Pacific
11-72

THE
VOICES
OF
TIME

THE VOICES OF TIME

A COOPERATIVE
SURVEY OF
MAN'S VIEWS OF TIME
AS EXPRESSED BY
THE SCIENCES AND
BY THE HUMANITIES

EDITED BY

J. T. FRASER

George Braziller / New York

Acknowledgments

Funds for this study, including the sponsorship of the necessary conferences with the authors, were derived entirely from financial support by Microcraft Consulting Service of Pleasantville, New York.

I want to thank GPL Division, General Precision, Inc. of Pleasantville, New York, for their endorsement of my labors and for their assistance in preparing the illustrations. Of great help has been my appointment in 1962 to the Department of Physics and Astronomy, Michigan State University, for the purpose of completing this work.

For their valuable opinions, critical comments, organizational assistance, and general good will I am grateful to the many "Friends of Time," and especially to the following men and women: Prof. A. Cornelius Benjamin, Mr. Arthur C. Clarke, Prof. J. L. Cloudsley-Thompson, Prof. John Cohen, Dr. G. R. Gamertsfelder, Mrs. Alice Grafflin, Mr. Ivan A. Greenwood, Jr., Prof. Robert H. Knapp, Mr. Arthur Koestler, Prof. Donald J. Montgomery, Dr. Joseph Needham, F.R.S., Rev. D. J. K. O'Connell, S.J., Prof. David Park, Mrs. Wendelyn A. Reymond, and Prof. Richard Schlegel.

Finally, for that measure of inspiration without which no sustained commitment can appear rewarding, I am indebted to the memory of a remarkable friend.

TO THOSE

WHO HAVE SHARED MY RESTLESS SEARC

FOR THE MEANING OF TIME

Contents

CONTENTS

PART II   TIME AND MAN
Communications, Rhythm, and Behavior

CONTENTS

PART III TIME AND LIFE
Rhythm, Life, and the Earth

PART IV TIME AND MATTER
Clocks, Man, and the Universe

CONTENTS

Introduction

Slowly but quite perceptibly, the young child learns the rhythm of life. With each sunrise and sunset he adapts himself more intimately to the enduring and to the changing aspects of his environment. As the seasons pass he begins his lifelong search for personal identity through a complex mental process involving expectation and memory. In due course he is likely to beget other human beings resembling himself and at least one other person. Ten or twelve scores of seasonal changes later he will be motionless, his body will begin its disintegration into the chemical elements making up the earth. We say that he was conceived and was born, that he learned, loved, struggled, aged and died. Through the life of this man, as through the life of all men, personal identity becomes intelligible and communicable to others because of the existence of a subtle private and communal understanding of an ordering principle. The same understanding, or type of knowledge, is also essential for the description of man's relation to the universe. This principle, or knowledge, couched in terms of an idea called *time*, is the central concern of *The Voices of Time*.

The concept of this book evolved from beginnings which cannot be precisely dated but which go back to the Second World War. In that upheaval, more than in the routine existence of organized society, ideas related to time, such as inevitability and the freedom to dispose of one's time appeared in no uncertain terms as the fundamental aspects of existence; reminders of the irreversibility of death (a knowledge of temporal nature) were the order of the day. Watching the clash of cultures and the attendant release of primordial emotions stripped of their usual niceties, I could not help observing that man is only superficially a reasoning animal. Basically he is a desiring, suffering, death-conscious and hence, a time-conscious creature.

Temporal experience, it seemed, more than any other aspect of existence is all-pervasive, intimate and immediate; and life, death and time combine in a dialectical unity which is hard to comprehend but which, nevertheless, is symbolically stated in all great religions. Furthermore, time also seemed to be a constituent of all human knowledge, experience and mode of expression; an entity intimately connected with the functions of the mind; and a fundamental feature of the universe. No other properties of reality, such as space or space-time, seemed to bear the same pertinence to the basic concerns of man as did the idea of time. In short, it appeared to me that time must and should occupy the center of man's intellectual and emotive interest. These thoughts turned out to be far from novel. But they were sufficiently stirring to lead to the present volume.

1 The Unity and Diversity of Time

During the summer of 1946 I began studying the literature in search of some generally acceptable view of time. Soon it became evident that the concept of time is curiously evasive when looked at from the separate and necessarily limited viewpoints of individual disciplines. For example, the physicist's symbol t is a deceivingly simple representation of "what we mean by time." It is useful in formal expressions and its meaning need not be questioned. If we ask, however, how it is supposed to relate to what we all intimately know as our existence in time, we shall be referred to psychology. Psychology, a science dealing mainly with mental processes, says hardly anything about the physicist's symbol t. Seeking for something common to the psychologist's and the physicist's time, we might inquire what psychology says about how our feeling of continuous becoming relates to the continuous becoming of the physical world, including the crucial questions of the beginning and end of time. For an answer we are likely to be referred to religion. But religion and theology speak mainly of purpose and history and not about feelings and physical symbols. We may proceed to the philosopher and inquire about the relations, if any, between the feeling of duration, apparent purpose in nature and the useful time of the physicist. Philosophy, "the most general science,"[1] has a great deal to contribute to the clarification of the problem of time. But, because of some unavoidable limitations, possibly intrinsic in the speculative method, it comes up against certain antinomies which, so it seems, may not be resolvable without the aid of the specialist. Perhaps, by observing with Aristotle that time is the "number of motion in respect of 'before and after,' " we may proceed to the people who make machines which measure time by motion. But the horologer is not interested in epistemology or the feeling of enduring and will probably send us back to the physicist.

One may object that we have consistently and willfully questioned the wrong man. This, however, is not strictly true, for there is no one who can give a satisfactory answer as to just how these numerous manifestations of time are interconnected; scientific concepts of time and the feeling of duration are seldom spoken of in the same context. Clearly our experience is rich and varied. Yet, it would be disappointing if this were to force us to live in a world of fragmented knowledge and accepted chaos.

To make possible some organized thinking about the numerous aspects of time, this volume of twenty-six essays includes discussions by leading experts from most major fields of knowledge, arranged according to a plan that will be unfolded. It includes views and opinions concerning time and studies of the role of time in philosophy and religion, history, music, linguistics, the sciences of the mind, the life sciences and the physical sciences. The essays relate to the dual problem: What is time as seen from special fields of learning, and what is the significance of time to the use of those disciplines?

This, then, is the primary purpose of *The Voices of Time*: to offer material and encourage the search for new knowledge related to time. To do this requires experimentation and much probing thought. We should not be surprised, then, that the essays include such varied subjects as Mohist causality, symbolization of time in psychoanalysis, construction details of various escapements and problems of the Lorentz transformation in Relativity Theory.

The question of "what is time?" is not emotionally neutral. Views held by individual scientists and scholars regarding which domains of knowledge are equipped to deal with this question tend to be dogmatic and often contradictory. Therefore, another purpose of the book is to present a variety of views in the hope that their survey will help reveal which fields will contribute to an understanding of time and what they might contribute.

It is also hoped that the interdisciplinary nature of *The Voices of Time* will be useful in epistemological studies relating to the organization of knowledge. We might approach this task by noting that a person's view of time is a method of discerning his personality. We may almost say: Tell me what you think of time and I shall know what to think of you. Experimental support for such observation appeared in 1958, when two psychologists from Wesleyan University in Connecticut[2] discovered that personality traits, especially the achievement motive, correlate significantly with attitude toward time. The investigators found it meaningful to divide their subjects, based on their responses to time imagery, into three different clusters. The three motivational traits were named Dynamic-Hasty, Naturalistic-Passive and Humanistic.[3] They were tentatively related to the Newtonian sense of time and to high need for achievement, to oriental and mystic thought and to classical Mediterranean thought respectively. The validity of these assigned areas is not here of interest. Important is that the groups are identifiable on the basis of attitude toward time.

That personality traits are decisive in the choice of occupation is a truism. Individuals have expressed their personalities in different ways by creating the principles and techniques of the various branches of intuitive and rational knowledge. It should follow that knowledge so created bears the personality marks of its creators, including their views of time. Continuing our speculation we might suspect that knowledge can be meaningfully classified according to some uniform test of attitude toward or of relation to time.[4] No attempt is made in this book to devise such a test. But we do note that there is a more profound connection among the varied expressions and views represented in these essays than just their use of the same English word.

This connection suggests that the idea of time be used to hold together separate areas of knowledge in the face of the fragmentation brought about by two major causes. One cause is the immense increase of available data and analytical information in the sciences. The other is the diverse experimentation in methods of thought and modes of expression in our age of profound social change.

An interdisciplinary study of time might also help us learn more about the nature of time by providing a direct confrontation of a multitude of views. Desire for the widening of the base on which the problem of time is to be studied is indicated by an interdisciplinary trend in recent publications.[5]

Insofar as knowledge itself is created by man, a wholistic attitude toward the structure of knowledge appears to be justified. But it seems that there exists a real duality between knowledge felt and knowledge understood. These two types of knowledge exhibit in each individual, as well as in the aggregate of individuals, a share of mutual incomprehensibility. Their seeming incompatibility may stem from the same source of incomplete perceiving power as does "the conflict between *lived time* apparently understood and the *idea of time* as an entity which when critically examined is found to be replete with obscurities and unsolved problems."[6] The idea emerges that C. P. Snow's "two cultures" are basically not those of the humanities and the sciences, but those of knowledge felt and knowledge understood. This division, then, might relate to the duality of the two *times* mentioned above. Yet we have learned to live with time and developed a working dialogue between our rational and emotional faculties with respect to time. This condition suggests another purpose for the essays: to provoke, if possible, communication between the humanities and the sciences, using time as a common theme. Such communication is important, for in my view, the humanities will retain their influence only if they are forced to interact with all matters that are of interest to man in his present world. Creativity in an individual, at least in the vocabulary of theoretical psychoanalysis, relates to his ability to resolve the instinctual conflict between Eros and aggression. By a possibly dangerous analogy, the total creativity of society may depend on the effectiveness of a harmonious dialogue between the two great branches of knowledge.

The *Weltanschauung* of an individual and of an age, that is, the perception of life and concept of things preferred, is essentially a view of time. Consequently, the reader may use, if so inclined, the organized material in this book in his own pursuit of a coherent world view, based on a combination of ideas about time prevalent today.

To make this or any cooperative venture directed to the problem of time meaningful and any systematization based on time possible, the following basic assumptions are made:

1) When specialists speak of time, they speak of various aspects of the same entity.

2) This entity is amenable to study by the methods of the sciences, it can be made a meaningful subject of contemplation by the reflective mind, and it can be used as proper material for intuitive interpretation by the creative artist.

These assumptions, which we shall call the *unity of time*, amount to saying that all of us, even working separately, are nevertheless headed toward the same central idea.

The unity of time had to be defined by assumptions. The *diversity of time*, that is, the existence of its multiple manifestations, hardly needs proof; it is all too apparent. According to Giordano Bruno, the creative activity of the mind consists of the search for the one in the many, for simplicity in variety. There is no better and more fundamental problem than the problem of time in respect to which such search may be conducted. It is always present and always tantalizing, it is the basic material of man's rational and emotive inquiries.

Just as individuals, if mature, should be capable of containing within themselves the unity and diversity of day-by-day existence, so, if we are to obtain a mature view of the nature of time, we should be able to assimilate and eventually make sense of the unity and diversity of time.

2   Arrangement of the Material

In an attempt to impose order on and, perhaps, to extract coherence from the material, it has been arranged in a generally evolutionary sequence. Historically, the flow of ideas is from the philosopher to the scientist. To avoid oblivion and oppose the destructive power of time, man first invented philosophy and theology, and only much later did he discover the methods of the natural sciences. The basic problems of existence remain, however, inscrutable, if for no other reason than their endlessness. From the bottomless well of problems, the speculative and reflective mind selects topics of interest to it (often for reasons unknown) and passes them on for interpretation to the vanguards of the method of discovery. In our age, these are the sciences. So we begin with the most general ideas about ideas, proceed to man's views of himself and terminate in man's most precisely formulated views of the material universe.

As we proceed from speculative philosophy to experimental science there is a fine gradation with regard to the idea of progress and duration of acceptability. Philosophy is individualistic, often accepted only within well defined geographic and ethnic borders. Its progress cannot be very well defined and change is likely to be slow. In the physical sciences, acceptability and universality are compulsory and results are acceptable only if they are not limited by geographic and ethnic borders. Progress is not only definable in terms of theoretical and practical improvements, but is required for general acceptability, and changes are apt to be sudden. There is a complete interdependence between the disciples, and individual results are cumulative. In a highly oversimplified sense, then, we progress in *The Voices of Time* from the individualistic and slowly changing to the communal and rapidly changing knowledge.

The sequence of papers permits a degree of cyclic permutation. A linguist reader, for instance, may start with Dr. Dürr's discussion on rhythm and time in music, proceed to Professor Whitrow's and Professor Benjamin's presentations of man's view of time and the universe and reach the conclusion of the book with Professor Stutterheim's pensive words:

"From of old, in lyrics about transiency and death, the painful realization has been expressed that time exists . . . whatever it may be."

Because of this possibility of cyclic permutation, the essays could have been arranged, for example, so as to conclude the book with philosophy. The view that potentially each of the fields represented has matters of equal importance to contribute to the central subject is contained in the idea of the unity of time. However, the concluding chapters of the book were left in the hands of the physical scientists because of my high regard for the "scientific method," which is most explicitly represented in the physical sciences. The scientific method is a manner of approaching the world and searching for order. It has been discussed many times but never successfully defined. This is not surprising for the "scientific method" itself is not entirely within the domain of science, that is of clearly definable ideas, but is essentially a mixture of logic and intuition. It depends on the *Gestalt* of the scientist and not on his argumentative power alone. It seems to me that if the problem of time is ever to be clarified, such clarification is most likely to come by applying to the complete spectrum of human experience that mixture of clear logic and unwritten superstition which produced Western science.

Finally, if all these arguments in favor of the arrangement chosen turn out to be unconvincing, *The Voices of Time* may still be looked upon as a long poem written in strange verse.

3 *"The Voices of Time" as a Project*

In a very true sense, the development of this volume was a research project. I had no prior example to guide me, for the book is not an anthology, not an encyclopedia, not a record of a symposium—it is not even a series of comments on an established field. I had to discern the outlines of a possible new field of study and measure the contents of the papers against the contents of a not yet delimited discipline. Therefore, to a great extent, the essays are biased by my personal likes and dislikes.

Concerning the depth of approach, the two obvious alternatives were: making all the articles elementary, or addressing each article to the specialist reader. The seriousness of the purposes enumerated above, even without the claim that any of them has been accomplished or even proved feasible, counseled the second approach. But, a series of highly specialized articles would have defeated the unifying purpose of the volume. The approach selected was neither of these obvious ones, but a pragmatic one: some topics are discussed in greater depth, others are stated and referenced. In all but Part I, the essays are somewhat in order of increasing complexity. The last paper in each Part is an overview of its subject, although not an attempt for a summary.

My Notes and Comments discuss selected topics, offer speculations, or just emphasize certain details. Although references to time are dominant in the belles-lettres, we do not have an essay on this subject. As a partial remedy, I have drawn liberally on the intuitive, poetic representation of time.

The titles of the four Parts are useful and descriptive; they do not imply undue dichotomy of our subject beyond the obvious. It was necessary to keep each paper sufficiently complete so as to assure its acceptance in terms of its own rationality and logic. We have kept professional jargon to a minimum, but had to permit the use of mathematics for some of the papers, for their messages cannot be conveyed through any other means. Since a universally acceptable definition of time does not exist, it was not possible to specify what the "thing" is that we all are to keep in focus.

The variety of subjects in the book is unusually great, and commenting on so many special fields is dangerous. I felt, however, and received ample support for such a view, that the undertaking of this task is timely, necessary and of broad interest.

Studying the literature, editing the articles and writing my own sections, I could not suppress a sense of awe regarding the dimensions of man's existence, both in its horizons and in its limitations. This type of reaction is not currently widespread, for the success of the sciences tends to submerge the acknowledgment of these limitations. The forces of speculation and reverence which were so essential in creating the natural sciences and Western technology are not always attractive because one must have courage to fear and wonder.

No shocking insights or revolutionary discoveries have come my way while completing the work; neither were they expected. The problems of time are too profound to hope for more than a modest advance and for the enjoyment of search.

This book is submitted in the deep belief that the search for the nature and the meaning of time is a challenge second to none.

J. T. FRASER

Pleasantville, N.Y.
August 23, 1964

But pardon, gentles all,
The flat unraised spirits that hath dar'd
On this unworthy scaffold to bring forth
So great an object. . . .
Piece out our imperfections with your thoughts:
Into a thousand parts divide one man
And make imaginary puissance.
Think, when we talk of horses, that you see them
Printing their proud hoofs i' the receiving earth.
For 'tis your thoughts that now must deck our kings,
Carry them here and there, jumping o'er times,
Turning the accomplishment of many years
Into an hourglass; for the which supply,
Admit me Chorus to this history;
Who, Prologue-like, your humble patience pray,
Gently to hear, kindly to judge our play.

HENRY V

PART I
TIME IN THOUGHT
Philosophy, Religion, and

Man's Attitude Toward Change

Faust in his study watching a magic disk.
Etching by Rembrandt.
Collection of J. T. Fraser.

Ideas of Time in the History of Philosophy

A. CORNELIUS BENJAMIN

1   *The Problem*

Man is confronted by a world which he commonly describes through two characteristics: on the one hand, objects are spread out in *space*; on the other hand, events succeed one another and endure for shorter or longer periods of *time*. The two structures, space and time, constitute the spatio-temporal framework within which the individual lives and dies and the human race develops and preserves the wisdom of the ages.

The words commonly used to describe the nature of space and time suggest that the two entities are alike in some respects. Contemporary physics claims that they are, indeed, in a very fundamental sense interdependent. Because of the unique pertinence of time to the basic concerns of man, already emphasized in the Introduction to this book, I shall concentrate neither on the idea of space nor on the idea of space-time but on the idea of time. Specifically, I shall attempt to depict through representative examples from the history of philosophic thought the conflict between *lived time* apparently understood and the *idea of time* as an entity which when critically examined is found to be replete with obscurities and unsolved problems.

Man exhibits behavioral responses which are identified as memory and expectation. The existence of such responses is not unique to man, however, but appears in organisms which are below him in the evolutionary scale. I shall begin by referring to man's unique ability to talk about time; this, I believe, is something he does not share with other creatures. In speaking of time he commonly uses words such as "when," "hour," "past," "present" and "future." These seem clear and are seldom confused; they are straightforward descriptions and are readily combined in comprehensible sentences which designate aspects of time as we experience it. Hence they appear to pose no problem.

But if we expand this vocabulary and add other expressions which are frequently used when we talk about time, we begin to perceive the difficulties. "Time is like an ever-flowing stream." (The *stream* of consciousness, the *flow* of an electric current, the *flow* of words of a great orator?) "Time unrolls like a carpet." (*Unrolls* in the sense of uncovering something which was previously hidden but now lies exposed to view; and will it continue to be displayed or will the carpet begin to re-roll from the other end and thus hide something again?) "Clocks keep time." (As we *keep* our possessions, *keep* our moral principles, *keep* a house?) "Time passes." (As we *pass* an automobile on the road, *pass* a course in a university, *pass* from life to the hereafter?) "Time is ever coming into being and passing out of being." (Where was it before it *came into being* and where does it *go* when it *passes out of being?*) "Time is all-embracing." (If it is all-embracing does it also *embrace time?*) "We tell time." (To whom, in what language, and *what* do we tell about time?) "We expect the future, experience the present, and remember the past." (Is time then merely a subjective image created by our mind, and having no counterpart in the world?) "Time is the relation of before and after." (But *before* and *after* refer *only* to time; hence we are saying literally that time is time.)[1] Does this not show that what I have called the straightforward descriptions of time contain metaphors and analogies, ambiguous words, subjective terms, hidden contradictions, and definitions which are purely verbal? And does it not mean that before studying time we must study the *language* of time?

There is a strong movement in recent philosophy and linguistics which would argue for precisely this point of view. It calls itself variously "significs," "linguistic analysis," and "ordinary language," and it argues that our view of the world is much more influenced by the language which we employ in talking and writing about that world than we have heretofore realized. Thus we should not be surprised to find that peoples speaking different languages have different conceptions of time, and that philosophers, scientists and laymen, even though they may all be using English, have views of time which are reflections of their own specialized, technical languages.

While this is all quite true, and has been generally recognized by philosophers for many years, I do not feel that it justifies prefacing our study of time with a detailed and careful analysis of the language of time, or, indeed, of any language. For to study a language requires another language (a meta-language) in terms of which this study may be carried on—and one immediately finds himself on the way to an endless regress. Furthermore, the language of time may prove to be a subject matter which is even more obscure and difficult to study than is time itself; hence we only replace a difficult problem by a more difficult one. Finally, one cannot possibly study any language of time without knowing something about what time itself is, and this can be best acquired simply by looking at time, while recognizing that one's own native language may have contributed more than he realizes to the conception of time at which he finally arrives. For these reasons, and because, to my knowledge, none of the linguistic analysts has produced any original, over-all view of what time is, I shall not make further reference to this approach to the problem of time. This is a volume about time, and should concern itself with what there is to learn about time through the specific tools and methods characteristic of the multitude of approaches represented in the book and only then, if it appears useful, perform a significal analysis of the results.

One of the best tests of our understanding of a concept is our ability to give a definition of it. The mathematician Louis Poinsot has implied by illustration what seems to demonstrate the deep-seated difficulty which arises when one attempts to answer the question "What is time?" Poinsot stated that if he were asked to define time he would in turn ask the questioner whether he knew what he was asking about. If the reply was "Yes," Poinsot would answer, "Very well, let us talk about it." But if the answer was "No," Poinsot would say, "Very well, let us talk about something else."[2]

In order to provide the matrix within which I believe man's idea of time develops, I shall point out what seem to be the outstanding characteristics which most philosophers who think about the problem of time recognize, more or less explicitly, and try to incorporate into their views.

The first and most obvious fact is that time, so far as our immediate experience is concerned, is relative to happenings or things *in* time. We are aware of time simply by virtue of changing events or enduring objects with which we come into contact in our daily living. We therefore speak of time as *relational*. But on occasions we feel that time, measured in terms of events, seems to "speed up," and on other occasions, to "slow down," and this leads us to believe that there must be some objective time, independent of what is *in* time, which flows at an even rate and in terms of which we recognize these varying times. This content-independent time we tend to characterize as *absolute time*. But these two views of time, the relational and the absolute, are clearly incompatible and must be reconciled if we are to gain a satisfactory understanding of this aspect of the world.

The second fact about any temporal series is that it exhibits two ordering characteristics. One is the *earlier-than* (the *before-after*) relationship which holds between any two instants, and is completely independent of the time when the selection is made. Although events may be simultaneous, instants cannot be. Consequently, given any two instants, x and y, either x is earlier than y or y is earlier than x; and once the choice between these alternatives has been made, the passage of time can never change it. The second ordering characteristic of a temporal series is the *past-present-future* relationship. This is a three-term, "between" relationship, *i.e.*, the present must always be between the past and the future. Its properties are such that *if* it holds for any three instants, it holds uniquely; for the "now" is continually changing.[3] Whether this "now" is the "knife-edge" between expectation and memory or the "saddleback" of passing time makes no essential difference.[4]

To the uninitiated the reconciliation of these two apparently inconsistent relational properties of time seems to offer no problem. Of course, if x is before y then it remains eternally before y, and nothing in the passage of time can change this. But the peculiar position of the "now" can be easily expressed simply because our language has *tenses*. The future *will be*, the present *is*, and the past *was*; the light will be red, it is now yellow, and it was green. But do we, in these terms, really describe the "processional" character of time? We sometimes say that an event *is* future, then it *is* present, and finally it *is* past; and by this means we seem to dispense with tenses, yet we portray the passage of time. But this is really not the case; for all that we have done is to translate our tenses into the words "then" and "finally," and into the order in which we state our clauses. If we were to omit these words or their equivalents, and mix up the clauses, our sentence would no longer be meaningful. To say that the future, the present, and the past *are* in some sense is to dodge the problem of time by resorting to the tenseless language of logic and mathematics. In such an *atemporal* language it would be meaningful to say that Socrates is mortal because all men are mortal and Socrates is a man, even though Socrates has been dead many centuries. But if we cannot describe time either by a language containing tenses or by a tenseless language, how *shall* we symbolize it?

The third fact about time is that it is closely tied up with the notions of change and constancy. Perhaps the best way to indicate this is to consider the verb "to change." Now this, like all verbs, requires a subject; if there is change *something* must change. If the entire universe changes it

6

must remain a universe while undergoing change; if I myself change I must in some way retain my identity in order that I—or anyone else, in fact— can know that I have changed. All change, therefore, is relative to a constant background, and without this element of permanency change would be meaningless. The simplest explanation of this constancy is that change is always the replacement of at least one specific quality, such as greenness, by another, such as redness, which is incompatible with it but shares a generic identity. The universe remains a universe but changes in form, or in size, or in structure; I remain a person but change in appearance, or in habits of living, or in age. What we mean in these cases is that whatever changes, retains a generic permanence (as a universe, or as a person). Or, very simply expressed, when a green light changes into a red light, the light remains the same but the color changes.

This explanation of temporal change has led many philosophers to deny that there can be any such temporal process as *coming into being* or *passing out of being*. To come into being involves changing from nonbeing into being. But what is the generic identity between nonbeing and being? A green light can change into a red light since each is a light. But how can nothing change into a light? Nonbeing is nothing, and being is something, and there is no generic identity between nothing and something. Hence there can be no creation out of nothing or disappearance into nothing. This is the philosophical equivalent of the scientific principle that mass-energy can be neither created nor destroyed, but only changed from one form into another. Evidence for this principle of conservation seems so widespread that we are tempted to speak of it as self-evident.

Yet it runs counter to a fact well established by the evolutionary philosophers. Creativity is obviously present in the universe. Life is more than bare matter, yet emerged from it in the temporal process; mind is more than life, yet appeared in the evolutionary scale. Time must therefore be conceived of not merely as change, but as *creative* change; not merely as replacement of specific qualities by other specific qualities, but as the origination of new genera. Hence something comes from nothing.

I have pointed out that time, as commonly understood, exhibits three puzzling properties. First, it is relational in character and yet somehow independent of happenings in time. Second, it involves a strange combination of two series of instants, differing in fundamental character. Third, it demands some sort of harmonious reconciliation between change and permanence, and in its changing aspects seems to forbid the emergence of novelty in a universe which is characterized by the continued appearance of higher and more complicated forms.

These are but a few of the paradoxes that arise when we try to penetrate beneath the superficial description of time, and achieve a more complete and fundamental explanation of it. How can we attain this more basic understanding?

C. D. Broad faces the difficulty squarely. When the logic of time leads to a contradiction, he says, so much the worse for our logic.[5] Jacques Maritain confesses that time is "a place where we put our blackest contradictions out to pasture."[6] J. E. Boodin almost rebukes the universe for exhibiting time, when he says that we describe the universe and then "time *creeps into* our world of description and negates it."[7] Schopenhauer goes further and actually *defines* time as the "possibility of opposite states in one and the same thing."[8] Whitehead, in a somewhat less disparaging mood, remarks that "it is impossible to meditate on time and the creative passage of nature without an overwhelming emotion at the limitations of human intelligence."[9]

I shall now turn to a chronological consideration of a few of the many attempts which philosophers have made to resolve these paradoxes and to achieve a better understanding of time. Because of the extreme limitations of space at my disposal I shall, with a few exceptions, confine myself to exposition, and omit any criticism of the views expressed. Occasionally I shall raise brief objections in the notes.

2   The Pre-Socratics

Of the many controversies carried on among the early Greek philosophers, that of *becoming* or *change* versus *being* or *permanence* was one of the most prominent. As representative of the emphasis on time and change I have selected Heraclitus (*fl.* 500 B.C.), and as advocates of the stress on static being I have chosen Parmenides (*fl.* 540 B.C.) and Zeno (*fl.* 460 B.C.). Our knowledge of the views of these thinkers is based on mere fragments of their writings, frequently lists of aphorisms, or (as in the case of Parmenides) a poem expressing his philosophy, together with commentaries by other philosophers, particularly Plato and Aristotle. Hence we can only guess at what they really had in mind. Space prevents me from raising controversial issues, and I shall state my interpretation of their views more or less dogmatically.

Heraclitus believed that he had discovered a truth about the universe which was contrary both to the ideas of his predecessors and to those of the common man of his day. He insisted that reality is flux and change. "You cannot step twice into the same river; for fresh waters are ever flowing in upon you" (fr. 41).[10] "We are and are not" (fr. 81). Plato attributes to him the views that "nothing really is, but all things are becoming"[11] and that "all things flow and nothing stands."[12]

As the substance and symbol of this eternal flux Heraclitus chose *fire* (fr. 26) because it so clearly exhibits change. The flame continues steadily and appears to be the same, yet it passes constantly into smoke, and the flame which takes its place must be fed repeatedly by new fuel. Thus fire seems to be a *thing,* but it is eternally undergoing change. The principle of all change is the law of opposites or of strife (fr. 62); everything tends to become its contrary and in this way change is produced.

Thus Heraclitus solved the paradox of time by showing that flux and becoming alone are real, and that permanence and constancy are merely apparent. Everything that seems to be static involves, when properly understood, continuous movement, and should this cease the universe would collapse into nothingness. Presumably Heraclitus did not realize that in adopting a *principle* of change (strife) he had unconsciously inserted a basic constancy into his system; for the *law* of change cannot itself change.

Parmenides and Zeno disagreed basically with Heraclitus. Their belief was that only the permanent and enduring are real, and all time, flux, motion and change are unreal. Parmenides refuted Heraclitus by arguing that in accepting change he presupposed that something can both be and not be; fire would have to be and not be, simply because the flame at one instant would have to be replaced by a flame which does not yet exist. But a flame which is not, is nothing, and it cannot be or even be thought about. Consequently, we can never say of anything that it becomes; for it would have to come from nothing, and this is impossible. If anything is, it is now, all at once.[13]

Zeno, in his famous paradoxes of motion, came to the same conclusion. This was inevitable since for him motion is change and change is time. For convenience, I shall describe his paradox of the race-course. "You cannot get to the end of a race-course. You cannot traverse an infinite number of points in a finite time. You must traverse the half of any given distance before you can traverse the whole, and the half of that again before you traverse it. This goes on *ad infinitum,* so that there are an infinite number of points in any given space, and you cannot touch an infinite number one by one in a finite time."[14] The story is told that when Zeno presented this argument to Diogenes the Cynic, the latter silently rose to his feet and walked away.[15] He was convinced that the best way to refute Zeno was to move. But Zeno was obviously not disturbed by this presumed contradiction, for he did not deny that things *appear* to move.

9

Logic proves that they cannot move; and if our senses tell us that they do, so much the worse for our senses.

Here, then, we have two typical solutions to the problem of time. If time requires both change and permanence, we have only to make one aspect real and the other apparent. For Heraclitus change was real, and constancy was unreal; for Parmenides and Zeno constancy was real, and change was unreal. Thus the problem has been "solved." But note that the universe has been bifurcated. What is only apparent has not been eliminated, since its apparent status must still be accounted for. If either change or constancy is unreal, we have the duty to explain why reality chooses to deceive us in either of these ways.[16] The classic exchange of thought between two small boys can well illustrate this point. "There ain't no sky," said one. "But what is it that ain't?" asked the other.

3 *Plato (428–347 B.C.)*

To discuss Plato's view of time adequately without putting it into the larger context of his total view of the world is impossible. Yet there is a sense in which we can say that if we disregard his moral and political philosophy, for which he is probably best known, the ontological view which remains has as one of its essential ideas that of the relation between *being* and *change*. It is possible to state with a certain measure of truth that Plato tried to resolve, or at least clarify, the opposition which remained after Heraclitus had established that all is change, and Parmenides and Zeno had to their satisfaction refuted him by showing that nothing changes. Plato saw more clearly than his predecessors that the problem of being and change is essentially the same as the problem of the one and the many, and he made a noble attempt to devise a theory of the universe which would solve both problems. He could not, of course, avoid bifurcating the universe, since time, as we have seen, must combine constancy and change.

The essence of Plato's view is that an ordinary object of everyday experience consists of two "parts," one called variously its "form," its "Idea," its "essence," and the other its "matter," its "individuality," its "sensible manifestation." All individual objects of a given class possess the same form or essence, which exemplifies itself in these objects with varying degrees of adequacy. All men, say, are human, not in themselves but by participation to some degree in the very essence of humanity. Forms are absolute, eternal and unchanging; particular objects appear and disappear, undergo changes, and therefore reside in the world of time and space. Men come and go, but the form of man persists throughout time and never alters. There are, therefore, two "realms" into which the world is divided. One contains only forms, some, but not all, of which are exemplified in the actual objects of our sensory experience; this world is outside of time, never exhibits change, and endures forever. The other contains only sensible objects, which exemplify the forms, though not all with the same degree of obviousness (hence our difficulty in identifying many objects of our experience). This world is in time, manifests the characteristic changes of the temporal process, and is for the common man, who has no curiosity about really understanding the universe, the only world there is.

The myths and arguments which Plato employed to clarify and support his belief in the existence of these two realms are many and various; but they were all designed to show that the world of forms (the intellectual world) is real and permanent, while the world of everyday objects (the world of opinion) is a world of mere appearance and change. In the allegory of the cave,[17] he supposed men in a cavern, chained with their backs to the opening so they cannot see what passes outside. The sun, shining into the cave, produces on the back wall shadows of objects which move back and forth across the opening, and for the inhabitants of the cave these are the only realities of which they have any knowledge. But when one of the prisoners is released from his chains and goes out into the world, he sees for the first time the forms which produce shadows. Although he is at first blinded by the Ideas in their purity, he recognizes that the shadows are ephemeral and but pale, changing images of the true reality exhibited by the forms.

In an analogy, the figure of the divided line,[18] Plato again attempted to clarify the relation between particular objects and forms. He compared the object-form relation to the relation between the mirror images of an object seen in a pool of water and the object itself, and between the objects which exemplify forms and the pure forms apart from their exemplification. The dialectical process of arriving at true knowledge consists in mounting from images of objects to the objects themselves; from these to the objects exemplifying the forms; and from the objects embodying the forms to the bare forms themselves. For example, we may go from the mirror image of a beautiful object to the beautiful object; from the object itself to the object as exemplifying beauty; and from beauty as exemplified to the pure form of beauty. In spite of the elusive character of these forms their being must be accepted, since they are presupposed in all sensory experience.

This cognitive role of the forms was brought to light by Plato in the *Meno*,[19] where Socrates talks with an ignorant slave, and by well-directed questioning enables the boy to prove how long the side of a square must be if its area is to be twice that of a given square. Since the slave had never studied geometry he must have acquired knowledge of pure forms in some previous existence, and Socrates had only to draw it forth by asking the proper questions. In much the same way, Plato would say, our ability to recognize cases of approximate equality, examples of imperfect justice and varying degrees of beauty in objects presupposes that we must have had, prior to birth, some awareness of eternal forms in their purity. For to identify approximate equality, imperfect justice and varying beauty, one must know what pure equality, perfect justice and ideal beauty are, and the changing world of sensory experience nowhere offers instances of these.

The relation of time to changelessness was beautifully expressed in the *Timaeus*, where the creator is explaining how he transformed the original chaos into a lawful universe. To bestow eternity directly upon the universe was impossible. "But he resolved to make a moving image of eternity, and as he set in order the heaven he made this eternal image having a motion according to number, while eternity rested in unity; and this is what we call time."[20]

From this brief sketch of Plato it should be clear that while he could not avoid dividing the universe into the temporal and the nontemporal, he succeeded in showing in much greater detail than the pre-Socratics did what characteristics the transient, multiple and changing portion of the universe possesses, and what properties the eternal, unchanging and timeless part exhibits. Most important of all, he explained, in terms of the figure of the divided line and other metaphors, how these two realms fit together without contradiction and thus constitute a universe rather than a pluriverse.

4   *Aristotle (384–322 B.C.)*

Aristotle's examination of time was much more analytic than that of his predecessors. He was more deeply concerned with its nature and properties, particularly its measurability, than with its status in the universe as real or apparent.

His famous definition of time as the "number of motion in respect of 'before' and 'after' "[21] contains many subtleties and many sources of confusion. In the first place, his restriction of time to that particular form of change which he calls motion (κίνησις) is designed to exclude such cases as *coming into being* and *passing out of existence*, on the grounds that movement is a single process whose beginning and ending can be positively identified. In the case of generation we can set a time when the object generated *is not* and a time when it *is*, but we cannot establish an instant when it *begins to be*; similarly we cannot set a time when a perishing object *ceases to be*. In the second place, there are cases where motion is more or less accidental; for example, if a man has a musical quality we might loosely say that when he moves the musical quality moves. And there are instances where an object may be said to move when one of its parts moves, as in the case of a man at rest moving his arms. These are omitted from Aristotle's treatment because he believes them to be more or less misleading ways in which motion constitutes the measure of time.

Motion is therefore restricted to three kinds:[22] qualitative, quantitative and local. Qualitative motion is simply alteration, *i.e.*, change from black to white, or change from cold to hot. Quantitative motion is change in size by increasing or decreasing. Local motion is change in place; it may be called locomotion, and is presumably the only one of the three kinds which we should be inclined today to call motion. It is for Aristotle the most basic of all forms of motion.

Proceeding with his analysis, Aristotle suggested that time is either identical with motion or it is something closely related to it. But since the movement of each thing is only *in* the thing which moves, or *where* the thing which moves happens to be, time is present equally in and with all things which move. Furthermore, we cannot say that time is fast or slow, for "fast" and "slow" are defined by time, "fast" being that which moves much in a short time and "slow" being that which moves little in a long time; consequently only motion can be fast or slow. This means that "time" cannot be defined by time, by being either a certain amount or a certain kind of it.[23] The conclusion is that time is not identical with motion, but is something closely related to it. What is this relation?

Time cannot exist without change, for in a deep, dreamless sleep we are totally unaware of the passage of time. On the other hand, if any movement takes place in the mind we suppose that some time has elapsed, and the time that has passed is always thought to be in proportion to the movement. Thus we perceive time and movement together, and they always correspond to each other in magnitude. The application of magnitude to motion may seem clear to us when Aristotle speaks of quantitative and local motion. But what can it mean in the case of the motion which is qualitative change? The answer is that change from black to white is a process of whitening and change from cold to hot is a process of heating, and these can readily be understood quantitatively as *rates* of whitening and *rates* of heating. Now time is a quantifiable change from "before" to "after." Hence we arrive at Aristotle's definition of "time" as the number of motion in respect to "before" and "after."

The introduction of number suggests measurement. Aristotle went on to say that "time is a measure of motion and of being moved, and it measures the motion by determining a motion which will measure the whole motion, as the cubit does the length, by determining an amount which will measure the whole."[24] Taken by itself this statement is misleading, for it suggests that measurement of time is essentially the same thing as measurement of space. But considered in the context of the *before-and-after* character of time, it makes quite clear that the cubit requires no distinction between its two ends, while the *passing* character of time demands that its unit of measure be *arrow-like* as well as *durational*. Hence the unit of time must be in the form of a vector whose direction is fixed and whose length indicates lapse of time.

Perhaps the most noteworthy point which Aristotle recognized in his discussion is that to which I called attention in my introductory section on time, *viz.*, that time is a combination of change and permanence. Aristotle said that just as motion is a perpetual succession so also is time, for the *now* as a subject is an identity which accepts a succession of predicates. In motion there is an identical substratum. A body which is carried along is different insofar as it is at one time here and at another there; nevertheless it is the same body at each stage in its motion. Similarly, "the 'now' is in one sense the same, in another it is not the same. Insofar as it is in succession, it is different (which is just what its being now was supposed to mean), but its substratum is an identity."[25] Both time and motion, therefore, as processes, imply *something* which proceeds; this itself must be unchanging.[26]

In further support of the ultimacy of some sort of permanence Aristotle asked whether there can be motion of motion, or, in general, change of change. His answer was in the negative, and he offered several arguments designed to prove his position.[27] One of them was as follows: Suppose there can exist a change of change. Let *A* be a change, and let *B* be that of which *A* is a change. Now by hypothesis *B* will be a change of change. Let us call this new change *C*. But this process will go on to infinity, and we shall never arrive at a stage of "simple becoming," i.e., a change which presupposes no further change. "And since in an infinite series there is no first term, there will be no first stage and therefore no following stages either. On this hypothesis, then, nothing can become or be moved or change."[28] But this is absurd; hence there can be no change of change.

While this discussion of Aristotle's view of time is very much truncated by the omission of his consideration of the prime mover, of the simplicity and completeness of rotary movement, and of the nature of infinity, it may, perhaps, indicate that Aristotle seems to have advanced beyond Plato in the understanding of time by introducing the concept of measurement, and by recognizing the directional character of all temporal processes.[29]

5   *Locke (1632–1704)*

To leap from Aristotle to Locke (about 2,000 years), without mention of intervening events, would be quite unjustifiable from the point of view of historical scholarship. Three figures, in particular, stand out for possible consideration: Plotinus (A.D. 204–69), St. Augustine (353–430), and Thomas Aquinas (1225–74). Plotinus, reminiscent of Plato, had a time-less and spaceless world of thought, which was a model of the phenomenal world. But Plotinus was a mystic, and his views are phrased in the obscure language of that point of view, which defies simplification and condensa-tion. St. Augustine, in Book IX of his *Confessions*, discussed the nature of time in a brief but penetrating passage which has become almost a classic in the history of philosophy. But the *Confessions* was an auto-biography, not a philosophical treatise, and to include St. Augustine while omitting A. N. Whitehead (whom I excluded only after long and serious deliberation) would destroy the effectiveness of an essay which claims to present only the most representative ideas of time which have been de-veloped by the great philosophers of history. Similar considerations led me to pass over Thomas Aquinas, whose eminence in the history of thought is unquestioned. But does St. Thomas belong in the history of philosophy or in the history of theology? To summarize, then, my omis-sion of the early Christian philosophers, and of those who made their con-tributions during the Dark Ages (which modern scholarship has shown to be not so dark as we had previously supposed) is based solely on the fact that my allotment of space in the volume was necessarily limited, and I had therefore to restrict myself (often arbitrarily) to those whom I considered to have made the most significant contributions to the prob-lem of time.

For the British empiricists in general, and for Locke in particular, the problem of time was quite differently conceived. Locke[30] was not primarily a metaphysician and was not concerned with the ontological status of time nor with the properties it possessed, supposing that its existence had been demonstrated. He was essentially an epistemologist and expended his efforts in showing how we build up the idea of time out of elements given to us in experience. He was opposed to the view, formerly argued by Plato and Descartes, and later to be developed by Leibniz and Kant, that there are ideas which are presupposed by experience and have to be taken for granted before any sensory knowledge is possible. His attack on the theory of innate ideas laid the foundation for his view that the mind is initially a blank tablet on which experience writes and thus provides us with the ideas by means of which we understand the world.

For Locke our ideas come from two sources—sensation and reflection.[31] Ideas of sensation, such as yellowness, heat, shape, softness and bitterness, come from extra-mental objects; ideas of reflection, such as perception, thinking, reasoning and willing, come through our internal sense and are those which the mind gets by attending to its own operations. Simple ideas[32] enter the mind distinct from one another even though they may be caused at the same time by a single thing, as in the case of the idea of motion and the idea of color produced by a moving object. Complex ideas[33] are those which arise when the mind acts on simple ideas by combining them, relating them and abstracting from them. Our idea of time is complex,[34] and is obtained when we reflect on the appearances of several ideas one after another in our minds, producing the idea of *succession*, and when we reflect on the distance between any parts of that succession, giving us the idea of *duration*. Thus time is a sort of quantified change in happenings, and a man who sleeps soundly, as Aristotle pointed out, has no conception either of succession or of duration. If anyone, *per impossibile*, could hold only one idea in mind without variation he could form no idea of duration. Motion, since it exhibits itself always as a succession of changes, provides an important source for our idea of time. Change which is so rapid as to take up the time of only one idea in our mind produces no notion of succession, and we call it an *instant*.

Measurement of duration differs sharply from measurement of spatial extension. In the case of space all that is required is the application of some arbitrary standard to the space to be measured, as in the measurement of cloth by a yardstick. But in the measurement of duration this is impossible because no two different parts of succession can be put together in such a way as to determine which is greater; we cannot set aside and have available at any time an unvarying, standardized duration, like a yardstick, which we can "apply" to a given lapse of time. Instead, we measure duration by motion, such as that of the earth on its axis or the movement of the earth about the sun. But we have no method for determining the equality of either the diurnal or the annual movements of the earth. Consequently, while the *origin* of our idea of a duration which goes on in one constant, equal and uniform course can be located, this does not guarantee, as Newton will argue, that there *is* such an objective, absolute time.

Furthermore, although, in spite of our abilities to abstract, we never achieve the capacity to conceive of time independently of events happening in time, there is nothing which prevents us from imagining a duration which we can freely populate with events other than those actually experienced by us. We can therefore extend time into the past, through memory and into recorded history, and into the future beyond the point when any of us will be alive. Locke stated that "though the Julian period be supposed to begin before there were either days, nights, or years, marked out by any revolutions of the sun,—yet we reckon as right, and thereby measure durations as well, as if really at that time the sun had existed, and kept the ordinary motion it doth now."[35] From this it is but a short step to the idea of *eternity*, which is produced when we "add such lengths or duration to one another, as often as we please, and apply them, so added, to durations past and to come."[36]

In summary, it can be seen that Locke, while providing no metaphysical analysis of time, and telling us nothing about the real nature of time in the structure of the world, did show that from the simple experiences of succession and duration, recognized by the most unsophisticated of us, we can easily build up the complex notions of the measurement of time, and of the past, the future and eternity. All that is required, in addition to the empirically given phenomena of succession and duration, is our mental capacity to abstract and relate simple ideas in such a way as to produce from them complex ideas. There is nothing in our minds that *compels* us to see the world in a temporal way.[37]

6   Newton (1642–1727)

Though Newton is remembered more for his science than for his philosophy, his views of time and space played such a significant role in the history of thought that they cannot be omitted from our discussion. Indeed, one should not say that he was a scientist rather than a philosopher, since the title of his great work, *Mathematical Principles of Natural Philosophy*,[38] together with many of the methodological problems which he examined in his *Optics*, and the role which he assigned to God in his system of the world, clearly indicate that he was at once physicist, philosopher and theologian.

Because Newton stated his conception of time so clearly and so concisely I shall quote him directly. "Absolute, true, and mathematical time,
of itself, and from its own nature, flows equably without relation to anything external, and by another name is called duration: relative, apparent,
and common time, is some sensible and external (whether accurate or un-
equable) measure of duration by means of motion, which is commonly
used instead of true time; such as an hour, a day, a month, a year."[39]
Newton then went on to add that in the same way absolute space, in its
own nature, without relation to anything external, remains always similar
and immovable, and that relative space is the measurement of absolute
space as determined by the sensible determination of bodies. From this it
follows, of course, that absolute motion, in contrast to relative motion,
which is the translation from one relative space to another, is the translation of a body from one absolute place to another.

That Newton should have introduced the notion of absolute time (together with the correlated ideas of absolute space and absolute motion)
seems surprising in view of the strong emphasis, expressed elsewhere in
his writings, on the empirical character of the scientific method. Near the
end of his *Principles* he stated definitely, "I frame no hypotheses; for
whatever is not deduced from the phenomena is to be called an hypothesis;
and hypotheses, whether metaphysical or physical, whether of occult qualities or mechanical, have no place in experimental philosophy."[40]

There is evidence throughout Newton's writings that he did not adhere
rigidly to this principle. But the deliberate introduction of absolute time,
space and motion, which are clearly unobservable, seems such a flagrant
violation of the empirical spirit of his entire philosophy that it cries out
for justification. He prefaced the introduction of these absolutes by admitting that he did not define time, space and motion in accordance with the
way in which they were ordinarily understood. "The common people conceive these quantities under no other notions but from the relation they
bear to sensible objects."[41] Even Locke, as we saw in the previous section,
emphasized the relativity of all ideas of time to sensible experience. But
Newton believed that such conceptions produced certain prejudices which
could be removed only by distinguishing sharply in each case the absolute
from the relative. Consequently, in spite of his warning that "Nature is
pleased with simplicity, and affects not the pomp of superfluous causes,"[42]
he insisted that "in philosophical disquisitions, we ought to abstract from
our senses, and consider things themselves, distinct from what are only
sensible measures of them."[43]

Apparently Newton was convinced, at least partly, of the reality of
absolute time by virtue of an experiment which seemed to him to show

that we can demonstrate the difference between absolute and relative motion in terms of their sensible effects. If a pail is filled with water, the water's surface is flat. Now if the pail is spun rapidly on its axis, the pail will gradually communicate its motion to the water, which will recede little by little from the center, and gradually take on a concave surface until it acquires the same rotational velocity as that of the pail. At first the surface of the water is flat, indicating that "its true circular motion had not yet begun." But the increasing concavity of the water surface suggests that the "real circular motion of the water" increases until such time that the water "had acquired its greatest quantity," meaning, the greatest quantity of real circular motion. This, Newton interpreted as experimental evidence for the existence of absolute rotational motion. Consequently, he argued, if an object can have absolute, though not always easily observable rotational motion, there must be both an absolute homogeneous space in which it moves, and an absolute, smooth-flowing time during which it carries on its motion.

Whether this constitutes a justifiable ground for belief in an absolute time is not here my concern. But even if it does not, there is strong evidence that Newton might have posited the existence of this unobservable entity on a further and quite different basis.[44] Newton was a profoundly religious man. Presumably disturbed by some of the criticisms leveled against his view (*e.g.*, that of Berkeley, who argued that the notion of absolute space led to atheism), Newton added to the second edition of the *Principles*, twenty-six years after the first one had appeared, the famous *General Scholium*. In this he stated his belief in a God—eternal and infinite, omnipotent and omniscient. "He is not eternity or infinity, but eternal and infinite . . . He endures forever, and is everywhere present; and by existing always and everywhere, he constitutes duration and space. Since every particle of space is *always*, and every indivisible moment of duration is *everywhere*, certainly the Maker and Lord of all things cannot be *never* and *nowhere*."[45] He must therefore be in absolute space and absolute time, and the demands of the religious experience thus conveniently provided Newton with the guarantee for these entities which his empirical science could not conclusively establish.

7   *Leibniz (1646–1716)*

Leibniz stood between Locke and Kant in his conception of the role which the mind plays in our knowledge of the external world.[46] He agreed with Locke in his attack on innate ideas, but he argued that Locke had misconceived the notion of an innate idea. "There are," he insisted, "ideas and principles which do not come to us from the senses, and which we find in us without forming them, although the senses give us the occasion to become conscious of them."[47] He believed that Locke had not sufficiently distinguished the origin of necessary truths, whose source is in the understanding, from that of factual truths, drawn from the experiences of the senses and the activities of reflection. On the other hand, he was not willing to go so far as Kant, who, as we shall see in the next section, projected certain subjective forms on sensory contents which they in no way possess in and of themselves. The task of Leibniz, therefore, in solving the problem of time, was to show which part of the temporal aspects of the world is contributed by perceived objects and which part by certain ideas implanted in the human understanding.

Like Locke he began by saying that a succession of perceptions awakens in us the idea of duration; but he then quickly added that the sequence of perceptions does not *make* the idea, as Locke would be inclined to say. "Our perceptions never have a train sufficiently constant and regular to correspond to that of time, which is a uniform and simple *continuum*, like a straight line. The change of perceptions gives us occasion to think of time, and it is measured by uniform changes; but if there should be nothing uniform in nature, time would not cease to be determined, just as space would not cease to be determined also if there should be no fixed or immovable body."[48] It appears, then, that Leibniz, like Newton,[49] conceived of two different kinds of time—one which is potential and ideal, and the other which is actual and real. Thus when he distinguished space from time by saying that the former is the order of coexistence and the latter is the order of succession, he meant only that these are potential orders which may or may not be exemplified in actual occasions, and the ideas which they embody contain more than is involved in any existent situation. "Time and space are of the nature of eternal truths which concern equally the possible and the existing."[50] Space and time, in themselves and outside the world, are imaginary.[51] Space comprehends all places, just as time comprehends all durations; but places in space and durations in time, unless occupied, are as ideal as space and time.[52] "Time . . . is metaphysically necessary, and the same in all possible worlds; whereas the *existence* of time is contingent, since it depends upon God's free resolve to create the world."[53]

One might object that space and time cannot be relational; for they are quantitative, and relations cannot be. But Leibniz argued that relations exhibit distance or interval. For instance, ratios and proportions in mathematics have quantity, which is measured by logarithms; yet they are relational.[54] One part of time can be greater than another because the former can contain more successive and like states than the other.

Since time and space are ideal and potential, if there were no created things there would be no actual time and space. The immensity of God is independent of space, and his eternity is independent of time. This does not mean, as it did for Newton, that an infinite space and an eternal time are required for the residence of God, but only that if there were an actual world, God would be present and coexistent with it. To ask whether God might have created the universe at a different time than he actually did— say, one day later—is nonsense; for if one supposes that all the parts of the two universes have exactly the same relations among themselves, there would no longer be two universes but only one, since they would be identical. It is true that a given succession of events, say *A* preceding *B*, would be different in the two universes in the sense that in the one *A* would be present while *B* would be future, and in the other *B* would be present while *A* would be past, nevertheless the ordinal characteristics of the two universes would be the same.[55] This indicates that for Leibniz the essential feature of time is *succession*, and that the *uniqueness of the now* in coming out of the future and moving into the past is of no great significance.[56] But to disregard, or even to minimize, this feature of time is, as we saw in Section 1, to overlook one of its essential characteristics.

8   Kant *(1724–1804)*

Kant's view of time, drawn mainly in what follows from his *Critique of Pure Reason*,[57] reflected the obscurity and confusion of this important classic of philosophy. Written within only a five-month period, as Kant confessed in a letter to Moses Mendelssohn,[58] it represented the hasty condensation of twelve years of the most concentrated thought. And, to make confusion worse confounded for our purposes, at least one well-known authority on Kant argues that his notion of time is not really justified and is, in fact, the most vulnerable concept in his entire system.[59]

Kant can be best understood in terms of his reactions to Newton, Leibniz and Hume. While he agreed with Newton that absolute time is independent of content, he was unwilling to go along with the great scientist in making it into an entity having reality outside the mind of the individual. Leibniz, he argued, was wrong in believing that time is an idealization of existing relations between things. Kant felt that the uniqueness of time prevented it from being an abstraction from experience. Hume, drawing the skeptical conclusions to which Locke's empiricism led, denied that the fundamental concepts of our knowledge have any intrinsic necessity or absolute authority, and argued that they are merely useful principles for organizing our experience. Kant's problem was to show that time has the *a priori* character which Newton's view implied. But he believed that it is not so closely tied to the actual relations of things as Leibniz supposed, yet it has a status which, in contrast to Hume's skeptical view, makes science possible as a legitimate study.

Since Kant's view of time is in many respects similar to his view of space, I shall restrict my discussion to the former. But some preliminary definitions are needed. By *intuition* Kant meant the direct experience of sensory contents, which he calls *phenomena,* and the capacity of receiving representations of these contents he calls *sensibility.* Every phenomenon contains both *matter* and *form.* Matter is that which corresponds to the contents of sensations, but form is that which affects the contents in such a way as to order them into a manifold. Form cannot itself be a sensation. However, there are pure representations in which the matter of the phenomena is not present and only the forms are given. These are called *pure intuitions* and constitute the *forms of sensibility.* They exist *a priori* and have no correspondence with anything in the sensations themselves.[60]

Kant stated his arguments that time is a form of intuition very briefly.[61] First, time cannot be an empirical conception since its essential characteristics (coexistence and succession) cannot be perceived by us unless we have some prior notion of time in our minds. In other words, sensations cannot be observed as temporal if we do not already know what is meant by "coexistence" and "succession." Second, we cannot think of phenomena as outside of time, yet we can readily think of empty time. Objects, therefore, can be annihilated from thought, but time cannot. This makes time logically prior to phenomena. Third, only on the supposition that time is a form of intuition can we explain why it is impossible to think of a two-dimensional time or of two coexistent times. This incapacity of our minds is due not to the fact that experience reveals no such notions but to the unthinkable character of the notions. Fourth, time is not a generalization from different times, for different times are merely *parts* of one and the same time; hence time is an *a priori* form which interrelates phenomena into a temporal manifold. Fifth, conceptions of time-segments, *i.e.,* limited durations, are possible only on the assumption of an unlimited or infinite time; but this cannot be an empirically derived notion and must consequently be given as a prior form of intuition under which phenomena are perceived.

Kant seemed to be unsure whether time is the form of internal sense or of external sense. Space is obviously the form of our external sense, because we attribute extension and figure to phenomena in space. But time seems to be a form of our internal experiences since, when we dream or carry on imagination, the contents change and occur in succession. On the other hand, Kant noted that one of the best examples of change is motion, and this is certainly a spatial phenomenon requiring the use of our external senses. Furthermore, because time possesses no shape or figure, we try to supply this want by analogy, and we find that we can successfully represent time in space by a straight line extending to infinity. Hence time seems to be really a form both of the internal and of the external sense. But in either case it remains a form contributed to reality by the mind, and not a property of objects in themselves.

Kant's general conclusion was that time is not something that subsists by itself (as Newton argued), nor does it inhere in things as an ideal relational structure which is presupposed whenever we observe them (as Leibniz insisted). It is merely a form of intuition "built into" the mind in such a way that we *must* see phenomena as temporal very much as we *must* see things as red when we are wearing red glasses. Time is not a property of things but a property of the instrument by which we view things. And since we have no instrument other than mind for observing them we are compelled to see the world as temporal. This was the answer to Hume's skepticism. It provides time with an objective status with reference to all objects that can ever be presented to our senses, yet it saves its subjectivity because apart from the mind time is nothing.[62]

9   *Bergson (1859–1941)*

The period between Kant and Bergson witnessed the development of the theory of evolution, and Bergson's conviction that time is the key to the understanding of reality can be at least partly explained by his belief that a mechanical theory of evolutionary change cannot account for the emergence of novel forms. Furthermore, Bergson was strongly interested in psychology, and wrote important works on memory, the comic, perception and dreams; in fact, the subtitle of his great work, *Time and Free Will*, was *An Essay on the Immediate Data of Consciousness*. This suggests that although his position harks back to Heraclitus, he believed that time is not merely change into opposites within a generic identity, for this would permit no novelty ever to emerge. Furthermore, he argued that time is not best exemplified in spatial movements of the outer world but in man himself, in the overlapping of mental states and their gradual transition into succeeding states as we experience the *passing* of time.

There are, said Bergson, two possible conceptions of time. One is that of duration as "the form which the succession of our conscious states assumes when our ego lets itself *live*, when it refrains from separating its present state from its former states."[63] The other conception of time arises when "we set our states of consciousness side-by-side in such a way as to perceive them simultaneously, no longer in one another, but alongside; in a word we project time into space, we express duration in terms of extensity, and succession thus takes the form of a continuous line or chain, the parts of which touch without penetrating one another."[64] For the purpose of characterizing these two forms of time I shall use Bergson's word, *durée*, to describe the first, and, adopting the terminology of Miss Cleugh,[65] employ the phrase *spatialized time* for the latter.

I shall consider the second form first. Objects in space may be said to derive their boundaries from the fact that they have relations of coexistence and are separated from one another by intervals in a homogeneous medium. Now suppose we try to think of time as a similar homogeneous and unbounded medium. In order to succeed we should have to abstract from the flow of duration and imagine the successive stages in time as being given all at once. Furthermore, and this is very significant, we should be forced to think of the parts of time as possessing sharp boundaries, as do material objects in space.

But if we examine time as it really exhibits itself in the unfolding of states of consciousness, we discover that it has neither of these properties. We never experience our psychic states as simultaneous, existing alongside one another. Nor do we ever find such states set off sharply from one another as discrete elements in the succession; they blend into one another in such a way that the present contains both memories of the past and anticipations of the future. "We can thus conceive of succession without distinction, and think of it as a mutual penetration, an interconnection and organization of elements, each one of which represents the whole, and cannot be distinguished from it or isolated from it except by abstract thought."[66]

Time, in this latter sense, is *durée*. *Spatialized time* is that which we obtain when we arbitrarily convert temporal order into simultaneity and project it into space. By this act we lose qualitative heterogeneity, and pure homogeneity remains. This is clearly indicated when we try to measure *durée*. Strictly speaking, the measurement of a lapse of time is simply the counting of simultaneities; for outside of the individual there is never anything but, say, the position of a hand on a clock when the past positions have disappeared and the future positions have not yet come into existence; only because the individual endures is he able to picture both the past and the future positions of the hand at the same time that he perceives the present one. This permitted Bergson to say that within the ego there is succession without mutual externality and outside the ego, in pure space, mutual externality without succession.[67] *Durée* cannot therefore be measured, because numbering implies that the numbered elements be completely excluded from one another. And experienced time, if described quantitatively at all, must be characterized as an *intensive* quantity, while spatialized time must fall in the category of *extensive* quantities.

Once we have realized the artificiality of spatializing time, the problem of the freedom of the will has been settled. Determinism and mechanism claim that, given certain antecedents, only one resultant action is possible. *Dynamism*,[68] which is the term Bergson employed to describe the action of the self, implies that the very indecision of the self in making a choice between alternative future acts transforms the choice, when it finally occurs, into something quite different from the mechanical effect of a mechanical cause. Suppose that I am trying to decide between two possible actions, X and Y. As I reflect on these alternatives my self grows, expands and changes. Just as in *durée* the present contains memories of the past, so my self "lives and develops by means of its very hesitation, until the free action drops from it like an over-ripe fruit."[69] My act, therefore, springs from my whole personality, and this is what we mean by saying, when we speak of a free act, that the contrary action was equally possible.

10   *Samuel Alexander (1859–1938)*

Without doubt the most comprehensive and painstaking analysis of the idea of time ever made was given by Alexander in his *Space, Time and Deity*, which constituted the Gifford Lectures for 1916–18. In the opening sentence of Chapter 1 he stated, "All the vital problems of philosophy depend for their solutions on the solution of the problem what Space and Time are and more particularly how they are related to each other."[70] Beginning with space and time, or, more accurately, with the whole which consists of a hyphenated space-time, he showed that this is the fundamental stuff of which the universe is constituted, the basis for the categorial properties which all things possess, and, by virtue of the creativity of time, the origin from which matter, life, mind and, ultimately, deity itself emerge. One is not surprised when he stated that he and Bergson were the first philosophers to "take time seriously."[71]

His method, he claimed, is empirical throughout; he simply looks at the universe and tells us what he finds. That this process of "looking" involves analysis of what is given, and frequently abstraction from what is concretely presented, he readily admitted. But he claimed to use no arguments; in fact he stated in one place[72] that he had a distaste for arguments. As a result, the comprehensiveness of his system and the dogmatism in some of his formulations of its features make a brief review of his position almost impossible. I shall confine myself, therefore, to an examination of the manner in which he united space and time into a single whole, and to the way in which he conceived of this fundamental stuff as possessing a power to emerge into ever higher and higher forms.

Since we have no special sense organs for the perception of either space or time, we must become aware of them in one of two ways. Either we determine their character by what they contain, *e.g.*, space by figures and shapes, and time by successions of happenings, in which case we are led into the error of believing, as Leibniz did, that both space and time are merely relations between events and objects; or we rid them of their contents and think of them as empty extents or undifferentiated processes, in which case we end with Newton's unexperienceable absolute space and time. Alexander proposed that we examine them by a method which permits us to recognize the empirical or variable aspects of the world, and at the same time to employ abstraction and construction by the intellect in order to detect its constant and pervasive features. This method he called *intuition.*[73]

In the use of the word "intuition" the transition from Bergson to Alexander is an interesting one from the point of view both of method used and outcome achieved. That Alexander called his method *intuition* is indeed unfortunate, for it is quite unlike Bergsonian *intuition.*[74] Alexander's method is analytic and conceptual. Bergson believed that this method is not illuminative but destructive of its subject matter; it renders static that which is flowing, and breaks into parts that which is an organic whole. Hence it is poorly adapted to the understanding of time. Furthermore, the view of time at which Alexander arrived is quite different from that of Bergson. Alexander found such a close relation between space and time that he set up a combined space-time as the basic stuff of the universe. Bergson, as we have seen, argued that time is fundamentally different from space, and loses its uniqueness if any spatial representation is joined with it or substituted for it. Yet, in spite of these basic disagreements, Bergson and Alexander both sought to find in time the explanation of the creativity in the evolutionary process.

By use of the method of intuition, according to Alexander, we grasp space and time as infinite and continuous wholes of parts. They are infinite because each finite space and finite time is part of a wider space and time, and they are continuous because between any two apparently adjacent parts of space or of time another can always be found. Alexander's essential contribution to the understanding of space and time lay in his conception of their necessary interrelationship into an integrated space-time. He was obviously acquainted with the work of Minkowski[75] and agreed that the latter's conclusions were "in spirit" the same as his own; but he felt that Minkowski had merely given a mathematical representation of the universe, which did not justify asserting that the world actually was a four-dimensional one.[76] Alexander insisted that "space is in its very nature temporal and time spatial,"[77] and he used the analogy of the mind-body relationship to clarify what he meant.

There are roughly four considerations which, in my opinion, seem to be of major importance in Alexander's position.

1) If time existed independently of anything else, there could be no continuity in it; the past would be gone, the future would not yet be, and there would remain only a series of "nows," perpetually renewed. But in this case the temporality of time would disappear. What is needed, therefore, is some continuum other than time which can secure and sustain the togetherness of the earlier and the later. This continuum is space.

2) If space existed independently of anything else it would also lose its continuity. For space, taken merely as a whole of coexistence, has no distinction of parts; and a continuum without elements ceases to be a continuum. A part of space, consequently, is selected by the passage of time through it; a point is something that endures. The conclusion is that there is no instant of time without a correlated point of space, and no point of space without a correlated instant of time. We can say that a point *occurs* at an instant and an instant *occupies* a point, and since neither can exist without the other the world is a complex consisting of point-instants, or what Alexander called *pure events*.[78]

3) The correlation between space and time is even more intimate since the three dimensions of space are related to the three dimensions of time. Now Alexander's use of the word "dimension" to characterize the three properties of time is a source of great confusion. By the three-dimensional character of space he meant what is ordinarily understood by both the geometrician and the man-in-the-street, *i.e.*, the capacity of space to be represented by a coordinate system of three lines intersecting at right angles. But the tri-dimensionality of time is something quite different; the three dimensions of time are successiveness, irreversibility and transitivity

27

(or betweenness). Alexander's correlation of the successiveness of time with a one-dimensional space seems clearly warranted. But he then went on and, in a very obscure passage, claimed to show that the irreversibility of time demands a two-dimensional space, and its transitivity a three-dimensional space. My only comment here is that he seems to me to be demonstrating not the correlations of the so-called dimensions of time with the dimensions of space, but merely the difficulty of *representing* the dimensions of time spatially.

4) Still another way to clarify the relation between space and time is to point out that time is the mind of space and space is the body of time. But the route which Alexander asked us to follow in order to achieve this recognition is a long one, and we must travel it gradually. The basic stuff of the universe, we found, is pure events. These acquire qualities by the manner in which space and time are interrelated. One of the simplest qualities which can be attached to pure events is motion. This is clearly a way of correlating space and time. And since space-time is the stuff of the universe, motion, which is a special type of correlation, will also be a substance. Now in much the same way matter is a complex of motions,[79] life is a special quality which emerges when matter takes on a certain complexity, consciousness appears in the higher forms of life,[80] and there are indications that the universe is now engaged in bringing forth a still higher quality which will characterize what we now call *deity*.

We thus arrive at a hierarchy of forms of existence—a pyramid with a broad base which is space-time, and a pinnacle which is God. Each level rests upon the one immediately below it, and thus differs from it both in its height from the base and in its size. The new is obtained by simple addition to the old; but with each higher level there is an increase in complexity which causes the emergence of new properties; life is not just matter but matter plus reproduction, growth and adaptation; consciousness is not just life but life plus self-awareness. The appearance of novelty at each level is due to the role which time plays in the space-time complex. Alexander said that it possesses a *nisus*, much like Bergson's *élan vital*, which "as it has borne its creatures forward through matter and life to mind, will bear them forward to some higher level of existence."[81] "The existence of emergent qualities thus described is something to be noted . . . under the compulsion of brute empirical fact, or, as I would prefer to say in less harsh terms, to be accepted with the 'natural piety' of the investigator. It admits no explanation."[82]

But while it admits of no explanation it can be clarified by analogy. Of the original unity of space and time it would be impossible to state that an instant can contemplate a point, or one pure event can contemplate another. But after time has progressed sufficiently long, and through its creative activity has produced first matter, then life, and finally mind, we ourselves, as combinations of life and mind, find that we can contemplate both our minds and our bodies—the former, through introspection, and the latter, especially through neural processes but also through observation. Alexander then argued that this body-mind complex, which we ourselves exhibit and can therefore understand, provides an analogy in terms of which we may better grasp the structure of space-time. That body is spatial no one would dispute. That mind is temporal is not so clear, even after Alexander's analysis. To be sure, the *recognition* of time is mental because its three dimensions—succession, irreversibility and transitivity— are disclosed to us through memory, expectation and the direct apprehension of the present. But since Alexander admitted that the analogy between body-mind and space-time is not perfect, we may pass over this difficulty. To the extent to which the analogy is successful we may conclude that time is the mind of space and space is the body of time. "In any point-instant the instant is the mind or soul of its point; in a group of points there is a mind of those points. . . . In space-time as a whole the total time is the mind of the total space."[83]

11   *Conclusions*

As we reach the end of this historical sketch of the attempts on the part of the philosopher to understand time, can we say that the problem has been solved? Certainly not. Much of the language employed, though suggestive, is too vague; unresolved paradoxes are still present; and many "solutions" are achieved through the generous use of analogies and metaphors. Has the problem been clarified? Probably, to a certain degree.

What about the difficulties which stubbornly remain? They might be met in several ways.

The easiest method would be for the philosopher to admit frankly that the problem of time is insoluble because time is one of the ultimate irrationals of the world. It is something whose presence must be admitted as a brute fact but whose understanding is forever beyond his grasp. This would be an admission of defeat and would discourage, if not forbid, all further investigation into the subject.

A second escape might be to claim that the problem of time is insoluble because the philosopher has not yet developed a logic which is adequate to handle it. If this is the case, then he will have to extend or broaden his logic so that it may be capable of resolving at least some of the apparent contradictions which now seem to be inherent in our understanding of time.

The most promising possibility, I think, lies in an extension of our vistas. I have necessarily restricted my view of time, with certain exceptions, to that seen through the eyes of the metaphysician, the rational man, trying to *understand* time without including his religious hopes, ethical ideals or aesthetic aspirations. Even if I had been more critical and analytic in my approach, and even if I had presented a wider and more diversified range of theories about time, I should still have given a distorted picture since man is an imaginative and valuational, as well as a rational, creature. Furthermore, I have of necessity, again with some exceptions, excluded the specialized information which the methods and tools of technical science have now made available. This omission also encourages the drawing of unwarranted conclusions since science has greatly expanded the area of available data concerning time, and has enabled theorizing activities based upon such data to proceed in ways never imagined by the abstract metaphysician.

The interdisciplinary character of the ensuing essays will assist in the removal of these restrictions.

Time as Succession and the Problem of Duration

FRIEDRICH KÜMMEL

Translated from the German by Francesco Gaona

The problem concerning the nature of time cannot be separated from the many forms of its interpretation in the history of thought. Even a purely analytic method is necessarily linked to the historical problem, for the very character of our contemporary conception of time is but the result of a long intellectual development. This fact could suggest that time, having no reality apart from the medium of human experience and thought, is relevant only in its various historical interpretations, for no certain criteria of the truth of such interpretations can be accessible to us. True, no single and final definition of the concept of time is possible since, far from representing a unitary self-contained form, such a concept is always conditioned by man's understanding of it. It would however be useless just to review the great variety of time theories or to limit our effort to the observation of a few of their abstract characteristics. The problem of time can, in our opinion, be solved only by progressive specification and differentiation of *all* questions related to it. In order to attain this end an interpretation of several theories of time is unavoidable. But we must remember how closely intertwined with the thought of a given epoch such theories have always been, and that as a whole they may be adequately understood only in the context of that thought. In Greek philosophy, for example, the main concern was the nature of the present, for only in time as present were the Greeks able to root their particular cosmic sense of life. In contrast, the far greater interest of Christian philosophy in past and future, in the beginning and the end of time, as well as in the course of history as such, and in the relation between time and eternity, depended wholly on the character of salvation as an historical event and the consequent valuation of time as a consecrated time of "salvation."

An analytical investigation into the nature of time, however, cannot consist merely of an attempt to synthetize various points of view since the problem is not only an historical one, but also one immediately vital to man. In addition to the historical theories we must also study more direct time forms, such as are found in various life situations; forms as varied and divergent as the theoretical concepts themselves. Here too, only by means of close attention to concrete phenomena and by avoidance of formalistic, foregone conclusions may the question be solved. A philosophical contribution to the present interdisciplinary study will not, therefore, seek to give an all-inclusive frame of reference within which the time concepts of the other disciplines are to be contained, but rather must, in its own way, seek to strike out in as concrete a manner as possible.

In order to limit our subject we shall start with the assumption that the question of time is primarily and essentially an "anthropological" question, that is, inseparable from the problem of man. This does not, however, make it a purely "subjective" problem. It means simply that although a great variety of temporal phenomena have a reality independent of man, man is nevertheless the only being on earth with an *awareness* of time, a being for whom the problem of time is not merely one of theory but one which is supremely and intimately related to the conduct of his life. Man, unlike other beings, is not merely chained to time: he is the one being who can determine the order and content of his time. Freedom means to him essentially freedom to dispose of his time. The indeterminate and unformed time which lies before him appears to him as an unbounded possibility. Yet it is also time that gives him the sharpest sense of his own limitation. In this double consciousness of power as well as of impotence in respect to time man apprehends a challenge he must overcome. This continual struggle of life against the *destructive* power of time causes us to forget, however, that it also has a positive aspect, that it represents a *sustaining* power as well, without which life could not exist. In this profound *ambiguity* of time is rooted the antinomical nature of its determinations.

Within the limited space at our disposal we shall discuss the following two questions: 1) The problem of duration. If time is understood as a succession of present moments, the question inevitably arises as to the possibility of a duration in time; 2) As the characterization of time as a simple before-after relation is not adequate, the problem of the specific nature of the relationship between past, future and present must be investigated in reference to this first relation.[1]

1   *Succession and Duration*

We conceive of time either as flowing or as enduring. The problem is how to reconcile these concepts. From a purely formalistic point of view there exists no difficulty, as these properties can be reconciled by means of the concept of a *duratio successiva*. Every unit of time measure has this characteristic of a flowing permanence: an hour streams by while it lasts and so long as it lasts. Its flowing is thus identical with its duration. Time, from this point of view, is transitory; *but its passing away lasts*.

This alleged identity and permanence encounters difficulties, however, as soon as we cease to consider time under its formal aspect (free from all content), and direct our attention to things which exist *in time*. They also take part in the movement of time and are altered in it. The concept of change seems at first sight to contain within it precisely such an identity of incessant alteration and permanence, since all that changes is necessarily altered while at the same time it persists.[2] This formulation proves, upon closer examination, to be merely a superficial covering up of difficulties which soon reappear when we examine the question more closely. All that is, does not remain the same even as it alters. Very special conditions must be satisfied before one may assert the identity in time, the duration of a given existent. The permanence of marble, over which time passes almost imperceptibly, differs so essentially from the permanence of a living being that one questions the usefulness of joining them under the same concept. Permanence in change may in strictness be predicated only of an existent which is centered in itself, of a living being, and even then a vast difference remains between the "self-centeredness" of an animal organism and that of a conscious human subject.

Without entering into the question of these differences, let us at first consider generally the nature of the permanence or duration of life forms. Organic existence is involved to a much greater degree than is inorganic in the stream of time and must be in a state of constant transformation and renewal in order to persist within it. The more highly organized life becomes, the greater the significance that time acquires for it in a negative as well as in a positive sense. Whereas time taken abstractly is an unfathomable continuum in endless flux, life *in* time exists both in the infinite chain of a species and in individual, finite forms. In spite of this double existence each one of these individual forms of life (and not merely the species) has a tendency to persist as it appears in two directions, in the continual transformation and renewal of the organic substance and in the tendency to ossification. For the individual form of life the relation of its duration to the flow of time may no longer be expressed by means of an abstract identity. The life impulse of the individual form may, of course, be spoken of as a form of duration in change, but only while the form lives. Its finitude means that while the unceasing flux of time is the basis of its duration, it at the same time brings with it its end. Even if for a formalistic argument duration and succession are resolved into one abstract identity, the persistence and the passing away of a concrete individual can in no way be considered as constituting a synthetic unity, but rather a perpetual opposition of contraries. The individual as a unitary form of life must endure by opposing its permanency to its transitoriness in time. Although the course of the individual's life corresponds to the course of time, this happens only in the sense of the struggle of the one attempting to annul the other. The antagonism is reflected in the contrary direction of the two "motions": The flux of time is experienced as a movement from the future toward the past; life itself, however, is directed toward the future, either pressing forward against the flow of time or, if incapable of doing so, being carried away by the flow of time and perishing.

This antinomy of transitoriness and duration and with it the ambiguity of time itself, emerges as an original form of life experience and is not merely the product of abstract reflection on time. If reason finds itself caught in the contradictions of time, that is because life itself is already entangled in them. By contradictions of time, however, we do not mean abstract paradoxes such as the Eleatic puzzles concerning motion. These mere conceptual paradoxes arise only in a very special context; they do not have any concrete relevance to the experience of motion. The antinomies of time, however, are rooted in life itself and therefore they affect its course.

The fundamental kinship of the problem of time to the problem of life is yet to be fully appreciated. The usual question concerns the nature of time in itself, either as duration or as succession. While, abstractly considered, duration and succession may appear to be identical, attentive investigation will reveal incompatibilities. Yet, the concepts remain inseparable. Duration without succession would lose all its temporal characteristics and would have to be conceived as an unchanging present. Pure succession without duration is equally inconceivable. A theory of time cannot, therefore, impose an either/or alternative between these two characteristics, but must seek instead to understand them in their necessary correlation. This means that only the two taken together describe time as a whole and that each aspect can be grasped only by reference to the other. Duration arises only from the stream of time and, conversely, only within the background of duration is the emergence and our awareness of succession possible. This easy correlation is illustrated by clock time whose flow remains indifferent to any particular content, and appears to contain within it no further problem.

The solution suggested by the last sentences, however, is valid only from a purely abstract point of view. As soon as one attempts to do more than merely comprehend conceptually the flow of time, and begins to seek for the concrete conditions which make an individual duration within the flow of time possible, the apparent abstract correlation between duration and succession at once reveals itself as an immanent contradiction. Abstract time, we have said, cannot adequately support the concept of duration, for real duration is predicated on the existence *in* time of individual, self-centered forms of life. The unity of life as an inner form or entelechy, is the decisive trait of each form of duration presupposing always, as it does, that something remains the same even as it alters in time. If time, however, in itself possesses no such entelechy and in its absolute fluidity signifies only the dissolution of things, it can manifest itself only in its transitoriness and can never possess real duration. In the relation between the flux of time and the duration of life in time, there emerge a contradiction and a disharmony which remained necessarily hidden to a purely abstract formalistic investigation. It would be merely sophistic (and in any case pertinent only to clock time) if we were to seek the solution to this problem by saying: time lasts *while* it passes. One would then have to ask: does time "last" indeed? does it possess duration only while and so long as it passes, as the measure of the clock shows us? Or is it not true that the clock registers a mere succession, but never a real duration? Does not real duration express principally a real *abiding* quality of time, and therefore inevitably a real *unity* in the *coexistence* of the past with the present and the future at one and the same time?

We intend to show that such an abiding quality, such a concrete duration, has a real existence in time. To accomplish this, however, we cannot take the definition of time as succession for our point of departure. If something is to abide, to endure, then its past may never be simply "past," but must in some way also remain "present"; by the same token its future must already somehow be contained in its present. Duration is said to exist only when the "three times" (put in quotation marks when used in the sense of past, present and future) not only follow one another but are all at the same time conjointly present. (The coexistence and integration of the "three times" does not mean simultaneity, for in their interrelation

35

succession is not excluded at all.) Thus formulated, the opposition between duration and succession is no longer to be overlooked: to define time as succession signifies that not all of its moments are present conjointly as parts of a whole, but that they follow and thus exclude one another. Conversely, the coexistence of the "times" means that a past time does not simply pass away to give way to a present time, but rather that both as *different* times may exist conjointly, even if not simultaneously. With this, however, we have already anticipated future considerations. We had found that, taking succession as a starting point, it was not possible to account for duration, and that recourse to an apparently alien subject matter (the self-centered organic individual) had to be taken in order to discover a form of unity in time which was not to be found in a mere before-after relation. It now appears possible, however, that the *basis* for such an individual unity may be found *in time itself* within the relation of temporal coexistence between past, future and present, a relation which can be conceived not only as one of succession but as one of conjoint existence, too. We are thus in a position to describe time as enduring insofar as the past does not wholly disappear and the future is in some way always already present. As, however, this is never valid of abstract but only of concretely realized time, we may also say that the living form is a necessary presupposition of duration and that, conversely, *real duration itself makes possible the living form.*

We chose as our point of departure the vain attempt to comprehend duration by the concept of a flowing duration from the point of view of succession. If life signified only incessant flow as successive time, it could not last; its form gives life the capability of enduring in the face of a time continually passing away. When one speaks in this manner of the effort of life to assert its power against time, he appears to be saying that time, in the final analysis, must retain the upper hand. It is therefore necessary to ask for the significance which this attempt of life to hold its own has, being confined as it is between a beginning and an end. Can we consider this act of enduring a mere vehicle for the preservation of the life of the species? Although this could be true of animal life, it is not adequate to describe the striving for permanency on the part of man.

What is now sought is a time which is itself duration and which within itself contains all concrete duration. This is qualitatively different from the mere incessant prolongation of time as a successive stream. To be sure, life in its striving after duration also strives for temporal prolongation. If, however, this were the only possible way of realizing duration it would have to follow that the higher the form of life, the higher also would be the degree of its resistance and consequently of its duration. Clearly, however, this is not the case. Time, for an organic existent, is to such a great extent immanent, that it must constantly alter itself in order to be able to endure at all. Vital duration and change exclude one another so little that a permanent fixation or a pure vanishing successivity would mean the end of the individual form. Life can, in other words, as little extricate itself from the stream of time as it can give itself entirely up to it. Neither abstract duration nor the stream of succession can be adequate to it, and it is not possible adequately to derive and describe the process of life from these concepts.

The conflict in which the individual form of life finds itself consists in the fact that these two possibilities are indeed open to it, and being unable to accept either alternative it finds itself continually tossed from one to the other. It will be our task to describe another conception of time which, being in conformity with life, not only makes its vital duration possible but also realizes and sustains it. This vital duration cannot exist in a time which is always passing away: life must either strive to escape this vanishing succession or be lost in it. On the other hand, as life in fact is realized within this process of succession, the conditions which make duration possible must already be contained within it. Time must, therefore, be understood as correlation of succession *and* of the coexistence of the "three times" which is the basis of vital duration. This correlation will allow of no other description than as a continual opposition or conflict between its elements, a "harmony of opposites" in Heraclitus' sense.

In order that we may discover the nature of time in vital duration as the coexistence of the "three times," and likewise discover that which distinguishes it from time as a succession of present moments, we must first orient ourselves by examining the nature of man's life, where the problem of duration is not only manifest but represents the sense of a conscious task. Moreover, just as in the life of mankind the significance of the individual transcends his usefulness as a mere element in the preservation of the race, so whenever man equates the duration of his life with a simple prolongation, he demonstrates that he has not yet grasped the meaning of true duration: for in the genuine experience of duration the flow of time seems to stand still. Not that there is then no more time: time goes on, but it loses all its negative externality and retreats to the intrinsicality of a latent and sustaining reality. Only when man steps out of the circle of this lived duration, of this *temps vécu*, does time as abstract succession emerge as an intended substitute.

True duration has a temporal character peculiar to itself, distinguishing it essentially from the experience of the flow of time; although neither aspect of time ever appears in an isolated form, one usually passes imperceptibly into the other. Vital duration is always concrete, fulfilled duration, so that time from this aspect is always immanent to life. As the sustaining basis of life itself, it loses its separate character, remains hidden and, as it were, stands still. By means of this contact with the intrinsic character of life, time loses all rigidity and determinateness; indeterminate and malleable, it opens to man a free horizon and all thought concerning its beginning or its end ceases. Vital duration, then, is experienced as an "always," and as an eternal present. Where, on the contrary, time emerges in our consciousness, the realization of life has always been interrupted and a certain emptiness makes itself manifest. It is at this point that time begins to be experienced as a flux and is transformed into a succession of mutually displacing moments, its wholeness having been destroyed. This means that only the inner, intrinsic, "fulfilled" time of vital duration can act as a sustaining time, while the extrinsic, superficial time is always negative in its effect wherever it alone has the dominant role.

In so distinguishing vital duration from the flow of time, we run the risk that it may then no longer be considered as forming part of time at all, and that in speaking of it one will be speaking of eternity. Whenever it is interpreted as eternity, however, the term is used only to indicate that one's experience of it differs radically from that of transitory time but is, nevertheless, an experience of time.

The description of duration as an experience of eternity is too far removed from the daily realization of life and too much limited to very rare situations for this description to amount to more than a hint, that succession alone cannot explain the complete essence of time. The extreme nature of this experience may serve as an example to clarify certain aspects of duration; these, however, must thereafter prove essential to all of life, that is to say, prove of importance also *within* the stream of time no matter how faint the mode of their appearance in this element may be. We have, therefore, spoken of duration not as an exceptional experience "beyond" time but rather as a *sustaining* element of time and as wholly "given" *in* temporal experience, although in its *pure* state it does not appear to be experienced as time at all but rather as eternity. We must, therefore, now go one step further and ask: what must the nature of time be if it is to be considered as the bearer and sustainer of life and not only as its destructive element?

The obvious solution to this problem would be an analysis of time by means of the distinction between an *internal* and an *external* time such as Schelling suggested.[3] Here the *external time* is understood as the framework for the possibility of duration but not as its foundation; *internal time*, on the other hand, is described as providing foundation for concrete duration in all its articulation and fulfillment. Internal time is never abstract and indeterminate but forms always an immanent aspect of each individual organism. The whole complex of problems centering on an inner clock of plants and animals is to be understood in terms of such inner time. One may oppose to inward time the successively flowing, formless, and form-negating external time. The internal and external aspects are not to be isolated and opposed, however, but always considered as participating actively together in each organism. If only internal time had reality, death would have no meaning and, conversely, where only external time ruled, life would come to an end.

This distinction, however, does not suffice: if internal time as a foundation of individual duration is able to prescribe to life, in a manner peculiar to each species, its temporal order, we must go further and ask whether or not for man, given the free conduct of his life, a quite different aspect of reality arises. The question must be referred to man, not only because for him alone is time "open" in his consciousness, but also because time plays, in a positive as well as in a negative way, a far greater role in his life than in that of other organic beings; a role in which he himself takes a decisive part. The fact that he can put to himself the problem of the possibility of his duration means at once that time is no longer predetermined and fixed in its form and order. Contrary to other organic beings who are fully subjected to the law of their specific time-form, man is able to condition the order of his time himself. This is possible because he is able to relate himself freely to both his past and his future, and by means of these relations he is able to actively mediate his present. His peculiar form of life thus represents both the task and the result of his free activity. Time-consciousness is in this connection a constant presupposition, for only such time-awareness makes the realization of man's life possible.

If man, therefore, not only grasps his duration through his consciousness of time but also is active in the construction of his duration the question arises whether it is not now possible, within the horizon of man's life, to understand duration no longer as an isolated unchanging present or an internal time, but as a past-present-future relation which is an integral part of time. To achieve this we refer to our earlier statement that in order for something to endure, the past cannot be merely "past," but in an important sense must still remain to be rediscovered, must, in other words, some-how remain "present." This coexistence of future and past *in* the present cannot be predicated of time as a succession of present moments. On the other hand, the concept of succession need not be rejected by this coexist-ence, since it is already contained in the very concept of a relation between past and future. The concept of time as the interrelation and coexistence of future, past and present rejects the idea that at any one given moment only one time (the present) may be exclusively real, whereas other times (past and future) are either no longer or not yet. Only when such a false preconception, which as we shall show is the source of the contradictions in the traditional theories of time, is abandoned, can any progress be made in the investigation of the nature of duration.

When time is considered as a succession of present moments, thus deny-ing the reality of both past and future, then duration is excluded from temporality. If the past is thought of as being merely a "no-longer-present" and the future a "not-yet-present," the real basis of duration, that is the presupposition that past and future are inherent in the present, is removed. As the concept of duration, however, cannot be wholly eliminated from time, one is forced to symbolize duration by other concepts which are meant to represent it, such as the image of time past conceived as having entered into *eternity* and there to have stopped; or the image of an abiding *space*-time continuum; or, finally, by the conception of past and future, as non-present times, being contained in the remembering or foreseeing *con-sciousness*. Clearly, any theoretical attempt to explain duration which is conceived under the presupposition that time is a succession of present moments must necessarily go beyond this conception in order to embrace something foreign to it, namely, to make the idea of duration expressible. The recourse to other realms of reality in order to conceive the temporal aspects of real processes and relations thoroughly is legitimate and neces-sary for all concrete investigations, because time is inseparably inter-twined in the whole framework of the world. But when asking for the nature of duration, we must remain within the limits of the properties of time itself and explain their specific temporal (not their spatial, organic and spiritual) conditions. The problem, as Section 2 of this paper will show, consists in the understanding of time not from a point of view which is essentially alien to time, such as eternity, space-time or consciousness mentioned above, and not as mere succession, but as a *coexistent inter-relation of future, past and present*.

Through the interpretation of the structural relation between past, future and present, however, we are enabled to give a temporal interpretation not only of duration but also of those very concepts—eternity, space, consciousness—which are used to represent it; so that the objection against the introduction of extra-temporal elements no longer holds. Clearly a temporal interpretation of these realities as symbols of duration was always intended; since, however, succession had to be excluded wholly from eternity or space, duration, as a result of this delimitation, was transformed into something purely static.[4] In order to be contrasted with a *duratio successiva* it was made into a *duratio permanens* (*i.e.*, eternity). To the vanishing instant of time, an eternal present *nunc stans* was opposed. The repose and the movement of time were seen as radically excluding each other.[5]

This sundering of elements is made especially clear by the conception of time as a line. The image of a point in movement along a line expresses in a concrete manner the succession as well as the coexistence of all points in time. The correlative relation between coexistence and succession is made evident by the fact that the progress of the present can be expressed only against the background of a static line, while the line itself symbolizes a time-line when it is understood as marking the course of a point in motion. But while the image thus symbolizes time as a whole, it perpetuates the disparateness of its elements, so that repose and movement exclude one another. Either time as a whole flows and then the present is conceived as motionless, or else time as a whole is motionless and only the present progresses. More importantly, the image of time as a line requires its representation by means of a spatial image so that the succession of time-points (their existing one *after* the other) is transposed into a mere spatial contiguity. The spatial image is thus used to express duration as the presence of all past and future times in one single motionless present. Space as a symbol of time is therefore adequate only to express a static duration. The usefulness of the spatial image rests consequently only in the case in which its most disparate elements may be visualized within it, including those symbolized temporal interconnections which are otherwise not to be visualized. While in space both the "here" and the "there," plus the whole range of relations between them, are clearly visualized; in time, since only the "now" is given, neither duration nor the relation of that "now" to the other "nows" is immediately evident. It is the coexistent *presence* of space as a whole which qualifies space to represent a duration precisely as such a presence, at one and the same moment, of past and future time. This simultaneity of

41

place, however, is not identical with the coexistence of moments in time. Space is, in its extension, always actually present as a whole. The symbolic space of time, on the other hand, does not signify the simultaneity of places but embraces past and future times. An actual spatial presence is thus transformed into a temporal omnipresence as a symbol of duration, an omnipresence from which all temporal succession has been effaced. In the spatial image of duration, past, present and future are leveled into one all-inclusive present. Likewise, time as a succession of present moments can also be duration only if it is thought of as the coexistence of succeeding moments, and only in just such an all-inclusive present, containing all successive "presents." Duration as an omnipresence is arrived at by a negation of all succession, and conversely the succeeding moments exclude every duration from themselves. The conceptions of duration are unable to express the vital element which must always be associated with it, while the conception of succession loses all continuance.

If one rejects both of these one-sided theories of time (as either an omnipresence or as a succession of present moments) and takes as a point of departure the view of time as a *correlation* of future, past and present (in which past and future can coexist with the present without negating temporal succession and thus the necessary precondition of vitality), he may arrive at a more adequate description of time, and of the structure of the time-awareness and the nature of permanence. The spatial analogy, as we saw, while possibly satisfying the imagination, can only lead thought astray by eliminating the specifically temporal character of time. Recourse to spatial analogies represents, therefore, nothing less than the death knell of thought concerning time. The Greek heritage, in other respects so fruitful, has in this instance been more of a hindrance than a stimulus to thought. Reality, we had mentioned, was for Greek thought rooted in the spatially visualized order of the universe and therefore in the present-ness of time. But as this present was made up of vanishing moments, two sorts of presents were postulated: a world of transitory passing away of the present and a world of eternally present Being, resulting in the consequent elimination of all time. Later on, when it became impossible for Christian philosophy to deny the existence of time while it at the same time remained attached to the Greek tradition, an all-inclusive present was necessarily assumed as a foundation of time, and recourse was once more taken, although in a modified way, to the spatial analogy from which the problem had in the first place arisen. The circle was thus completed, although it represented for the philosophy of time, only a vicious circle.

2 *The Relation Between Future, Past and Present*

From the point of view of the philosophy of time, examined in Section 1, the nature of the relation between past, present and future is essentially one of succession: while a particular time exists as present, there is a time which "not yet" is but which will sometime come into being, as well as a time which, already having been, "no longer" exists. Time is therefore never present as a whole but is divided into the elements of a succession: two periods, delimited by the present and continually passing into one another, so that what was previously a future is "now" a present and will soon be a past.

In so describing the interrelation of the periods of time, however, we cannot help but feel that something essential has been left out. To be sure, succession is such a vital element of time that any state from which it is absent can no longer be experienced or conceived as temporal. If, however, time is defined exclusively as succession, then, as we saw above, duration can no longer be understood in reference to it, and an alien form is required (such as living organism or space), which can make the idea of duration explicit in the assertion of a permanence opposed to the successive stream of time. A similar difficulty arises when, in defining time as an actual present, the reality of past and future are called into question. Further, since it is possible in the continuity of time always to proceed further in the division of the present until one finally arrives at a limit which is a point without extension, the doubt can arise, in spite of the unmistakable evidence of experience, as to the reality of time itself.[6] If in spite of all difficulties we were to insist on the reality of time, we should be compelled to go beyond it and transform it into a transcendental time-consciousness or into an eternal present, neither of which would exclude past and future but would contain time as a whole.

Only a genuine dilemma could lead us to such an apparently unavoidable and at the same time unsatisfactory conclusion. Under the influence of the traditional preconception, the historical investigation of time was led into this difficulty. The contradiction arises from one fundamental presupposition: that one could conceive the relation between the "times" *only* as a succession and that time therefore could have reality only as a transitory present. This preconception must be questioned if we hope to make any progress in the study of time. Our problem will be to describe the interrelation of past, present and future not only as an obvious succession, but also and more importantly as a real coexistence. The spatial image of a time-line discussed in section 1 can no longer help us in distinguishing between time as coexistence and time as succession. The course of time as symbolized by the line is seen as an addition of successive present moments; that is, as a substratum of possible points of time. On this view, past and future are no longer to be distinguished from any other time periods which, relative to particular points of time, could be designated as "earlier" or "later." Whatever marks the specific difference of the present from other points of time,[7] such difference remains hidden in this diagram and is ignored in the calculation of time. Similarly, if time is viewed as successive, all essential difference between past and future is also leveled, and the question as to whether a relation of "earlier-later" point-moments is to be understood as lying in the past or in the future or coinciding with the real past-future relation, cannot be answered by means of this line image. All that can be said about past and future from this point of view is that neither one is a present. If one insists further on a positive determination, then they must be designated as a future- or a past-present. The only difference between them lies in their not being both "at the same time." Whenever the three time periods are thus conceived as forms of the present, time can only be understood as either a succession of present moments or else as an all-inclusive present. The second alternative leads to the region of the extra-temporal, however, since in time the "past" and "future" present and the immediate actual present are not to be reconciled.

There is in spite of this a possibility of escaping from this dilemma in that one can ask for the reality of the past and the future *as such*, and not merely in terms of the present. Only from the point of view of the independence of the periods of time can the possibility of their coexistence be conceived without contradiction. When their relationship is conceived as a row of "coming," "actual," and "elapsed" present moments, they inevitably exclude each other. Granted their independent individual nature and determination their coexistence can no longer be in contradiction with the present; all periods being conceived as existing at one and the same time.[8] Thus arises the apparent paradox that the amalgamation of time periods into a mere succession inevitably results in their mutual exclusion, whereas their distinction makes possible their harmony within an articulated unitary structural interrelation. Further, their coexistence may be considered the condition for a real continuity of time which is always based on a concrete relationship between past, present and future, whereas mere succession expresses only the abstract addition and relation of discrete units. Moreover, only a fully independent past (a past as past) can effectively influence the present. All this means that time is never a splintered series of successive moments but always a whole from the point of view of which the possibility of its fragmentation may be understood. Time is traditionally described as a *fragmentation* of successive vanishing moments; one can, however, just as logically assert the *integrity* of time based on the inner correlation and coexistence of its parts. *Only the two definitions taken together can fully describe the nature of time*. Since, however, as we have seen, they are contradictory, there is no question of there existing only one form or aspect of time but a variety of time structures and their corresponding modifications. The peculiar relation between past and future, their harmony or discrepancy with each other and with the present, will itself cause time to manifest itself always in a different manner.

With this we have at last arrived at a point of view from which the variety of temporal phenomena may be approached. The traditional dilemma arose, as we saw, from the view that past and future were merely latent presents, and that the question of their individual reality could be ignored, so that the last answer concerning the nature of time was the doubt of its reality. No positive determination could be given to the concept of time, with the result that all other essential concepts, such as form, substance, duration, freedom, had to be thought of as divorced from temporality and as existing in a purely ideal world. This separation of reality into a temporal-empirical and an intelligible world may be considered the result of the one-sided conception of time as a succession. All aspects of reality which are fulfilled within a temporal relation—the duration of the vital form, the consciousness as memory, freedom in connection with the possibilities of the future, the realization of the self in the integration of the personality—are removed from the context of time and assigned an extra-temporal reality. It is not possible, within the limits of this paper, to document these far-reaching questions. We must limit ourselves here to noting that it was the idealism of the nineteenth century with its "dissubstantialization"[9] of the concept of man's consciousness, which first opened the way for the consideration of his cultural-historical world in its temporal-historical significance and permitted, after a long period of stagnation, a deeper understanding of the problem of time. This new approach was made possible only because the attitude toward time itself had changed. Not only did time become considered as a fundamental aspect of man's nature but the new positive attitude toward it has brought a re-evaluation of history, by making history an instrument of existential analysis designed to replace metaphysics. Indeed, metaphysical thought must now be understood as conditioned by history and thus by time itself.[10]

The concept of time as traditionally developed is not adequate to the new point of view which we have tentatively designated as that of an inward, sustaining and integral time, and described as a structural correlation of future, past and present, understood not as a succession but as coexistence. In this new view of time the focus of interest has shifted from the present and the stream of time to the past and future; they are revealed as being essential for man's knowledge of himself.

It is not surprising that the renewed interest in the problem of time tended to rely on traditional concepts essentially foreign to it, concepts which remained in circulation because of the influence of the natural sciences and the philosophy of science, which in turn influenced the methodology of historical investigation and related subjects. This affinity with the scientific philosophy as well as with the old philosophical tradition whose influence had not yet been overcome, accounts for the initial interest of the new philosophy of time in investigations centered around the problem of flux. We shall attempt now to show briefly how more harm than good was done by the persistent influence of this concept which has been adopted, since Bergson, primarily by the psychologists in studies concerning the so-called experienced time (*temps vécu*).[11] We shall also examine briefly Bergson's concept of *durée* which is the source of these investigations.

The main basis of comparison for the description of *temps vécu* experience was the flux of time as exemplified by the common, "objective" clock time. The experiences themselves were considered as subjective departures from the experience of this time measure and with it the "experienced speed" of the time flux diverged greatly from that of the "objective" course. The time of concrete experience, it was held, could extend or contract itself, with memory very often providing a radically different, if not a totally opposite, impression from that of the actual experience. "Lived" time may speed up or be retarded, in extreme cases it may even appear to stand still. The cause behind these various time experiences remained, however, unexplained.

The changes themselves were thought to depend on the intensity with which life itself was lived, its relative emptiness or fullness, as well as on the most varied psychological factors. A definite correlation of the factors themselves with specific time experiences was not achieved and their origin remained obscure. The fact that a "full" as well as an "empty" time can either be extended or pass away rapidly, depending on the consciousness which experiences it, was not taken into account. Equally ignored were such determining factors as the degree of ratiocination (exact thinking) or immediacy, as well as desires, intentions, fears, etc., on the part of the experiencing subject. The quality of the experienced time was also seen to depend on whether the attention is directed to a past, a present or a future time, whether the subject "has" time or not; or finally, whether the experienced time is actual or exists only in the memory. Clearly, experienced time is conditioned by the whole of man's constitution and it reveals something of the human situation without, however, allowing any inferences concerning the conditions which determine it. The modifications of experienced time clearly reflect the most varied human attitudes to past, future and present, and can only be fully understood in terms of the interrelation of these attitudes.

It seems to us, therefore, a mistake to attempt to understand psychological time experiences from the point of view of a modifiable flux of time, for any intentional relation[12] to past and future on the part of the subject is incapable of being explained by this scheme. As we have seen, these relations can never be understood from the point of view of succession, as it is evidently something quite different to understand the intentionality of human conduct within the context of past, present and future than it is to describe merely the external process of its fulfillment. When this latter is understood simply as an event in a set of successive events, then its specific time structure in its intentionality has been overlooked and has been identified with the process of its occurrence. The incongruity between the temporal character of human action which is always directed toward the future, and the course of the time process, which is always toward the past, is made manifest, whereas in reality not their disparateness but their interrelation is what makes the peculiar dynamics of the life process understandable. The conflict between these two times (that is, the temporal character of human action and the time process) is best expressed in the experience of the latter as an alien, threatening power and of the first as an inwardness which we identify with the self itself in its intentionality. Thus the question once more arises whether a better attempt at understanding of the process of life cannot be made by grasping the conditions on which it really depends; that is, the various relations to future, past and present which serve as its foundation and therefore also as a condition of the manifold aspects of experienced time.

A similar objection must be made to Bergson's concept of duration, developed in contrast to the concept of quantitative time.[13] The context and interrelationship of the "three times" cannot be understood from the point of view of a *durée,* for the problem is on the contrary, how to understand a concrete duration in its presuppositions from the point of view of the temporal context itself.

It would at first appear as if in his concept of *durée* the realization of a coexistence of past and future *in* the present would be fulfilled. Bergson not only rejects any fragmentation of duration into a before and after, but also insists on the point that only an interrelation of all its moments may properly be considered duration. Past and future may not be dissociated from the present without destroying the unity of duration itself. Whatever could be wholly independent of its past and at the same time have no future, could not possibly endure, since the future represents the possibility and the past the sustaining basis of every concrete duration. Thus far Bergson's view contains no difficulties, although it makes a difference whether one expressly speaks of these concrete conditions of duration or, as in Bergson's case, one speaks simply of duration as a form of the present. To be sure, he describes this present *durée* as an indivisible temporal extension within which "before" and "after" are in a relation of vital interpenetration. An *integral* inclusion of both past and future *as such* is however no longer possible in his formulation of the problem: they are both fused into the one indivisible present. The reason for this sacrifice of the independent significance of past and future to the one indivisible *present durée* resides in Bergson's interpretation of past and future as succession which threatens duration.

The difficulty in the formulation of the Bergsonian *durée* has its origin in its being rooted in the traditional conception of time as a succession of present moments. Attempting to give expression to a new insight, but not yet having eradicated this hidden preconception, that is to say, not yet having grasped the idea of a coexistence of past and future, the Bergsonian *durée* is led to their negation and to a paradoxical formulation. For, the *durée* must extend, without disintegrating into a succession of moments; it must constitute a time magnitude, without being accessible to a purely quantitative reflection; it must, finally, contain a manifold which cannot, however, be separated and counted. This indissoluble interpenetration of all times in the present does not allow the past its own significance as a past. Imperceptibly, it extends into the present, giving it content and continuity and preventing it from becoming an abstract vanishing point in a flux. An effective differentiation of either, however, is now impossible. The effort to avoid the definition of the relation of past, present and future as "successive sum" has thus led to a definition of a non-differentiated duration. For Bergson, past, present and future may not be carefully distinguished for fear of destroying a duration which he has from the beginning defined as a lasting present contrasted against a background of successive time. A past *as* past, therefore, a future *as* future, may no longer be understood from this point of view.

Herein lies the limitation of Bergson's theory: a true succession, a ceasing to be and a new beginning, are as little to be understood from the point of view of his *durée* as is the real lasting presence of the past in the present or the vital relation of a present to a future in action, expectation or hope. The abstract relation of past to present and future which distinguished them within a successive flux had, at least to some extent, maintained their distinction. In the Bergsonian formulation, even this vestige of specification must be rejected as conflicting with the identity of an interfusion of the "times" in the present. If, in the traditional view, the specific differences were leveled, at least their distinction was maintained; with Bergson not only does the leveling remain, but all effective delimitation disappears in one enduring flux. Both for the traditional view and for that of Bergson, past and future are mere modifications of the present—either separately as the abstract moments of a present "gone by" or a present "still to come," or interfused in the one lasting present. In spite of the advance which the philosophy of time underwent as a result of the attempt to understand duration in its *vitality*, Bergson's solution remained bound to the traditional antithesis which had its source in one fundamental prejudice: Time is either a lasting *or* a flowing present. This accounts for the difficulty Bergson's theory had in freeing itself from the framework of the abstract-objective time as well as for its inability to grasp past and future in that part of their aspects which transcends their nature as a past- or future-present.

With these considerations as a background, we are now able to define our starting point for the analysis of the complex relations between man and time.[14] Any new beginning must necessarily concern itself with the great variety of concrete phenomena, since as we saw, the relations of past and future to the present can no longer fruitfully be analyzed from a purely abstract point of view. These relations are so varied in their content as well as in their structure that no simple scheme of a time sequence can possibly do justice to them. Within them the "three times" may show the most varied values. It is, therefore, of vital importance, for example, to ascertain whether the future, within a given concrete situation, is open as a possibility or closed, whether the past represents only a burden to the present moment or has been assimilated and become easy to bear; whether, finally, the present itself is full of content or only an empty thing. All these aspects are further to be investigated in close relation to such temporal phenomena as fear, hope, expectation, desperation, guilt, depression, etc., etc.—feelings which decisively influence the course of human life.[15] Only within the framework of these temporal relations is man able to realize his freedom: the openness of future and past is, in other words, the vital condition for the conduct of man's life and all his actions. Indeed, the main difference between animal and human life is the complete lack of time consciousness in the former. For man, the mediation of time transforms his environment into a world in which he can freely act; the animal, on the other hand, remains always limited to his spatial situation, living the strict correlation of organism and environment, incapable of choosing its world or the form of its individuality, constantly reliving the impersonal past of the race without ever being able to assimilate it. Unable to put a distance between the past and its own being, it is forever bound to the present, without possessing any future which could enable it to transcend itself.

Man, on the contrary, is able, by means of the latency of his future, to place himself at a "distance" from his immediate past. Existing within an open relation between past and future, he can dispose of his own time and condition of the mode of his existence. He creates the context of his life and is able to make the past his own by bringing it into a free and positive relation with the present. Wherever this mediation and integration of possibility and actuality is not fulfilled, future and past necessarily conflict with each other, and freedom, which is based on the presupposition of an assimilated past, is turned into mere impotent impulse. When there is no such radical antagonism, the natural discrepancy of future and past constitutes a productive tension which forms the real medium for new action and new mediation. No act of man is possible with reference solely to the past or solely to the future, but is always dependent on their interaction. Thus, for example, the future may be considered as the horizon against which plans are made, the past provides the means for their realization, while the present mediates and actualizes both. Generally, the future represents the possibility, and the past the basis, of a free life in the present. Both are always found intertwined with the present: in the *open circle* of future and past there exists no possibility which is not made concrete by real conditions, nor any realization which does not bring with it new possibilities. This interrelation of reciprocal conditions is a historical process in which the past never assumes a final shape nor the future ever shuts its doors. Their essential interdependence also means, however, that there can be no progress without a retreat into the past in search of a deeper foundation.[16]

Within the limits of this paper we can only outline the general structural relationship between the elements of the temporal context. The real task, which is still to be accomplished, lies in the exposition and interpretation of man's life—his freedom, memory, action, character, history—in its essential temporal character within this context. We can here only point out a few general aspects of this temporal context which differ radically from those which are derived from the conception of time as succession.

Within the context of the circular relation of past and future, time loses its tendency in one exclusive direction, so that man—Prometheus and Epimetheus at once—is always moving in both directions. Not tied to an insurmountable present from which he would at best be able to cast a fleeting glance at future and past, he lives mostly in the future or in the past, and leaves them only to return to a present which he must continually transcend in order to make it his own. This incessant interweaving of the "times" does not however, as in Bergson, imply their fusion. For only the past *as* past and the future *as* future are able to make the present, entering into it and giving it a foundation. To remember something, for example, means that we "re-present" it, but always at the same time suppose it as a past, and separate from the present.[17] Memory thus accomplishes a double task in one act: it is able to "represent" the past while understanding it *as* past. Only if the past is removed to a "distance" is memory able to enter into a free relation with the present. The mechanism of repression has exactly the opposite effect. Whatever is merely repressed without having been previously freely assimilated is never truly past but breaks through once again at the slightest pretext. As something which has been merely negated, it transforms itself into a negative value. It is only in giving up all opposition to it and accepting it that a free relation to it can be achieved, a relation characterized by the fact that from that moment a distance has been created between the consciousness and the event. Only then does the event become past. Only by means of such "re-presenting" assimilation does the repressed event cease to be a burden and able to enter into a free relationship with the present. Negation cannot therefore be the primary attitude of man to reality for it cannot create the distance vital to life, which thus becomes itself negated. Only when man stands in a positive relation to reality does reality become accessible to him. He is thus able to come in contact with reality while he at the same time creates a distance between reality and himself. Only an originally affirmative act can place reality objectively before him; negation remains always a subjective self-imprisoning act. The affirmative act of memory and the negative reaction of repression thus constitute the two possible attitudes of man toward the past. Memory alone is able to subdue and integrate the past into a free sense of the present; repression, by reinforcing the antagonistic power of the past, causes the engulfing of the present by the past.

As this example shows, the concept of an integration of the "times" does not in any way annul their succession, but presupposes it in such a way as to make possible a free relation between them. Precisely because each time has its own function are they able to be integrated into an organic whole, whereas a past which, for example, always remains present, is only a hindrance to the emergence of a new situation in the present. In the differentiation of the "times" by means of memory, however, the consciousness already assumed that independently of the will of man time is fragmented and is in a state of flux; this is an experience which forms the steady background of all human life. In its independence from man, time becomes a burden only when he either rejects or insists on holding on to it forever. When rejected it never becomes a true past but only a mere negation and man discovers that the experienced time, by not really becoming a past, does not pass away even when he wishes it to do so. When, as in the second case, he wants to hold on to it so that it does not pass away, he discovers that he cannot stop it and that it disappears before his eyes. In both cases the permanence and the flux of time become negative and unfree experiences; in both cases man is the instrument of a compulsive reaction rather than a freely acting agent and is overcome by the counteraction of the tyranny of time. A free relationship is achieved only when one allows the past by means of its assimilation to be a true past, which at the same time creates the possibility of its recall. No longer does there exist a contradiction between the enduring and the successive aspects of time. Time returns to itself and brings the present into a state of vital repose without either making it rigid or canceling it. A true reconciliation then takes place in which duration and succession are constantly creating each other. Succession loses the inexorability of its passing, while duration gains through succession both flexibility and vitality; scattered time is collected and rigid time made malleable within the free mediation of the "times."

This last image remains, of course, a purely ideal one. Man's life is fulfilled within the most varied modifications between the extreme situations of either total integration or total disintegration of time into fleeting moments. Whenever it has not found its vital equilibrium it will always remain in danger of either ossification or disintegration, and precisely from the harmony and opposition of these simultaneous possibilities do the variety and the intensity of life as a hazardous enterprise arise.

It would be one-sided to fix one's attention exclusively on man's activity of time integration and to forget that a material of time must first of all be "given" to him to integrate; a material which at once forms the basis of his action and threatens it through an incessant stream which he cannot control. Equally one-sided, on the other hand, and indeed impossible, would be the establishment of a law of flux from the point of view from which life is to be understood. In spite of the partial truth contained in this view, man cannot possibly orient himself exclusively within it. The concept of the irreversibility of time, derived from it, can serve as a brief illustration of this. The idea of irreversibility cannot be completely rejected because it is an indisputable fact that time vanishes; it cannot, on the other hand, be unconditionally accepted, since time not only vanishes but also has a "lasting" effect and it is then no longer wholly irreversible. Whenever time is considered only from the point of view of its successive motion in one single direction, the past is then truly not evocable and stationary, and there is no longer any possibility of influencing it in any way. Such fatalism, however, does not take into account the real relation of man to his past, a relation in which the past emerges as not wholly irrevocable but capable of transformation from the present.

If one considers that the past is never wholly so, but always remains partially present in the context of life, one must then reject the idea of its irreversibility. In life one does not think of the irreversibility of time but hopes rather that what has happened may still be "made good"; he hopes, further, that no one decision has committed him forever and that he still "has time" to make up what he has missed; that, in short, what is once done can at least partly be undone. In such expectation not only the future, but the past also lie before man as an open field of possibilities for transformation. The whole of time must remain open to him in order that he may fulfill his life in it, and in this sense the past also must represent for him a potentiality so that he may then assimilate it into the process of his life. This possibility is not to be thought of as existing only in acts of repentance or forgiving—on the contrary, every return to the past by means of memory signifies already in some way a transformation of it. It is in this sense that one can say that history must be rewritten by each period, becoming in the process something new. This implies no subjective transformation of the "facts" of history; they remain constant, yet at the same time always receive a new significance from the present and acquire the possibility of having a new influence upon it; indeed, history itself, whenever it still maintains a vital relation to life, may be said to have lost its finality and to have become latent with possibility.[18]

53

3   *Conclusions*

What here must remain a mere fragmentary sketch requires a full exposition. Our purpose here has been to show briefly how, by means of an archetypal form of time different from the traditional one, new aspects of human life come to light. Time as a succession of present moments has not thereby lost its validity; we have only proved that it cannot express the whole truth of time, and that it must be transcended if human life in its temporality is to be fully understood. For this purpose, succession alone is not sufficient. An extensive analysis of the complex structural context formed by the interrelations of past, future and present in its most varied manifestations must be undertaken.

That time as succession could hold the attention of philosophy for such a long period in the history of thought, is not surprising. Time as flux, with its essentially destructive power, is always confronting man, whereas the positive "sustaining" time, being largely immanent to him, remains always partly hidden. It is not easily externalized or abstracted for its very inwardness constitutes a necessary precondition of the free form of life which is realized within it. This aspect of time was already known to St. Augustine and was the subject of the masterly studies of Schelling and Baader.[19] This aspect remained largely forgotten until Heidegger, in *Being and Time*, once more brought it to the attention of modern philosophical thought. This new, complex, and extremely productive philosophy of time, to which intensive investigations in other fields of the humanities show a parallel development, cannot yet be discerned in the totality of its results and influences. It would have taken us beyond the limits of this paper to discuss even a few of its special problems. Our task has been only to develop the new point of view by showing the way out of the traditional contradictions in which the philosophy of time so long had been caught. This appeared to us as especially necessary since the traditional concepts, inadequate for a description of the human temporal situation, are still in vogue, particularly in the natural sciences. Clearly, time as a succession of present moments, in its quantitative flux, is more adequate to a mathematical formulation than the structural interrelation of the "times," for their structural relation cannot be easily abstracted and made formal. Added to this is the difficulty that there are two temporal structures, one applying to nature and one applying to man's free conduct and actions, and that the structure in one domain cannot be carried over to the other. The main task cannot therefore be the formulation of a single, all-embracing concept of time. Instead, the temporal structure of the diverse spheres of reality must be differentiated as far as possible. Such a differentiation, as we have seen, is possible only against the background of the temporal interrelation of past, present and future, since the concept of a successive time flux tends, in its abstract universality, to level all differences and can at best only be modified. Only when this task of differentiation has been accomplished can the question as to the general characteristics of time as a whole be allowed to arise. Even then it is doubtful that a unitary definition of time can be attained, and perhaps the best we can hope for is an insight into the basic temporal structure of reality in its innumerable transformations. In our opinion, the greatest speculative progress toward such a unitary but differentiated theory containing within it the most contradictory aspects of time has been made by Franz von Baader, who not only called for but himself greatly contributed to the task of "not only of making clear that the fundamental problem of philosophy is to provide us with an integral theory of time, but also of facilitating its solution."[20]

Notes Concerning
Some Properties of Time

Philosophical speculation, represented by Dr. Kümmel's essay, uses as its tool, introspection, or in popular terms, reflection. The validity of introspection as a source of learning about reality has been questioned by some of the logical positivists who maintain that any acceptable description of reality must be capable of leading to propositions which are observationally confirmable. Unfortunately, the nature of time, perhaps because it may represent an instinctual ambivalence related to life and death, has so far defeated all attempts to achieve an exclusively scientific-empirical description of it independent of man.

From Dr. Kümmel's formulations of the properties of time, I have selected two for brief discussion because they are important for what we are going to say later. One relates to the succession-duration antinomy, the other to man's attitude toward time.

In the first part of his paper Dr. Kümmel examines what Professor Benjamin called "the third fact about time," that is, the reconciliation of succession with duration. In this respect one of two extreme views may be taken. On the one hand we may claim with Heraclitus that time is but a succession of present moments, change is real, permanence is not. We may label duration "not time" but since we cannot eliminate it, we must relegate it to extratemporal concepts such as *space*-time, *eternity*, or *consciousness*. This is an example of what was called in the Introduction the evasiveness of the idea of time. For, the explanation is passed on to physics, religion, psychology or, in any case, some field of knowledge other than philosophy.

Alternatively, we may claim with Parmenides that time is duration, change is a mirage and only permanence is real; or, as some contemporary physicists would put it, everything is "already written" and we just come upon events. This argument can easily lead to the banishment of the temporal from the problem of time, as is exemplified by the spatialization of time. Dr. Kümmel's point that the spatial nature of spatial figures ruins some of the most important aspects of reality contained in time is quite relevant to our study. The evasiveness of time would manifest itself here if we were to imagine that freedom of choice or awareness of the past could naturally be embodied in geometric relations, as indeed has been alluded to by certain interpretations of Relativity Theory. For, in this case, the problem of time would be passed on to the geometricians.

If we reject these extreme views we must attempt to understand time as a correlation of future, past and present. To find this correlation and solve the problem that Professor Benjamin raised in his "second fact about time," Dr. Kümmel proposes a theory of metaphysics, in which the self-conscious individual proves to be the instrument for unifying past, future and present by creating a special type of coexistence.

In the introduction to his essay, Dr. Kümmel speaks of man's simultaneous awareness of his power and of his impotence in regard to time. This ambivalent and uniquely human attitude has been keenly sensed by poets. The belles-lettres, in general, speak frequently about the puzzlement and tyranny of time. "Time—you know that—" writes Remarque, "is diluted poison, administered slowly in harmless doses."

Time in Shakespeare's powerful words is destructive for nature at large, for individuals—for everyone.

> O God! that one might read the book of fate,
> And see the revolution of the times
> Make mountains level, . . .
> The happiest youth, viewing his progress through, . . .
> Would shut the book, and sit down and die.
>> *—Henry IV, Part 2*
>
> Ruin hath taught me thus to ruminate,
> That Time will come and take my love away.
>> This thought is as a death, which cannot choose
>> But weep to have that which it fears to lose.
>>> *—Sonnets LXIV*
>
> Golden lads and girls all must,
> As chimney-sweepers, come to dust.
>> *—Cymbeline*

Opposing the destructive aspects of time is the overpowered man raising his stubborn voice.

> Do not go gentle into that good night,
> Old age should burn and rave at close of day;
> Rage, rage against the dying of the light.
> —Dylan Thomas
>
> There is not a woman turns her face
> Upon a broken tree,
> And yet the beauties that I loved
> Are in my memory;
> I spit into the face of Time
> That has transfigured me.
> —W. B. Yeats

Dr. Kümmel maintains that time represents not only destructive but sustaining power as well. For symmetry of illustrations, we may search the literature for poetic expressions of the sustaining power of time, but such statements are very hard to find. "Chronos begets his children before eating them, thus uniting in his person two functions, the creative and the destructive, which we attribute to time. But his cannibalism is more impressive than his procreative activities."[1]

We often blame time for what it ruins but seldom praise it for what it maintains or creates. This latter activity is commonly considered to be the numinous task of a Supreme Being, and is generally reserved for such a Creator. It is His messenger who declares victory over time in many languages and in innumerable forms:

> O death, where is thy victory?
> O death, where is thy sting?
> —St. Paul

To celebrate creation, to perpetuate what each epoch thought to be noble and in general to spite the destructive powers of time, religious, political and professional institutions have evolved and combined so as to form the scaffoldings of our gregarious as well as individualistic desires and aspirations. They embody our many attitudes toward change and form the subjects of inquiry of the following three essays.

We shall inquire into the Christian, Hindu and Japanese views of time, and question, using the example of China, their intricate roles in the development of scientific civilization.

J. T. F.

Time in Christian Thought

J. L. RUSSELL, s.j.

The theme of this essay is not easy to delimit with any precision. There is no "official" theory of time, defined in creeds or universally agreed upon among Christians. Christianity is not concerned with the purely scientific aspects of the subject nor, within wide limits, with its philosophical analysis, except insofar as it is committed to a fundamentally realist view and could not admit, as some Eastern philosophies have done, that temporal existence is mere illusion. Nevertheless, Christianity has a legitimate interest in time. It asks, for instance, whether there is a purpose or pattern in history indicative of God's love and care for mankind. Is the series of events in time oriented towards a goal? Does it show forth the progressive unfolding of some Divine plan for the world whose nature we can, to some extent at least, hope to understand, or does the whole process appear rather as a blind and arbitrary succession of one thing after another?

Such questions can be asked at three different levels relating respectively to cosmic time, historic time and individual time. Cosmic time relates to the ordering of events in the universe as a whole, and we may ask whether there is here any indication of a developing pattern which is relevant to Christianity. Similarly, at the level of human society, we may ask whether history can be recognized as subject to God's Providence or as manifesting His designs. And the same question can be asked of the individual human being: to what extent does or should his life reflect a single purpose, so that all its stages are ordered and integrated into a harmonious pattern? If history has a meaning, whether at the cosmic, social or individual level, then the time series will, to that extent, have a religious significance for the Christian.

All Christians would probably agree that God's Providence does in fact operate in human history, but there is room for wide divergences of opinion on the way in which this occurs and the extent to which we, in this life, can hope to recognize and understand His designs. Philosophies and theologies of history have been many and various; it would be of little interest to try to catalogue them all. The problem cannot, however, simply be dismissed, since it has had important repercussions on theology, philosophy and social thinking during the past two thousand years. I shall select three of its more significant aspects for discussion. These will be: 1) the significance of time as it appears in the New Testament; 2) modifications and developments of the New Testament view which occurred during the Middle Ages, chiefly under the influence of Greek philosophy; and 3) the influence of scientific discoveries and theories on Christian ideas of time from the eighteenth century onwards.

1 *Time in the New Testament*

Christianity stands out in sharp contrast to the pagan religions which preceded it by its firm insistence that it is a historical religion. From the beginning it claimed to be based on a particular series of events which really happened at a definite time and place. Jesus was an actual person, born in Bethlehem during the reign of the Emperor Augustus, while an imperial census was being taken in Judea. John the Baptist began to preach "in the fifteenth year of the reign of Tiberius Caesar, Pontius Pilate being governor of Judea and Herod being tetrarch of Galilee . . . under the High Priests Annas and Caiaphas" (Luke 3:1–2). The earliest official Christian creeds, which limited themselves to absolute essentials, stated that Jesus Christ "suffered under Pontius Pilate," not because Pilate in himself had any theological significance but in order to emphasize that the Crucifixion was a real historic event and not merely, for example, a nature myth couched in terms of death and resurrection, such as are to be found in some pagan mythologies. St. Paul asserted most emphatically that the actual physical resurrection of Christ was of the essence of Christianity: "If Christ be not risen again, your faith is vain" (I Cor. 15:17). And he gives a list of witnesses to the fact, many of whom were still alive when he wrote.

This concern with time and history was to be found also in Judaism but was generally absent from the other religious systems of the ancient world. The Greek or Egyptian devotee had no interest in establishing the exact date when Zeus overthrew Cronus or when Osiris was murdered by Set. Such events occurred in another age, at another time, even in a different sort of time from our own workaday world. His attitude was similar to that of an unsophisticated child towards the "once upon a time" of a fairy tale.

The question—when precisely?—would hardly have entered his conscious-ness.[1]

The historic reality of the Incarnation—of the birth, death and resurrec-tion of Jesus within a specific and identifiable historical context—is the central fact of Christianity, but it has never been viewed in isolation, as an event without antecedent or consequent. It is seen, rather, as the culminat-ing point of a long historical process ordered, from the beginning, to that particular climax. The record of that process is to be found in the Old Testament. I shall be concerned here with the Old Testament, not as understood by a modern higher critic or by an orthodox Jew, but as inter-preted by the New Testament writers and the Early Church. Making due allowances for subsequent advances in knowledge, it is still, in substance, the generally accepted Christian view.

The Old Testament is seen, then, as the history of God's dealings with His Chosen People, designed to prepare the world for the coming of Christ. It shows us the working out of God's purposes in time. The *mise en scène* of this history is sketched out in the early chapters of the Book of Genesis, where we are given a brief picture of the creation of the world, the creation of Adam and Eve and their establishment in an earthly para-dise where, if they obeyed God's commandments, they and their descend-ants would live in a state of primeval innocence, enjoying the continuous friendship of God. Adam failed to fulfill the conditions laid down. In consequence he, and with him the whole human race, forfeited the privi-leged relationship with God to which he had been called, and became subject to all the penalties of sin.

Human history begins, therefore, with man in a fallen state, subject to labor and hardship, suffering and death; caught up in the toils of sin; in a state of rebellion against God and incapable of returning to Him or of achieving the happiness of Heaven by his own efforts. However, the motif, even in the Book of Genesis, is not one of despair but of hope. Even as Adam and Eve are being expelled from paradise a promise is given that one of Eve's descendants would crush the head of the serpent which had caused their downfall (Gen. 3:15). From the beginning, God intended to restore to mankind the essentials of what had been lost, though in a differ-ent way and under different conditions. The plan involved a long period of preparation during which God protected, by His special Providence, one line of descent from Adam through Seth to Noah and thence to Abraham, Isaac and Jacob.[2] Finally, when Jacob's descendants, the twelve tribes of Israel, were led out of Egypt by Moses, He ratified a formal treaty, or Covenant, with them by virtue of which they were to be His Chosen

People. They would serve Him and worship Him as the one true God; He in turn would establish them in the Promised Land, would protect them from their enemies and eventually would send them a Messiah under whom they would become the religious leaders of the world—always assuming they fulfilled their part of the Covenant.

The remainder of the Old Testament shows God's designs working themselves out against a background of the weakness and instability of the Chosen People. The nature of the Messiah and the vocation of Israel were progressively clarified by the Prophets. As time went on, it became clearer that the Messiah was to be the saviour not only of the Jews but of all mankind; he was to undo the tragic effects of Adam's fall and restore mankind to the friendship of God. At the same time, the monotheism of the Jewish people was being purified and strengthened so that they could be effective mediators of God's plan for the Gentiles. By the close of the first century B.C., the preparation was complete. The Prophets had given their message; the Jews were firmly, even fanatically, worshipers of the one true God; almost the whole known world, as a result of Roman organization, was united and at peace, so that it would be possible to spread the good news throughout the world with a minimum of delay. Then, "when the fulness of time was come, God sent His Son" (Gal. 4:4). The Incarnation was the focal point of time. The history of Israel had been orientated towards it from the beginning; even that of the Gentiles had foreshadowed it (for example, in the mystery religions of the Greeks) and had cooperated in various ways to bring it about.

Thus we find the New Testament writers vividly conscious of their role as inheritors of the old dispensation. The Old Testament was always pointing beyond itself to the New. Not only were the obviously Messianic prophecies fulfilled in Christ, but many other passages suddenly acquired a new and unsuspected significance in the light of subsequent events: "God who at sundry times and in divers manners, spoke in times past to the fathers by the prophets, last of all, in these days, hath spoken to us by His Son, whom He hath appointed heir of all things" (Heb. 1:1–2).

Hence, up to the time of Christ, history, or at least that part of it which concerned the Chosen People and the plan of redemption, had a definite purpose and direction. It found its explanation and fulfillment at a particular moment of time when God became man in Bethlehem. However, it did not come to an end at this moment. The Incarnation was not only the end of the old dispensation; it was, much more importantly, the beginning of a new order of things. The problem of time and history arose once more, within the new context of Christianity.

2   *Patterns of the Future in the New Testament*

The most definite characteristic of historic time in the New Testament is its closure. Human history had an absolute beginning with our first parents and it will have a conclusion. The whole present order of things will come to an end with the Second Coming of Christ and the Last Judgment. When this will be it is impossible to say. Some of the sayings of Jesus seem to imply that it would occur within the lifetime of some of his hearers (Matt. 24:34), but elsewhere he asserted that the time was known to no man but to God alone; that it would come when least expected; that it might be in the evening, at midnight, at cock-crow or in the morning (Mark 13:32–36), and that the gospel must first be preached to the whole world (Matt. 24:14).[3]

Whether the interval between the First and Second Coming of Christ was to be long or short, the question arose whether we could still look for a significant pattern of history during this period. To this question the New Testament gives a somewhat qualified affirmative answer. Evidently there could be no very obvious structure with an exactly predictable climax since the date of the Last Judgment was unknowable. Nevertheless, some indications are given of a divine plan for the future as for the past, insofar as Christ commissioned his Church to preach the gospel to all nations (Matt. 28:19–20) and promised that the forces of Hell would never prevail against it (Matt. 16:18).

The idea of a Christian pattern of history is to be found more specifically in the writings of St. Paul, although here also it is only sketched out in very general terms. He likened the growth of the Church to that of a human body with Christ as its head and the Holy Spirit as its soul (Rom. 12:4–8; I Cor. 12:12–31; Eph. 4:11–16). Just as a body has many members—head, heart, hands, feet, etc.—which have different functions but are all necessary or useful, cooperating for the good of the whole, so the Church consists of many different members with diverse gifts—teachers, missionaries, healers, prophets, etc.—who all contribute to the well-being of the Church as a whole. And just as it is the nature of the body to become organized and thus to grow to its full maturity, so the Church, the Mystical Body of Christ, must develop and grow to its ultimate perfection by incorporating new members and so acquiring a richer, fuller and more varied life of its own. St. Paul clearly regarded this growth towards organic unity as a historical process: the actual visible Church on this earth would develop into a universal, unified society in which all types of men and all diverse human abilities would find their fulfillment in their union with Christ and, through Christ, with one another. St. Paul apparently envisaged the growth of the Mystical Body as involving not only human but also cosmic history. The passages in which he expressed this idea are obscure (Rom. 8:19–23; I Cor. 15:24–28; Eph. 1:9–12, 22–23) but it would seem that he regarded the whole of nature as being in some way involved in the fall and redemption of man. He spoke of nature as "groaning and travailing" (Rom. 8:22)—striving blindly towards the same goal of union with Christ to which the Church is tending, until finally it is re-established in that harmony with man and God which was disrupted by the Fall.

St. Paul's view of time is, therefore, on the whole, optimistic and progressive. History has a meaning and a direction in the future as in the past. Although he is far from painting a picture of automatic and inevitable progress all along the line, he insists that God has a plan and he appears confident that its fulfillment will become progressively manifested.

The earliest post-Apostolic writers shared, on the whole, St. Paul's optimism and his conviction that there is a pattern in history.[4] In the second century A.D., St. Justin, Athenagoras, Theophilus of Antioch and Clement of Alexandria regarded the classical Greek culture, especially its philosophy, as a *preparatio evangelica*—intended and providentially guided by God to prepare men's minds for the Gospel. This view was strongly supported by Eusebius of Caesarea (died *c.* A.D. 340). St. Augustine, in his *City of God*, saw the whole of pagan history, especially the Roman Empire, as subserving the designs of Providence.[5]

Many of the early writers accepted an intelligible pattern of history for the future as well as for the past. Irenaeus (died *c.* A.D. 202) regarded the whole of creation as being summed up or recapitulated in Christ. He saw the course of history up to the present as a growth of the human race from childhood (at the time of Adam) to maturity—but a growth which, as a result of sin, had fallen away from its true ideal. So Christ came on earth as a child, and grew to manhood, in order that he might sanctify every stage in human life and every period in human history. He recapitulated them in his own life and, as it were, presented them to God purified and divinized. The Incarnation gives to human life a new meaning and a new orientation. Henceforth all history—past, present and future—will be understood in its light. In union with Christ as its head, humanity will grow, according to God's plan, to new levels of perfection.[6]

Origen (*c.* A.D. 185–*c.* 254) put forward a somewhat similar idea in his theory of a universal restoration (apocatastasis) of all things in Christ. He held that, although at the present time there is disharmony between God and man, and between God and nature, nevertheless at the end of time a complete unity and harmony will have been re-established. Eventually all men, and even the fallen angels, will have become reconciled to God and restored to happiness.[7] A similar view was accepted by other early writers, including St. Gregory of Nazianzus and St. Gregory of Nyssa. In its extreme form, however, the theory of an inevitable universal restoration was later condemned by the Church.

3   *Time and the Individual*

We have so far been considering the movement of time as it affects the human race in general and the Chosen People in particular. But the human race is composed of individual persons, each living his own life and experiencing the temporal succession of events from birth to death. What, then, is the significance of this personal time in Christian thought? The Old Testament does not give much help here. It is concerned primarily with the history of the Israelites as a community whose vocation it was to bear a corporate witness to the True God. The destiny of the individual, at least up to the time of the Babylonian exile (586–538 B.C.), was left vague. In earlier times there was no clear teaching concerning a life after death; it was enough for an Israelite that he was a member of God's People, able to love and serve Him in this life and perpetuating his own name in his offspring. His life was inextricably bound up with that of his race. Such indications as there are of a life after death are indefinite and not inviting: a shadowy existence in Sheol without hope or purpose.[8]

In the later books of the Old Testament a more definite and cheerful picture of a future life emerges, especially in the deuterocanonical Book of Wisdom, written probably in the first century B.C. However, it is in the New Testament that the value and significance of the individual person are most clearly expressed. Every man is in one sense mortal and in another immortal. He is mortal insofar as his life on earth had a definite beginning in time and will come to an end when he dies. Death, however, is not the end of all things, since he will rise again at the Second Coming of Christ and will have unending life either in Heaven or Hell.

A man's destiny in the next life depends on his actions in this, which in turn depend upon God's grace freely offered to him and freely accepted or rejected by him. Human life in this world is therefore essentially orientated towards the future. Its purpose and value can only be judged in relation to eternity: "We have here no abiding city, but we seek that which is to come" (Heb. 13:14). Time has, moreover, a twofold aspect in the life of a Christian. In the first place he must order his whole life and every part of it with a view to growing in the love of God and union with

65

God which will come to full fruition in the life to come. Secondly, he cannot choose for himself how and when he will accept God's grace. Christianity has been called a religion of crisis, and this is a valid description insofar as every man must be prepared for critical moments when he finds himself, as it were, at a crossroads, faced with a specific choice between acceptance and rejection of God.[9] We cannot choose our own time for turning to Him. We must continually watch and pray in order not to be found sleeping when the moment of crisis arrives. This is a theme which recurs constantly in the New Testament: "Watch and pray, for you know not when the time is" (Mark 13:33); "The kingdom of God will come like a thief in the night" (I Thess. 5:2); "Behold, now is the acceptable time; behold, now is the day of salvation" (II Cor. 6:2).[10]

For the most part, early Christian writers approached the question of individual time from a predominantly Scriptural point of view; there was little attempt to analyze the nature of time as such. St. Augustine was, however, a notable exception. In his *Confessions*[11] he discussed some problems associated with the relations of past, present and future time from a purely philosophical point of view. But even here the main interest was theological: to contrast the transitoriness and contingency of created things with the eternal existence and perfection of God. This theme was later to be taken up by many medieval writers.[12]

The New Testament reflects a predominantly Hebrew cultural background. The time series is linear, not in the sense that we should picture it as a straight line of steady and inevitable progress, but in the sense that it had a definite beginning, at the creation of the world; it will have a definite end when Christ returns at the Last Day; and between these two limits it manifests, at least in a general way, the working out of a consistent Divine Plan. The plan may not always be clear to us; it may suffer, to our way of thinking, serious setbacks or modifications. Nevertheless, the movement towards its fulfillment does give a direction and purpose to the time series; it makes sense in a way in which it would not if, for instance, we thought of it as unfolding in the reverse order. Irenaeus, Origen, Augustine and many of the early Fathers of the Church belonged, in their own way, to this tradition.

4 *The Greek View of Time*

From its earliest days, Christianity began to be influenced by another view of time which derived from Greek philosophy. This originated with Plato and was developed or modified in various ways by his successors:

Aristotle, the Stoics and the Neoplatonists. Plato took as his starting point the eternity of God. God is eternal, not in the loose sense of existing for an infinite time, but in the strict sense of transcending time altogether.[13] There is no change and succession and, therefore, no before and after in God. He has His full and complete being in an eternal "now." The classical definition of eternal existence was given by Boethius nearly nine centuries later, but it would have been fully accepted by Plato: "*Interminabilis vitae tota simul et perfecta possessio*"; ("the totally simultaneous and perfect possession of unending (or 'unendable') life."[14] No other being except God can be strictly eternal, but the physical world is modeled upon this pattern and reflects it so far as it is possible for a created and mutable being. Every created thing strives to participate in and show forth God's perfection so far as it is capable of doing so. The higher and more perfect it is, the more perfectly it will do so. The most perfect physical being is the outermost heaven, *i.e.*, the sphere of the fixed stars, which Plato regarded as being everlasting and absolutely incorruptible though not strictly eternal. It is in a state of perpetual uniform rotation, since uniform rotation, inasmuch as it has no intrinsic beginning or end and no variation, is the nearest approach to an eternal activity which can characterize any created thing. Plato identified the rotation of this first heaven with time, which he called "the moving image of eternity."[15]

Besides the first heaven with the fixed stars there are the seven planets (including the sun and moon). These are also everlasting and incorruptible, but their movement is more complex as it is compounded of at least two distinct circular motions. Finally, below the sphere of the moon, are the terrestrial regions where there are no everlasting substances; everything is subject to generation, change and decay. These are furthest from the eternal nature of God.

The heavenly bodies exert a profound influence on the earth. This is especially true of the sun, whose changes of position relative to the earth are responsible for night and day, and for the succession of seasons throughout the year. But the moon and the other five planets also produce their effects, though in less obvious ways. Terrestrial processes at any particular moment are more or less completely determined by the configuration of the heavenly bodies at that moment. It was believed that the periods of the planets were commensurable with each other; hence at constant intervals the same configurations would reappear. The interval between two successive returns was known as the Great Year; it was analogous to an ordinary solar year but on a vaster scale. Like an ordinary year, the Great Year had a Great Summer when the planetary influences would combine to destroy the earth (partially or completely) by fire, and a Great Winter when it would be overwhelmed by floods. The length of the Great Year was variously estimated by Greek astronomers; 36,000 years was a fairly typical value and this may have been accepted by Plato.

Plato's theory of the Great Year was carried to extravagant lengths in later Greek times by the Stoics and Neoplatonists. According to the more extreme Stoic view, the whole history of the earth is exactly, and even identically, repeated in each cycle. Socrates has already drunk the hemlock infinitely many times in the past and will do so infinitely many times in the future; and so for every event in the universe. Others took a less extreme view but nevertheless accepted the general principle that world history substantially repeats itself over and over again, in endlessly recurrent cycles.

Aristotle was in general agreement with Plato concerning the links between celestial and terrestrial processes, but he approached the question from a rather different angle. Like Plato, he believed the celestial spheres to be everlasting and incorruptible—the nearest approach to God's eternal being. Like Plato also, he thought that they all exerted influences of some sort on terrestrial processes. But he was not much concerned with the theme of eternal recurrence. He referred rather casually to the Great Year in his work on meteorology,[16] but he seems to have regarded it as a long-term weather cycle of a less drastic type, not involving a complete destruction and renewal of the world and not producing an exact repetition of history. With his great interest in biology, he was much more concerned with the observable cyclic influence of the sun on the seasons and hence on the growth and breeding cycles of animals. All generation and decay were linked, in his view, with solar and, to a lesser extent, other planetary movements. It is in this way that corruptible terrestrial matter shares in the everlastingness of the heavenly bodies and is thus assimilated to the eternity of God. Although any particular terrestrial substance decays and ceases eventually to be, the cycles of generation are everlasting since they are dependent on the everlasting heavenly cycles. Hence in the terrestrial sphere it is not the individual but the species which endures forever. This was true for all species, both living and non-living. All have existed and will exist for an infinite time; none ever came into existence for the first time or will ever cease to exist.[17]

It will be clear from what has been said that the Greek view of time was radically different from the Judaeo-Christian. For the Greeks, cosmic and historic time were infinite in both directions; the human race had no beginning[18] and no end. History had a long-term cyclic pattern imposed by the planetary cycles, but it had no purpose beyond that of reflecting God's eternity. It was not fulfilling God's designs progressively nor moving to any final, irreversible consummation; man was there simply to be born, to perpetuate his species and to die.

In practice, the picture was not quite so negative as it might appear. Side by side with this arid philosophy of history there was a mystical tradition, both among the philosophers and in the more popular mystery religions, which taught that the individual soul can find its salvation and permanent fulfillment in the vision of God. The Stoics also held that there is a definite pattern of history within each cycle, determined by the divine *Logos*. But for the human race as a whole there was no long-term progress; only the endlessly repeated round of rise and decline, birth and death.[19]

For the first Christians, on the other hand, the world and the human race were both created at definite moments in the past; the present cosmos and the present order of human history were both destined to come to an end at a definite moment in the future. Between these two limits history had a purpose and pattern imposed by Divine Providence. It was orientated in the first place to the coming of a redeemer and in the second to the final establishment of the Kingdom of God.

The contrast can perhaps best be expressed by saying that the Greeks were primarily concerned with the eternity of God; the Hebrew Christians with the Providence of God. For the Greeks, God was the Eternal Being, beyond time and space, beyond all change and succession. Since he is also the ultimate ground and reason of the world, the world itself must reflect this eternity in whatever way a finite, changeable thing is able to. It will best do so if it is infinite in duration and if the fundamental processes which determine its history are cyclic, so that they can continue uniformly forever. Against such a background there can, in the long run, be no systematic evolutionary processes.

The Hebrew-Christian tradition did not, of course, deny the eternity of God. It is true that the Old Testament, in speaking of Him, does not clearly distinguish between eternity in the strict sense and infinite temporal duration, although there are a few passages in which the former seems to be implied.[20] But the New Testament explicitly asserts that God, and His Son, Jesus Christ, exist eternally.[21] Hebrew thought, however, was more immediately concerned with God's dealings with mankind than with His intrinsic nature. Inevitably, therefore, it tends to speak of Him in anthropomorphic language. He is represented, especially in the earlier books of the Old Testament, as deliberating, changing His mind, being angry and appeased, because we find it difficult to visualize His concern for mankind except in such terms. Under these circumstances, the concept of eternity will have a less dominating role than it did among the Greeks. It then becomes reasonable to expect that the time series will have an orientation towards some goal whose nature and purpose we can hope to discern, even if only dimly and uncertainly.

Christianity, from its infancy, has come under the influence of both the Greek and Hebrew ideas of time. It was the task of the Middle Ages to work out a cosmology which should, as far as possible, do justice to both: to see God as the ultimate ground of a universe which shows forth His perfections in the measure of which it is capable and at the same time to recognize the workings of His Providence as revealed in the Scriptures. The theory that cosmic and, still more, human history, are infinite in past and future was decisively rejected. The Great Year, in its most extreme forms, was equally unacceptable. It was inconceivable that God should repeat the same plan over and over again; that Adam should fall and Christ be crucified innumerable times in past and future. Origen[22] and Augustine[23] both found it necessary to attack the theory of eternal recurrence in their controversial writings. In spite of this, occasional echoes of the theory are to be found in the later Middle Ages.[24] It evidently had supporters in thirteenth-century Paris, since it was explicitly condemned by Stephen Tempier, Archbishop of Paris, in his famous decree of 1277.[25]

The Biblical teaching on history limited the extent to which Greek theories of time could be accepted, but within these limits it exerted a powerful influence on medieval thought. In particular, it tended to inhibit any idea of systematic progress either on the cosmic or human scale. The belief that terrestrial cycles of generation and decay, both in the inorganic and biological realms, were controlled by the eternally repetitive cycles of the heavenly bodies left no room for large-scale evolution either in the cosmos as a whole, in the organic realm, or in human history. In modern terminology the universe, including the earth, was regarded as a steady-state system, not as an evolving one. It is true that man was recognized as being, by virtue of his free will, partially exempt from the tyranny of the planets. Nevertheless there seemed to be no ground for supposing that human society could go against the universal law of the cosmos by achieving a systematic progress. Certainly there was nothing in the concrete situation in which medieval man found himself which could lend support to such an idea. He was surrounded by the ruins of a civilization which had been much more highly developed than his own. The histories of the Homeric, Egyptian, Babylonian, Greek and Roman empires must have strongly suggested that decline and degeneration are the normal lot of society, interrupted only for relatively brief periods by a transient flowering of civilization.

As a result, the idea that human history, taken as a whole, has a significant structure, reflecting a Divine Plan, disappeared largely from view. It was recognized that God is working out His purposes but we cannot hope to understand what these are. Generally speaking, the medievals had no philosophy of history.[26]

5   *Modern Science and the Christian View of Time: The Extent and Finitude of Historic Time*

The scientific revolution of the seventeenth century influenced religious beliefs in many ways, but did not have any immediate important repercussions on the Christian view of time and history. However, the rise of geological science in the early nineteenth century eventually necessitated a radical change in perspective. The work of James Hutton (1726–1797), Charles Lyell (1797–1875) and others gradually made it clear that the earth must have existed for very much longer than the six or seven thousand years suggested by a literal reading of the Old Testament. Some of the more fundamentalist Christians were deeply disturbed by this conclusion and stood out resolutely against it, but on the whole the new ideas were assimilated without much difficulty. In the 1830's, for instance, soon after the publication of Lyell's very influential *Principles of Geology* (1830–1833), two well-known Catholic theologians—Nicholas (afterwards Cardinal) Wiseman[27] and the Jesuit Giovanni Perrone[28]—were both teaching, in Rome, that the six days of creation need not be taken literally and that the universe may be very much older than had been supposed. Perrone added that this view was held by many Catholic theologians and Scripture scholars of his time. It was also held by many Anglican writers.

Evidence that not only the earth but also the human race is much older than seven thousand years was cogently presented by Lyell in his *Antiquity of Man* (1863). It met with more resistance than his earlier work, but caused no major upheaval.[29] More recently, of course, this particular horizon has receded much further. The date of the first appearance of man has been pushed back to perhaps a million years ago or more. The universe as a whole is at least several thousand million years old and it seems unlikely that purely scientific evidence will ever be able to establish with any certainty whether it had an absolute beginning or not. It is clear, however, that the earth and its living organisms, including man, have had a finite duration only. In principle the Hebrew-Christian finitist view of human history still stands, as against the Greek infinitist view. If the historic perspective has been profoundly modified, this would be regarded by most Christians nowadays as an enrichment rather than an impoverishment of the older tradition.

With regard to the future terminus of history there is little that science can tell us. Christian tradition has generally assumed that the end will come extrinsically as the result of a direct intervention by God and that it is, therefore, unpredictable in principle.[30] If this is the case, then it is outside the range of scientific investigation altogether. Abstracting from this possibility, however, it seems certain that the earth will ultimately, after some thousands of millions of years, cease to be inhabitable owing to long-term changes in the sun—always supposing that it has not been previously destroyed by some man-made or unforeseen natural catastrophe. Even so, the human race could still survive if it had by then colonized a planet in some younger solar system. If an evolutionary theory of the universe is correct, life will ultimately become extinct everywhere, but this would not necessarily be the case on a steady-state theory. For the present at least, the question whether human history will be subject to closure in the direction of the future as in that of the past, is theological rather than scientific.

6   *Modern Science and the Pattern of History*

Human History   Seventeenth-century advances in astronomy discredited the medieval belief that terrestrial processes are determined by the stars, but did not call in question the essentially static or nonprogressive view of history, although one well-known Catholic writer, Jacques Bossuet, did propose a linear theory of history similar to St. Augustine's.[31] He did not, however, attempt to lay down any precise pattern which it was to follow. On the other hand, G. B. Vico put forward a cyclic theory in his *Scienza Nuova* (1725), according to which civilizations more or less inevitably rise and decline in a series of stages conforming to a constant general pattern.

The theory that the structure of history is essentially progressive was characteristic of the latter part of the eighteenth century and was at first associated mainly with anti-Christian writers. Some of the French Encyclopedists looked upon human nature as readily perfectible; they tended to suppose that a few quite simple and straightforward reforms, such as better education or the abandonment of revealed religion, would suffice to produce a perfect society. But it was Condorcet (1743–1794)[32] who propounded the first systematic philosophy in which the whole of human history was seen as a steady and inevitable progress towards a perfect society which, he believed, would be achieved in the near future.

Christians, generally speaking, were very reluctant to accept any systematically progressive view of history. This was partly due, no doubt, to the continuing influence of Greek ways of thought and partly to the association, in men's minds, of theories of progress with atheism. The main objection arose, however, from current ideas concerning the nature of Original Sin. Official Protestantism, following Luther and Calvin, held that human nature has been totally corrupted by the Fall, so that we can never hope to see any general or consistent improvement in human society. Catholic teaching was compatible with a more hopeful view of history, but in fact was strongly inclined, at this period, to a pessimistic assessment of the situation owing to the Jansenistic climate of opinion then prevailing. The problem did not, however, become acute until the middle of the nineteenth century, when it became inextricably linked with that of biological evolution.

The Theory of Evolution   The theory that biological species are permanent and unchanging is logically independent of the theory that human history is nonprogressive, but there is a fairly close connection between the two. Both followed from the Greek view of time. Both were therefore accepted without serious question during the Middle Ages and, in the absence of any contrary evidence, continued to hold the field until the second half of the eighteenth century. The theory of evolution began to take shape in the writings of Linnaeus, Buffon, Kant and, more positively, in those of Erasmus, Darwin and Lamarck, at about the same time as the corresponding theory of human progress.[33]

Evolutionary ideas were widely known and discussed during the first half of the nineteenth century but did not arouse any great interest or hostility among Christians. It is true that Robert Chambers' *Vestiges of the Natural History of Creation* (1843) caused some indignation; it was, however, so speculative that it did not need to be taken seriously. It was not until the publication of Darwin's *Origin of Species* in 1859, followed by his even more controversial *Descent of Man* in 1871, that the problem became acute. Darwin's ideas were criticized by Christians on two main grounds. First, his theory of natural selection seemed to exclude design and purpose from the physical world. Secondly, his views on the evolution of man seemed to reduce human beings to the level of animals and to leave no room for the immortality of the soul. Some of the more extreme Bible Christians rejected the whole principle of evolution as being contrary to the Book of Genesis, but these were never more than a minority.[34]

Today the controversies which raged around Darwin's work have largely died away. Few Christians have any difficulty about accepting evolution, and the objections to natural selection can be seen to have been based upon a misunderstanding of God's relation to the created world. Many would still hold that Darwin's treatment of human evolution unduly minimized the distinction between man and the animals, but would nevertheless accept that there is a genuine genetic connection between the two.

However, even if the fact of evolution is accepted, it can hardly be said as yet to have been sufficiently integrated into the Christian world picture. The habits of thought of many centuries are not easily shaken off; we are perhaps only beginning to assimilate the fact that the Greek static view of time has gone for good. The way is open for a return to the more dynamic, unidirectional view of world history implicit in the Bible, but enriched and particularized by the findings of modern science. Among those who, in recent years, have attempted to reinterpret the Christian world picture in this way, the most influential has probably been the Jesuit paleontologist Père Teilhard de Chardin. I will therefore conclude with a brief survey of his leading ideas.[35]

7   The Theory of Teilhard de Chardin

Teilhard regarded the whole universe—inorganic, biological and human; past, present and future—as conforming to a single developing pattern, implicit from the very beginning and directed to a final goal of union with God. This pattern is manifested in different ways at different stages of world history, but it always conforms to the same general plan. The process is intrinsic throughout. It does not proceed by means of external interference or manipulation on God's part; He has made the world in such a way that it moves to its final goal by virtue of the innate tendencies of matter. Teilhard did not, of course, exclude the possibility of *de facto* interventions by God, *e.g.*, in the performance of miracles or in the raising of man to the supernatural order. But such interventions are concordant with the general tendencies of nature, enriching them and increasing their significance rather than simply negativing them.

The most characteristic feature of the pattern of history (using this term in its widest sense to cover cosmic as well as human events) is a tendency to "complexification." Simpler units tend to combine together to form more complex structures, showing a unity in diversity which was not present before. At the most primitive level, for instance, a proton and an electron show a very strong tendency to combine with each other to form a hydrogen atom. This represents the first stage of complexification. A hydrogen atom has a unity and structure which are not to be found in either constituent separately, nor in the pair of them before combination. It is a new sort of ordering—a unity in diversity—which opens up a whole new range of potentialities.

Hydrogen is the simplest type of atom but it is not, of course, the only one. Under suitable conditions of temperature and pressure, such as occur in the stars, more and more complex atoms may be built up until ultimately, by a variety of different processes, all the known elements have been synthesized. Under other less extreme conditions, such as are found on the earth, a new and higher level of structural unity appears when atoms combine together to form molecules. These again are ordered structures with their own mathematically intelligible unity, more complex if less stable than that of their constituent atoms. Simpler molecules, in turn, can combine to form more complex ones. There are good reasons for supposing that conditions on the primitive earth a few thousand million years ago may have been favorable to the building up of progressively more complex carbon compounds until, eventually, the first and most primitive living organisms appeared.

Life represents a new level of organization transcending that of a collection of lifeless molecules, just as a molecule has a higher type of organization than a collection of atoms or an atom than a collection of subatomic particles. And just as there is nothing miraculous or unintelligible about the transition from a proton and an electron to a hydrogen atom or of a sodium and a chlorine atom to a molecule of sodium chloride, so it was with the production of the first organism. In each case it was the result of a tendency to complexification within the bodies themselves, created by God with the intention that the ensuing effect should be produced.

Once life was established on the earth, the ordinary processes of biological evolution came into effect and ensured that there should be a progressive movement towards further complexification. This has occurred most clearly and significantly in the vertebrate phylum, where it is characterized by an increasingly complex brain and nervous system. Finally, when a sufficiently high level of organization had been reached, the stage was set for the appearance of man. Man differs from the animals by his power of reflecting upon himself, his actions, his motives and his ideals, thus enabling him to achieve a new and higher type of psychological unity, opening up unlimited further possibility of development.[36]

At the present time, the tendency to complexification has not been exhausted but it will no longer operate primarily at the physical level. In the future its most significant manifestations will be in the direction of an ever closer unification of the human race at the social level until ultimately there is produced a new "hyper-personal" unity of all men (or at least of all those who do not deliberately cut themselves off from the process). This is the climax of history and is called by Teilhard the "Omega Point."

Teilhard presents us, therefore, with a philosophy of history in which the whole process—from the first union of a proton with an electron in the primeval cosmic gas to the ultimate unification of the human race—conforms to a single pattern and plan. His theory differs from superficially similar Marxist and humanist theories of progress by its strongly Christian orientation. God is the origin and first cause of the process insofar as He made matter with these tendencies, with the intention that they should work out in the way they have. He is also the final goal since the ultimate super-society is not, and could not be, a purely human organization. The reason why God became man in the person of Jesus Christ was so that the human race could become united to God. The ultimate unity will represent the union of man with God (and, through man, of the whole physical world with God) in the Mystical Body of Christ. Essentially, Teilhard's theory is a development of St. Paul's vision of the Mystical Body, worked out within the framework of a modern scientific world picture.

Teilhard's ideas have their critics both among scientists and theologians. They will, no doubt, need modification on certain points. But in its broad general lines, his synthesis of the Biblical Hebrew-Christian view of linear time moving towards a climax with the scientist's insight into the innate capacity of matter to complexify would seem to provide a richer and more fruitful context for Christianity than did the static, structureless view of history which dominated Western thought during the Middle Ages and for so many years afterwards.

Time in Indian and Japanese Thought

HAJIME NAKAMURA *

1 The Notion of Time in India

The Indian conception of time is very different from what the Western mind regards as intuitively obvious. In Indian thought, time, like other phenomena, is conceived statically rather than dynamically. It is, of course, recognized that the things of this world are always moving and changing. But the substance of things is seen as basically unchanging, its underlying reality unaffected by the ceaseless flux. The Indian does not concede that we never step into the same river twice; he directs our attention not to the flow of water but to the river itself, the unchanging universal. Indian thought places a high value on universality, and the connection between this, and the static conception of phenomena, is of course not accidental. "The one remains, the many change and flee."

This static conception of time permeates Indian thought. It could hardly fail to do so, for it is present in the very forms of language itself, conditioning all philosophical thinking. In the classical Indian languages, there are no words which corresponded to the concept "to become." The verb formed from the root *bhū* can be translated as both "to become" and "to exist." These two aspects of perceived reality, conceived as antithetical by the Western mind, are not even distinguished. "To become" is merely an aspect of "to exist." The noun *bhava*, formed from the same root, can mean either "being born" or "existing";[1] in other words, to become is to be born. To express the idea of change at all, Indians had to make shift with the words *anyathā bhavati* or *anyathābhāva*—"being otherwise." Becoming is expressed in terms of being, dynamic is seen as a phase of static.

* The author and the editor wish to thank Clara Park of Williamstown, Massachusetts, for her valuable assistance in revising the manuscript.

The point of view permeates the language. The noun, which expresses the more stable and unchanging aspects of a thing, is in Sanskrit more likely to be used than the verb, and correspondingly adjectives are more frequent than adverbs. In classical Sanskrit,[2] indeed, especially in prose writings, it became usual to employ verbal nouns or participles instead of finite verbs. For example, the sentence "Because of the rain, the food appears" is expressed in classical Sanskrit as "Because of the rain, appearance of the food (is possible)." It has been the practice since ancient times to use the participial form instead of the finite verb to express the past tense, and it became a common expression in colloquialism of the later periods.[3] Sanskrit will also use an adjective, which is static in feeling, to express an idea which might take a verb in the languages of the West. The classic Western expression of the sense of flux uses a vivid and specific verb. "All things flow" (πάντα ρει). The corresponding idea is expressed in Sanskrit as *sarvam anityam*, "all existences are impermanent."

We find the same habit of mind conditioning the use of periphrastic forms. The periphrastic perfect, though seldom found in the Vedas, appears frequently in the literature after the Brāhmaṇas. "He went" becomes *gamayām cakāra* (literally, "he did going"). Again, the periphrastic future may be used to express future action.[4] For example, the word *gantāsi* (you are the one who goes) is used to express the meaning "you will go," thus directing the attention away from the action to the stable state of the actor.

The primacy of the noun is illustrated in the Sanskrit denominative, a category of verb not found in the classical grammar of the West. For example, the denominative *putrīyati* is formed from the noun *putra* (son) and means "to desire to have a son," and *svāmīyati*, from the noun *svāmin* (master) means "to regard as a master." Generally speaking, the denominative connotes the meaning of "to be . . . ," "to work as . . . ," "to regard as . . . ," "to desire . . . ," but the real emphasis of the word is on the noun.

Similarly, the meaning "to be able to," expressed in Western languages by verbs or auxiliary verbs, is expressed in Sanskrit by an adjective, *śakya*, or an indeclinable, *śakyam*. For example, *na devāsuraiḥ sarvaiḥ śakyaḥ prasahitum yudhi* (*Rāmāyaṇa* II, 86, 11) = *non potest proelio superari a cunctis dis daemonibusque*[5] (he cannot be conquered in battle by all the gods and spirits).

In Sanskrit, then, finite verbs are seldom used; the verb appears mainly as a verbal noun, and the nominal sentence is more often used than the verbal sentence. Usage of the infinitive of the verb is also limited; it is never used as subject[6] or as object. When it seems necessary to use the-infinitive as an object, an abstract noun formed from the root of the verb is used instead, thus directing attention from the changing aspect of the action to the unchanging universal: "to appear" does not equal "appearance."

The centrality of the noun is further illustrated by the absence in Sanskrit of the adverbial suffix which is common to all Western languages. Adjectives are converted into adverbs by adding ὡς in Greek, -*ment* in French, -*ly* and -*lich* in English and German. In Sanskrit, however, the accusative case of the adjective is used if it is necessary to modify the verb. Ablative and locative cases of adjectives may also be used adverbially. The adverb itself is not even acknowledged as a part of speech in Sanskrit.

There are other curious illustrations of this tendency to comprehend things through their static aspects. To connect two ideas, Western languages use such conjunctions as *and* or *then*; Sanskrit, in contrast, will express the same idea by adding the demonstrative pronoun *sa* to the subject of the sentence, as if "John runs and jumps" were to be expressed as "John running he jumping." The conjunction emphasizes the separateness of events; the demonstrative focuses on the subject, unchanging through time.

On the whole, then, Western people comprehend action through its changing aspects, while Indians tend to comprehend it attributively. In particular, many Indians consider that action is an unchanging aspect, even an attribute, of existence. Westerners tend to regard action as an active phenomenon while Indians tend to look upon it statically. In the sentence *sabbe saṅkhārā aniccā* (all things are impermanent), a basic idea of Indian Buddhism, *aniccā*, is an adjective. For an Indian, even the statement that "all things of this world are changing and moving" is not, as it was for Heraclitus, the expression of the changing aspects of existences, but the expression of a static and unchanging state.

In Indian philosophy the Absolute is generally explained as a Being beyond all temporal appearances. These exist and change in time; the Absolute, in contrast, is essentially static. In the Upaniṣads, the Absolute is repeatedly expressed as "Imperishable."[7] "Ātman is imperishable for it cannot be destroyed. . . . It is unfettered, it does not suffer, it is not injured."[8] "This is that great unborn Self who is imperishable, incorruptible, eternal, fearless, Brahman."[9] Early Buddhism does not lay emphasis on a metaphysical Absolute as such, but the same habit of mind is found in the principle of *pratītyasamutpāda*, later developed in Mahāyāna Buddhism, which states that nothing can disappear or arise. In Indian thought, as in the Sanskrit language, it is the idea of Being which receives central consideration.

Indian philosophers in general replace the concept of Becoming by three aspects of temporal existence: Appearance, Extinction, and Continuance. All three states are clearly conceived as static. They are referred to early in the Upanisads and are generally accepted by the orthodox schools of Brahmanism and Jainism. Buddhism also designates these as the three aspects of the conditioned or phenomenal being.[10] Other words which are considered equivalent to "becoming" (*vikāra, vikriyā, pariṇāma, vipariṇāma,* etc.) in fact express the specialization of the simple into the complex and should be understood as meaning "evolution" or "development," rather than "becoming." Indian philosophy contains a number of variations on the three basic states, and the Sarvāstivāda school, the most eminent of Abhidharma Buddhist schools, added a fourth, namely *jarā* or "decaying," which was interpreted as "changing to the other" (*anyathā-bhāva, anyathātva*).[11] This might seem to come close to "becoming"; the theory, however, was not accepted by all Buddhist schools, and Decay is no real analogue of Becoming as the idea appears in Western philosophy.

There are evident similarities here to ancient Greek thought, at least in its Platonic and Parmenidean aspects. Plato formulated the antithesis between Being and Becoming; he saw the true essence of reality as consisting of changeless, timeless forms. Geometry, as an investigation of the fixed forms of material bodies in space, was the typical pattern of science in ancient times, and in the physical sciences only statics was developed. Interest in the changing world of phenomena, however, was also an important element in Greek thought; "all things flow" is after all as Greek as Plato's ideal forms. Modern thought has concerned itself increasingly (though not exclusively) with Becoming; kinetics has replaced statics in the center of the physicist's attention, and mathematics has turned to analytics and algebra, in which *variable* quantities are examined. Modern thought is described as "progressive," "dynamic"; the unique contribution of Indian thought, in contrast, can be a kind of rest and joyfulness which may be very welcome to those who are tired of the frantic movement of their culture.

The persistent Indian conception of a transcendent reality as more important than the phenomenal world it underlies and sustains results in a kind of paralysis of the individual's sensitivity to time, if we understand "time" to mean the passage and flow of specific events in our experience. This paralysis manifests itself in a characteristic lack of time concepts which non-Indians regard as common sense. (Indian thought may show an intense preoccupation with other, more metaphysical senses of time; in the Vedic period time was seen as the fundamental principle of the universe, "Time, the steed, [who] runs with seven reins, thousand-eyed, ageless, rich in seed. The seers, thinking holy thoughts, mount him, all the worlds are his wheels. . . . With seven wheels does this Time ride, seven navels has he, immortality is his axle. . . . Time, the first god, now hastens onward"[12] But this is hardly the time in which human beings carry on their common concerns.)

Language, as usual, is where this lack of common-sense concepts is most clearly seen: the Indian people did not have a clear awareness of the discrimination of tense. Although in Sanskrit, as in Greek, there are five kinds of tenses, they are not sharply discriminated in meaning.[13] To indicate past time, the imperfect, perfect, past participle active, aorist and historical present are used almost indiscriminately,[14] and the frequency with which a given tense is used varies not according to meaning but according to historical period. The aorist is often used in the sixth century B.C., for instance, but in classical Sanskrit is no longer common. The discrimination between absolute past and relative past is not clearly made in the ancient Indian language.

In modern Hindustani as well, we find similar linguistic phenomena. The adverb *kal* means both "yesterday" and "tomorrow." *Parson* means "the day after tomorrow" as well as "the day before yesterday"; *atarson* means equally "three days ago" or "three days from now." The meaning of these terms can be determined only through context.

Since the lack of common-sense time concepts is built into the language of India, both ancient and modern, it is not surprising to find it manifested in Indian religion and historiography. The Buddha was born under a tree in the park at Lumbini, attained Enlightenment under a tree at Gaya, and entered Nirvana under a tree at Kusinagara. These three events, according to common-sense notions, must have taken place on different dates, yet they are all celebrated by Indians and South Asiatics on the same Wesak day in May. Indians have not exerted themselves to grasp the concept of time quantitatively, and have never written historical books with accurate dates. According to the Indian world view, the universe, world, and social order are eternal; personal life, however, is only one sample of a succession of lives existing repeatedly in limitless time. If one's life is conceived as infinitely repeated, it becomes meaningless. The idea of the transmigration of souls, the perpetual self-revolution of rebirth, has appeared only occasionally in the West, but in India it is a basic assumption of the common people as well as of philosophers. Passing phenomena, whether the events of the individual life or of more generalized history, have no real significance. It is natural enough that no importance is given to providing them with accurate dates.

We should thus be prepared to find the Indian conception of history very different from our own. Indian books of history are few in number, and these few are tinged with a fantastic and legendary color. They are not products of historical science but rather works of art. Usually they are written in verse. Indians are not satisfied with the simple description of facts in the language of daily use. They beautify the past and try to idealize it. They ignore precise figures, exact sequences of events, and other details of time and place. Far from exerting themselves to give exact sizes of armies, say, or expenditures, they exaggerate astronomically with magnificent and brilliant hyperbole.

As an example, consider the *Mahāvaṃsa*, the most reliable work of history produced in ancient Ceylon. Even this book, though highly informative from the modern historian's point of view, is saturated with a mysterious and legendary atmosphere. For instance, though Mahānāman, the author of the *Mahāvaṃsa*, lived in the fifth century A.D., in an age not too distant from the time of King Duṭṭhagāmani, his descriptions of this greatest of Ceylon's rulers are already full of fantastic elements, and the reader must make a careful distinction between myth and that which is historically true. The histories or "chronicles" of medieval European monks and the biographies of eminent Buddhist monks in China and Japan have a similar style, but the *Mahāvaṃsa* stretches historical truth to an incomparably greater degree.

Another example is Kalhaṇa's *Rājataraṅgiṇī*, the chronicle of a Kashmiri dynasty and one of the best historical works ever written by an Indian. In it Kalhaṇa details the social situation of his time and the activities of the various personages in it with an accuracy that no other Indian book of history has attained. Yet Oldenberg can still describe it in these terms:

If one removes all the poetic elements from Kalhaṇa's story, and compares it with events of the time, he will find that the account is in essence on a level no higher than that of a more or less accurate article in a newspaper or a cartoon in a political comic paper. The process of formation that this story has undergone is not that of historical thinking but that of poetry— poetry in the Indian sense with its brilliant quality and also with its weakness. And Kalhana himself has a very distinct idea on this point; he feels himself as a poet and he is a poet.[15]

It is worth pointing out that Kalhaṇa scarcely pays heed to causal sequence when considering historical events. His dates are inaccurate and sometimes clearly the products of pure imagination.

The Indians themselves have attached little significance to their books of history; most Indians have been much more interested in religion and poetry than in historical documentation. For the Indians, a minor error in the recitation of the Vedas has been a serious matter. But they have been thoroughly indifferent to the erroneous recording of dates or facts in their books of history.[16] This lack of historical consciousness is distinctly observable in the Buddhist attitude to the rules of their order. In the period after the death of the Buddha, Buddhists had to establish new precepts in order to meet changing social conditions. As some of the new rules were not compatible with the older ones, they hesitated to include them in the traditional books of ordination (*pāṭimokkhas*), and instead attached them to the *paṭimokkhas* as supplements. Although they would not alter the traditional books, however, they were not afraid to claim the authority of the Buddha's own teaching even for these supplementary precepts of their own creation, completely ignoring the historical facts. Their concern for the proper observance of the precepts was far stronger than their regard for historical accuracy.

This lack of interest in history is very different from what we find in China. The Chinese derive their rules of social conduct from the examples of their ancestors as set down in their books of history. The Indians, on the other hand, gain their principles of behavior from their religious books, and at the same time fables and parables such as the *Pañcatantra* and *Hitopadeśa* contribute toward the diffusion of practical morals into daily life. These books, embodiments of the enduring spirit of folk-tale, present for contemplation eternal paradigms of human experience—paradigms which are by their nature timeless and in that sense, outside history.

The concentration on the universality behind and beyond the variety of concrete phenomena of our experience is in its essence contemplative. Language again provides a key to thought; the meditative character of Indian thought is forcibly illustrated in the concept of causal relations as expressed in the forms of Sanskrit itself. To indicate the causal relation between two notions, *Sanskrit forms a compound which suggests that the natural order of thought is to begin with the effect and trace it back to the cause.* Accordingly, the expression "effect and cause" (*phalahetu*) occurs instead of the familiar "cause and effect." The contemplative attitude thus erases time: one can only speak of "effect and cause" if the effect is already known and both effect and cause present to the contemplative mind *sub specie aeternitatis*. Although the Latin phrase suggests that this habit of thinking is not wholly foreign to the West, the natural order of Western thought is clear: it is to proceed temporally from cause to effect. Even though the relationship is seen, it is seen in time. In Sanskrit, in contrast, many expressions emphasize this meditative view in which progressive phenomena are seen as already complete. *Kāryakāraṇabhāva* means, not "the relation of cause and effect," but of "effect and cause." What would in Western languages appear as the "relation of the knower and the knowable" is in Sanskrit "the relation of the knowable and the knower (*gamyagamakabhāva*)."[17] We find similar reversals of Western order in the "relation of the generated and the generative (*janyajanakabhāva*)"; "the relation of the proved and the prover (*sādhyasādhakabhava*)"; "the relation of the established and the establishing (*vyayasthāpyavyavasthāpakabhāva*)";[18] "the relation of the activated and the activator (*pravartyapravartayitrtva*)."[19] Each of these expressions appears reversed to Western minds, and even to other Orientals. Accordingly, when scholars translated the original texts into Chinese they changed the word order. Tibetan scholars also understood the causal relationship differently from the Indian; they translated *phalahetu* ("effect and cause") into *rgyu daṇ hbras-bu* ("cause and effect"). This way of thinking, in which the notion of effect is formed first and that of the cause inferred and stated afterward, is retrospective, and is basically different from the approach which starts from the cause. The retrospective, contemplative attitude is in further contrast to the thinking processes of natural science, through which, with the help of inductive and deductive reasoning, the cause of an effect is investigated and ascertained by functional correlation without giving primacy either to cause or effect.

Even when Indians do investigate the relation of two phenomena from cause to effect, they generally do not take the view that a single effect is caused by a single active movement, but prefer to consider that effects are produced by the combination of various causes. Therefore, most Indian thinkers do not employ the term which corresponds to the Aristotelian notion of efficient cause. While *nimitta-kāraṇa* is linguistically the equivalent of *causa efficiens*, it is also used to express the Western notion of *causa occasionalis*. The Sanskrit expression in fact describes a final cause or aim, that is, a teleological relation. And such relations, East and West, are traditional subjects of contemplation.

It would be incorrect to infer from the foregoing that the Indian people have no concept of abstract time. On the contrary, the view of the uncertainty and transiency of life which is at the center of both Buddhism and Jainism demonstrates that they understand from their heartfelt experience the changing phases of the world. Buddhism from the outset emphasized the transience and impermanence of human existence. All things pass away. On account of our fragility we are subject to disease and death. From transience comes suffering. The Buddha asked his disciples: "That which is transient, O monks, is it painful or pleasant?" "Painful O Master!"[20] Our dreams, our hopes, our wishes—all of them will be forgotten as if they had never been. This is a universal principle. "Whatever is subject to origination is subject also to destruction."[21] Necessary and inexorable is the death of all that is born. The difference is only in the degree of duration. Some things may last for years, others for a brief while only. But all must vanish. For the ignoble craving for worldly things must be substituted the noble aspiration for the "incomparable security of Nirvāṇa free from corruption":

> *O transient are our life's experience!*
> *Their nature 'tis to rise and pass away*
> *They happen in our ken, they cease to be.*
> *O well for us when they are sunk to rest!*[22]

There is no substance which abides forever. All matter is force; all substance is motion. The state of every individual is unstable, sure to pass away.

In later days the sentiment of impermanence became more peculiarly Indian than Buddhist. Suffering is seen as one with transience. Craving causes suffering since the impermanence of what we crave causes disappointment and sorrow. The Buddhist beatitude lies in our realization that all things are transient and we should not cling to them.

2 *The Japanese Attitude Toward Time and Change*

Japanese Buddhism also emphasized the transience of the phenomenal world. But the Japanese attitude toward this transience is very different from the Indian. The Japanese disposition is to lay a greater emphasis upon sensible, concrete events, intuitively apprehended, than upon universals. It is in direct contrast to the characteristic Indian reaction to the world of change, which is to reject it in favor of an ultimate reality, a transcendent Absolute in which the mind can find refuge from the ceaseless flux of observed phenomena. The Japanese reaction is rather to accept, even to welcome, the fluidity and impermanence of the phenomenal world.

This way of thinking, far from positing a changeless Absolute, regards the phenomenal world itself as the Absolute and explicitly rejects the recognition of any ultimate reality beyond or above it. What is widely known among post-Meiji philosophers in the last century as the "theory that the phenomenal is actually the real" has deep roots in Japanese tradition.

Master Dōgen, the thirteenth-century thinker who is said to have founded the Sōtō-Zen sect, asserts the transience of things as strongly as any Indian Buddhist. "Time flies more swiftly than an arrow and life is more transient than dew. We cannot call back a single day that has passed." But his emphasis is positive, not negative: "A man may live as the slave of the senses for one hundred years but if he lives one day upholding the Good Law, it will favorably influence his coming life for many years." And Dōgen stresses the primacy of the phenomenal world: "We ought to love and respect this life and this body, since it is through this life and this body that we have the opportunity to practice the Law and make known the power of the Buddha. Accordingly, righteous practice for one day is the Seed of Buddhahood, of the righteous action of All the Buddhas."[23]

What we see and experience is thus recognized as itself the ultimate reality. There is no greater reality, changeless and invisible; there is nothing to be apprehended that is not already exposed to us.[24] For Master Dōgen, impermanence is itself the absolute state, and this impermanence is not to be rejected but to be valued. "Impermanence is the Buddhahood. . . .[25] The impermanence of grass, trees, and forests is verily the Buddhahood. The impermanence of the person's body and mind is verily the Buddhahood. The impermanence of the country and scenery is verily the Buddhahood."[26]

In other places Dōgen says: "Death and life are the very life of the Buddha," and "These mountains, rivers and earth are all the See of Buddhahood." In the Lotus Sūtra, Dōgen finds the same vein of thought: "Concerning the Lotus Sūtra . . . the cry of a monkey is drowned in the sound of a rapid river. [Even] these are preaching this sūtra, this above all." He who attains the purport of this sūtra, says the Master, will discern the preaching of the doctrine even in the voices at an auction sale, for even in the mundane world "our Buddha's voice and form [are] in all the sounds of the rapid river and colors of the ridge."[27]

One is reminded of the words of the Chinese poet Su Tung-p'o: "The voice of the rapids is verily the wide long tongue [of the Buddha]. The color of the mountains is no other than [his] pure chaste body." This way of thinking is Japanese Zen Buddhism. In the words of Master Mujū, "Mountains, rivers, earth, there is not a thing that is not real."[28]

Starting from such a viewpoint, Dōgen gives to some phrases of Indian Buddhist scriptures interpretations that are essentially different from the original meaning. There is a phrase in the *Mahāparinirvāṇa-sūtra* that goes as follows: "He who desires to know the meaning of Buddhahood should survey the time and wait for the occasion to come. If the time comes, the Buddhahood will be revealed of itself."[29] To this concept of Buddhahood as something possible and accessible, Dōgen gives a characteristic twist. He reads the phrase "survey of time" as "make a survey in terms of time," and the phrase "if the time comes" as "the time has already come." His interpretation of the original passage becomes, in this way, something like the following:

> Buddhahood is time. He who wants to know Buddhahood may know it by knowing time as it is revealed to us. And as time is something in which we are already immersed, Buddhahood also is not something that is to be sought in the future but is something that is realized where we are.[30]

We see here Dōgen's effort to free himself from the idealistic viewpoint held by some of the Indian Mahāyāna Buddhists. In Dōgen's unique philosophy of time, "all being is time";[31] the ever-changing, incessant temporal flux is identified with ultimate Being itself.

Again and again Dōgen emphasizes that the true reality is not static but dynamic. "It is a heretical doctrine," says Dōgen, "to think the mind mobile and the essence of things static. It is a heretical doctrine to think that the essence is crystal clear and the appearance changeable."[32] Again, "It is a heretical doctrine to think that in essence water does not run, and the tree does not pass through vicissitude. The Buddha's way consists in the form that exists and the conditions that exist. The bloom of flowers and the fall of leaves are the conditions that exist. And yet unwise people think that in the world of essence there should be no bloom of flowers and no fall of leaves."[33]

Dōgen criticizes the Chinese Zen Buddhist Ta-hui (1089–1163), who taught that mind and essence are not caught up in the world of birth and death. According to Dōgen, Ta-hui was wrong in teaching that "the mind is solely perception and conceptualization, and the essence is pure and tranquil."[34] Here again a static way of thinking is rejected, and this rejection makes Dōgen's emphasis very different from anything which Indian or Chinese Buddhism has prepared us for.

This characteristic willingness to accept the phenomenal world as given and to live contentedly in it is not confined to Buddhism in Japan. It appears in modern Shintoism as well. The founder of the Konkō sect teaches: "Whether alive or dead, you should regard the heaven and earth as your own habitation."[35] And as Dōgen criticized and metamorphosed Chinese Zen, so Jinsai Itō (1627–1705) changed the form of Chinese neo-Confucianism. To Jinsai, the true reality of both earth and heaven is strongly active in a way which we would call evolutionary. Nothing but eternal development exists. Jinsai completely denies what is called death.

The *Book of Changes* (*I Ching*) says, "the great virtue of heaven and earth is called life." It means that living without ceasing is nothing but the way of heaven and earth. And in the way of heaven and earth there is no death, but life, there is no divergence, but convergence. That is because the way of heaven and earth is one with life. Though the bodies of ancestors may perish, their spirits are inherited by their posterity, whose spirits are again inherited by their own posterity. When life thus evolves, without ceasing, into eternity, it may rightly be said that no one dies.[36]

According to Jinsai the world of reality is nothing but change and action, and action is in itself good. "Stillness is the end of motion, while evil is the change of good; and good is a kind of life, while evil is a kind of death. It is not that these two opposites are generated together, but they are all one with life."[37]

Jinsai Itō's younger contemporary Ogiu Sorai, though a rival of Jinsai's, admires the latter's activities as "the supreme knowledge of a thousand years," and denounces the static character of the Chinese School of *Li*. In fact it can be said that all of the characteristically Japanese scholars believe in phenomena as the fundamental mode of existence. They unanimously reject the quietism of the neo-Confucianists of the Sung period.[38]

The way of thinking that recognizes absolute significance in the temporary, phenomenal world seems to be culturally related to the traditional Japanese love of nature. The Japanese love mountains, rivers, flowers, birds, grass and trees, and represent them in the patterns of their kimonos; they are fond of the delicacies of the season, keeping edibles in their natural form as much as possible in cooking. Within the house, flowers are arranged in a vase and dwarf trees are placed in the alcove, flowers and birds are engraved in the transom and painted on the sliding screen, and in the garden miniature mountains, streams and lakes are created. Japanese literature is deeply involved with nature and treats it with warm affection. Typical are the essays in the *Pillow Book* (*Makura no Sōshi*), which describes the beauties of the seasons. The loving concern with the particularities of nature is familiar to us through Japanese art; it is just as marked in Japanese poetry. If the poems on nature were to be removed from the collections of Japanese poems, how many would be left? Haiku, the characteristic Japanese seventeen-syllable short poems, are unthinkable apart from natural objects and the changing seasons, but the differences in attitude are as instructive as the similarities. Here is a poem by Master Dōgen:

> *Flowers are in spring, cuckoos in summer,*
> *In autumn is the moon, and in winter*
> *The pallid glimmer of snow.*

The meaning of the above poem is very close to that of the Chinese verse by Wu-men Hui-k'ai:

> *A hundred flowers are in spring, in autumn is the moon,*
> *In summer is the cool wind, the snow is in winter;*
> *If nothing is on the mind to afflict a man,*
> *That is his best season.*

Similar as the poems are, the Japanese substitution of "cuckoos" for the Chinese "cool wind" has produced an entirely different effect. Both cuckoos and cool wind are sensible phenomena, but while the wind gives the sense of indefinite, remote boundlessness, the cuckoos give an impression that is limited, almost cosy.

An even better example is the poem composed on his deathbed by Ryōkan:

> *For a memento of my existence*
> *What shall I leave (I need not leave anything)?*
> *Flowers in the spring, cuckoos in the summer*
> *And maple leaves*
> *In the autumn.*

"Maple leaves" are felt to be far closer to ourselves than "the moon," which Wu-men chose to associate with autumn. Enjoyment of nature is common to both China and Japan, but whereas the Chinese prefer the boundless and distant, the Japanese prefer the simple and compact. Dōgen took a Spartan attitude toward human desires, but he had a tender heart for seasonal beauties:

> *The peach blossoms begin*
> *To bloom in the breeze of the spring;*
> *Not a shadow of doubt*
> *On the branches and leaves is left.*
> *Though I know that I shall meet*
> *The autumn moon again,*
> *How sleepless I remain*
> *On this moonlit night.*

89

What is the origin of this tendency of the Japanese to grasp the absolute in terms of the world as it exists in time? Probably in the mildness of the weather, the benign character of the landscape, and the rapid and conspicuous change of seasons. Since Nature appears to be relatively benevolent to man he can love it rather than abhor it. Nature, as it changes in time, is thought of as at one with man, not hostile to him. Man feels congenial to his world, he has no grudge against it. This is at least a partial explanation for what is a basic tendency in Japanese thought.

Whatever its source, this willingness to accept the human being's situation in time has many manifestations in Japanese philosophy. Most of the Buddhist sects in Japan teach that doctrines should always be made "à propos of the time." Later Mahāyāna Buddhism employs the concept of the Three Times, the three periods which follow the demise of Lord Buddha. The first thousand years is called the Period of the Perfect Law, when the religion of the Buddha was genuinely and perfectly practiced. The second thousand years is the Period of the Copied Law, when the religion of the Buddha was practiced only in limiting the practices of the sages and monks of the past. The last period, the Period of the Latter Law, is seen as a time of open degeneration.

These ideas took deep root in Japan. The idea, in particular, of the third, degenerate age penetrated deep into the core of the doctrines of various sects. These admitted that they were in the age of degeneration, but instead of exhorting a return to the Perfect, or even the Copied Law, they claimed that the exigencies of the time should be considered and religious doctrines made suitable to them. The sects even vied in claiming the superiority of their respective sūtras (or doctrines) *because* they were most suited to the corruption of the age. Nichiren, the Buddhist prophet, claimed that one could be saved only by the spiritual power of the Lotus Sūtra, whose gospel he alone was entitled to spread. The corruption of the age is no handicap: "The Adoration of the Lotus of the Perfect Truth shall prevail beyond the coming ages of ten thousand years, nay eternally in the future." It is indeed an advantage:

Is it not true that one hundred years' training in a heavenly paradise does not compare with one day's work in the earthly world, and that all service to the Truth during the two thousand years of the ages of the Perfect Law and the Copied Law is inferior to that done in the one span of time in the age of the Latter Law? All these differences are due, not to Nichiren's own wisdom, but to the virtues inherent in the times. Flowers bloom in the spring, and fruits are ripe in the autumn; it is hot in summer and cold in winter. Is it not time that makes these differences?[39]

Nichiren here welcomes the processes of time, even if they bring corruption; he sees them as an opportunity for service to the truth. Time provides the opportunity for a turning point from degeneration to regeneration.

Nichiren laid special emphasis upon the particularity and specificity of the truth of humanity. The Japanese unfriendliness for universals is plain in this passage:

> The learning of just one word or one phrase of the Right Law, if only it accords with the time and the propensity of the learner, would lead him to the attainment of the Way. The mastery of a thousand scriptures and ten thousand theories, if they should not accord with the time and the propensity of the one who masters them would lead him nowhere.[40]

Nichiren evaluates doctrines by five standards, all specific in character. These are: the teaching of the sūtra, the spiritual endowments of the learner (what he calls the "propensity"), the country in which the doctrine is practiced, and the temporal order of circumstances affecting the practice of the doctrine. Saichō, an ancient Buddhist teacher, also regarded the time and the country as important factors, but it was Nichiren who established them as basic principles, presented in a clear and distinct form. Such a method of evaluation of religious truth in terms of social and individual particularities, would hardly be found in the Buddhist thought of India or China. It is clear that even where India and Japan have shared a set of religious assumptions, the characteristic national habits of thought have led to entirely different conceptions both of Time and of Ultimate Reality.

Time and Knowledge in
China and the West

JOSEPH NEEDHAM, F.R.S.

The *philosophia perennis* of Chinese culture was an organic naturalism which invariably accepted the reality and importance of time. This must be related to the fact that although metaphysical idealism is found in China's philosophical history, and even enjoyed occasionally a certain success, as when Buddhism was dominant in the Liu Chhao and Thang periods, or among the followers of Wang Yang-Ming (A.D. 1472–1529) in the sixteenth century,[1] it never really occupied more than a subsidiary place in Chinese thinking. Subjective conceptions of time were therefore uncharacteristic of Chinese thought. Although we are speaking here, of course, of ancient and medieval or traditional thought, and not of sophisticated modern ideas, it may also be said that clear adumbrations of relativism occur in the ancient Taoist thinkers. But whatever happened in time, or times, whether flourishing or decay, time itself remained inescapably real for the Chinese mind. This contrasts strongly with the general ethos of Indian civilization,[2] and aligns China rather with the inhabitants of that other area of temperate climate at the Western end of the Old World.

1   Time in Chinese Philosophy and
Natural Philosophy

Time and its content were often the subjects of discussion and speculation in the philosophical schools of the Warring States period contemporary with Aristotle.[3] We may take a brief look at what interested them. The expression which is now used for "the universe," *yü-chou*, has essentially the meaning of "space-time." In a text of 120 B.C. we read: "All the time that has passed from antiquity until now is called *chou*; all the space in every direction, above and below, is called *yü*. The Tao (the Order of Nature) is within them, yet no man can say where it dwells."[4] The original meaning of both these two ancient words was "roof," of house, cart or boat, so that the semantic significance is that of something stretching over an expanse to cover it. So indeed we still in English say that such and such an exposition "covers" ten or fifteen centuries. The word for duration (*chiu*) is explained by the Han lexicographers as derived from the character *jen*, man, a man stretching his legs and walking "a stretch," just as a roof stretches across a space, and time stretches from one event to another.

Interesting definitions of time and space are contained in the writings which have come down to us from the Mohist school, the followers of Mo Ti (*fl.* between 479 and 381 B.C.). This was the group of ancient Chinese thinkers most interested in the philosophy of mathematics and science. The *Mo Tzu* book is of different dates, for the systematic account of Master Mo's doctrine, including that of universal love (*chien ai*), cannot be much later than 400 B.C., while the Canons and their Expositions (a kind of corpus of definitions explained by a commentary) are not much earlier than 300 B.C., and the technological chapters lie within half a century later than that. Let us look at some of the definitions.

Duration
(Canon)   Duration (*chiu*) includes all particular (different) times (*shih*).
(Exposition) Former times, the present time, the morning and the evening, are combined together to form duration.[5]

Space
(Canon)   Space (*yü*) includes all the different places (*so*).
(Exposition) East, west, south and north, all are enclosed in space.[6]

Movement
(Canon)   When an object is moving in space, we cannot say (in an absolute sense) whether it is coming nearer or going further away. The reason is given under "spreading" (*fu*) (*i.e.*, setting up coordinates by pacing).
(Exposition) Talking about space, one cannot have in mind only some special district (*chhü*). It is merely that the first step (of a pacer) is nearer and his later steps further away. (The idea of space is like that of) duration (*chiu*). (One can select a certain point in time or space as the beginning, and reckon from it within a certain period or region, so that in this sense) it has boundaries, (but time and space are alike) without boundaries.[7]

Movement

(Canon)   Movement in space requires duration. The reason is given under "earlier and later" (*hsien hou*).

(Exposition) In movement, the motion (of an observer) must first be from what is nearer, and afterwards to what is further. The near and far constitute space. The earlier and later constitute duration. A person who moves in space requires duration.[8]

Thus motion, as Forke[9] said, led genetically to the idea of time as well as space. The distances left behind by a moving body, an observer, constitute space, and the changes of position of an observed moving body, such as the sun or moon, awaken the conception of time. Of course man, like all other animals and plants, had his own built-in biological clocks, sensitive to intrinsic needs and the rhythmic cycle of light and darkness.

Great debate went on among the philosophers concerning relativity and infinity. In the fourth century B.C. under the influence of Hui Shih, one of the School of Logicians (Ming Chia), many strange sayings were mooted which bear much resemblance to the Eleatic paradoxes of Greece. Hui Shih said, for example, "The sun at noon is the sun declining, the creature born is the creature dying," and again, "Going to the State of Yüeh today, one arrives there yesterday."[10] The brief moment of noon seems illusory, and if observed from different places on the earth's surface the sun is always declining; senescence begins from the moment of conception, and indeed goes on faster the younger the organism. "Going to Yüeh" is probably a recognition of the existence of different time scales in different places. The Mohists held not only that time was constantly passing from one moment to another but also that particular locations in space were constantly changing; it may be that they had recognized the movement of the earth.

Space and time

(Canon)   The boundaries of space (the spatial universe) are constantly shifting. The reason is given under "extension" (*chhang*).

(Exposition)   There is the South and the North in the morning, and again in the evening. Space, however, has long changed its place.

Space and time

(Canon)   Spatial positions are names for that which is already past. The reason is given under "reality" (*shih*).

(Exposition)   Knowing that "this" is no longer "this," and that "this" is no longer "here," we still call it North and South. That is, what is already past is regarded as if it were still present. We called it South then and therefore we continue to call it South now.[11]

Perhaps the Mohists envisaged something like what we should now speak of as a universal space-time continuum within which an infinite number of local reference frames coexist, and guessed that the universe would look very different to different observers according to their positions in the whole.

Then there was the infinitely long-enduring, the infinitely brief, the infinitely large and the infinitely small. Other paradoxes of the school of Hui Shih were concerned with atomism, just as those of Zeno had been. An interesting passage occurs in the *Lieh Tzu* book,[12] cast as so often in the form of a conversation between semi-imaginary characters.

Thang (the High King) of the Shang asked Hsia Chi saying, "In the beginning, were there already individual things?" Hsia Chi replied, "If there were no things then, how could there be any now? If later generations should pretend that there had been no things in our time, would they be right?" Thang said, "Have things then no before and no after?" To which Hsia Chi answered, "The ends and the origins of things have no limit from which they began. The origin (of one thing) may be considered the end (of another); the end (of one) may be considered the origin (of the next). Who can distinguish accurately between these cycles? What lies beyond all things, and before all events, we cannot know."

So Thang said "What about space? Are there limits to upwards and downwards, and to the six directions?" Hsia Chi said that he did not know, but on being pressed, answered, "If they have none, there can be an infinitely (great). If they have, there must be an indivisibly (small). How can we know? If beyond infinity there were to exist a non-infinity, if within the infinitely divisible there were to exist an indivisible, then infinity would be no infinity, and the infinitely divisible would contain an indivisible. This is why I can understand the infinite and the infinitely divisible, but I cannot understand the finite and the indivisible. . . ."

This shows the kind of arguments that were current about time and space. A little later in the same chapter Hsia Chi goes on to tell of the immense variations of the life-spans of plants and animals, pointing the moral of the relativity of time as it must seem to different living creatures. The *Chuang Tzu* book says:[13] "Man has a real existence, but it has nothing to do with location in space; he has a real duration, but it has nothing to do with beginning or end in time." One could not restrict the influence or knowledge of a man to the physical space which his body happens to occupy, nor is it limited to the time span between the moments of his conception and his death, even if these could be precisely identified. But though Hsia Chi favored an infinite space-time continuum, the Mohists were more inclined to atomism, at least with regard to the definition of the geometrical point and the instant of time.

Instants of time

(Canon)   The beginning (*shih*) means (an instant of) time.

(Exposition)   Time sometimes has duration (*chiu*) and sometimes not, for the "beginning" point of time has no duration.[14]

In spite of this, however, atomism in the physicochemical sense never played any role of importance in traditional Chinese scientific thinking, which was wedded to the ideas of the continuum and action at a distance.[15]

How advanced the conceptions of the Mohists and logicians were may be seen from surveys of the scientific thought of the Greeks.[16] The Mohists were very near the formulation of "functional dependence" in the relation of motion to time. Although the Stoics, with their great emphasis on a continuous rather than an atomistic universe, developed the first beginnings of multivalued logic and grasped one of the elements of the concept of function, the continuous variable, they could not go much further for they could not think of time as an independent variable with phenomena as its function. The description of motion by analytical geometry as change of place functionally dependent on time had to await the mathematization of physics at the Renaissance. For the Peripatetics, as we shall later see, time was cyclical rather than linear; they could not think of time as a coordinate stretching to infinity from an arbitrary zero, like the abstract coordinates of space, in fact a geometrical dimension mathematically tractable, as Galileo did. The Mohists had no deductive geometry (though they might have developed one), and certainly no Galilean physics, but their statements often give a more modern impression than those of most of the Greeks. How it was that their school did not develop in later Chinese society is one of the great questions which only a sociology of science can answer.[17] For most of the Peripatetics, moreover, there was something unreal about time,[18] and they were followed in this by most of the Neo-Platonists. In China the Buddhist schools shared this conviction as part of their general doctrine of the world as illusion, but the indigenous Chinese philosophers never did.

The Mohists also discussed causality. Take for instance the following:

Causation

(Canon)   A cause (*ku*) is that with the obtaining of which something becomes (comes into being, *chhêng*).

(Exposition) Causes: a minor cause is one with which something may not necessarily be so, but without which it will never be so. For example, a point in a line. A major cause is one with which something will of necessity be so (*pi jan*) (and without which it will never be so). As in the case of the act of seeing which results in sight.[19]

This distinguishes between necessary conditions and efficient causes. The former is like competence to react to a stimulus in modern biology. An enlightened understanding of the relation of causality to time is also implicit in some of the other Mohist propositions already mentioned. But this was not always so in ancient Chinese thought. Just as ancient and medieval Europeans spoke of Aristotelian final causes, so in ancient China we hear that such and such a lord, in his lifetime, was not able to obtain the hegemony of the feudal states because, after his death, human victims were sacrificed to him.[20] Both facts were felt to be part of a single pattern, not exactly timeless, but with time as one of its dimensions, in which causation could operate backwards as well as forwards. Elsewhere, within the framework of that primary natural philosophy of China which we shall discuss immediately below, there was scope for ideas of causation distinctly different from the Indian or Western atomistic picture in which the prior impact of one thing is the cause of the motion of another. A causal event might not be strictly prior in time to its effect, bringing the latter about rather by a kind of absolutely simultaneous resonance.[21] Although this conception was quite congruent with the highly organicist tendency of Chinese science and philosophy in general, it was never very thoroughly worked out. One may doubt therefore whether in itself it played any great role in the inhibition of the rise of modern science and technology in Chinese civilization.

The ancient Confucian school, occupied always with human affairs, was of course not interested in all these speculations, and even disapproved of them. Time entered into their considerations only in relation to the appropriate times of action of the sage in society. The "mean" or "norm" (*chung*) was the guide for emotion and action, but it must be flexible in application, for circumstances alter cases and no fixed rules of duty could be laid down, hence in the classic called the "Doctrine of the Mean" (*Chung Yung*), it is a "timely mean" (*shih chung*) that one must follow.[22] In the "Book of Changes" (*I Ching*) too, this conception of the right timeliness in everything is very prominent. The Neo-Confucian school of the Middle Ages, however, had an altogether different approach. These Scholastic thinkers were by that time (eleventh to thirteenth centuries) aware not only of all the Mohist and Taoist speculations of old, but also of the numerous philosophies of Buddhism, and they adopted eclectically whatever suited the purpose of their new synthesis. It would not be possible here to examine in detail their various attitudes to time, but most of them accepted it as real, objective and infinite. So, for example, Shao Yung (A.D. 1011–1077).[23] Some, however, such as his son Shao Po-Wên (A.D. 1057–1137), regarded time as subjective, since for the eternal Tao there could be no past, present or future.[24] Most, as we shall see, believed in cycles of recurrence within time.[25]

In this they were drawing on ancient Taoism. Nothing could be more striking than the appreciation of cyclical change, the cycle-mindedness, of the Taoists. "Returning is the characteristic movement of the Tao (the Order of Nature)" says the *Tao Tê Ching*.[26] "Time's typical virtue," wrote Granet, "is to proceed by revolution."[27] Indeed time (*shih*) is itself generated, some thought, by this uncreated and spontaneous (*tzu-jan*) never-ceasing circulation (*yün*).[28] The whole of Nature (*thien*), the Taoists felt, could be analogized with the life cycles of living organisms. "A time to be born and a time to die," a time for the founding of a dynasty and a time for its supersession. This was the meaning of destiny (*ming*), hence the expressions *shih-yün* and *shih-ming*. The sage accepts; he knows not only how to come forward but also how to retire.[29] This preoccupation with cycles had interesting results in later times when Chinese scientific thought appreciated the existence of natural cycles, sometimes before other civilizations; for example, the meteorological water cycle,[30] or the circulation of the blood and pneuma in the human and animal body.[31] It was prevalent in nearly all the schools, not only the Neo-Confucians, and in this case as in others a close relation may be perceived between the cyclical world view and that other paradigm of Chinese scientific naturalism, wave theory as opposed to atomism.[32]

An important question here is how far the individual cycles, or particular parts of cycles, were compartmentalized, separated off from one another into discrete units, for the ancient and medieval Chinese thinkers. In an influential book, Granet concluded that time in ancient Chinese conception was always divided into separate spans, stretches, blocks or boxes, like the organic differentiation of space into particular expanses and domains.[33] *Shih* (time) always seemed to imply specific circumstances, specific duties and opportunities[34]; it was essentially discontinuous "packaged" time.[35] This conclusion was based, not on the study of the philosophical schools already mentioned, but (as was Granet's way) on the mythology, folklore and general world outlook of the classics and other ancient writings, including Chhin and Han literature. This world outlook was systematized by yet another group of thinkers, the proto-scientific

School of Naturalists (Yin Yang Chia), headed by Tsou Yen (*fl.* 350–270 B.C.), oldest of the Chi-Hsia Academicians.[36] The Naturalists elaborated the theory of the two fundamental natural forces, Yin and Yang, the theory of the five elements, and the system of the symbolic correlations, in which a great number of objects and entities were classified by fives in correspondence with the elements Wood, Fire, Earth, Metal and Water.[37] Since the seasons were prominent in this classification,[38] as also the double hours of the day and night (because of the use of the twelve cyclical signs to denote them), time was to that extent "boxed"; and since this differentiation of time was duly extended to states, dynasties, rulers and reign periods, one can understand the exceptional political power of Tsou Yen and his followers in the late feudal and early imperial period. Success or failure in peace and war might well hinge, it was thought, on adherence to the appropriate element and all its corresponding entities in the symbolic correlation system, hence the prestige of the proto-scientific prognosticators. All this formed the primary Chinese world view on which was based the later traditional natural philosophy of alchemists and acoustic experts, geomancers and pharmacists, smiths, weavers and mastercraftsmen throughout the centuries. In my own description of it I wrote, "For the ancient Chinese, time was not an abstract parameter, a succession of homogeneous moments, but was divided into concrete separate seasons and their subdivisions."[39] The idea of succession as such was subordinated to that of alternation and interdependence.[40]

This was assuredly true but it was not the whole story.[41] First, although the theories of the Naturalists were in general widely accepted, and even to some extent developed for proto scientific purposes, they were much less powerful in some realms of Chinese society than in others. The competing schools of Mohists and Logicians of course were never interested in them, and they played relatively little part in the long evolution of astronomy and cosmology; furthermore, as we shall see, the historians on the one hand and the mass of the people on the other, when they engaged in long-term sociological study and speculation, found no use for the compartmentalized time of the Naturalists. And secondly, while cyclical recurrence was indeed prominent in the natural philosophy, it was almost entirely the cycles of the annual seasons—months, days, hours, etc., and of those which present themselves in biological or social organisms—long-term astronomical periods played an insignificant part, and "Great Year" conceptions (see Section 8), with their consequence of temporal recurrence, none at all. Thirdly, it may be noted that the political applications of the natural philosophy were more and more doubted as time went on.

In the Chhin and Han periods there were intense and anxious debates about the proper color, musical notes and instruments, sacrifices, etc., appropriate to a particular dynasty or emperor, and such questions could still be a live issue in the sixth century A.D.,[42] but after that the political significance of the symbolic correlations seems to have played a steadily decreasing part in men's minds.[43] In sum, there was both compartmentalized time and continuous time in Chinese thinking.[44] Both were important in different ways, the former for some of the sciences and technology, the latter for history and sociology.

Here we may be putting a finger on one of the keys for the answer to the question why modern science did not develop spontaneously in China. Insofar as the traditional natural philosophy was committed to thinking of time in separate compartments or boxes, perhaps it was more difficult for a Galileo to arise who should uniformize time into an abstract geometrical coordinate, a continuous dimension amenable to mathematical handling. But then Chinese astronomy does not prominently show this compartmentalization of time—for example, one never finds a particular planetary revolution associated with an element or a color, though the planets themselves were named after the elements. And in the story of the associated invention of mechanical clockwork (entirely Chinese, as we shall see), there is nothing to indicate inhibition by any ideas of sharp boundaries between stretches of time; the clocks were tended continually, and they ticked away continually decade after decade.

Truly, the cyclical does not necessarily imply either the repetitive or the serially discontinuous. The cycle of the seasons in the individual year (*annus*) was but one link (*annulus*) in an infinite chain of duration, past, present and future. By the use of two interlocking sets of cyclical characters, one of ten and one of twelve, the Chinese were also accustomed to measure time in sexagenary periods. These were used for the day-count from the fifteenth century B.C. onwards, and for the year-count from the first century B.C., thus giving a system independent of celestial phenomena, with a "week" (*hsün*) of ten days. In a civilization primarily agrarian the people had to know exactly what to do at particular times, and so it came about that the promulgation of the luni-solar calendar (*li*) in China was the numinous cosmic duty of the imperial ruler (the Son of Heaven, Thien Tzu). Acceptance of the calendar was the demonstration of fealty, somewhat analogous to the authority of the ruler's image and superscription on the coinage in other civilizations.[45] Since celestial magnitudes are incommensurable and subject to slow secular change, continual work on the calendar was necessary through the ages, and few were the mathematicians and astronomers in Chinese history who did not work upon it. Between

370 B.C. and 1742 A.D. no less than one hundred "calendars" or sets of astronomical tables were produced, embodying constants of ever greater accuracy, and dealing with the determination of solstices, day-, month- and year-lengths, the motions of sun and moon, planetary revolution periods, and the like.[46] Metonic, Callippic, and Saros-like eclipse cycles were early recognized.[47] The calendar had indeed a central role in the history of Chinese science and culture. And since one calendar system covered many years or decades, it blended into history itself.

2   *Time, Chronology, and Chinese Historiography*

The closeness of the relation between calendar systems and history can be seen strikingly in the title of a high official, the Thai Shih (or Thai Shih Kung, or Thai Shih Ling). Today we translate this, from all medieval and late texts, as "Astronomer-Royal," for he was the head of the Bureau of Astronomy in the civil service, and less and less occupied with State astrology as time went on. In the early Han, however, "Astrologer-Royal" would not be a wrong rendering, but what is really significant is that "Historiographer-Royal" would not be wrong either. This was in fact the title borne by the first of China's great historians, Ssuma Chhien, and by his father Ssuma Than before him. Someone has suggested that perhaps the term "Chronologer-Royal" would be the best, for the office was certainly thought of as combining an earthly archivistic with a heavenly uranographic function. Whether the primary function of the Thai Shih was terrestrial or celestial is not yet decided;[48] certain at least it is that by the middle of the first century A.D. the Bureau of Astronomy and the Bureau of Historiography were two entirely separate branches of the bureaucratic organization. The second of these was charged with conserving the records and archives of the current dynasty, and with writing the history of the preceding one, and in theory it was free from manipulation by the reigning monarch or the officials of the day.[49]

It was quite characteristic, in view of Chinese realism about time, that China should have possessed perhaps the greatest of all ancient historical traditions.[50] One can say without hesitation that the Chinese were the most historically minded of all ancient peoples; a quality which makes the dating of events in their civilization comparatively easy. Archaeology too proves this daily, for objects and inscriptions were meticulously dated. No other culture has given us so great a mass of historical writing as that constituted by the twenty-five official dynastic histories, starting with the *Shih Chi* (Historical Records) of Ssuma Chhien already mentioned, which was finished about 90 B.C., and ending with the *Ming Shih* (History of the

Ming Dynasty), completed in A.D. 1736. That customary expression "the Chinese annals" so often encountered, shows how little Western writers have understood the pattern of the dynastic histories. To be sure, these contain the basic annals of the successive reigns, but they also contain a great quantity of treatises on special subjects such as astronomy, economics, civil service organization, administrative geography, hydraulic engineering, taxation and currency, law and justice, court ceremonial, etc., etc. Lastly they embody a vast wealth of biographies of individuals, and this is among the most valuable of all the material in the dynastic histories. Fortunately, the greatness of the Chinese historical tradition, the work of a people who took time seriously, is now more and more appreciated by Western scholars.[51]

The question has been raised, however, whether the time of Chinese historians was not "boxed time" rather than continuous time.[52] It is quite true that the idea of a single era-count, such as the Olympic dating from 776 B.C., or the Seleucid from 311 B.C., or our own Christian Era defined in the early sixth century A.D., did not spontaneously originate in China.[53] Years were counted in terms of dynasties and reigns and (from about 165 B.C. onwards) special regnal periods (*nien hao*) within reigns. But the historians worked out a coherent "single track" theory of dynastic legitimacy and made great efforts to correlate the chronology of events in concurrent minor dynasties, overlapping kingdoms and barbarian peoples, with the graduations on the main time scale adopted. One of the greatest astronomers who thus joined Joseph Scaliger (A.D. 1540–1609)[54] and Isaac Newton[55] in the field of chronology was Liu Hsi-Sou (*fl.* A.D. 1060) whose *Liu shih Chi Li* (Mr. Liu's Harmonized Calendars) embodied the results of his *chhang shu* (art of reconciling long-period data) by means of the sexagenary cycles, identification of intercalary months, solstice dates, etc.[56]

Furthermore, Chinese historiography was by no means confined to the set framework of the dynastic history, for as time went on there grew up various forms of "continuity history-writing" which dealt with long periods of time, including the rise and fall of several dynasties. Ssuma Chhien himself had set a pattern for this, since his *Shih Chi* began with the remotest antiquity and came down to about 100 B.C. in the Earlier Han dynasty, but he did not theorize much about the work of the historian. The philosophy of history was brilliantly studied, however, in the Thang period with the *Shih Thung* (Generalities on History) of Liu Chih-Chi (A.D. 661–721), finished in A.D. 710—the first treatise on historiographical method in any language,[57] quite worthy of comparison with the work of the European pioneers Bodin (1520–96) and de la Popelinière (1540–1608) eight and a half centuries later.[58] It was Liu Chih-Chi's own son, Liu Chih (*fl. ca.* A.D. 732), and another Thang scholar Tu Yu (A.D. 735–812) who invented a new form of encyclopedic institutional history, the former with his *Chêng Tien* (Governmental Institutes) the latter with the famous *Thung Tien* (Comprehensive Institutes; a Reservoir of Source Material on Political and Social History) of A.D. 801. But the climax of this type of work was not reached until the Yuan period, when in A.D. 1322 the *Wên Hsien Thung Khao* (Comprehensive Study of the History of Civilization), by Ma Tuan-Lin, saw the light.[59] His lucid and outstanding treatise, in 348 chapters, was essentially a general history of institutions, which, together with the social structures and economic situations implied by them, seemed to Ma a much more important form of history than any chronological catalogue of contingent events. This search for causal sequences in history more fundamental than dynastic and military mutations and permutations was remarkably advanced for its time; indeed it paralleled the sociological history initiated by Ma's near contemporary, the great Ibn Khaldūn (1332–1406),[60] and the history of institutions later to be achieved by Pasquier (1529–1615), Giannone (1676–1748) and de Montesquieu (1689–1755).[61]

The first move for narrative continuity history came early in the sixth century A.D., when Emperor Wu of the Liang (Hsiao Yen) commissioned Wu Chün to produce a *Thung Shih* (General History); this he did, in 620 chapters, but it has not come down to us.[62] Still preserved, however, is the *Thung Chih* (Historical Collections) of Chêng Chhiao (A.D. 1104–1162), *ca.* A.D. 1150. Chêng Chhiao was more successful in his theory of "synthesis" or "interrelatedness" than in his practice as a historian, and the Lüeh or monograph section of his work, a topically arranged historical encyclopedia, is the only part now used and admired.[63] He had indeed been preceded by the grandest of all the Chinese continuity histories, the *Tzu Chih Thung Chien* (Comprehensive Mirror (of History) for Aid in Government), finished in A.D. 1084 by Ssuma Kuang and a team of collaborators.[64] It covered the whole period from 403 B.C. to A.D. 959 in 354 chapters, and during the following centuries was constantly commented upon, abridged, digested, imitated and extended. It generated yet another form of history, the *chi shih pên mo* (narratives of major sequences of events from beginning to end). This again gave rise to a whole genre of historical writing in subsequent centuries. Thus did the Chinese overcome the "compartmentalization" of time.

It is worth while to notice particularly here the title of Ssuma Kuang's masterpiece—"for aid in government." There was a seeming paradox in the Chinese conception of history. In China good history was considered

a) objective, b) official and c) normative. Confucius himself, in his doctrine of the "rectification of names" (*chêng ming*), had insisted on a spade being called a spade, no matter how powerful the interests which wanted it to be called a shovel;[65] and it was the duty of the historian, even though he was what we should call a civil servant, an eater of the bread of authority, to render judgments on the acts of the past without fear or favor, "for the punishment of evil-doers, and the praise of them that did well."[66] Government in China could and did bestow titles and honors on the dead as well as the living (again illustrating an attitude somewhat different from our own to the dimension of time), so it was natural that the making of a just and definitive record of the past should be its function also. Finally history served an essential moral purpose "for aid in government," for guiding administrative action, encouraging virtue and deterring vice. Such was the basic "praise-and-blame" (*pao pien*) theory of Chinese historiography, a high endeavor of the human spirit, however displeasing to the Tory historians of the modern West.[67] Anything apparently paradoxical in this combination was resolved by a profound if tacit conviction which ran through all generations of Chinese historical writers, namely that the process of social unfolding and development had an intrinsic logic, an indwelling Tao, which rewarded "human-heartedness" (*shan hsing, pu hu jen chih hsin, tshê yin chih hsin*)[68] with good social consequences in the long run and when all balances were struck, while its opposite brought irretrievable evil.[69] This induction was felt to have overwhelming empirical justification. Thus history is the manifestation of the Tao, and has its origins in Heaven.[70] How could anyone ever have imagined that the time sense of the Chinese was inferior to that of Europeans? Almost the opposite could be said; for the incarnation of the Tao in history was a continuous process, ever renewed.[71]

It would really be true to say that in Chinese culture, history was the "queen of the sciences," not theology or metaphysics of any kind, never physics or mathematics. History thus even helped to inhibit the growth of the sciences of Nature, confined as they were to hypotheses of medieval type down to the end of their autochthonous development, and never achieving that mathematization which generated modern science in Europe and in Europe only. Some indeed have gone so far as to urge that the preeminence of history is almost alone sufficient to account for the failure of Chinese culture to develop systematic logic on Aristotelian and scholastic lines out of the brilliant beginnings of the Mohists and Logicians.[72] Insofar as explicit syllogistic logic would have helped the growth of the sciences (a somewhat debatable point),[73] here was another limiting factor, for it was not available in medieval China. Seeking for the concrete causes of this great difference, Stange has contrasted the social conditions of the ancient philosophers of China and Greece, pointing especially to the proto-feudal bureaucratic character of the one and the city-state democracy of the other.[74] The Greeks had relatively little history of their own, they looked back with curiosity rather than reverence to the long ages of Babylonia and Egypt, the morality of which did not particularly apply to them; but they were very interested in proving a point by rigorous logical process *coram publico* in their assemblies, where every man was an equal citizen and every man could argue back. It was natural therefore for formal logic to develop among the philosophers of the Greek democracies. The Chinese philosophers were in a rather different position; they had indeed some important academies and societies for discussion among themselves,[75] but for the most part they frequented the courts of the reigning feudal princes as advisers and ministers.

The Chinese philosopher [wrote Stange] could not, like his Greek counterpart, discuss his ideas on a political situation with an assembly of men of equal rights on the same level as himself, he could only bring his thoughts to fruition in practice by gaining the ear of a prince. The democratic method of logical argumentation was not feasible in discussions with an absolute ruler, but an entirely different method, the citation of historical examples, could make a great impression. Thus it was that proof by historical examples prevailed very early in Chinese history over proof by logical argument.

There can be little doubt that from the social differences between the slave-owning city-state democracies of the ancient Western world on the one hand, and the feudal and proto-feudal bureaucratic states and empires of China on the other, far-reaching divergences in cultural development will be explainable in this sort of way.[76] Essentially the Chinese method was analogical—like causes bring like effects, as it was then so it is now, and so it will be forever. This faith was profound. Hence the great dominance of history (and all its ancillary sciences) throughout Chinese history. The idea that Europe was the only really history-minded civilization is in fact untenable—Clio was at least as much at home in Chinese dress.

3   *Mechanical and Hydro-mechanical Time Measurement*

In view of the appreciation of concrete time, celestial and terrestrial, in Chinese culture, it is perhaps not so remarkable, though the relation of China with chronometry has only recently been properly appreciated, that we owe to medieval artisans and scholars there the first solution of the problem of mechanical timekeeping. Six centuries of mechanical clockwork in China preceded the appearance of clocks in the European West.[77]

Sundials and clepsydras (dripping water clocks) were of course developed in ancient Babylonia and Egypt, spreading out from the Fertile Crescent in high antiquity all over the Old World. In China the former were generally equatorial, never developing the complexity of Arabic and Western gnomonics because of the absence of Euclidean deductive geometry, but they gave rise to many a proverb similar to our own—"an inch of gold will not buy an inch of time (*lit.* light-and-dark)," "a foot of jade is no treasure, but one should struggle for an inch of shadow."[78] The clepsydra evolved much further; adopting the inflow type, the Chinese stabilized pressure heads by multiplying the number of superimposed vessels and then by using constant-level overflow devices. They also had the receiver weighed continuously on a steelyard, and ended by weighing the intermediate vessel. This was probably the prelude to mounting a whole series of such vessels on a rotating wheel, and so the great breakthrough in accurate time measurement came about.[79]

The invention of the mechanical clock was one of the most important turning points in the history of science and technology, indeed of all human art and culture.[80] The problem was to find a way of slowing down the rotation of a set of wheels (an escapement, in fact) so that it would keep step with the great clock of the skies, that apparent diurnal rotation of the heavens which star-clerks and astronomers had studied since the beginning of civilization. This machine, so familiar to all of us today, was truly at its birth a cardinal triumph of human ingenuity; it constituted perhaps the greatest tool of the Scientific Revolution of the seventeenth century, it trained the craftsmen who were needed for making the apparatus of modern experimental technique, and it furnished a philosophical model for the world picture which grew up on the basis of the "analogy of mechanism." But when exactly was it born? Until lately books on the history of timekeeping used to begin with a couple of chapters on sundials and clepsydras, then passing with a mortal leap to the invention of the verge-and-foliot escapement of the mechanical clocks of early fourteenth-century Europe. Some link was very obviously missing. And indeed we now know that an effective escapement was first invented at least six hundred years earlier, at the Eastern, not the Western, end of the Old World. The fact that it was for a hydro-mechanical clock shows precisely the nature of the link.

This knowledge came forth from the recent study of a book written by one of the greatest Sung statesmen, also a naturalist and an astronomer, Su Sung (A.D. 1020–1101). In the year 1090 he took up his brush to give a monographic description of an elaborate astronomical clock tower which during the previous two years had been erected under his supervision and with the collaboration of an engineer, Han Kung-Lien, at the capital Khaifêng. It is entitled *Hsin I Hsiang Fa Yao* (New Design for an Armillary Clock). The first two chapters deal with the sphere and globe while the third describes the horological machinery in great detail. Since the observational instrument on the top storey was mechanized, as well as the globe on the first floor, and all the jackwork which manifested itself at each storey of a pagoda-like time-annunciator, this was the first astronomical clock-drive in history. Necessarily also it embodied the first solar-sidereal conversion gear. Power was derived not from a falling weight as in later Europe but from the torque of a water wheel with scoops like a mill wheel or Pelton turbine. The escapement which checked the forward motion of this wheel was a device of weigh bridges and linkwork which remained stationary while each scoop was filling, but then operated instantaneously so as to open a gate and release one spoke, the next scoop being brought into position under the constant-flow water-jet.[81] Steady motion was thus secured by intersecting the progress of a powered machine into intervals of equal duration—an invention of genius.

But it was not Su Sung's own. Once his technical terminology had been understood it was possible to trace back through earlier literature the records of the building of such hydro-mechanical clocks. A key point was found in the year A.D. 725 in the Thang dynasty, when the first of these escapements was devised, by a Tantric Buddhist monk I-Hsing, probably the greatest mathematician and astronomer of his age, and a military engineer Liang Ling-Tsan. The presence of the linkwork escapement in their clocks is certain, as also is that of luni-solar orrery gearing. The possibility still remains open that the escapement may go even further back, for from the time of Chang Hêng (A.D. 78–139) in the Han onwards, many texts tell of the automatic rotation of celestial globes in time with the heavens. They are not explicit, unfortunately, on the mechanisms of these "proto-clocks." It is not at all difficult to understand why these developments should have taken place in China so long ahead of Europe, for the expression of star positions on equatorial coordinates (equivalent to modern declination and right ascension) was standard practice there, not the ecliptic coordinates of the Greeks. All stellar motion follows the former, while along the latter nothing moves, so that it was quite natural to wish to reproduce the motion in model form, "for aid in computation."

What we still do not know is whether the Chinese escapement system was known and used in Europe during the century or so preceding the appearance of the verge-and-foliot falling-weight clocks, but there are some grounds for thinking so. At the least, the knowledge that the problem of mechanical timekeeping had been successfully solved elsewhere may have been inspiration enough for the first makers of mechanical clocks in Europe.[82]

Western writers have had much to say about the "timeless Orient," but whatever other civilization their words may have applied to, it was not China. It is impossible to imagine that the vast works of erudition in Chinese literature, only a very few of which can be mentioned in this article, could have been brought to completion unless their authors and team-leaders had had "an eye on the clock." A book ascribed to the great literary critic Liu Hsieh (*d. ca.* A.D. 550), the *Hsin Lun* (New Discourses), has an interesting section entitled Hsi Shih (Sparing of Time).[83] Here we read:

The worthies of old, wishing to spread abroad benevolence and righteousness in the world, were always struggling against time. They set no value on whole foot-lengths of jade, but a tenth of an inch of shadow (on the dial) was as precious as pearls to them. Thus it was that Yü the Great[84] raced with time to finish his work and paid no attention to the enquiries of Nanjung.[85] Thus it was that Tao Chung never stopped walking till the soles of his feet were as hard as iron. Confucius grudged every moment lost from reading, and Mo Ti was up and about again before his bed had had time to get warm. All these applied their virtue and genius to relieve the miseries of their times, so that they have left a good name behind them through a hundred generations.

4 *Biological Change in Time*

We have now said something of philosophy, history, chronology and horology. We must next inquire what went on in this endless chain of time which the Chinese took so seriously. First, what of the position of biological change and evolution? As soon as one looks at the ideas of traditional Chinese culture on living things, one finds that they never had any belief in the fixity of species. This followed because they never had any conception of special creation, and that was because creation *ex nihilo* by a Supreme Deity was itself unimagined by them; consequently there was no reason to believe that different kinds of living things could not turn into each other quite easily, given sufficient time. Careful observation would show what did or did not happen. Thus on one side the Chinese view of life was far more open than that of medieval or even eighteenth-century Europe, but on another side their Stoic-Epicurean non-creationism precluded them from a conception which in the West proved (at least in certain periods) favorable to the growth of the natural sciences, namely that of Laws of Nature laid down by a supreme celestial lawgiver.[86] Without a more or less personal Creator one could not think of a divine Legislator for animals, plants and minerals as well as men; the operations of the Tao were in a way more mysterious, even though certain clear regularities (*chhang tao*) would certainly reveal themselves to "faithful and magnificent" observers and experimenters, of whom there was no lack.

Thus the lore of metamorphoses was even more prominent in Chinese literature than in that of the West.[87] Numerous texts may be cited to show the acceptance of the possibility of slow evolutionary modifications and interconversions.[88] Recognition of the *scala naturae* developed among Warring States philosophers contemporary with Aristotle (fourth century B.C.), and there was an independent elaboration of the "ladder of souls" theory with different and less animistic terminology, as in the *Hsün Tzu* book (third century B.C.). Recognition of the animal (and even vegetal) relationships of man led during the first millennium A.D. to a resolution of those controversies about human nature which had so occupied the philosophers of the late Chou and Han periods, in terms of animal-like components within man. Thus Tai Chih about A.D. 1260 saw that the more highly social tendencies of man were peculiar to him, while his antisocial tendencies had to do with those elements of his nature which he shared with the lower animals.[89] Among the Neo-Confucians this led to a marked interest in comparative animal psychology, where "gleams of righteousness" (*i i tien*) might perhaps be perceived.[90] A direct statement of evolutionary transformation is found in a famous passage of the *Chuang Tzu* book (fourth century B.C.), though several of the species mentioned there are not now identifiable. In this book too we have a view of biological changes arising as adaptation to particular environments, and an adumbration of the idea of natural selection in passages which point out the "advantages of being useless."[91] Copious biological discussions occur in the book by the great sceptic Wang Chhung entitled *Lun Hêng* (Discourses Weighed in the Balance) and written about A.D. 83. He insists that man is an animal like other animals, though the noblest of them, rejects mythological birth stories but not spontaneous generation, maintains that all transformations, however weird, are fundamentally natural, and speaks of "sports," genetic inheritance, animal migrations and tropisms. After the spread of Buddhism in China, interest in the philosophy of metempsychosis led to renewed study of embryological and metamorphic time processes. Early in the twelfth century A.D. Chêng Ching-Wang tried to analyze certain believed natural transformations, linking them with the ladder of souls and interpreting them in the light of Buddhist migrations between the lower and higher levels of being. Good actions authorized some spirits to rise in the scale, while soteriological virtue (as well as evil actions) impelled others to descend.[92]

Evolutionary naturalism came fully into focus in the thought of the great Neo-Confucian school, a movement of systematization quite close in date to that of the scholastic philosophers of Europe, with whom these Chinese thinkers are often compared. Just as the Europeans sought to harmonize Greek philosophy with the doctrines of Christian theology, so the Neo-Confucians drew upon all the older philosophies, Confucianism, Taoism and Buddhism for their own synthesis. But their spirit of organic materialism was so different from that of the European scholastics that Chu Hsi (A.D. 1131–1200), their greatest figure, has been termed with at least equal enthusiasm the Herbert Spencer as well as the Thomas Aquinas of China. To understand the universe, as man sees it, the Neo-Confucians worked with two fundamental concepts only, *chhi*, or what we should now call matter-energy, and *li*, the principle of organization and pattern in all its forms. It was extraordinary that they could reach this economy of principle, this world view so congruent with modern science, in a civilization which not only had not developed modern science, but was destined not to be able spontaneously to develop it. For the Neo-Confucians the universe was essentially moral, not because there existed beyond space and time a moral personal deity controlling his creation, but because the universe was one which had the property of bringing to birth moral values and ethical behavior when that level of organization had been reached at which it was possible that they should manifest themselves. Organization in the animal kingdom begins to approach this, very incompletely and one-sidedly (hence the gleams of righteousness), but it is only with the fully developed nervous system of gregarious social man that the universe manifests ethical values. Thus long before the Darwinian age evolutionary naturalism was very clearly stated by Chinese philosophers. But they envisaged a whole succession of these phylogenetic unfoldings rather than one single evolutionary series.

This was doubtless a legacy from Indian thought mediated through Buddhism, envisaging successive time spans, finite but enormously long, which included the *kalpa* and the *mahākalpa*.[93] All the Neo-Confucians accepted the idea that the universe passed through alternating cycles of construction and dissolution. It seems to have been first systematized by a Taoist precursor of Neo-Confucianism, Shao Yung (A.D. 1011–1077), who applied the duodenary series of cyclical characters (see Section 1) to its various phases.[94] Chu Hsi was probably led to his remarkably correct views on the nature of fossils, and other Sung scholars such as Shen Kua (A.D. 1031–1095) to their penetrating insights into mountain-building and erosion, foreshadowing the "plutonic" and "neptunian" ideas of the early nineteenth century, by their meditations on the recurrent world catastrophes or cataclysms in which they believed.[95] Other thinkers, such as Hsü Lu-Chai (A.D. 1209–1281) applied the hexagrams of the *Book of Changes* to the phases of the evolutionary cycle, and Wu Lin-Chhuan (A.D. 1249–1333) estimated its length in time as 129,600 years.[96] Like the Thang calculations of astronomical periods in millions of years,[97] these world views were immensely more spacious than those of seventeenth- and eighteenth-century Europe, with its fixing of the date of creation at October 22, 4004 B.C., at six o'clock in the evening.[98] Biological and social evolution were thus conceived in a cyclical setting, and would forever continue to recur, each cycle being separated by a kind of Ragnarök, a twilight of the gods, the reduction of everything to a disordered and chaotic state, after which all things slowly evolved anew. One might say that the single action of the world's drama, as we think of it today, was replaced by a whole series of repeat performances. And while the rise was slow, the downfall was rapid. The Neo-Confucians would have appreciated the words of William Harvey on individual beings: "For more, and abler, operations are required for the fabrick and erection of living beings, than for their dissolution and the plucking of them down, for those things that easily and nimbly perish, are slow and difficult in their rise and complement."

5   Time and Social Devolution or Evolution, Ta Thung and Thai Phing

We have been speaking of social evolution, but it was implicit in the Neo-Confucian world view rather than clearly defined. Chinese thinkers were rather divided on the question of what had happened to human society in time, and there were two sharply contrasting attitudes. On the one hand there was the conception of a Golden Age of primitive communalism or of sage-kings from which mankind had steadily declined,[99] while on the other there was a recognition of culture-heroes as progenitors of something much greater than themselves, with an emphasis on development and evolution out of primitive savagery.[100]

The first of these views was characteristic of the ancient Taoist philosophers, and in them it was closely connected with a general opposition to proto-feudal and feudal society.[101] They harked back always to the ancient paradise of generalized tribal nobility, of cooperative primitivity, of spontaneous collectivism ("When Adam delved and Eve span, Who was then the gentleman?"), before the aëneolithic differentiation of lords, priests, warriors and serfs.

Of course there are many European parallels for the Taoist idea of a Golden Age—the Cronia and the Saturnalia of Rome commemorating the vanished ages of Cronos and Saturn, the repudiation of over-civilized life by the Stoics and Epicureans, the Christian doctrine of the Fall of Man (perhaps deriving from the ancient Sumerian laments for lost social happiness in lordless society), the stories of the "Isles of the Blest," and finally the eighteenth-century admiration for the Noble Savage, stimulated by the first contacts of Westerners with the real-life "paradises" of the Pacific.[102]

By some literary accident the most famous statements of the Taoist theory of regressive evolution occur in books of other schools, the second-century B.C. *Huai Nan Tzu*, and the first-century A.D. Confucian *Li Chi* (Record of Rites). Here we shall quote the passage from the Li Yün chapter of the latter.[103]

When the Great Tao prevailed, the whole world was one Community (*thien hsia wei kung*).[104] Men of talents and virtue were chosen (to lead the people); their words were sincere and they cultivated harmony. Men treated the parents of others as their own, and cherished the children of others as their own. Competent provision was made for the aged until their death, work was provided for the able-bodied, and education for the young. Kindness and compassion were shown to widows, orphans, childless men and those disabled by disease, so that all were looked after. Each man had his allotted work, and every woman a home to go to. They disliked to throw valuable things away, but that did not mean that they treasured them up in private storehouses. They liked to exert their strength in labour, but that did not mean that they worked for private advantage. In this way selfish schemings were repressed and found no way to arise. Thieves, robbers and traitors did not show themselves, so the outer doors of the houses remained open and were never shut. This was the period of the Great Togetherness (Ta Thung).[105]

But now the Great Tao is disused and eclipsed. The world (the empire) has become a family inheritance. Men love only their own parents and their own children. Valuable things and labour are used only for private advantage. Powerful men, imagining that inheritance of estates has always been the rule, fortify the walls of towns and villages, and strengthen them with ditches and moats. "Rites" and "righteousness" are the threads upon which they hang the relations between ruler and minister, father and son, elder and younger brother, and husband and wife. In accordance with them they regulate consumption, distribute land and dwellings, raise up men of war and "knowledge"; achieving all for their own advantage. Thus selfish schemings are constantly arising, and recourse is had to arms; thus it was that the Six Lords (Yü "the Great," Thang, Wên, Wu, Chhêng and the Duke of Chou) obtained their distinction. . . . This is the period which is called the Lesser Tranquility (Hsiao Khang).

The Mohists undoubtedly sympathized to some extent with this account of the ideal cooperative, even socialist, society, which had, it was thought, existed in the remote past, but it was certainly not part of Confucian ideology at all. Nevertheless in spite of the later universal dominance of Confucianism in Chinese life, the idea of the Ta Thung society enjoyed a certain immortality, for if it had really once existed upon the face of the earth, it might perhaps be brought into existence again.[106] Indeed, Confucianism itself, with its emphasis upon development and social evolution, contributed to this very end. And although the innumerable peasant rebellions through Chinese history rarely pushed their thinking beyond the establishment of a new and better dynasty,[107] at the same time their more visionary elements were often inclined to reverse the time dimension of the regressive conception and turn it into a progressive one. Nineteen centuries later than the Han, in our own time, these two little words (Ta Thung) had vastly gained, not lost, in numinous, emotional and revolutionary force.[108]

There was indeed a parallel (or rather, inverse) ascending Confucian sequence, but before examining it we must take a look at another, related, conception, that of Thai Phing (the Great Peace and Equality).[109] This was another "phrase of power," but widely varying in interpretation.[110] The Golden Age and the realizable Utopia are here not very clearly dissociable; it is hard to find definite statements in ancient texts that this was an era only in the far past which could never come back, or that it was purely something to look forward to in the future. Undoubtedly many imperial reigns were consciously trying to attain it. The term appears first in 239 B.C., in the *Lü Shih Chhun Chhiu* (Master Lü's Spring and Autumn Annals), a famous compendium of natural philosophy, where it denotes a state of peace and prosperity which can be brought about magically by music in harmony with the cyclical operations of Nature.[111] During the following centuries the emphasis was sometimes upon social peace springing from the harmonious collaboration of different social classes each contented with its lot, sometimes upon a harmony of natural phenomena (which man could perhaps induce) leading to an abundance of the kindly fruits of the earth, and sometimes upon the idea of equality, with undertones of reference to that primitive classless society which might in the last day be restored. Some thought that the Great Peace had existed under the sage-kings of high antiquity, others that it was attainable by good imperial government here and now, and others again that it would come to pass at some future time. It is worth while to cite some of these different opinions.

The mysterious social magic of Master Lü appears again in the chapter on rites in the *Chhien Han Shu* (History of the Former Han Dynasty), *ca.* A.D. 100, where it is said that the full application of the rites of the former kings will bring about the Thai Phing state.[112] This had particular reference to the economic chapter of the *Chhien Han Shu* which goes so far as to apply the term Thai Phing to record harvest years.[113] One of the sections of the *Chuang Tzu* book which is probably a Han interpolation says that the highest aim, good government, Thai Phing, is to be attained not by human skill and planning, but only by following the Tao of Heaven. The *Chuang Tzu* book (*ca.* 290 B.C.) contains some of Chuang Chou's clearest keys to Taoist epistemology, scientific world outlook and democratic social thinking.[114] "The great highway (of the Tao of justice and righteousness)" says the *Tao Tê Ching* "is broad and level (*ta tao shen i*)" —one of those pregnant sayings which recalls Hebrew prophecy, "Make straight the way of the Lord," and touches the mystical poetry of road engineering in all ages and peoples, "the valleys shall be exalted, and the mountains shall be made low."[115] Its equalitarian meaning cannot be in doubt for the poem goes on to castigate the feudal lords for amassing wealth and oppressing the peasant-farmers; "these are the riotous ways of brigandage, these are not the great highway."[116]

Many ancient texts, however, speak of Thai Phing only as the Golden Age of the sage-kings of high antiquity. Others make it clear that in certain prosperous periods the Great Peace was considered as already having been attained.

We come now to the incorporation of the Thai Phing concept into a temporal sequence analogous to that of the Ta Thung. It arose out of the exegesis of the *Chhun Chhiu* (Spring and Autumn Annals) by the scholars of the Han. This book was a chronicle of the feudal State of Lu between 722 and 481 B.C., and there was a persistent tradition that Confucius himself had edited it. It has come down through the ages accompanied by commentaries in three traditions known as the *Tso Chuan*, the *Kuliang*

Chuan and the *Kungyang Chuan*.[117] Master Tsochhiu's Enlargement carried the history down a little further, to 453 B.C., and was compiled from ancient written and oral traditions of several States (not only Lu) between 430 and 250 B.C., though with many later changes and additions by Confucian scholars of the Chhin and Han. Master Kuliang's Commentary and Master Kungyang's Commentary differed from this in that they were not formed partly from independent ancient historical writings, but restricted themselves to word-for-word explanations of the chronicle text.[118] The importance of this was the belief, already mentioned, that great moral weight attached to the precise terms which Confucius had used in each given historical circumstance. During the second and first centuries B.C. the scholars of the Han formed groups which specialized in the study of one or other of these traditions, indeed separate chairs were established for them in the imperial university.[119] Among those learned in the tradition of Master Kungyang was that remarkable philosopher (who made his mark in many other ways), Tung Chung-Shu (179–104 B.C.). Tung developed a theory of the San Shih or Three Ages, grouping the events in the *Chhun Chhiu* into a triple classification, those that Confucius himself had personally witnessed (541–480 B.C.), those that he heard of from oral testimony (626–542 B.C.), and those that he knew only through written records (722–627 B.C.).[120] Then in the Later Han this was converted into an ascending social evolutionary series, first applied to the Confucian redaction, and afterwards extended to a universal application. Here the key mind was Ho Hsiu (A.D. 129–182), whose work became the standard commentary on the *Kungyang Chuan*.[121] He wrote:

In the age of which he heard through transmitted records, Confucius saw (and made evident) that there was an order arising from Weakness and Disorder (Shuai Luan),[122] and so directed his mind primarily towards the general (scheme of things). He therefore considered his own State (of Lu) as the centre, and treated the rest of the Chinese oikoumene as something outside (his scheme). He gave detailed treatment to what was close at hand, and only then paid attention to what was further away. . .

In the age of which he heard through oral testimony he saw (and made evident) that there was an order arising of Approaching Peace (Shêng Phing). He therefore considered the Chinese oikoumene as the centre, and treated the peripheral barbarian tribes as something outside (his scheme). Thus he recorded even those assemblies outside (his own State) which failed to reach agreement, and mentioned the great officials even of small States. . .

Coming to the age which he (personally) witnessed, he made evident that there was an order (arising) of Great Peace (Thai Phing). At this time the barbarian tribes became part of the feudal hierarchy, and the whole (known) world, far and near, large and small, was like one. Hence he directed his mind still more profoundly to making a detailed record (of the events of the age), and therefore exalted (acts of) love and righteousness. . . .

Here then we have the formal simulacrum of a process of social evolution in time, ready to be taken over into the general thought of the people as applicable to the whole of civilization.

Already before the time of Ho Hsiu, religious Taoism fermenting among the people had adopted the idea of Thai Phing in this way.[123]

Much study is now being given to a corpus of ancient documents of which the chief is a book entitled *Thai Phing Ching* (Canon of the Great Peace), difficult to date because probably written at different times between the Warring States period (*ca.* fourth century B.C.) and the end of the Later Han (A.D. 220).[124] Though the greater part of this is concerned with religious and superstitious practices, revelations and prophetic warnings, there are passages which link up with the revolutionary Taoism of the great national uprisings—the "Red Eyebrows" of A.D. 24 led by Fan Chhung, and the "Yellow Turbans" (A.D. 184–205) under Chang Chio.[125] But the popular religious Taoism of the Han was millenarist and apocalyptic; the Great Peace was clearly in the future as well as the remote past. In the "Canon" we hear of rural social solidarity, sins committed against the community and their forgiveness, an anti-technology complex, the overcoming of village feuds, and the particularly high place accorded to women. We also find a theory of cycles opposite in character to those of the Neo-Confucians already mentioned. As the sins of mankind's evil generations increase to a climax, world catastrophes, flood and pestilence sweep all away—or nearly all, for a "holy remnant" (a "seed people," *chung min*), saved by their Taoism, win through to find a new heaven and a new earth of great peace and equality, under the leadership of the Prince of Peace (Ta Thai-Phing Chün), of course Lao Tzu. Then everything slowly worsens again until another salvation is necessary. Thus, unlike the cycles of the Neo-Confucians which rose extremely slowly and ended in a flash, those of the religious Taoists issued fresh from chaos "wie herrlich als am ersten Tag" and then fell slowly till the day of doom. But whether or not time was thought of as boxed this way in cyclical periods, the Thai Phing ideal was now forever inscribed upon the banners of the Chinese people in one rebellion after another. It was clearly stated to be the aim of the Ming revolutionary Chhen Chien-Hu (*ca* A.D. 1425); and gave of course the name to the great Thai-Phing Thien-Kuo movement which between 1851 and 1864 nearly toppled the Manchu (Chhing) dynasty, and which is regarded in China today as the closest forerunner of the People's Republic.[126]

But this was not at all the end. One of the greatest reformers and representatives of modern Chinese thought, Khang Yu-Wei, who lived (1858–1927) throughout the period of intellectual strain when China was absorbing and digesting the new ideas which contact with the modern scientific civilization of the West had brought, drew greatly upon these age-old dreams and theories of progress. His thought was deeply influenced both by the seemingly Mohist Great Togetherness (Ta Thung) and the Taoist Great Peace and Equality (Thai Phing). Interpreting them both in the ascending evolutionary sense, he chose the former as the title of an extraordinary Utopia, the *Ta Thung Shu* (Book of the Great Togetherness),[127] conceived and first drafted in 1884. An abridged English translation appeared in 1958,[128] so that Western readers now have access to a magnificent description of the future, visionary perhaps, but extremely practical and scientific, which it would not be inappropriate to call Wellsian in its authority and scope, a vision which no Chinese scholar could have been expected to create if his intellectual background had been as timeless and static as Chinese thought has only too often been supposed to be. Khang Yu-Wei predicted a supranational cooperative commonwealth with world-wide institutions, enlightened sexual and racial policies, public ownership of the means of production, and startling scientific and technological advances including the use of atomic energy. In our own time the charismatic phrases of old became the nation-wide watchwords of the political parties, *Thien hsia wei kung* (let the whole world be One Community) for the Kuomintang, and *Thien hsia ta thung* (The world shall be the Great Togetherness) for the Kungchhantang.

Enough has surely now been said to demonstrate conclusively that the culture of China manifested a very sensitive consciousness of time. The Chinese did not live in a timeless dream, fixed in meditation upon the noumenal world. On the contrary, history was for them perhaps more real and vital than for any other comparably ancient people; and whether they conceived time to contain a perennial fall from ancient perfection, or to pass on in cycles of glory and catastrophe, or to testify to a slow but inevitable evolution and progress, time for them brought real and fundamental change. They were far from being a people who "took no account of time." And to what extent they often visualized it in terms of progress we can see by following another line of thought.

6   The Deification of Discoverers and the Recognition of Ancient Technological Stages in Time

No classical literature in any civilization paid more attention to the recording and honoring of ancient inventors and innovators than that of the Chinese, and no other culture, perhaps, went so far in their veritable deification so late in historical times.[129] Texts which might be termed techno-historical dictionaries, or records of inventions and discoveries, form a distinct genre.[130] The oldest one of the kind is the *Shih Pên* (Book of Origins), most of which simply recites the names and deeds of the legendary or semi-legendary culture heroes and inventors, often dubbed "ministers" of the Yellow Emperor, systematizing thus a body of legendary lore more copious than that of the "technic deities" of Mediterranean antiquity. Thus Su Sha invented salt-making, Hsi Chung invented carts and carriages, Chiu Yao invented the ard, Kungshu Phan the rotary millstone, and Li Shou computations. Five or six classes of these names have been distinguished: the clan patrons and ancestors, the gods of antiquity demoted to heroes, the patron deities of trades, the mythical heroes euhemerized to inventors, then certain made-up names of transparent etymology (like the first example above), and lastly the inventors who were undoubtedly historical personages, such as the fourth of the above examples. The history of the text, which we now have in eight versions, is complicated, but there can be no doubt that while Ssuma Chhien used one form of it, it never had anything to do with Master Tsochhiu the historian, to whom third-century A.D. scholars attributed it. The most probable view is that it was first put together by somebody in Chao State between 234 and 228 B.C., thus just a little later than the *Lü Shih Chhun Chhiu*. From the post-Han centuries one could find a dozen or more books to place in this category, and writers were still not tired of it as late as the Ming, when Lo Chhi wrote his *Wu Yuan* (On the Origin of Things) some time in the fifteenth century A.D.

So greatly prized was the lore of the traditional inventors of old that a list of them was incorporated into one of the greatest arcana of Chinese naturalistic philosophy, the *I Ching* (Book of Changes). This is a very strange classic; it took its origin from what was probably a collection of peasant omen texts, accreted a large amount of material concerned with ancient divination practices, and ended as an elaborate system of symbols with their explanations—sixty-four patterns of long and short lines in all

118

possible permutations and combinations. Since to each of these was assigned a particular abstract idea, the whole system played the part of a repository of concepts for developing Chinese science, the symbols being supposed to represent a gamut of forces actually acting in the external world. The continuing additions to the book made by many profound minds through the ages in the form of appendices and commentaries turned it into one of the most remarkable works in all world literature, and gave it immense prestige in traditional Chinese society, so that philosophical sinologists are still today studying it with great interest.[131] One indeed wrote only a few years ago on the concept of time in the *I Ching*, showing how inescapably this is bound up with its theme—"Change: that is the unchangeable."[132] Nevertheless others have felt that on the whole the *I Ching* exerted an inhibitory effect on the development of the natural sciences in China, since it tempted men to rest in schematic explanations which were not explanations at all. It was in fact a vast filing system for natural novelty, a convenient mental chaise longue which discouraged the need for further observation and experiment.[133]

The dating of the *I Ching* is a highly involved question, but we shall not go far wrong if we place the canonical text (a compilation of omens) mainly in the eighth century B.C., though not complete until the third century B.C., while the principal appended writings (the "Ten Wings") must date from the Chhin and Han, not finalized until the second century A.D. One of these appendices makes now a curious correlation between the great inventions and a select number of the symbols.[134] Precisely from these, it is alleged, the culture-heroes got their ideas. In other words, the scholars of the Chhin and Han found it necessary to adduce reasons for the inventions from the corpus of symbols in the concept-repository. Nets, textile-weaving, boat-building, houses, the crafts of the archer, the miller and the accountant, all are derived ingeniously from Adherence, Dispersion, Massiveness, Cleavage, the "Lesser Topheaviness," the "Breakthrough," and the like. What this teaches us here is chiefly, I think, the honor that was done to the venerated technic sages by incorporating them in the sublime world system of the "Book of Changes."

There was also more concrete liturgical veneration. Everyone who spends time in China and travels about in the different provinces is deeply impressed by the many beautiful votive temples dedicated not to Taoist gods or to Buddhas and Bodhisattvas but to ordinary men and women who conferred benefits upon posterity. Some keep up the memory of great poets, such as the Tu Fu Tshao Thang at Chhêngtu, others that of great commanders such as the Kuan Kung Ling south of Loyang. But the technicians have a most eminent place. Twice in my lifetime I have had the privilege of burning incense (literally or metaphorically) at Kuanhsien in the temple of Li Ping (*fl.* 309–240 B.C.), the great hydraulic engineer and governor of Szechuan province, which stands and has for centuries stood beside the great cutting made under his leadership through the ridge of a mountain. This work divides the Min River into two parts and irrigates still today an area fifty miles square supporting some five million people who can till the soil free of the danger of drought and flood. Every branch of science and technique is represented in these temples of doers and makers deified by popular acclamation. The great physician and alchemist of the Sui and Thang, Sun Ssu-Mo (*ca.* A.D. 601–682), has such a temple, and the custom did not cease even in the Ming, for Sung Li (*d.* A.D. 1422), the engineer who made the summit levels of the Grand Canal a practical proposition, was given a votive temple posthumously beside its very waters.[135] Nor was incense burnt only to men. Huang Tao-Pho (*fl* A.D. 1296) was a famous woman textile technologist, instrumental in the propagation of cotton-growing, spinning and weaving, which she brought to the Yangtze Valley from Hainan. The towns and villages of the cotton areas all honored her, and built many votive temples to her after her death.[136] It is thus impossible to maintain that the Chinese people had no recognition of technical progress. It may have proceeded at a leisurely rate very different from what we have been accustomed to since the rise of modern science, but the principle is clear.

We can also see it in quite another, and rather unexpected way. The conception of the three major technological stages of man's culture, the ages of stone, bronze and iron following each other in a universal series, has been called the cornerstone of all modern archaeology and prehistory.[137] In its modern form this was crystallized in 1836 by the Danish archaeologist C. J. Thomsen.[138] Though sometimes criticized in recent years, it remains the basic classification of periods of high antiquity, and a permanent part of human knowledge. For its general acceptance there were several limiting factors. Thus it had first to be acknowledged that stone-tool artifacts had indeed been made by man (and this was only slowly accepted after the Renaissance as acquaintance with existing primitive peoples grew).[139] It was necessary also to understand the correlation of orderly series of geological strata with time, and to escape from the prison of traditional Biblical chronology so as to recognize the archaeological evidence of man's true antiquity.[140] Furthermore it was necessary to link archaeological findings with some knowledge of the distribution of metallic ores and some reconstruction of the most primitive techniques of copper, bronze and iron production. Nevertheless Thomsen was only the nucleus of crystallization, for the general idea had been "in the air" since the middle of the sixteenth century A.D., a time when curious inquirers into "fossilia" were, as humanists, well acquainted with Greek and Latin texts. They certainly knew of the passage in Lucretius which distinguishes the three ages:

> arma antiqua manus ungues dentesque fuerunt
> et lapides et item silvarum fragmina rami,
> et flamma atque ignes, postquam sunt cognita primum.
> posterius ferri vis est aerisque reperta.
> et prior aeris erat quam ferri cognitus usus,
> quo facilis magis est natura et copia major.[141]

This has been called "just a general scheme of the development of civilization, and based entirely on abstract speculation."[142] I am not so sure that Lucretius never picked up a flaked arrowhead. At any rate his contemporaries in China were saying exactly the same thing, with no less appreciation of the rise of man in time from primitive savagery, and with perhaps more sure and certain reason for what they averred.

Lucretius' words may have been written in the neighborhood of 60 B.C. The *Yüeh Chüeh Shu* (Lost Records of the State of Yüeh), a feudal princedom absorbed by Chhu State in 334 B.C., is attributed to Yuan Khang, a scholar of the Later Han, whose work, which certainly made use of ancient documents and oral tradition, was completed by A.D. 52. Here, in the chapter on the work of the swordsmiths we find the following passage.[143] The Prince of Chhu (Chhu Wang) is engaged in a discussion with an adviser named Fêng Hu Tzu.

The Prince of Chhu asked: "How is it that iron swords can have the wonderful powers of the famous swords of old?" Fêng Hu Tzu replied: "Every age has had its special ways (of making things). In the time of Hsien-Yuan, Shen Nung and Ho Hsü weapons were made of stone, (and stone was used for) cutting down trees and building houses, and it was buried with the dead. Such were the directions of the sages. Coming down to the time of Huang Ti, weapons were made of 'jade,' and it was used also for the other purposes, and for digging the earth; and it too was buried with the dead, for jade is a numinous thing. Such were the directions of the (later) sages. Then when Yü (the Great) was digging (dykes, and managing the waters), weapons were made of bronze. (With tools of bronze) the I-Chhüeh defile was cut open and the Lung-mên gate pierced through; the Yangtze was led and the Yellow River guided so that they poured into the Eastern Sea— thus there was communication everywhere and the whole empire was at peace. (Bronze tools) were also used for building houses and palaces. Was not all this also a sagely accomplishment? Now in our own time iron is used for weapons, so that each of the three armies had to submit, and indeed throughout the world there was none who dared to withhold allegiance (from the High King of the Chou). How great is the power of iron arms! Thus you too, my Prince, possess a sagely virtue." The Prince of Chhu answered: "I see; thus it must have been."

Here then, apart from the intercalation of a "jade" subperiod, possibly meaning stone of better qualities, but also perhaps referring to worked stone as distinct from unworked stone, we have a sequence just as clear as that of Lucretius. And Yuan Khang had a double advantage.

First, he belonged to a distinct tradition.[144] If we read the books of the Warring States philosophers we find time after time a lively appreciation of the stages which mankind had passed through in attaining the high civilization of the late Chou period.[145] The Taoists and Legalists from the fifth century B.C. onwards worked out a highly scientific version of ancient history and social evolution.[146] They had at their disposal the ancient epics of Yao and Shun enshrined in chronicles, they had the lists of culture-heroes and inventors, and they had plenty of oral mythological traditions. From these they made their culture-stage sequence with conscious reference to the customs of the primitive peoples around them. They spoke of men living in nests in trees (pile-dwellings perhaps) or holes in the ground (including cave-dwellings), of the food-gathering stage and the origin of fire and cooked food, of the first making of clothes, the development of the art of the potters (whose neolithic Yangshao and Lungshan wares are now so well known), and of the first writing on bone and tortoise shell. A passage in the *Han Fei Tzu* book (*ca.* 260 B.C.) relating a speech of Yu Yü to the Prince of Chhin, suggests strongly that the writer had seen neolithic pottery both red and black, and also the bronze vessels of the Shang cast in deep relief.[147] Wood, stone, "jade" (worked stone), bronze and iron were regularly associated, as in the passage just quoted from the *Yüeh Chüeh Shu*, with one or other of the mythological rulers.[148] A whole book could well be written on this antique "proto-archaeology."[149]

Secondly China differed from Europe in that the three technological ages had succeeded each other rather faster, and were thus almost parts of history rather than prehistory. Stone tools were still in general use in the Shang kingdom (fifteenth to eleventh centuries B.C.) and continued so down to the middle of the Chou, probably till the coming of iron, for it seems that bronze was very little used for agricultural tools at any time. It is revealing that the physicians maintained a persistent tradition that in ancient times their acupuncture needles had been sharply pointed pieces of stone (possibly obsidian).[150] Copper, tin and bronze metallurgy, however, quickly reached great heights of expertise under the Shang, and the "beautiful metal," as it was called, remained in use for weapons and for marvelous sacrificial and commemorative bronze vessels down to the middle of the Chou.[151] The introduction of iron then occurred in perfectly historical times, a little before the life of Confucius, towards the middle of the sixth century B.C.,[152] and it is not difficult now to trace many of the profound economic and social effects which it brought about.[153] Those, therefore, who have sought to dismiss Yuan Khang's generalization as cavalierly as that of Lucretius, have had even less justification. "This is not a case," someone wrote, "of genius forestalling science by two thousand years; an alert intelligence is simply juggling possibilities without any basis in fact or any attempt to test them."[154] Actually, neither of these alternatives is applicable. The scholars of the Chou and Han did not make stratigraphical excavations, but they had a far more secure basis for their conviction of the truth of the three technological stages than such a criticism could conceive. For the very tempo of development of their civilization had made them historians rather than prehistorians.

7   *Science and Knowledge as Cooperative Enterprise Cumulative in Time*

It is possible to follow the conception of a progressive development of knowledge a good deal further, far beyond the level of ancient techniques. It would be quite a mistake to imagine that Chinese culture never generated this conception, for one can find textual evidence in every period showing that in spite of their veneration for the sages Chinese scholars and scientific men believed that there had been progress beyond the knowledge of their distant ancestors.[155] Indeed, the whole series of astronomical tables ("calendars," mentioned earlier) illustrates the point, for each new emperor wanted to have a new one, necessarily better and more accurate than any of those that had gone before.[156] No mathematician or astronomer in any Chinese century would have dreamed of denying a continual progress and improvement in the sciences which they professed. The same also may be said to be true of the pharmaceutical naturalists, whose descriptions of the kingdoms of Nature grew and grew. It may be worth while to look at the diagram (analogous to Figure 3 of Mr. Lloyd's essay in this volume, which shows the gradual increase in the accuracy of mechanical timekeeping).[157] In the diagram here the number of main entries in the major pharmacopoeias between A.D. 200 and 1600 are plotted so as to show the growth of knowledge through the centuries; the unduly sharp rise just after A.D. 1100 is probably referable to increasing acquaintance with foreign, especially Arabic and Persian, minerals, plants and animals, with a synonymic multiplication which subsequently righted itself.

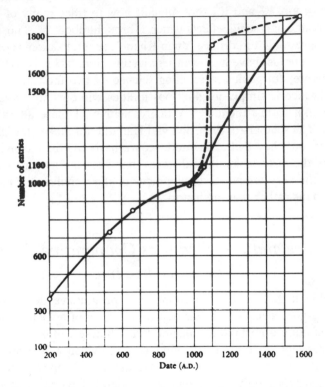

Chart showing the increase in the number of entries in the Chinese pharmacopoeias through the centuries (graph by J. Needham, based on the census of Yen Yü).

The position in China would be well worth contrasting in detail with that in Europe. In his great work J. B. Bury showed long ago that before the time of Francis Bacon only very scanty rudiments of the conception of progress are to be found in Western scholarly literature.[158] The birth of this conception was involved in the famous sixteenth- and seventeenth-century A.D. controversy between the supporters of the "Ancients" and those of the "Moderns," for the studies of the humanists had made it clear that there were many new things, such as gunpowder, printing and the magnetic compass, which the ancient Western world had not possessed. The fact that these (and many other innovations) had come from China or other parts of Asia was long overlooked, but the history of science and technology as we know it was born at the same time out of the perplexity which this discovery had generated.[159] Bury had dealt with progress in relation to the history of culture in general; Zilsel enlarged his method to deal with progress in relation to the ideal of science.[160] The "ideal of scientific progress" included, he thought, the following ideas: (a) that scientific knowledge is built up brick by brick through the contributions of generations of workers, (b) that the building is never completed, and (c) that the scientist's aim is a disinterested contribution to this building, either for its own sake, or for the public benefit, not for fame or private personal advantage. Zilsel was able to show very clearly that expressions of these beliefs, whether in word or deed, were extremely unusual before the Renaissance, and even then they developed not among the scholars, who still sought individualistic personal glory, but among the higher artisanate, where cooperation sprang quite naturally from working conditions. Since the social situation in the era of the rise of capitalism favored the activities of these men, their ideal was able to make headway in the world.

Thus "science," said Zilsel, "both in its theoretical and utilitarian interpretations, came to be regarded as the product of a cooperation for non-personal ends, a cooperation in which all scientists of the past, the present and the future have a part." Today, he went on, this idea or ideal seems almost self-evident—yet no Brahmanic, Buddhist, Muslim, or Latin scholastic, no Confucian scholar or Renaissance humanist, no philosopher or rhetor of classical antiquity ever achieved it. He would have done better to leave out the reference to the Confucian scholars until Europe knew a little more about them. For in fact it would seem that the idea of cumulative disinterested cooperative enterprise in amassing scientific information was much more customary in medieval China than anywhere in the pre-Renaissance West.

There are quotations to be given, but first one must recall that the pursuit of astronomy throughout the ages in China was not the affair of individual star-gazing eccentrics;[161] it was endowed by the State, and the astronomer himself was generally not a free-lance but a member of the imperial bureaucracy with an observatory often located in the imperial palace.[162] Doubtless this did harm as well as good, but at any rate the custom of cumulative teamwork was deep-rooted in Chinese science. And what was true of astronomy was also true of the naturalists; we know, for instance, of large groups who worked together at pharmacognosy and the taxonomic sciences. In these respects the medieval scientists of China, building on the knowledge of their forebears, resembled quite closely the historians, who also came together in teams to produce some of the splendid large-scale works which have already been mentioned.

Some voices from the past may now give color to this perhaps unexpected attribute of Chinese culture. Science is cumulative in that every generation builds on the knowledge of Nature acquired by previous generations, but always it looks outward to Nature to see what can be added by empirical observation and new experiment. "Books and experiments," wrote Edward Bernard in 1671, "do well together, but separately they betray an imperfection, for the illiterate is anticipated unwittingly by the labours of the ancients, and the man of authors deceived by story instead of science."[163] This theme of empiricism was extremely strong in the Chinese tradition. "Those who can manage the dykes and rivers" says the *Shen Tzu* book (probably third century A.D.), "are the same in all ages; they did not learn their business from Yü the Great, they learnt it from the waters." "Those who are good at archery," says the *Kuan Yin Tzu* book (eighteenth century A.D.), "learnt from the bow and not from Yi the archer. . . . Those who can think, learnt from themselves and not from the Sages."[164] This is in part the message of that splendid story of Pien the wheelwright in *Chuang Tzu*, who admonished his feudal lord, the Prince of Chhi, for sitting and reading old books instead of learning the art of government from personal knowledge of the nature of people, just as the artisan learns from personal knowledge of the nature of wood and metal.[165] Thus always alongside the Confucian veneration of the sages, and the Taoist threnodies about the lost age of primitive community, there flourished these other convictions that true knowledge had grown and could yet grow immeasurably more if men would look outward to things, and build upon what other men had found reliable in their outward looking.

There is no Chinese century from which one could not cite quotations to illustrate the conception of science as cumulative disinterested cooperative enterprise. In the field of astronomy and geophysics Liu Chhuo appealed to the throne in A.D. 604 for the authorization of new research on solar shadow-measurements, proposing the geodetic survey of a meridian arc. He said:

> Thus the heavens and the earth will not be able to conceal their form, and the celestial bodies will be obliged to yield up to us their measurements. We shall excel the glorious sages of old and resolve our remaining doubts (about

the universe). We beg your Majesty not to give credence to the worn-out theories of former times, and not to use them.

His wish was not granted, however, till the following century, when a remarkable sixteen-hundred-mile meridian arc survey was accomplished between A.D. 723 and 726 under the superintendence of I-Hsing and the Astronomer-Royal, Nankung Yüeh. This did give results different from those previously accepted, and their descriptions show an enlightened recognition that age-old beliefs about the universe must necessarily bow to improved scientific observations, even though the "scholars of former times" (*hsien ju*) were discredited thereby.[166] Again, at the end of the eleventh century A.D., the idea of cumulative advance came up against the superstition that each new dynasty or reign-period must "make all things new," when a new prime minister wanted to destroy the great astronomical clock tower of Su Sung (mentioned earlier). There was doubtless an element of party politics here, but two scholar-officials, Chhao Mei-Shu and Lin Tzu-Chung, who warmly admired the clock and regarded it as a great advance on anything of the kind that had previously been made, exerted themselves to pull strings to save it. This they succeeded in doing, and the great clock continued to tick on until the year of doom, A.D. 1126, when the Sung capital was taken by the Jurchen Chin Tartars. They transported it to their own capital, near modern Peking, and re-erected it there, after which it still ran for some decades.[167] It is in connection with these astronomical clocks that we often find the expression "nothing so remarkable had ever been seen before," and though it may be considered a stock literary phrase, it nevertheless reveals the fact that Chinese scholars were very conscious of scientific and technical achievements, by no means always trivial in comparison with the works of sages of old.[168] It remains to be seen whether, when all the information is in, pre-Renaissance Europe was as conscious of the progressive development of knowledge and technique as they were.

In the light of all this, the widespread Western belief that traditional Chinese culture was static or stagnant turns out to be a typical occidental misconception. But it would be fair to call it "homoeostatic" or "cybernetic." For there was something in Chinese society which continually tended to restore it to its original character (that of a bureaucratic feudalism) after all disturbances, whether these were caused by civil wars, foreign invasions, or inventions and discoveries. It is truly striking to see how earth-shaking were the effects of Chinese innovations upon the social systems of Europe, when once they had found their way there, yet they all left Chinese society relatively unmoved. We have mentioned gunpowder, which in the West contributed so powerfully to the overthrow of military aristocratic feudalism, yet after five centuries' use in China left the mandarinate essentially as it had been to start with. At the other extreme, the beginnings of Western feudalism had been associated with the invention of equestrian stirrups, but in China, their original home, no such disturbance of the social order resulted. Or one may take the mastery of iron-casting, achieved in China some thirteen centuries before Europe obtained it—there it was absorbed into customary usage for a great variety of purposes both peaceful and war-like, here it furnished the cannon which destroyed the feudal castle walls, and it formed the machines of the industrial revolution.[169] The fact is that scientific and technological progress in China went on at a slow and steady rate which was totally overtaken by the exponential rise in the West after the birth of modern science at the Renaissance. What is important to realize is that although Chinese society was so self-regulating and stable, the idea of scientific and social progress and real change in time was there. Hence, however great the forces of conservatism there was no ideological barrier of this particular kind to the development of modern natural science and technology when the time was ripe.

8 Time and History in China and the West

We come now to what is perhaps the greatest question of all that have here been raised; could there have been a connection between differences (if any) in the conceptions of time and history characteristic of China and the West and the fact that modern science and technology arose only in the latter civilization? The argument set up by many philosophical writers consists of two parts, first the supposed demonstration that Christian culture was much more historically minded than any other, and secondly the view that this was ideologically favorable to the growth of the modern natural sciences at the Renaissance and the scientific revolution.[170]

The first half of the argument has long been familiar ground for occidental philosophers of history.[171] Unlike some other great religions, Christianity was indissolubly tied to time, for the incarnation, which gave meaning and a pattern to the whole of history, occurred at a definite point in time.[172] Moreover Christianity was rooted in Israel, a culture which, with its great prophetic tradition, had always been one for which time was real, and the medium of real change. The Hebrews were the first Westerners to give a value to time, the first to see a theophany, an epiphany, in time's record of events. For Christian thought the whole of history was structured around a center, a temporal midpoint, the historicity of the life of Christ, and extended from the creation through the *berith* or covenant of Abraham to the *parousia* (παρουσια; second coming of Christ), the messianic millennium and the end of the world. Primitive Christianity knew nothing of a timeless God; the eternal is, was, and will be[173] ἀιωνων των ἀιωνων, "unto ages of ages" (in the sonorous words of the Orthodox liturgies); its manifestation the continuous linear redemptive time process, the plan (*oikonomia*, οἰκονομία) of redemption. In this world outlook the recurring present was always unique, unrepeatable, decisive, with an open future before it, which could and would be affected by the action of the individual who might assist or hinder the irreversible meaningful directedness of the whole. A moral purpose in history, the deification of man, was thus affirmed, significance and value were incarnate in it, just as God himself had taken man's nature upon him and died as a symbol of all sacrifice.[174] The world process, in sum, was a divine drama enacted on a single stage, with no repeat performances.

It is customary to contrast this view sharply with that of the Greek and Roman world, especially the former,[175] where cyclical conceptions were generally dominant.[176] We have mentioned the description of successive ages in Hesiod, and their eternal recurrence is one of the few doctrines which it is certain that Pythagoras taught;[177] the other end of Hellenism saw the Stoic doctrine of four world periods[178] and the fatalistic pietism of Marcus Aurelius.[179] Eudemus, Aristotle's pupil, envisaged a complete return of time so that once again, or many times again, he would be sitting talking with his students; Aristotle himself,[180] and Plato too,[181] were wont to speculate that every art and science had many times developed fully and then perished, or that time would return yet again to its beginning and all things be restored to their original state. Such ideas were often combined of course with the long-term recurrences of observational and computational astronomy, hence the notion, probably Babylonian, of the "Great Year." Now cyclical recurrence precluded all real novelty, for the future was essentially closed and determined, the present not unique, and all time essentially past time. "That which has been is that which shall be, and that which has been done is that which shall be done, and there is no new thing under the sun."[182] Salvation therefore could only be thought of as escape from the world of time, and this was partly what led, as some suppose, to the Greek fascination with the timeless patterns of deductive geometry and the formulation of the theory of Platonic "ideas,"[183] as well as to the "mystery-religions."

Deliverance from the endless repetitions of the wheel of existence at once recalls the world outlook of Buddhism and Hinduism; and indeed it does seem true that non-Christian Greek thought was extremely like that of India in this respect.[184] A thousand *mahāyugas* (four thousand million years of human reckoning) constituted a single Brahmā day, a single *kalpa*; dawning with re-creation and evolution, ending with dissolution and re-absorption of the world spheres with all their creatures into the absolute.[185] The rise and fall of each *kalpa* brought ever-recurring mythological events,[186] victories of gods and titans alternately, incarnations of Vishnu, churnings of the Milky Ocean to gain the medicine of immortality, and the epic deeds of the *Rāmāyaṇa* and the *Mahābhārata*. Hence the innumerable reincarnations of the Lord Buddha told in the *Jātaka* birth stories.[187] The dimension of the historically unique was not really present in Indian thought, so that India remained by general consent the least historically minded of the great civilizations,[188] while in the Hellenic and Hellenistic situations uninfluenced by Israel, only a few remarkable minds broke through the prevailing doctrine of recurrence, Herodotus and Thucydides, and they but partially. Of course the hopelessness of this world outlook was greatly modified in India by the wisdom (more Hindu than Buddhist) of the duty of the householder and husbandman in his generation, in fact its own kind of Stoicism which gave to ordinary social life its honored place in part at least of every individual's life cycle.

Paul Tillich has brought together the characteristics of the two great types of world outlook into almost epigrammatic form.[189] For the Indo-Hellenic, space predominates over time, for time is cyclical and eternal, so the temporal world is less real than the world of timeless forms, and indeed has no ultimate value.[190] Being must be sought through the fleshly curtain of becoming, salvation can be gained only by the individual, of whom the self-saving *prateyeka buddha* is the prime example, not by the community. The world eras go down to destruction one after the other, and the most appropriate religion is therefore either polytheism (the deification of particular spaces) or pantheism (the deification of all space). It may seem this-worldly, concentrating hedonistically on the passing present, but it dares not to look into the future and seeks lasting value only in the timeless. It is thus essentially pessimistic. For the Judaeo-Christian, on the other hand, time predominates over space, for its movement is directed and meaningful, witnessing an age-long battle between good and evil powers (here ancient Persia joins Israel and Christendom)[191] in which since the good will triumph, the temporal world is ontologically good. True being is immanent in becoming, and salvation is for the community in and through history. The world era is fixed upon a central point which gives meaning to the entire process, overcoming any self-destructive trend, and creating something new which cannot be frustrated by any cycles of time. Hence the most appropriate religion is monotheism, with God as the comptroller of time and all that happens in it. It may seem other-worldly, despising the things of this life, but its faith is tied to the future as well as the past, for the world itself is redeemable, not illusory, and the Kingdom of God will claim it. It is thus essentially optimistic.

We may surely accept, then, as historical fact, the intense history-consciousness of Christendom. The second part of the argument, which appears to have been hinted at rather than worked out as yet by philosophers of history, is that this consciousness directly contributed to the rise of modern science and technology at the Renaissance, and may therefore rank with other factors in helping to explain it.[192] If it helps to explain it in Europe, perhaps its absence (or putative absence) elsewhere might help to explain the absence of the scientific revolution in those other cultures.

There can be no doubt that time is a basic parameter of all scientific thinking—half of the natural universe indeed, if only a quarter of the number of common-sense dimensions—and therefore that any habit of decrying it cannot be favorable to the natural sciences. It must not be dismissed as illusory; nor depreciated in comparison with the transcendent and the eternal. It lies at the root of all natural knowledge, whether based on observations made at different times, because they involve the uniformity of Nature, or upon experiments, because they necessarily involve a lapse of time, which it may be desirable to measure as accurately as possible.[193] The appreciation of causality, so basic to science, must surely have been favored by a belief in the reality of time. It is not at first sight obvious, however, why this should have been more favored by linear Judaeo-Christian rather than by cyclical Indo-Hellenic time, for if the time cycles were long enough the experimenter would hardly be conscious of them;[194] but it may be that what the recurrence theories really sapped was the psychology of continuous cumulative never-completed natural knowledge, the ideal that sprang from the higher artisanate but came to fruition in the Royal Society and its virtuosi. For if the sum of human scientific effort were to be doomed beforehand to ineluctible dissolution, only to be re-formed with endless toil eon after eon, one might as well seek radical escape in religious meditation or Stoic detachment. Psychological strength was certainly not always weakened in this way, for otherwise Aristotle would never have labored at his zoological studies, the very title of which is relevant to our thought, *Historia Animalium*, περὶ ζωων ἱστορίας, showing as it does the original undifferentiated meaning of "history," any knowledge gained by inquiry—hence the expression still in use, natural history.[195] Nevertheless it is probably reasonable to believe that in sociological terms, for the scientific revolution, where the cooperation of so many men together (unlike the individualism of Greek science) was part of the very essence, a prevalence of cyclical time would have been severely inhibitory, and linear time was the obvious background.

Sociologically it may have acted in still another way, it may well have strengthened the resolution of those who worked for a "root-and-branch reformation in Church and State," bringing into being thereby not only the "new, or experimental, science" but also the new order of capitalism. Must not the early reformers and merchants alike have believed in the possibility of revolutionary, decisive, and irreversible transformations of society? Linear time could not of course have been one of the fundamental economic conditions which made this possible, but it may have been one of the psychological factors which assisted the process. Change itself had divine authority, no less, for the new covenant had superseded the old, the prophecies had been fulfilled; and with the ferment of the Reformation, backed by the traditions of all the Christian revolutionaries from the Donatists to the Hussites, people dreamt again apocalyptically of the foundation of the Kingdom of God on earth. Cyclical time could contain no apocalyptic. In many ways the scientific revolution, however sober, however patronized by princes, had kinship with these visions; "That discouraging maxime, *nil dictum quod non dictum prius*," wrote Joseph Glanvill in 1661, "hath little room in my estimation; I cannot tye up my belief to the letter of Solomon; these last ages have shown us what antiquity never saw, no, not in a dream."[196] Perfection no longer lay in the past, books and old authors were laid aside, and instead of spinning cobwebs of ratiocination men turned to Nature with the new technique of mathematized hypotheses, for the method of discovery itself had been discovered. As the centuries passed linear time influenced modern natural science more deeply still, for it was found that the universe of the stars itself had had a history, and cosmic evolution was explored as the background to biological and social evolution.[197] Then the Enlightenment secularized Judaeo-Christian time in the interests of the belief in progress which is still with us, so that although today when "humanists" or Marxists dispute with theologians they wear coats of different colors, the coats (to an Indian spectator, at least) are actually the same coats, worn inside out.

This brings us to consider the position of Chinese civilization; where did it stand in the contrast between linear irreversible time and the "myth of eternal recurrence"? There can be no doubt that it had elements of both conceptions, but broadly speaking, and in spite of anything that has been said above, linearity, in my opinion, dominated. Of course European culture was also an amalgam, for although the Judaeo-Christian attitude was certainly dominant the Indo-Hellenic one never died out—one can see this in the Spenglerian view of history in our own time[198] and it has always been so. While Aurelius Augustinus (St. Augustine, A.D. 354–430) worked out the Christian system of one-way time and history in his *City of God*,[199] Clement of Alexandria (*ca.* A.D. 150–220), Minucius Felix (*fl.* A.D. 175) and Arnobius (*fl.* A.D. 300) were inclined to favor astral cycles like the *annus magnus*, the "Great Year." Similarly, in the twelfth and thirteenth centuries A.D., just when Joachim of Floris (1145–1202) was setting forth his evolutionary and apocalyptic theory of the Three Ages, inspired successively by the Three Persons of the Trinity, in his *Liber Introductorius ad Evangelium Aeternum*,[200] Bartholomaeus Anglicus (*ca.* 1230), Siger of Brabant (1277) and Pietro d'Abano (*d.* 1316) were prepared at least to discuss with calm, if not complete approbation, the theory that after 36,000 solar years history will repeat itself down to the minutest detail owing to the resumption by the planets and constellations of their original places.[201]

For China the case is very similar. Cyclical time was certainly prominent among the early Taoist speculative philosophers, in later Taoist religion with its recurring judgment days, and in Neo-Confucianism with its cosmic, biological and social evolution ever renewed after the periodical "nights" of chaos. The second and the third were undoubtedly influenced by Indian Buddhism which brought to China the lore of *mahāyugas*, *kalpas*, *mahākalpas*, etc., but for this the first was too early, and indeed we do not find in it any developed form of the doctrine, rather a poetic ataraxy based on acceptance of the cyclism of the seasons and the life spans of living things.[202] But all this leaves out of account both the mass of the Chinese people throughout the ages and also the Confucian scholars who staffed the bureaucracy, assisted the emperor in the rites of the age-old "cosmism" or Nature-worship, and provided the personnel for the Bureaux of Astronomy and Historiography.[203] Sinologists have appreciated for more than a century the linear time-consciousness of Chinese culture, but whatever they know takes at least as long as that to become the common property of occidental intellectuals.[204] Thus, in an interesting paper, Bodde wrote:[205]

Connected with their intense preoccupation with human affairs is the Chinese feeling for time—the feeling that human affairs should be fitted into a temporal framework. The result has been the accumulation of a tremendous and unbroken body of historical literature extending over more than three thousand years. This history has served a distinctly moral purpose, since by studying the past one might learn how to conduct oneself in the present and future. . . . This temporal mindedness of the Chinese marks another sharp distinction between them and the Hindus.

To the great historical tradition of China we have already referred. It envisaged love (*jen*) and righteousness (*i*) incarnate in human history, and it sought to preserve the records of their manifestation in human affairs. Its "praise-and-blame" (*pao pien*) bias, "for aid in government," though somewhat of a limitation and liable to crystallize into dead convention, had nothing to do with the *karma* of Buddhist faith. What it affirmed was that evil social results would follow evil social actions, and though these might lead to the personal ruin of an evil ruler, the effects might also, or only, be visited on his house or dynasty; but inescapable effects there would be. To this the system of rewards and penalties for good or evil actions, worked out through a series of reincarnations of a particular individual, was quite foreign, for the Confucian historians were much more concerned with the community than with the individual. If their time had not been linear it is hardly conceivable that they would have worked with such historical-mindedness and such bee-like industry. Moreover, we have seen that theories of social evolution, technological ages initiated by inventive culture-heroes, and appreciations of the cumulative growth of human science, pure and applied, are in no way missing from Chinese culture.

It would easily be possible, finally, to overestimate comparatively the Judaeo-Christian keying of time's flow to a particular point in space-time when an event of world significance occurred. The first unification of the empire by Chhin Shih Huang Ti in 221 B.C. was a never-to-be-forgotten focal point in Chinese historical thinking, all the more important because of the unity of secular and sacred which no schizophrenia of pope and emperor ever broke up. If one demands something still more numinous, the life of the Sage, the Teacher of Ten Thousand Generations, Confucius (Khung Chhiu, 552–479 B.C.), supreme ethical molder of Chinese civilization, the uncrowned emperor, whose influence is vitally alive today in the tenements of Singapore as well as the communes of Shantung, forming the inescapable background of the Chinese mind whether traditional, technical or Marxist; this life was at least as historical as that of Jesus. That the Confucian outlook was essentially backward-looking is a thesis which in the light of the evidence brought forward in this essay alone cannot

134

stand a close look—the Sage's Tao was not put into practice in his own generation but his assurance was that men and women could and would live in peace and harmony whenever and wherever it was practiced. When this faith, less other-worldly than Christianity (for Thien Tao, Heaven's Way, was not strictly speaking supernatural), joined with the revolutionary ideas implicit in Taoist primitivism, the radically apocalyptic dreams of Ta Thung and Thai Phing, dreams that men could, and did, fight for, began to exert their potent influences. Tillich wrote: "The present is a consequence of the past but not at all an anticipation of the future. In Chinese literature there are fine records of the past but no expectations of the future."[206] Once again it would have been better not to come to conclusions about Chinese culture while Europeans still knew so little about it. The apocalyptic, almost the messianic, often the evolutionary and (in its own way) the progressive, certainly the temporally linear, these elements were always there, spontaneously and independently developing since the time of the Shang kingdom, and in spite of all that the Chinese found out or imagined about cycles, celestial or terrestrial, these were the elements that dominated the thought of the Confucian scholars and the Taoist peasant-farmers. Strange as it may seem to those who still think in terms of the "timeless Orient," on the whole China was a culture more of the Irano-Judaeo-Christian type than the Indo-Hellenic.

The conclusion springs to the mind. If Chinese civilization did not spontaneously develop modern natural science as Western Europe did (though much more advanced in the fifteen pre-Renaissance centuries) it was nothing to do with her attitude towards time. Other ideological factors, of course, remain for scrutiny, apart from the concrete geographical, social and economic conditions and structures, which may yet suffice to bear the main burden of the explanation.

Comments on Time, Process, and Achievement

Dr. Needham's detailed arguments support the view that the historical-mindedness of China in its temporally-linear nature was much closer to that of the West than was previously suspected. In spite of such similarities, however, natural sciences spontaneously developed in the West but not in China. By a modification of Mill's method of difference, we might attribute the cause to something that was not common to China and the West. If the view of *time as history* was, indeed, common to both, then our view of time in its long-range and communal significance might be absolved from the main burden of explanation, regardless whether history is thought of as linear and directed, cyclic and eternally returning, or eternal and hence of a generally uniform nature.[1] We should not, however, also dismiss as a possible cause *man's personal and intimate view of time* as a medium in which achievement may be realized. We may inquire, for instance, whether there existed any attitudes or practices, secular or religious, in daily Western life but not in daily Chinese life which encouraged the sciences and facilitated the rise of industrial capitalism. We should look at the natural sciences and industrial capitalism as portions of a larger unit of historical development for, as Dr. Needham pointed out, "whoever would explain the failure of Chinese society to develop modern science had better begin by explaining the failure of Chinese society to develop mercantile and then industrial capitalism."[2]

We will not attempt to speak about China, but we will discuss briefly a few ideas taken from experimental psychology which suggest some relationship between the rise of the West and man's attitude toward change.

It is convenient to begin by considering the views and practices with regard to time in the United States, the foremost product of that scientific and industrial revolution whose historic origins Dr. Needham has sought to reveal. Time in the United States has become a commodity, but only present time. The past is useless, the future of interest only as a potentially better present. The pre-industrial sense of continuity with the past has been contracted into a narrow present.[3] In Max Lerner's opinion, "American culture cuts away the sensitivity to death and grief, to suicide and immortality, emphasizing the here-and-now as it emphasizes youth and action."[4] A steadily increasing concentration on the "now" has made America a country of the process that is, of continuous becoming. What symbolizes this better than the improvisation of jazz, an image of the unpredictable, *par excellence*?

Professor Kouwenhoven, a keen observer of the American scene, when writing about mass-production systems, has remarked that whereas in other societies and epochs the final product is important, in the United States "it is the process of production itself which becomes the center of interest rather than the product."[5] This attitude did not develop overnight: "Emerson, a century ago was fascinated by the way 'becoming somewhat else is the perpetual game of nature.' The universe, he said, 'exists only in transit' and 'man is great not in his goals but in his transitions.' " American industrialism is, indeed, a continuous transition, a permanent revolution directed toward a steadily increasing control of the environment. The time consciousness and open anxiety often attendant on this permanent revolution are commonly acknowledged in the United States.

These, then, are the hallmarks of the society brought about by the striving of Western man: the control of the environment as measured in terms of achievement, interest in the present, and interest in process as such. When thinking of these traits, one can hardly avoid the haunting image of Faust, the scientist and man of letters, ransoming the time and suffering the conflict between his concrete experience of Gretchen, his interpretation of society, his understanding of Nature, and his abstract knowledge of God. "Europeans suffered from a schizophrenia of the soul, oscillating for ever unhappily between the heavenly host on the one side and the 'atoms and void' on the other; while the Chinese, wise before their time, worked out an organic theory of the universe which included nature and man, church and state, and all things past, present and to come. It may well be that here, at this point of tension, lies some of the secret of the specific European creativeness when time was ripe."[6] But where did this anxiety-laden desire for achievement, this holy restlessness *en masse* spring from?

Some suggestions may be found in recent studies conducted by R. H. Knapp and his collaborators. As we have already noted in the Introduction to this volume, they have reported psychological tests showing significant correlation between emphasis upon achievement motivation and time imagery. Knapp noted[7] that "the rise of time measurement as a serious concern, the development of time pieces themselves, the first establishment of time-monitored industrialism occurred in exactly those Northern European cultures which . . . fostered entrepreneurship, the rise of capitalism,[8] and strong emphasis upon achievement motivation." Subsequently, Knapp's group has also found correlations between combinations of aesthetic asceticism, time judgment, achievement motivation and interest in the physical sciences. Later he put forth the view that "emphasis upon achievement motivation, asceticism in matters of taste and decor, interest in science and technology, and probably the preferred employment of repression as a means of coping with instincts and affects"[9] are part and parcel of what he calls the Puritan pragmatic syndrome. The association of these qualities in a common complex is not a simple historical accident, he argues, but a unit whose existence may be supported by coherent and persuasive reasoning.

The above speculative propositions are interesting, for they transfer our attention from time as history to time as something individualistic and intimate. Whether socio-cultural studies would substantiate causal relations between the psychological principles put forth by Knapp and the spontaneous development of natural science, remains to be seen. We must remember that where many other arguments can become suspect, psychological factors should not, because, as Dr. Needham quotes Bishop Monte Corvina's remark about the Chinese, all humans are basically "di nostra qualita."[10]

Be that as it may, the rise of the "Protestant pragmatic syndrome," if that be the case, probably would not have been possible were it not preceded by the Medieval Church which continuously reminded the faithful of the constant passage of time,[11] as Prof. Brandon phrases it in the following essay, "by the church bells that now regularly sounded across the countryside recording the course of the hours or the passing away of other Christian souls." Thus we might have to seek for earlier clues.

In examining creativity and problem-solving, E. R. Hilgard called attention to some qualities in creative people which reminded him of childhood traits.[12] This observation does not seem to run counter to common experience. Expressing the view of theoretical psychoanalysis, Norman Brown observed that it is the aggressive, dominating attitude toward reality which appears to have been useful in the rise of Western science.[13] But, it is well known that aggressiveness, desire for the feeling of omnipotence, of control of the environment, are pronounced in the child's make-believe world. Dr. Needham himself brings in a similar thought when he asks whether the "medievally conceived Laws of Nature in their naïve form" were not necessary for the birth of modern science.[14] Finally, according to St. Matthew, Christ Himself has said: "unless you be converted, and become as little children, you shall not enter into the kingdom of heaven." Perhaps a certain degree of naïveté and trust in the simplicity and the purpose of existence might be necessary before scientific laws of nature can be formulated.

Many objections may be raised about psychological approaches to what appear to be historical and philosophical problems relating to man's attitude toward change. There is, for instance, a great danger of oversimplification; also, it is clear that the validity and limitations of theoretical arguments such as those quoted can only be determined by careful experimental studies and continuous analysis. Our purpose here has not been to explore the immense problem of what made the West what it is vis-à-vis other cultures, but to point to the important role the individual's view of time, *his time*, might have played in the ascendance of the Western preoccupation with process and achievement. Significantly, this preoccupation is no longer limited to the West. Reflecting on the fate of Japan, Arthur Koestler[15] coined this jingle:

> If East is East and West is West,
> Where will Japan come to rest?
> In the restless West.

It is clear that in the cultural evolution of man, a variety of attitudes toward time form a dialectical give and take, a perennial interplay of views. The preceding three articles suggest the extent, the complexity and the importance of the problems involved. A summary and an analysis of the various responses to the knowledge of transience, as these responses shape man's destiny, are the subject of the concluding chapter of our survey of time as expressed in philosophy, in religion and in man's attitude toward change.

J. T. F.

Time and the Destiny
of Man

S. G. F. BRANDON

1   *The Challenge of Time*

It would appear that the complex nexus of the temporal categories of past, present and future grows in depth and clarity with the growth of the mental abilities of the individual; for mature rationality implies a keen appreciation of these categories and efficient ability to exploit their potentialities. It would, accordingly, be reasonable to suppose that in the process of the evolution of our species the human mind must gradually have grown in its awareness of the temporal categories, and that *pari passu* mankind became more effective in the common struggle for existence.[1]

The archaeological record, from the very appearance of *homo sapiens*, significantly documents man's consciousness of time, or, perhaps more accurately, man's exploitation of the potentialities of such consciousness. Two instances may be given by way of illustration. The making of tools presupposes not only their future use, but also the utilization of past experience for future benefit—in fact such production involves all three of the temporal categories; for it means that the toolmaker, instead of taking his ease in the present, occupies his leisure and his energies fashioning the hand ax or arrowhead, according to past experience with such objects; and he does so with the intention of using them on what he foresees as a future occasion. That this may not be just a kind of instinctive action, such as that of the nest-building of birds or the operations of ants, implying no real individual consciousness of time, would seem to be adequately attested to by a consideration of the variety of the products of paleolithic technology.

But more significant evidence of the time consciousness of Early Man, and of a very dramatic kind, is to be seen in his cave art. The wonderful frescoes found in caves in the Pyrenees and the Dordogne district of

France are justly famous, and it is generally known that expert opinion regards them as being designed to meet the requirements of hunting magic. What is not so often appreciated is the nature of mental processes that the creation of these pictures must surely imply. These pictures are mostly situated in the parts of the cave systems that are most difficult to approach; even today, with all the aids of modern technology, many of them cannot be made accessible to the ordinary visitor. To have painted these pictures, the paleolithic artist must have faced a terrible ordeal that not only called for high moral courage, but implied a powerful motivation of a rather complicated kind. For he accepted the menace of losing his way in the darkness, which his primitive lamp but feebly illuminated, and whatever other terrors the unknown might have held, not in response to some urgent present need, but to achieve an abstractly conceived purpose. This purpose was based upon the belief that the painted representation on the wall of a sacred cave of some animal transfixed with lances would help to make a future hunt successful. In other words, by depicting a wounded animal, as remembered from past experience, it was thought that some, as yet hypothetical, future occasion could be shaped in the way desired. But paleolithic art has even greater significance for man's evaluation of time.

In a cave known as that of the Trois Frères at Ariège, in its innermost recesses, there is a remarkable picture of a strange anthropomorphic figure, clad in an animal's hairy pelt and surmounted by the antlers of a stag.[2] From its posture, which would seem to be that of dancing, this figure is generally referred to as the "Dancing Sorcerer." Such a designation is based upon the reasonable assumption that in these paleolithic hunting communities magical dances were performed, in which the dancer was disguised as an animal and mimed its actions, with the intent of thus gaining power over such animals when hunted or to ensure their fertility for the future supply of food. The assumption that such dances did then exist does not, however, explain why it was deemed necessary to depict an example of them in this cave. Now, it is obvious that this picture must have had some definite purpose, and, on the analogy of the animal pictures, it would seem most probable that it was designed to achieve that purpose in terms of what is known as sympathetic magic, *i.e.*, that like will produce like. But, since the actual dance would itself have been of the order of sympathetic magic, it is surely strange that it was thought to be necessary thus to represent it.

To meet this difficulty there would seem to be one likely explanation, which is also of particular significance to us. It is that the necessary fact of the temporary duration of a dance was felt to constitute a problem, namely, how could its magical efficacy be conserved when the action ended and the dancer departed? In terms of paleolithic thought, insofar as we can surmise its nature, it would be intelligible to have sought to conserve such efficacy by a linear representation of the action that generated it—in other words, that it was believed that the depiction of a magical dance on the walls of a sacred cave would result in the constant generation of its supposed potency. Thus, in this picture of the so-called Dancing Sorcerer we may have a paleolithic prototype of a form of human faith and practice which finds expression in later ages in many religions, and which may be conveniently designated the "Ritual Perpetuation of the Past."[3]

We see, then, from this evidence of the earliest forms of human culture, that already there exists significant indication of man's disposition for intelligent planning, involving, as it does, anticipation of the future in the light of the past and the application of present effort to future ends.

There is still another aspect of paleolithic culture which we must consider, because it, too, provides us with significant witness to another side of man's involvement with the logic of time. As his cave art attests paleolithic man's concern to secure his future food supply, other archaeological data reveal his preoccupation with his personal destiny. Before interrogating this evidence, it will be helpful to reflect upon two cognate matters of our own common experience.

At some time during childhood each normal human being makes two discoveries of profound personal significance: they are those of the fact of birth and of death. The latter discovery is naturally calculated to cause the greater impression, although it has rather a delayed and gradual effect. Thus, the normal healthy child, learning usually by inquiry that people die, does not brood on the fact; but its personal significance does not escape him, and, although he may forget many other things, he will never forget this, and with the passing of years its significance inexorably deepens. Now, this becoming aware of the fact of one's mortality also involves a proleptic element, because the knowledge that human beings die at once causes the individual to anticipate his own death. This consciousness of one's mortal nature is a factor of fundamental significance for any evaluation of man, since it means that each individual can foresee the termination of his life, however far in the future he may hope it to be. Indeed, this ability to anticipate his own mortality may rightly be regarded as an attribute that definitively marks off man from all other forms of life. The discovery of the fact of birth, on the other hand, although it may more greatly intrigue the child, does not necessarily grow in terms of personal significance; however, it has its function, as that of death, in setting the terms of human life—in other words, the knowledge of birth and death invests the life of each individual with a temporal significance that is basic, and it ultimately affects the evaluation of all experience.[4]

These a priori considerations are of immense significance when we turn to study certain evidence of the thought and practice of those earliest representatives of our race with whom archaeology acquaints us. For the extant data of paleolithic culture appear to document what must have constituted the three basic concerns of human life at that remote period. One form of this evidence we have already noticed, namely, cave art as attesting to concern for the supply of food; the other evidence relates respectively to birth and death. Preoccupation with birth finds expression in representations of the human female form, with the maternal attributes grossly exaggerated, that have been discovered on many sites. Since most of the objects concerned are faceless, it would seem that they were designed to concentrate attention on maternity as symbolizing birth or the promise of new life. The careful burying of the dead, usually equipped with food, ornaments, tools or weapons, and sometimes covered with a red pigment or with the body tightly flexed, attests the fact that death constituted an event about which special action had to be taken. The provision of food and other equipment reasonably suggests that it was believed that the dead still continued in some form to exist and need such tendance. Now, this special concern for the dead, which is exclusive to mankind, must surely mean that already at the very dawn of human culture, as it is known to archaeological science, the phenomenon of death constituted a crisis which, in the light of our a priori considerations, must have caused the individual to anticipate his own demise. Moreover, such ritual burial also suggests that this anticipation was linked with some belief in, or hope for, a post-mortem existence. Hence we may conclude that, as soon as he emerged in the archaeological record, man was urgently aware of the temporal character of his life, knowing that it had a beginning and that it must also have an end.[5]

The evidence of paleolithic culture, accordingly, reveals that already man's sense of time, which is peculiar to his species, was operating in two distinctive forms that have been correspondingly distinctive in their effects. Thus, the ability to draw upon past experience in the present to plan for the future has conferred upon man an advantage which in the struggle for existence against other species has proved superior to their often greater muscular strength, swiftness of action, or ferocity. This ability is the very foundation upon which the immense structure of human civilization, with all its scientific and technological achievements, has since been built. In a very true sense all the planning that must necessarily precede each human enterprise represents an attempt to anticipate future contingencies and win some form of security against the menace of temporal change of which man's time sense makes him so urgently aware. But the advantages which this time sense thus confers in material benefits are offset by what may be described as a profound sense of spiritual insecurity which also stems from consciousness of time. To the tracing out of the various forms of reaction to this sense of insecurity the rest of this essay will be devoted.

2   *The Positive (Optimistic) Response to Time*

We have seen, from our brief reference to the evidence of paleolithic culture, that already at that time man was significantly concerned with the phenomena of birth and death. Further, we found reason for thinking that this concern would have had its personal aspect, so that already there existed some perception of the mystery of human destiny. The fact that provision was made in the burial of the dead for some form of post-mortem existence in turn suggests that the problem of destiny prompted certain practical action which was motivated by some belief or hope that death was not definitive in its personal consequences. In other words, we may reasonably say that at the very dawn of human culture man's awareness of time made him conscious of his own mortality, causing him to have a profound sense of insecurity that prevented him from living immersed completely in present experience as do the other animals. Instead, he became increasingly aware, although he would undoubtedly have been unable to define and describe the nature of his awareness, that the passage of time brought change that might often be unpleasant and would ultimately bring death to him. Faced with this disturbing knowledge, he believed or hoped that the dead, in whose company he would himself one day be numbered, would be helped by the careful tendance of the living. It is possible that in those instances in which the corpse was colored with red pigment some action was also taken, by way of sympathetic magic, for its reanimation by restoring to it the sanguine hue of life. Such tendance of the dead, moreover, meant that man was prepared to sacrifice his present resources for post-mortem security or well-being: in communities that must have lived often close to the margin of economic subsistence, the endowing of the dead with food or valuable implements could only have sprung from a profound conviction of the worthwhileness of such sacrifice.

We see, then, that in the paleolithic era, while man's time sense was already giving him success in the struggle for survival, it was, however, an ambivalent faculty in that it also gave him a deep sense of insecurity by causing him to anticipate his own mortality and seek for some assurance of existence beyond its dread event. It is, accordingly, reasonable to think that from this sense of insecurity, engendered by awareness of time, that complex reaction of hope and fear, which we call religion, originally stemmed. The continuing influence of this stimulus in history is evident, as we shall now endeavor to show, in the more significant religions, past and present.

The earliest written records that we possess afford impressive evidence that anticipation of death inspired the desire to achieve some form of security beyond the effacing flux of time. The Egyptian Pyramid Texts, which date from about 2400 B.C., comprise long series of religio-magical texts inscribed on the interior walls of the pyramids of certain pharaohs of the Fifth and Sixth Dynasties for the purpose of securing their safe passage from this world to the next. The material from which these Texts are drawn gives evidence of being even more ancient, and it often appears to have been originally intended for private use. Now, the significance of these documents from our point of view lies in the fact they offer the dead king various forms of post-mortem security that put him beyond the touch of the change and decay that time inevitably brings. For example, he is conceived as being associated with the circumpolar stars, which are called the "Imperishable Ones" because they never disappear from the night sky. Or the pharaoh is pictured as joining the sun-god in his celestial boat on his unceasing daily passage through the sky.[6]

This Egyptian evidence, affording our earliest documented insight into the human mind, is notable for its vivid attestation of man's aspiration, when faced with death, for some form of post-mortem survival. In Egypt this aspiration profoundly influenced the current *Weltanschauung*, and it will be useful to note it at some length because it constitutes a particularly illuminating pattern of human response to this challenge with which to compare other forms.

As we have seen, the Pyramid Texts reveal that by the middle of the third millennium B.C. it was believed that the pharaoh could achieve a blissful post-mortem existence which would be eternal. In process of time this royal privilege was gradually democratized until everyone who could afford the necessary obsequies could hope for such eternal well-being. In other words, the Egyptian believed that by the employment of a prescribed technique, which took the form of an elaborate mortuary ritual, time's menace of personal extinction could be surmounted and everlasting felicity attained. To accomplish this for himself much concentration of attention and expenditure of treasure were required of the individual, so that such mortuary provision consumed a significant part of the resources and energy of ancient Egyptian society. Generally this preoccupation with death did not make Egyptian life and thought excessively morbid. The wise man believed that it was for him to enjoy this life to the full, while taking the practical measures necessary for his eternal well-being; this enjoyment was to some degree conditioned by the belief that one would be judged in the next world for one's moral conduct in this. The Egyptian seems to have found no ultimate significance in this life, and he generally regarded it as a meaningless process; but he was saved from despair by his trust in the efficacy of the traditional mortuary ritual, which was naturally thought to have divine sanction.[7]

This ancient Egyptian reaction to what may be described as the challenge of time provides the earliest example of a certain pattern of response to that same challenge which has been made in subsequent cultures in various parts of the world. The pattern may be defined as positive or optimistic in that it connotes belief, and consequent action, that man can achieve an everlasting state of security from the evils of temporal change, particularly as manifest in decay, old age and death. Sometimes, as in Egypt, this ultimate security is thought to be achieved by the practice of a ritual technique; but always there is essential dependence upon divine aid. One aspect of the Egyptian response must particularly be described here, because not only was it its classic expression, but it also constitutes a very notable instance of the principle of the Ritual Perpetuation of the Past to which reference has already been made.

Already in the Pyramid Texts the resurrection of the dead (king) to a new post-mortem life, and thereby to immunity from the physical corruption of death, was being sought through the practice of ritual assimilation. It was believed that there once lived in Egypt a divine hero named Osiris, who, after being foully murdered by an evil being called Set, had been raised to life again by the devoted efforts of certain deities. This legend of Osiris formed the rationale of a mortuary ritual designed to represent what were thought to be the crucial acts in effecting the resurrection of the divine hero. The dead person, who was the subject of this ritual, was mystically identified with Osiris, with the intent that he should participate in the efficacy generated by the ritual re-presentation of what were believed to have been the original acts that had revivified Osiris. In other words, the Osirian mortuary ritual sought to perpetuate ritually on behalf of each person for whom it was performed, the saving efficacy of those acts that had once restored Osiris to life.[8] Accordingly, resurrection and security from the corruption of death depended ultimately upon divine aid, which was variously provided by a number of deities, but with essential reference to what was held to be a historical event.

In the light of this Egyptian prototype, we shall now briefly notice the more significant of the later examples of this pattern of response to the challenge of time. Of these various mystery cults those of ancient Greece and the Near East demand first a passing reference. We remain, unfortunately, still badly informed about certain essential aspects of the ritual and belief concerned in these cults; however, there is sufficient evidence for thinking that they each embodied in some form the principle of the Ritual Perpetuation of the Past in that, like Osirianism, salvation was achieved by the ritual assimilation of the initiates in some manner with certain acts or actors that had once won new life or the assurance of immortality.[9]

Pursuing our inquiry in chronological sequence, we come next to consider the evidence of Hebrew religion. Here we encounter an interesting variant of the pattern of reaction with which we are now concerned; it is a variant, moreover, that introduces us to another, and a very significant, aspect of man's evaluation of time. For reasons that are too complicated for exposition here, but which the present writer has set forth at length elsewhere,[10] the party in Israel that strove over many centuries, with ultimate success, to make Yahweh the unique god of the nation found itself obliged at first to propound a doctrine of human nature and destiny that precluded any hope of a significant post-mortem existence. However, in process of time, impelled by the very logic of its concept of Yahweh as both omnipotent and just, Yahwist orthodoxy accepted the idea of a post-mortem resurrection. But, owing to the essentially ethnic character of Yahwism, this post-mortem resurrection, although personal in its application, was integrated into a theology that was really concerned with only the relationship of Yahweh and Israel, as Yahweh's chosen people. Now, this theology, if as such it may be described, was based upon a philosophy of history, or interpretation of a particular series of past events. It had become customary for the Yahwist prophets to recall the Israelites to their allegiance to Yahweh by appealing to the memory of their deliverance from the Egyptian bondage and their settlement in the promised land of Canaan. These signal events constituted the mighty deeds that Yahweh had done for their fathers, and their memory was invoked as demonstrating the power and providence of Yahweh.

About the ninth century B.C. a Yahwist writer of genius elaborated this prophetic appeal into a long narrative that traced the providential action of Yahweh from the very creation of the world, through the early history of mankind to the birth of Abraham, the progenitor of Israel, and then on to the events of the Exodus and the conquest of Canaan. This narrative, which became part of the sacred literature of the people, gradually came to exercise a mighty influence on the Jews by causing them to regard the passage of time as the field in which their God manifested his power and his providence on their behalf. In other words, history was interpreted as the revelation of the divine purpose so that a teleological *Weltanschauung* emerged, according to which the whole cosmic process was regarded as the gradual unfolding or achievement of God's plan, the destiny of Israel being its central theme. Into this complex the destiny of the faithful Israelite was fitted, in that he had the assurance of Yahweh that, as a member of his elect people, he would finally be raised to life again to participate in the glorious vindication of Israel which Yahweh would in his own proper time accomplish. Thus time, on this Jewish reckoning, was replete with an awful, yet an inspiring, significance, for it was the medium wherein the divine purpose was most signally demonstrated.[11]

By way of the Hebrew evaluation of time, we are naturally brought to consider the evaluation implicit in Christianity, since it was from Judaism that this world-faith originally stemmed. Here we meet with one of the most thorough and comprehensive attempts on the part of man to assign a definitive significance to time both in terms of the destiny of the individual and of mankind as a whole, and even of the physical universe. This all-embracing scheme was achieved by the taking over of the Jewish sacred Scriptures by the Church on the theory that they constituted a kind of *praeparatio evangelica, i.e.,* they recorded the first great stage in the divine plan for man's redemption which ran from the Fall of Adam to the birth of Christ, the Second Adam: hence these Jewish Scriptures were significantly designated the Old Testament. Jesus of Nazareth was regarded both as the Messiah foretold by the Hebrew prophets and as the incarnated Son of God. The purpose of his incarnation and death by crucifixion was explained in terms of a *mythos* involving cosmology and an anthropology both esoteric in origin and concept. Accordingly, a series of eons or world ages was conceived, before the beginning of which God had planned to save mankind, whose members in some unexplained way had become fatally subject to the demonic powers that were associated with the planets and ruled the world. Man's deliverance from this bondage was effected when these powers were led, by ignorance, to crucify the incarnated son of God. This decisive event was held to mark the start of the new age of man's redemption, which was effected by faith in Christ and through membership of his Church.

Since the primitive Christian movement inherited the outlook of Jewish eschatology in connection with the coming of the Messiah, it was at first believed that the Risen Christ would return almost immediately in power and glory to bring the present world order to a catastrophic end, to redeem his own, and to judge mankind. When the first generation of the faithful passed away without the fulfillment of this expectation, the hope that it represented was not abandoned but projected forward into the future. Accordingly, a Christian *Weltanschauung* was constructed which envisaged the purpose of God as unfolding majestically through the ages. After man's frustration of the divine intention implicit in the creation by his fatal act of disobedience, God had His plan of redemption which, as we have already noted, found expression in the history of Israel until the birth of Christ. The second stage of that purpose, initiated by the divine Saviour's death and resurrection, would continue until "the gospel had been preached in all the world," and then the Second Coming of Christ would mark the end of the world, which would be tantamount to the end of time, and the final achievement of the divine purpose.[12]

Into this cosmic scheme the destiny of the individual was gradually fitted until the superb synthesis of the medieval form of Christianity was attained. According to this theological tour de force, each person who is born into this world, which is veritably the testing ground for eternity, inherits the original sin of a fallen race and a nature prone to actual sin. However, Christ had made such provision through his Church that, by the sacrament of Baptism, this original sin is cleansed away and the neophyte incorporated into membership of the Church, which is the mystical body of Christ. This membership rendered the new Christian eligible to receive the divine grace, mediated through the other sacraments, to strengthen him against the temptations of this earthly life and to restore him when he falls. Supreme among these sacraments was that of the Mass, whereby each day the saving sacrifice of Calvary was re-presented and communion made with Christ through his sacramental Body and Blood. The Christian thus lived out the days of his life in preparation for eternity, fortified by divine grace and ever mindful of the destiny of his soul—indeed, he was solemnly reminded of the constant passage of time by the church bells that now regularly sounded across the countryside recording the course of the hours or the passing away of other Christian souls. For death itself he was prepared by the prayers and rites of Holy Church, believing that he would go forth from this life to Purgatory, where he would expiate his sins, with all others who had died, until that awful Second Coming of Christ would end the present order of things and bring the terrible Last Judgment whereby the fate not only of every human being, but even of Satan and his demons would be irrevocably decided for all eternity.[13]

Therefore, time for the Christian was invested with the most profound significance. He was taught to see it in a solemn twofold aspect as the gradual revelation of the mighty purpose of God, in which his own personal destiny, here and hereafter, had its minute but essential place. But the awful meaning with which time was invested in this *Weltanschauung* does not sum up the whole of the Christian preoccupation with it, and we must now briefly note how the ancient principle of the Ritual Perpetuation of the Past has also its part in Christian faith and practice.

The soteriological interpretation of the death of Christ inevitably imposed a problem similar to that which must originally have confronted those who, more than three millennia before in Egypt, had regarded Osiris as a saviour-god. It was in essence the problem that, as we have seen, already confronted man in the paleolithic age, namely, how can the efficacy of an event be conserved and made available on subsequent occasions? The solution found then, and invoked in many religions since, has been in terms of what we have defined as the Ritual Perpetuation of the Past. Christianity in its turn conformed to this pattern in its solution of the particular form of the problem with which it was concerned. The means had to be found whereby the potency of the saving death and resurrection of Christ, which were events believed to have happened on specific occasions in the past, could be made available for the salvation of individual men and women who might be converted, generation after generation, in the future. The solution found was a ritual one, and it took two distinctive, yet complementary, forms.

The first may be stated in terms of its earliest formulation by St. Paul in his Epistle to the Romans (6: 3–5): "are ye ignorant that all we who were baptized into Christ Jesus were baptized into his death? We were buried therefore with him through baptism into death: that like as Christ was raised from the dead by the glory of the Father, so we also might walk in newness of life. For if we have become united with *him* by the likeness of his death, we shall be also *by the likeness* of his resurrection." According to this statement, then, the Christian neophyte is, in baptism, ritually assimilated to Christ in his death and resurrection, so that, in sharing in the death, he will participate also in the new life of resurrection. In other words, by the action of baptism, practiced on any given occasion, the saving efficacy of the Death and Resurrection of Christ, which were events that had happened at a definitive point in the past, was made operative in the present. We may note that this belief was dramatically interpreted in the early baptismal rituals, and it is still represented in the modified forms current in the Church today.[14] The other way in which the efficacy of the historical event of the crucifixion of Christ was made continuously available to the faithful was in the Eucharist or Mass, as we have already briefly noted. Here the principle of the Ritual Perpetuation of the Past finds expression in a very striking manner. In the classic forms of the liturgy what is termed the *anamnesis* constitutes the critical moment of the rite, when reference is made to the historic occasion of the Last Supper and to Christ's words concerning the significance of his broken body and poured-out blood, thereby instituting the rite. The ritual acts of the celebrant symbolize the death of Christ, and in the subsequent offering of the Eucharistic sacrifice the historical death of Jesus outside the walls of Jerusalem about A.D. 29 is ritually re-presented to God the Father. Thus, in solemn word and action, it is believed that at each celebration of the Mass what was in reality an event that took place in Palestine long ago is made a present reality for the faithful there gathered and efficacious for those for whom that particular presentation of the Sacrifice may be pleaded.[15]

Christianity is often described as a historical religion. The description is justified, although its connotation is too often left ambiguous. Christianity can indeed be regarded as a historical religion in two different ways which we have seen, each of which is significant for us in terms of this essay. Thus its soteriology is essentially based upon a philosophy or interpretation of history that sees the passage of time as the unfolding of the purpose of God. It is historical also in the sense that it stems from certain events, alleged to have occurred at a specific place and time, to which it assigns a superlative significance; moreover, it seeks to preserve what it regards as the essential efficacy of some of these events by repetitive ritual action.[16] Accordingly, Christianity may truly be evaluated as a reaction to the challenge of time which has made of that very challenge a most comprehensive and closely integrated interpretation of human destiny, on both the individual and the social plane, in terms of a divine purpose. We shall leave until the end of this essay consideration of the consequences which have ensued for Western society by its gradual abandonment of the Christian *Weltanschauung* which is based upon this estimate of time.

Continuing our survey of what we may describe as the positive or optimistic response to the challenge of time, we must briefly notice the form in which it has found expression in Islam. It would appear that in Arabia, before the preaching of Muḥammad, time was closely associated with a capricious destiny that determined the lives of men. Muḥammad's conception of God (Allāh) was closely akin to that of Judaism and Christianity in regarding the whole cosmic process as the expression of the divine will, which will ultimately terminate it. However, since the will of Allāh is inscrutable, and individuals are predestined thereby to heaven or hell, the faithful can only trust in Allāh's mercy and strive to obey his commands. In practice such a creed seems generally to have produced an attitude of fatalism toward life; but such fatalism has not been of an apathetic kind, because a real personal attachment to Allāh has fostered an encouraging sense of trust in his providence. By reason of its belief in an ultimate Day of Judgment that will mark the end of the world, Islam is teleological in its outlook; but this character has not produced a preoccupation with the significance of time that is in any degree comparable to that of Christianity.[17]

There remains one other notable example of the positive response with which we have so far been concerned. This occurs in pre-Islamic Iran. Its elucidation and exposition involve many very difficult textual, linguistic and historical problems, so that only the briefest summary may be given here. Zarathustra, who attempted to reform the dualistic polytheism of his people in the sixth century B.C., made Ahura-Mazda the sovereign lord of the universe, who would bestow upon his faithful everlasting felicity. He also pictured the universe as the battleground of two primeval forces which he designated respectively the "Good Spirit" and the "Lie," the one being identified with all that is good and the other with all that is evil. It would seem, however, that Zarathustra's dualistic conception of life was not absolute, since he held out the hope that the Good Spirit, which he closely associated with Ahura-Mazda, would ultimately prevail.

In the following centuries Zarathustra's teaching, which, with certain modifications, came to represent Iranian orthodoxy, began to raise certain problems for the more philosophically minded. One of these problems concerned the origin of the two opposing cosmic principles of good and evil, and in the Sassanian period apparently an attempt was made to find an answer by assigning their origin to a god called Zurvān. This deity is a mysterious being, concerning whose nature there has been much conflict of expert opinion. The word "Zurvān" means "Time." There is evidence of its use as the name of a deity as far back as the twelfth century B.C., and some scholars have maintained that Zurvān was an ancient Iranian "high-god." The information that we have of the theological conception of Zurvān dates, however, from the Sassanian period or later; but it is very striking. For example, Zurvān, as the personification of Time, was also associated with Space. Moreover, he seems to have been conceived under a twofold aspect, namely, as *Zurvān-akarana* and *Zurvān-dareghō-chvadhāta*, the former designation signifying "Infinite Time" and the latter "Time of Long Dominion." This second concept is particularly significant, because in this form Zurvān was regarded as the force that brings decay, old age and death to men. There is some reason for thinking that, under the form of "Time of Long Dominion," Zurvān had a place in the theology of Mithraism, being represented in the *mithraea* in the guise of a lion-headed monster, entwined about by a serpent and adorned with the signs of the zodiac. The subject is a fascinating one, and it awaits further elucidation. However that may be, from our point of view in this essay Zurvān, as the personification of Time in these two aspects, constitutes a truly remarkable, and it would seem a unique, concept in the history of religions.[18]

3   *The Negative (Pessimistic) Response to Time*

The various forms of reaction to the challenge of time which we have surveyed have been described as conforming to a pattern that has been defined as positive or optimistic. This definition has been chosen because in each instance, while time has been regarded with awe as associated intimately with death, the belief prevailed that in some way man could ultimately reach a state of security beyond the fatal touch of time. We have now briefly to consider what may be designated the negative or pessimistic reaction. As we shall see, such reaction has assumed two main forms of manifestation: either it has evaluated the human situation as completely hopeless, or it has denied the reality of this life and of the phenomenal world in its space-time continuum—in one sense this latter form of expression is not, however, ultimately pessimistic, since it encourages the hope that security can be found beyond this illusory world.

The most notable example of the completely pessimistic reaction is found in Mesopotamian culture, which equals that of Egypt in antiquity. Here a doctrine of man was taught, according to which it was held that the gods, in creating mankind to serve them, withheld the gift of immortality. However, death was not regarded as personal annihilation, which would surely have been better than the post-mortem fate that was believed to await the individual; for by dying a man was horribly transformed and departed to exist in dust and darkness in the grim realm of *kur-nu-gi-a*, the "Land of No Return." A similar hopeless fate awaited man according to the view of human destiny that is first attested in Homer and thereafter

153

comes to form the accepted *Weltanschauung* of ancient Greek society. Such views logically counseled a *carpe diem* philosophy of life, and they precluded any disposition to seek in time for some transcendental meaning—life was accepted realistically and lived existentially, for it held no other purpose.[19]

What may also be reckoned as pessimistic reactions, yet in a rather different sense from those generally current in Mesopotamia and Greece in that they were based upon premises that could promote a kind of willing acceptance of an inevitable destiny, are the ancient Greek philosophies of Stoicism and Epicureanism and the classic systems of China, usually designated Taoism and Confucianism. Basic to these interpretations of human significance was the conviction that man was an integral part of nature or the universe, and that he should not strive to set himself over against it as having a unique form of being or destiny. Accordingly, the individual was counseled to live in conformity to nature, cheerfully accepting ultimate annihilation as occasioned by the physics of the universe. Time was part of the fundamental pattern of things; it had no transcendental significance, and in Stoicism it was regarded as a cyclic process which continuously reproduced the same order of phenomena.[20]

It was in northern India during the so-called Upanishadic period (800–300 B.C.) that the foundations were laid of that evaluation of existence which underlies the great systems of Hinduism and Buddhism. The quintessence of this evaluation may be fairly defined as a rejection of the consciousness of self, or the sense of individual being, as an illusion of dangerous consequence. In turn the reality of the phenomenal world, apprehended by the self, is also regarded as equally illusory. From the sense of being a separate individual being, and the desire to preserve this sense of individuality, it was held that all the pain and suffering of life derived; moreover, this pain and suffering was unceasing, because the urge to preserve self-identity meant that the self or *ātman* at the end of one life was reborn to another, and so on and on through the dreary process of *saṃsāra*, with its ceaseless round of births and deaths with all the misery that these repeated incarnations entailed. The situation, however, was not hopeless; for *mokṣa* (salvation) could be attained by achieving a true understanding of the illusion of individuality, and, in turn, emancipation from its consequences. Such salvation meant, significantly, surrender of the desire for the persistence of one's sense of individuality and absorption into *Brahman*, the Cosmic *Ātman* or Self, by identifying one's self with this Cosmic Self. The belief is eloquently expressed in the *Chāndogya Upaniṣad* (III. 14. 54): "This is my self within the heart, smaller than a grain of rice, than a barley corn, than a mustard seed, than a grain of millet or than the kernel of a grain of millet. This is my self within the heart, greater than the earth, greater than the atmosphere, greater than the sky, greater than all these worlds . . . this is the self of mine within the heart; this is *Brahman*. Into him, I shall enter, on departing hence. Verily, he who believes this, will have no more doubts."[21]

The Buddhist solution was of the same order, but more radical. For it rejected the very existence of an individual self, maintaining that the so-called individual person is only a temporary assemblage of various constituents, material and psychical, which are dissipated at death. Consequently, the empiric concepts of "past" and "future" are bereft of meaning, and time really consists in the existential "instant": "instantaneous being is the fundamental doctrine by which all the Buddhist system is established 'at one stroke.' "[22] Subsequent speculation elaborated a *Weltanschauung* that accounted for every mode of being, past, present and future; but basically time was an aspect of the *māyā* or illusion of the phenomenal world. Time, however, could be said to have a real significance as "time past" through the doctrine of *karma*; for it is fundamental to Buddhist thought that "behind the natural causality which links events in the world of sense there are other, invisible, chains of a moral causality which ensures that all good acts are rewarded, all bad actions punished; and that this chain of moral sequences is not interrupted by death, but continues from one life to another."[23] But, whatever logical clash there may be between the insistence on the instantaneous nature of being and the adoption of the twin doctrines of *karma* and *saṃsāra* ("rebirth"), it is certain that the Buddhist form of salvation, *i.e.*, the individual's ultimate attainment of *nirvāṇa*, conceived of a state beyond the range of time.[24]

Out of the rich complexity of Indian thought, either Hindu or Buddhist, concerning human nature and destiny, perhaps the most significant evaluation of time that is to be found occurs in the theophany passage of the famous Bhagavadgītā. The hero, Arjuna, to whom the great god Vishṇu had manifested himself in his benign being, feels that there is another side to the deity and he begs that it might be revealed to him. The request is granted, and Arjuna, terrified by the awful revelation, exclaims:

Looking upon thy mighty form of many mouths and eyes, of many arms and thighs and feet, of many bellies, and grim with many teeth, O mighty-armed one, the worlds and I quake . . .

Thou devourest and lickest up all the worlds around with flaming mouths; filling the whole universe with radiance, grim glow thy splendours, O Vishṇu.

And the god replies, significantly explaining his awful form:

I am time that makes the worlds to perish away, waxed full and working here to compass the world's destruction.[25]

Here the supreme deity is represented as both the Creator and the Destroyer, and it is as Time that he destroys. We are reminded of the Iranian conception of Zurvān in the twofold form of *Zurvān-akarana* and *Zurvān-dareghō-chvadhāta*—possibly a dualistic conception of deity in these terms represented the deepest insights of both the Indian and the Iranian mind.

4   *The Dilemma of Western Thought*

We may here terminate our survey, having noticed the chief patterns of reaction to the challenge of time in the various historic cultures. We have seen how time, by making man conscious of his mortality, has caused him either to seek for some assurance of security beyond its reach or induced in him a pessimistic resignation to its logic. Accordingly, we may reasonably assert that the various religions and philosophies of life have stemmed from the sense of insecurity that man's awareness of time has inspired. We began our study by noting that the origin of civilization may also be legitimately traced back to this innate consciousness of time. We may now close with a brief comment upon the part which this consciousness has in the present malaise that afflicts Western culture.

As we have seen, the Christian interpretation of time's significance came to form the basis of the *Weltanschauung* of medieval Christendom, which was coterminous with Europe. With the gradual secularization of Western thought, owing especially to the progress of scientific research, since the eighteenth century, that *Weltanschauung* has been progressively abandoned. The process has been rather of the nature of a tacit relinquishing than a consciously violent rejection. Moreover, appreciation of the seriousness of the change was considerably dulled by the fact that the traditional Christian view of the passage of time as the achievement of God's purpose was substituted by belief in the progress that would inevitably follow as mankind learned to direct its affairs more rationally. However, the tacit abandonment of the Christian view was destined to have profound consequences. That view, which was essentially teleological, was inspired and sanctioned by the belief that the whole time process from the Creation was divinely guided, and through it God's purpose was gradually being fulfilled. Moreover, it was believed that that purpose was good, and in it both the individual and mankind, as a whole, had their significance. Further, the present world order, manifest in time, was seen as moving toward its ultimate *telos* or end, which would be accomplished by the Second Coming of Christ and the Final Judgment.

The gradual replacement of this essentially theological estimate of the temporal process as the drama of Man's destiny under the hand of God, by the evaluation authorized by science, has in effect been a tremendous revolution in orientation and value. For the scientific picture of the evolution of the space-time continuum is that of the operation of impersonal forces that only have a time-direction in terms of the Second Law of Thermodynamics, and in which such terms as "beginning" and "end" have merely a relative significance. In this complex, mankind, and its affairs, are dwarfed to an incidental insignificance, being doomed to extinction when the balance of natural conditions, which has permitted the appearance of man, changes as it must undoubtedly do in the course of time. The fundamentally impersonal character of the only *Weltanschauung* which science thus authorizes, and the essential relativity it assigns to mankind and its values is gradually making its chill logic felt. Yet, instinctively Western man still tends to think, with a kind of teleological optimism, in terms of an old-fashioned doctrine of progress, which is still recognizably inspired by the Christian tradition that "God is working His purpose out as year succeeds to year." Hence, to the hiatus that exists between the verdict of our science and that of our instincts, and of which we are becoming increasingly and disturbingly more aware, is surely to be attributed in large measure that spiritual malaise which afflicts our culture, and, inevitably, our personal lives. For Western thinkers there can be no more urgent task than that of resolving this dilemma, and, if possible, of producing an adequate philosophy of history, *i.e.*, of the meaning of man's life in time, in both its individual and its communal extensions.

PART II
TIME AND MAN
Communications, Rhythm, and Behavior

Introductory Note to Part II

There are reasons to believe that our apprehension of the future is intimately associated with the origins of human speech,[1] and that the imaginary reconstruction of events necessary for our recognition of the past is also linked with the invention of language.[2] Prof. Cohen has remarked that the emotive and intelligible aspects of sound symbols are distinct; furthermore, the former tend to refer to the present, while the latter may also refer either to the past or to the future.[3] Possibly this description may contain some oversimplifications, but it is certainly conceivable that, as he suggests, the separation of intelligibility from emotion in the evolution of speech has increased man's ability to plan ahead and made practical the growth of his interest in the past. The precise relations between the gifts of communicating thought and emotion, on the one hand, and the awareness of future, past and present on the other, are not sufficiently understood, but they do seem to be in some ways interdependent. Consequently, we are probably correct in subsuming the creation and enjoyment of music and the invention of language under the same basic and unique ability of communication.

Perhaps the most important element shared by speech, music and time is the general concept of rhythm, that is a complex apportioning of fractions of time. Rhythmic structures in speech and music have been discussed by Fraisse[4] and others; we only note here that to make communication between men possible, it has apparently become necessary to temporally subdivide the trains of signals at the source and develop suitable decoding ability at the destination, both according to mainly intuitive rules. The specific purpose of such rhythm makes it a richer concept than the cyclically repetitive behavior of all the living which is also called "rhythm."

TIME AND MAN

The intimate relation between language, the view of time and the way of life has been well illustrated by Professor Nakamura's essay. The following two papers relate to the techniques of communication in that they present the extreme sophistication of timing control necessary for the transfer of thought and evocation of feeling. The papers survey the methods of structuring human time, telling us in a subtle way with T. S. Eliot:

> Words move, music moves
> Only in time; but which is only living
> Can only die.

From communications and rhythm, we will shift our attention to rhythm and behavior. Beginning with an inquiry about time perception in children, we shall in subsequent steps speculate on psycho-physical isomorphism, consider the multitudes of meanings of time in psychiatry and conclude Part II with an essay on subjective time.

J. T. F.

Time in Language and Literature

C. F. P. STUTTERHEIM

Translated from the Dutch by Pieter Johannes Verhoeff

The title gives only an indication of the subject to be discussed in the following pages. It does not claim to be either exact or comprehensive.

The word "language" in the first place needs some explanation. Being coupled with "literature" as it is here, it is restricted in its modes of use. It can, for instance, no longer refer to gesture language, the language of the bees, to light signals, flag signals, telegraph signals or other similar codes. But its meaning is still too wide and vague, and will have to be further differentiated. The chief distinction to be made is that between what in modern linguistics is called "language" (in a narrower sense), and "speech." As the wider and the narrower concepts are both usually indicated by the word language, difficulties of terminology may arise, causing confusion and misunderstanding. However, this is not a serious danger as the term does not have the same syntactical valence in both cases. *A* language is a *particular* language as a complex whole of vocabulary and morphological, syntactic and other constituents. Speech we understand to mean the concrete utterances of a particular language, a sentence or a group of sentences. Such utterances, in spite of the etymology of the word speech, may be either oral or written. We have to make this distinction, as the relations to time of each of these two terms has its own character and its own problems.

Although the word language in the title may now have become a little clearer, it still remains a universal concept. It can refer to what all the languages of the world and to what the utterances of those languages have in common in relation to time. And here we may encounter great difficulties. No one can deduce these general characteristics from all the relevant empirical material, for no one knows more than a relatively small number of (usually cognate) languages. In any discussion of our subject, grammatical terms such as substantive, adverb, adverbial adjunct, verb, tense or verbal aspect will probably be introduced and used as if they belonged to the terminology of a scientifically acceptable *grammatica universalis*. But they were coined for the description of a particular type of language, and it is doubtful whether they can be applied for the description of other languages. Moreover, even between such cognate languages as I have at my command, there may be such enormous differences that it is impossible to make statements that apply to all of them.

"Time in language" will make some readers think immediately of the tense system of verbs. We need not go outside the Indo-European group to realize that there are various tense systems, and that, within the verb, time can be expressed in various ways. If we assume that all people experience the temporal aspect of reality in exactly the same way and that it leads them to make exactly the same distinctions in their world of experience, all we can say is that some of these distinctions may appear in one language and not in another. But may we assume this, and can we determine, excluding language, what this experience of time, common to all people, is? The danger remains that we declare whatever has been laid down in the verbal system of a particular language or group of languages to be in this respect "common to all people," and then we do nothing but compare other systems with it. The difficulties are increased by the fact that there is no separate verbal system whose description can as yet be said to be definite. And what has been said about the verbs also applies to everything else in language that has a relation to time, and for all the terms and notions that have to be handled when we wish to formulate our theories about the temporal aspect of language. Thus, there are the most divergent theories about the parts of speech and the various syntactical phenomena. In the past few decades, linguistics has been, more than ever before, a dynamic science, a science in gestation. The diversity of languages and the diversity of opinions together render a discussion of our subject a risky undertaking.

Speech shares in the problems of language, although in some respects it does not so easily give rise to unwarranted generalizations. There is a considerable difference between language and speech in their mode of being: they are realities of different orders. This is apparent especially in their relation to time. To mention only one instance: the duration of a concrete utterance (a sentence or group of sentences, spoken or read to us) can be measured accurately. But although we can use the term duration in the case of language, too, this is the sort of duration that cannot be measured with stopwatches.

The word "literature" in the title refers to utterances of a special kind. Much of what has been said about speech in connection with time also holds true for literature. But the latter has, in addition, sufficient individual characteristics to call for a separate discussion. That here, too, there is a danger of unwarranted generalization, and that here, too, there is still no *communis opinio* with regard to many a problem, are things I need mention only in passing. I shall not here enumerate the conditions an utterance has to fulfill in order to be regarded as literature.

The many time phenomena that are of importance for language, for speech in general and for literature in particular cannot be described systematically if no distinction is made between "form" and "content." The problems involved in formulating such a distinction will, for the greater part, have to remain undiscussed here.

The first word of our title, "time," is not a scientific notion but an English word, one of the tens of thousands of words of a particular language. As such it has a specific form, and a specific semantic unit with a number of semantic elements, and it has therefore specific semantic uses. Furthermore, it has a specific morphological and syntactical valence. Here, too, certain distinctions will have to be made if it is even partly to lose its character of "word" at the level of pre-scientific usage. Anyone reflecting about our subject will feel the need to juxtapose linguistic, psychological and physical time, and to consider as time aspects duration, isochronism and tempo, simultaneity, order and repetition, as well as change. He will also become aware that he will have to make use of the term "a-temporal."

Meanwhile, he might become somewhat doubtful as to the validity of the statements he makes and conclude that he cannot attain a high degree of exactness in defining these concepts. He will wonder why we should denote something linguistic, something psychological and something physical, three categorically different things, by one and the same word, time. And there is the harmless-looking preposition "in," which connects "time" with "language and literature" in our title. It could be argued that time *is* in language, just as well or just as unconvincingly as that language *is* in time. Compare, for instance, "an utterance takes place *in* time" with "time appears (as duration or order) in an utterance." A consideration of these and similar problems would lead to the question of what is the relation between language and thought, and what is the influence of a particular language on the thought of its speakers. A discussion of this question, however, is not within the scope of this essay.

Although in a discussion of time it is probably impossible to formulate everything in such a way that the result is proof against rigorous linguistic criticism, we can at least elucidate some of the more important formulas. The title of this article refers to all the possible relations. As to the titles of the various sections, "Language in Time" belongs to diachronic, "Time in Language" to synchronic linguistics. "Time in Literature" will be discussed in Section 4.

Within the scope of this article, a discussion of our comprehensive and complicated subject can only consist in a preliminary inventory of the various phenomena and a summary delineation of the main problems involved. These phenomena are extremely heterogeneous and they have never before been dealt with in a single essay.

One last remark: it would take up too much space to cite examples here. It would also lead to almost complete arbitrariness in view of the large number of languages, the even larger number of written utterances (literary or non-literary), and the unimaginably large number of spoken sentences. In order to avoid arbitrariness, we have not quoted from the work of others who have written about one or more of the relevant temporal aspects.

1   *Language in Time*

The proposition "Every language changes in the course of time" does not seem open to discussion. It is not really a proposition, it is a simple statement of fact. And yet, a critical analysis of this statement will bring to light a few problems.

Reflections about time always lead to reflections about reality. This is particularly true in the case of language. How are we to visualize a language? In an earlier period, people saw in it an organism, something that comes into being ("is born"), and perishes ("dies"), and then is "dead," or is liable to perish ("is transitory"), and which, between birth and death, "changes," "develops." This representation of things is quite natural; it has also been used for other entities—literary forms and currents, civilizations, communities, etc.—and it is still used occasionally. Nowadays, however, it is rejected by a great many scholars. The term "organism" has been exposed as a dangerous metaphor. It is not incorrect to speak of dead and living languages. Undoubtedly, these are metaphors, too, but they can easily be translated into words that are not metaphorical, and they refer to established facts. Dead languages are languages that used to be but are no longer spoken and/or written by members of a speech community.

There is nothing problematical in this. But how are we supposed to consider a changing or developing language as an organism? In this case the metaphor hides a real problem.

Nowadays, we no longer regard a language as something comparable to a plant or an animal. We see in it a system or, at any rate, something systematic in many respects. As long as we stay within the limits of synchronic linguistics the term "system" applies better than the term "organism." But as soon as we say (explicitly or implicitly) that a system changes or develops, we get into the same trouble.

Whatever a language may be, and with whatever else we may compare it, one thing is certain: it does not change of itself, it is changed. And the cause of what happens to it in the course of time is not difficult to find; that cause must be the people, the members of the speech community, the speakers of the language. In other words, a language changes because people change their speech.

Still, even this statement does not cover everything. How do we establish that a language changes? No consideration of what passes in our minds when we speak, write, listen or read will suggest to us that the sentences made by us or by others are utterances in a language that used to be different and will be different again in the future. To the speakers of a language, their language is an invariable quantity, it is something beyond the opposition variable/invariable. They are not even aware that they use "their," *i.e.*, "a" language (one of many). If we wish to establish a change, no matter how it came about, we must consciously compare utterances from different periods. From a change in speech we can deduce a change in language. The "being different" in time we interpret as a change in something which, in spite of that change, retains its identity. We have from the outset recognized this identity as such, for we have compared with each other utterances that not only lay apart in time, but also had something important in common. It is clear that the historiography of a language—just like the historiography of a literature, a civilization, and also that of a town, a community, a nation, etc.—leads us, not only to the problem of reality, but also to the problem of identity.

When facts belonging to the history of a language are established, speech is only in one respect primary as compared with language. In another respect the relation is just the opposite. For something has to precede the comparison of utterances from different periods, namely understanding. And this understanding presupposes a more or less complete command of the languages or phases of a language that form the basis of the utterances.

What is the relation between speech and language if we try to disregard the way in which facts belonging to the history of a language are established, and once more concentrate on the facts themselves? We have said that the changes in a language are only the result of changes in the use of that language, and that the former does not change of itself. Although this is true, it is not the whole truth. We still have to reckon with the following factors. Supposing a language is, if not a system, at any rate systematic (and this cannot be doubted). Then elements must be interdependent, which means, among other things, that changes in one element must cause changes in other elements. The freedom the speaker of a language is permitted in deviating from existing norms is restricted by the fact that *this* systematicalness, but not systematicalness itself, can get lost. In this respect language, even as a historical object, determines speech. In the field of diachronic studies, the systematicalness appears most clearly in the changes that have taken place in phonemes and word forms. This enables historical and comparative linguists to formulate sound laws. A sound law lays down the beginning and the end of a process in which many word forms, innumerable realizations of these forms and innumerable speakers are involved, and which does not allow of an empirical investigation.

Should these processes be thought of as continuous? "A language changes continuously" or "is changed continuously" sounds like an acceptable proposition. But what is continuous in this connection? From second to second, from day to day or from year to year? How is this continuousness to be measured? A language is such a complicated phenomenon that it must be considered impossible to describe it as it is "at a given moment." Composing a dictionary or a grammar that is as complete as possible takes many years, and this, if the object of investigation has not remained the same in a number of its elements, may have unpleasant consequences for the synchronic character of the description.

In the last few decades there has been considerable discussion about the relation between synchronic and diachronic linguistics. At first they were considered to be essentially different, mutually exclusive. One went in either for the one or for the other. The former was descriptive, the latter dynamic. This sharp differentiation has proved untenable. Establishing a change is impossible without a synchronic knowledge of two historical phases of a language. On the other hand, if we observe the use of language in our own days, we cannot but notice that phonemes and words show combinatory and individual variants, and that not all members of a speech community speak in exactly the same way. In this "being different" lie the germs of change.

And yet, attempts have been made at drawing up a formula, especially with relation to changes in vocabulary. The starting point for this was that the rate of preservation *r* of those words which belong to the most stable part of a vocabulary is constant in time. Once the number of related words *c* in the basic vocabularies of the two languages is known, the formula can be used to calculate the time *t* that has elapsed since those languages were split off from a common parental language (assuming that no disturbing factors have been at work, such as borrowing, migration, etc.). The formula is

$$t = \frac{\log c}{2 \log r}.$$

Here *c* is the percentage of cognates, *r* the percentage of cognates retained after a millennium of separation, while *t* is in millennia. The discussion about this new branch of diachronic linguistics, "glottochronology," which is connected with lexicostatistics, is still not closed.

2   Time in Language

In his object of investigation, the student of synchronic linguistics will come across the temporal aspects of duration (sometimes called length) and order. Duration plays a part especially on a phonemic level, but it can also determine morphological characteristics. As a distinctive feature of a phoneme it distinguishes pairs of minimum contrast. The members of such a pair have all their characteristics in common except one, and form the opposition duration/no duration (or length/no length). From this statement it appears that linguistic duration and therefore linguistic time does not coincide with physical duration and physical time. This matter will be further discussed in Section 3.

The distinction between long and short stem syllables is based not only on the difference in length, but also on the difference in number of the phonemes belonging to a stem syllable (this number is also regarded as a form of duration, as discussed in Section 3). In the description of a language this distinction may only be made if morphological phenomena render this inevitable. In some languages, for instance, the finite verb forms and the case forms of nouns depend (partly) on the length of the stem syllable, or the influence of this temporal aspect can be discerned in the ways in which diminutives are formed.

Whether duration has a phonological and morphological function, and if so, what this function is, will have to be investigated for each language separately. When in phonetics "continuants" are opposed to "checked sounds," this applies likewise to a difference in duration.

More important than duration is order, which also presupposes time.

Here the difference between the various languages is in certain respects less of an obstacle to the suggesting of general propositions. The order of the elements of which it is built up, as well as their number and their nature, is a constituent of a word form in any language. The distribution rules that hold good for a phoneme have to be expressed in terms such as "before" and "after" (other phonemes), "beginning" and "end" (of a word form). Furthermore, in the description of a word group, the chronological relation between the members of that group has to be stated. The same goes for the description of syntactic schemata. The term "place categories" is sometimes used in this connection, but this is a spatial metaphor for something temporal. In the theory of word formation, terms such as prefix, suffix and infix imply the idea of order. Some languages have compounds that are distinguished formally only by the order of their constituent parts or by their stress relations (prominent/not prominent as opposed to not prominent/prominent). This may suffice to show what functions this temporal aspect can have in the system of a language.

Special mention ought to be made of repetition, the occurrence one after the other of the formally identical or the formally and semantically identical. Also, a morpheme may be repeated in a variant form or only partly. The functions of this process differ from language to language, and are not homogeneous within one and the same language. It may intensify, it may typify onomatopoeic formations or it may be used in the formation of plurals or preterites. In the formation of preterites in some languages a reduplication syllable occurs. Although this seems to be in contrast with the meaning of the word "repetition," it is, from a linguistic point of view, certainly correct to say that in the reduplication syllable, which precedes, not follows, the stem, something is "repeated." This is paralleled in speech, where we also find relations "contrary to time."

With the notion of past tense we are already outside the field of purely formal phenomena, for in the past tense the temporal is also revealed at a semantic level. In all languages there are words that in some way or other indicate something temporal. This semantic characteristic may have a formal indicator. Of a series of word pairs, we can say that the same semantic difference indicating something temporal goes with the same formal difference.

But the situation is slightly more complicated than that. There may be more than two forms and therefore more than two tenses. Moreover, words that can be opposed as regards their meaning may differ from each other not just in one, but in two or three respects. And finally, in many sentences the function of the so-called present appears not to be to indicate something that happens at the moment of speech, and the same, *mutatis mutandis*, can be said of the preterite. This does not alter the fact that in certain cases there is a semantic opposition present/past, and that in certain cases this has a formal indicator. It is understandable why the term "verbal tense system" should have remained in use, and why the verb should have been called *Zeitwort* by German scholars. In this connection, we may also mention perfect presents, frequentatives, inchoatives, as well as perfective and durative verbal forms or verbs. This does not, of course, mean that all of them should occur in all languages.

In all languages the temporal can be discerned in the meaning of many words which are not part of a system of formal oppositions, in verbs, substantives, adjectives, adverbs and conjunctions (if I may use these terms quite unproblematically for the moment). It is a remarkable phenomenon that conjunctions and adverbs but not substantives or adjectives "of time" should have been distinguished. Such terms are not acceptable in a *grammatica universalis*. In the description of a particular language they can only be used if the adverbs indicating time, for instance, can be shown to have an individual phonematic structure or an individual syntactic valence in which they differ from, say, the adverbs of place. An inventory of all the words, belonging to different parts of speech, that have something to do with time, seems to have little or no linguistic value. Still, we may expect to find a main division or subdivision "time" in the layout of an ideological dictionary or in the construction of a so-called *lingua universalis*. We may also find this category in the division of word meanings into "spheres," which some scholars have needed for their theory of the metaphor.

Only a part of a language system is temporal as it appears in linguistic forms and contents. Many elements are related a-temporally, although the description of them (a case of speech) can only take place in time. Thus, although it is customary to say that word forms are "built up" of phonemes, phonemes do not precede word forms in time. And all the elements which the speaker of a language has at his disposal and which he combines into a variety of sentences exist "at the same time."

3   *Time in Speech*

An utterance consists of at least one sentence. A sentence consists of at least one actuated word and an actuated intonation. An actuated word consists of at least one actuated phoneme. Between a monoverbial and monophonematic sentence and a sentence comprising more than a page in printing, between the former and a family novel in twenty volumes, lie all the possibilities of length or duration in speech. Duration and the related tempo are the only temporal aspects that are objects of investigation in phonetics. What is there called the "time pattern of speech" is determined by the duration of the speech sound, the syllable, the word, the sound unit (that which is articulated without interruption) and of larger speech units, in which the pause, too, is considered. Phonetic duration is determined in two different ways. One may count the number of smaller units making up a larger unit, the number of words of a particular sentence, the number of syllables of a particular word, the number of phonemes of a particular syllable or of a particular word. Once a sufficient quantity of material has been investigated, it is possible to calculate averages comparing writers from different periods or speakers of different languages or, within one speech community, men and women, regarding for instance, the average length of the sentences uttered by them. The counting method is on the whole simple and precise. Yet, several problems may present themselves. It is not always clear whether we are dealing with one word or with two, nor is it always easy to determine the number of phonemes constituting a word. Furthermore, it is really incorrect to include the fact that a particular word consists of two syllables or five phonemes among the "time phenomena of speech," because a word, unlike a sentence, is a language element, and because we are dealing here with a quality it has as a language element rather than a quality it acquires because it is actuated.

We do find ourselves in the domain of speech when we try to ascertain the duration of actuated phonemes, syllables and words, and sentences in relation to a unit of time such as a millisecond. This method, however, is much more complicated and also less exact. Something may be achieved with a chronometer in the case of a sentence if it has a specific duration. For words, syllables, and *a fortiori* for sounds, this aid cannot be used. Spectrograms are made of words in which a speech sound occurs whose duration one wishes to determine. Each sound appears on them with its own figure. The difficulty is that these figures overlap; there are no clear boundaries. The phonetician is aware that an exact measuring of the duration of a sound (*i.e.*, one realization of a phoneme at a time) must therefore remain something illusory, even though he can express it in terms of thousandths of a second. One has to disregard this essential difficulty if, on the basis of a large number of realizations, one wishes to calculate the average duration of a speech sound—*i.e.*, the duration of the average realization of a phoneme—or if one wishes to investigate how much male and female speakers differ in this respect, or to compare realizations in speaking with those in reading aloud. When pauses or sentences marked off by pauses are measured, this difficulty does not arise.

In the preceding paragraph I have held to the traditional formulas. It is customary to say that a speech sound has a duration of, say, 0.184 second. This is an attempted quantitative evaluation or measurement of a quality, seen in arbitrary isolation as a bundle of air vibrations. Vibrations, however, are not sounds, and even less are they speech sounds. A speech sound is the correlative of air vibrations in the mind of the listener, not just any listener, but one who knows the language to which the word containing the sound in question belongs. And this does not necessarily mean that everything that is to be seen in a spectrogram or an oscillogram has a correlative in the mind of the listener. If the term "duration" is to be used in the linguistic-psychological sense, it refers to a duration that cannot be measured but is experienced. This may be the case in linguistically relevant differences of duration (Section 2). Also, the listener may observe that a speaker incidentally draws out a particular vowel in a particular word in emotional speech, for instance.

Instrumental phoneticians nowadays have at their disposal means by which a spoken word and its constituent sounds can be reproduced first in an increasing, then in a decreasing, number of very small fragments. From this, they try to draw conclusions as to duration as a characteristic of a phoneme. In the investigation of the so-called diphthongs, a similar fragmentizing artifice is applied to psychological time though not in connection with duration. Something that lasts about 0.334 second is successively broken up into fragments, each of 0.10 or 0.20 second, which at their beginning and at their termination differ one one-hundredth of a second from each other. That is to say, fragment $n + 1$ does not at its beginning contain the one one-hundredth of a second of sound with which fragment n begins; instead of that it gets at the end the penultimate one one-hundredth of a second of sound of fragment $n + 2$. There is an extremely great risk of categorical errors in the interpretation of perceptive data obtained in such a way, especially in the first-mentioned case.

The time that is needed to express a thought has been the object of inquiry of some linguists, and their opinions have influenced their theories of language and their descriptions of languages. The time referred to here is called "operative time."

Whether, and if so in what way, people's experience of time in all sorts of situations belongs to the mental processes cannot here be discussed. Time is certainly present in the minds of speaker and listener whenever there is speech. This appears in a peculiar way in the utterances of inhibited speakers, people who stammer. With them there lies between the words not simply a gap, but something positive that is not filled, something that continues in an irritating way, even, perhaps especially when the speaker tries to cover it up with some sort of basic sound (. . . er, er. . .). An unpleasant experience of duration (boredom) may result for the listener when the spoken word is unable to hold his attention.

Time has sometimes been called an implicit function of speech. Actuated language elements and their combinations must not only have duration, they must also follow each other in time. They appear, in the same tempo and in the same order as they left the mouth of the speaker, in the mind of the listener. What appears is, for the time being at any rate, preserved together with that order. Besides, all kinds of grammatical relations are established. Each relation is bilateral and is, consequently, also established "contrary to time." This is also true of the order relation. The moment there is a "following" word, another word becomes the "preceding" word. In general it is impossible to say what is linguistically functional in that order, as this differs from language to language, and may differ, within the utterances of one language, from sentence to sentence. Elements which belong together grammatically need not follow each other immediately. And phenomena other than linguistically relevant ones determine the grammatical relations. There is a method of syntactic analysis in which the investigator in the first instance follows the sentence from word to word along the time axis, and determines when a "connection" does and when it does not operate. This method is more or less a slow-motion picture of the process that takes place in the mind of the listener. Meanwhile, it is not unusual for the grammatical structure of a sentence to be represented in a two-dimensional scheme that is meant to be a-temporal. The characteristics of some languages enable two or more sentences consisting of the same words in a partially different order to have the same grammatical structure.

A word has, in addition to a specific form, a specific meaning and a specific valence. If a word does not function in a monoverbial sentence, what follows is, as soon as the word has been uttered, already partly determined for the listener. It makes him expect something, look forward to some connection. Expectation is a state of mind that presupposes an experience of time. Listening to and interpreting a sentence is too complicated a process, and the tempo in which it takes place is usually too rapid, for an introspection to grasp that expectation as part of the total consciousness. But the moment a connection fails to come, the listener becomes aware of the fact that this expectation plays a role. This is the case, for instance, when the speaker produces an anacoluthon, a change in grammatical construction in mid-sentence, which may (but need not) result in the listener being unable to understand him. For the rest, the listener and the reader may lose the connection even in a well-constructed sentence. One might speak of the critical length or critical duration of a sentence, but this depends on too many factors for it to be expressed in a fixed number of words.

A few other aspects of "time in speech" will be considered in the following section.

4   *Time in Literature*

The age-old, plausible division of literature into prose and poetry has in our century become problematical both because of the development of the theory of literature, and that of literature itself. Criteria of content, form and typography are apparently involved in this distinction, which has lent these terms a polysemical character and has thus led to differences of opinion. And yet it cannot be doubted that there are utterances that are characterized by a specific form principle which is revealed as such from beginning to end of these utterances through the repetition. This does not mean that everything that is called a poem or is reckoned among poetry should have this characteristic. Utterances with such a form principle consist of lines of verse, or verses. A line in printing corresponds with each verse. Although this line is in many cases not a completely adequate symbol, the boundary between the verses is then much more clearly indicated on the visual (typographical) level than on the auditory level. Apart from a boundary, a form principle also has an internal periodicity. Both boundary and periodicity exist only by virtue of the repetition of identical elements appearing simultaneously with different elements. The number and nature of these identical elements may vary from poem to poem, from literary period to literary period, and from language to language. As to the latter, the possibilities of metrical construction depend on the structure of the language in which the poem is written. This does not alter the fact that greatly varying form principles may be found within one national literature. Verse may be, among other things, syllabary or (in more than one way) metrical. Boundaries may be marked by a syntactical boundary, by a metrical group not allowing of variation, by a rhyme-word or by a combination of these elements.

Duration has a function in that type of metrical verse in which the regular alternation is based on the so-called quantitative accent, *i.e.*, on the difference between long and short syllables. Such verse is only possible in languages that have the opposition duration/no duration (Section 2). But in the metrical structure of other languages, too, duration can be of importance. A regular alternation in that case is based on other characteristics, on a different type of accent. In a group of two or three syllables, one will be more prominent because of higher pitch and/or greater loudness. Many form principles may be modified in the individual verses. If such a form principle is characterized, for instance, by four groups of syllables, each consisting of two non-prominent syllables followed by one prominent syllable, then in one of the lines the third group, for example, may consist of only two syllables, of which the second is prominent. When reciting (or when reading silently), one will make the distance in time to the preceding prominent syllable just as great as that between any other two prominent syllables. This can only be done by changing the tempo in such a place, by reading or speaking more slowly. In this case both duration and tempo are considered to belong to the poem itself and not only to the recitation. For, duration and tempo determine one characteristic of the members, isochronism. The correlative of this experience of duration, which only operates in tripartite meters, has been measured; it appeared to be about three-quarters of a second.

The verses of a poem based on only one form principle make, when read, the impression of being (roughly) the same length. Attempts have been made to prove with a watch that, irrespective of the experience, the duration of all the verses of such a poem was exactly the same and could be expressed in seconds. These attempts must be considered to have failed. As to the duration of the intervals, only in rare cases can it be considered to form part of the poem. It will then have to be shown that the interval in a metrical poem takes the place (*i.e.*, the duration) of some syllables.

The distance in time between two elements of the same order is limited to a certain length. In the case of rhyme, for example, there is a critical distance within which it can, and outside of which it can no longer, be experienced. This depends on the type of rhyme (full rhyme, alliteration, assonance), on the number of times a certain rhyme has already occurred in the utterance, and also on its function: it may be incidental, but it may also belong to the form principle, notably to the verse boundary.

What was said in Section 3 about the order of the words and about their grammatical relations also holds true for literary utterances. But other relations, too, are established both "along" with time and "contrary" to time. Thus, a word becomes a rhyme-word only at the moment when another word also becomes one. The matter is, however, rather more complicated. He who participates in a linguistic work of art participates in a growing structure. His mind not only receives what is given in an irrevocable order, retaining as much of it as possible, but is also actively directed towards what is coming but is not yet there. He expects something. And what he expects is partly determined by what he has already read. If he is already aware of the form principle, and if a rhyme-word in a fixed place is part of that principle, he will, for instance, after a certain number of verses expect that a particular word will rhyme. This expectation is then fulfilled. Similar observations can be made in the case of other formal elements.

Surprise may be aroused if something unexpected appears or if something expected fails to appear. Expectation and surprise are unthinkable without an experience of time. Without an experience of time, the participation in a linguistic work of art, too, is unthinkable. Such a work of art may evoke these experiences by its formal characteristics as well as by its content. The same may be said of suspense. At a formal level we generate suspense when the distance—that is, the duration experienced together with other moments of consciousness—between two elements is greater than normal, which means, greater than expected. If the relation between the elements is syntactical, we speak of phenomena of syntactical suspense. These, together with the release of the suspense, have an aesthetic function. In fact, everything mentioned above in connection with duration, order and repetition has an aesthetic function. Or rather, it may have such a function. It does not have this per se but only if it functions in a successful work of art.

In all the cases mentioned so far, and in many others, the occurrence one after the other of the elements has been symbolized univocally in printing. There are, however, forms of "typographical poetry" in which the order is no longer univocally given. The words or groups of words appear in such diverse ways on the paper that either one thing or another may be read first. In modern literature this is one of the many ways in which poets hope to free themselves from the "tyranny of time."

It may be asserted with some justification that an utterance, if it is understood, is present in the mind "simultaneously" with all its elements and with all the relations between them, among them the relation of order. However, there exists another type of simultaneity. Examples of this are the variations on a metrical scheme and enjambment. We cannot experience something as a variation if we are not able to be aware of the characteristics of the scheme at the same time. If the end of the form principle is characterized by, among other things, a syntactical boundary, then enjambment, the absence of that characteristic in one of the verses, is likewise experienced as a variation. Furthermore, there is often the simultaneity of a formal and a semantic element. In this connection we might think of the metrical division of a plurisyllabic word and lexical meaning. In general, however, this simultaneity is avoided in metrical poetry, and word boundaries do not coincide with metrical boundaries. At a semantic level, *e.g.*, when we experience a metaphor, our attention is often focused on something that shows two aspects at the same time. Experiments have even been made with a simultaneity of words that differ formally only in their vowels, have related meanings and fit, both of them, in the context. In the recitation of the poem, alternately one and two persons are then engaged.

When we speak of the content of a linguistic work of art, we usually disregard the fact that all formal characteristics correspond to something in the content. This content, the problems of which will have to remain undiscussed here, is sometimes a-temporal. Lyrical poems—however much they are structures in time, growing structures with a more or less definite duration—do not refer to successive events, that is to "facts" that stand in a chronological relation to each other. They are expressions of a feeling, a mood, an inspired thought through which the stream of time does not pass. This is different in plays and narrative literature. This does not mean that, insofar as they are expressions of a personality, a philosophy of life, drama and narrative should not at the same time partake of the a-temporal.

For a definition of what might be called the time pattern of a play, several criteria may be used. In the first place there is the relation between the time of action and the time of acting, that is, between the duration of the play and the duration of the whole of the events to which it refers. Sometimes (in one-act plays) the two coincide. When the play lasts one hour, not only the spectators and the actors but also the *dramatis personae* have grown older by one hour at the end of it. But usually there is a greater or smaller discrepancy in this respect. In many plays with an acting time of some hours, the action comprises a period of twenty-four-hours ("unity of time"). The time of action may also comprise days, months, years or even a whole life. This is connected with the division into acts. The interval, the caesura between one act and the next, can be filled by the spectator just as he sees fit. It remains functional as empty duration, usually representing passage of time within the play.

Apart from a duration, the acting time of a play has a degree of homogeneity or continuity. We could think here of a-temporal (*e.g.*, lyrical) interruptions of the dramatic action. Besides, the *dramatis personae* may narrate something, and then greater or smaller discrepancies between the time of narrating and the time narrated might occur. The degree of continuity is determined by something else, too. The order of events can be consecutive; what is said to have happened ten years later appears later in the play. But it is also possible that in a flash-back the past interrupts and breaks through the present, that, as stage reality, it occurs between what was first represented as the present and what will be represented as such again later on in the play.

However important all this may be as an expression of the manifold experiences of reality and time, far more important for the play as a piece of work are the dramatic aspects of action. Of these aspects only the retrospective, the prospective and the simultaneous can be mentioned in this connection. Some phases of a play refer back to the past, others point forward to what is yet to come. They raise expectation, suspense and surprise. The simultaneous has greater possibilities in drama than in other literary forms as drama is both seen and heard and as it is a combination of action and acting.

No matter how complicated the time of action is, all the various actions in the play form part of a rounded-off occurrence which unfolds itself in the tempo set by the speaking and acting of people, and which, however much the flash-back is used, passes simultaneously with our experience of time.

Much more complicated and much more heterogeneous is the time pattern of novels, especially modern ones, even if we disregard the complications that may arise from the use of the historical present. Even more important in this connection than the many possible relations between time of narration and time narrated are the problems of order.

The necessary linearity of all linguistic utterances forces the investigator to find an order for the moments of an essentially a-temporal complex. The writer of history has fewer problems to solve and will feel the tyranny of time less strongly than other writers do. Between the events he has to narrate there are relations, not depending on himself, of before and after, relations that influence the order of his sentences, paragraphs or chapters. But he does need other principles of arrangement which eliminate the first-mentioned principle. Besides, there are often a-temporal interruptions such as the descriptions of persons, situations and so on. We find more or less the same situation in historical novels. In other narrative literature however, everything depends on the author: the order of events in the imaginary reality and the order in which he narrates them. In modern novels there is often an enormous discrepancy between these two. There are several reasons for this discrepancy. One is the conviction that a human being as he is "at a given moment" is at the same time his past. Time, as a fascinating problem to the mind and as a frightening mystery to the feeling, has become a literary subject.

But if we wish to find it used as a "subject" of linguistic works of art (be it in a different way), we need not even turn to the literature of our own day. From of old, in lyrics about transiency and death, the painful realization has been expressed that time exists . . . whatever it may be.

Rhythm in Music: A Formal Scaffolding of Time

WALTHER DÜRR

Translated from the German by Conrad H. Rawski

Mr. Weir has told me of the case of a bullfinch which had been taught to pipe a German Waltz, and who was so good of a performer that he cost ten guineas; when the bird was first introduced into a room where other birds were kept and he began to sing, all the others, consisting of about twenty linnets and canaries, ranged themselves on the nearest side of their cages and listened with the greatest interest to the new performer.

Charles Darwin, *Descent of Man*

The great office of music is to organize our conception of feeling into more than an occasional awareness of emotional storm, i.e. to give us an insight into what may truly be called the "life of feeling," or subjective unity of experience; and this it does by the same principle that organizes physical existence into a biological design—rhythm.

Susanne K. Langer, *Feeling and Form*

In the temporal structure of the perception of melody, features can be discovered which appear irrational in any visual-mechanical model of physical reality: the primacy of events, the absence of infinite divisibility, the compatibility of novelty and mnemic causation, the compatibility of continuity and individuality, the fusion of becoming with its concrete content. Needless to say, such use of auditory models is purely propaedeutic; it only helps to free our mind from the exclusive sway of visual imagination, whose influence may be detected even in some apparently abstract mathematical habits. No epistemologically educated person would dream of reinstating auditory or any secondary qualities into the physical world. The positive significance of the auditory experience is in the fact that from it a certain imageless dynamic pattern may be abstracted which will probably offer a key to the understanding of the nature of the type of "extensive becoming" that seems to constitute the nature of physical reality.

Milič Čapek, *The Philosophical Impact of Contemporary Physics*

1   *Music and Time*

Music begins, progresses and ends; music moves in time; music is a temporal art.[1] Ever since Lessing's *Laokoon*, so it seems, the relation of an art to time and space has been regarded as its foremost characteristic:[2] poetry and music are grasped in relation to time—painting, sculpture and architecture in relation to space. There are also mixed art forms, such as pictorial narratives, which attempt to present an action in time in form of a strip of pictures, or the calligrammatic poem, the *carmen figuratum*, which seeks artistic effect not only through content and poetic language but also through the visual appeal of its contour lines.[3] Between the space-arts and the time-arts are those spatio-temporal arts in which artistic realization in space is combined with music, such as in the dance and the theater.

Music as a temporal art is thought to find realization within time. Yet, this realization is not without certain spatial concomitants: the sound waves, for instance, must expand in space. This spatial aspect of music has led in orchestral and choral performance to frequent experimentation with the arrangement of performing groups (as, for instance, the polychoral practices of the sixteenth century); it is also significant in musical performance on instruments such as the organ.[4]

Spatial aspects are also implied by the human capacity of retention. Our memory can transform time sequences into patterns in virtual space. It can retain pictorially, as it were, an overall impression of a composition and can even interconnect some of the musical details within this complex image. Melodies may become horizontal lines, harmonies vertical clusters. This metamorphosis of musical time into pictorial simultaneity finds expression in our musical notation which in most of its forms is nothing but the graphic record of our memory image, in which rhythms, melodies and harmonies appear projected upon a plane.

Similar processes seem to hold in the plastic arts where simultaneity is dissolved into a time continuum by the viewer's eyes as they seek out details, one after the other, rather than view the picture or sculpture as a whole. This, in turn, is acknowledged in art when the painter shows an action sequence simultaneously on one and the same canvas, as he relies on our ability to translate pictorial simultaneity into a virtual time sequence.

However, these possibilities of interaction do not render less valid the basic dependence of any single art mostly upon time or mostly upon space. Music remains music even when it sounds in earphones lacking the conventional space component. And musical memory must seek expression within a sonorous medium of a temporal order, if its spatial imagery is to become music again. Here too, musical notation serves as an instructive example: By looking at a score, we may obtain a general impression of the music notated on the page. But in order to actually transform the page into music we must perform it or read it, thus re-establishing temporal sequence.

This sequence of time in music is a very special time experience—time significantly arranged, formally grouped—time with a special content, so to say.[5] Time in music becomes a privately experienced order of occurrence—musical rhythm.[6] Rhythm as grouping of durations in a characteristic manner represents the time element of music. Without rhythm there would be no "before," no "after" in music—no evolvement, no form.

Rhythm is not only an element of the fine arts, of music, and of poetry. Rhythm emerges wherever we sense groups and patterns in the continuum of time. The world which surrounds us reveals itself in rhythmic forms. Within this general rhythmic framework specific rhythmic configurations emerge when groupings within time are interpreted as continuous recurrence "*eines als rhythmische Einheit der Rhythmuswelle bezeichneten ähnlichen Sachverhaltes in ähnlichen Zeiten*" ("at similar time intervals, of a similar event, which is identified as the rhythmic unit within the rhythmic wave").[7] This continuous recurrence is the property of waves (*i.e.*, undulating motion),[8] which is rhythmically interpreted by the eye (the breaking waves of the surf), the ear (the pounding of the waves), and the touch (the pulse beat). Also more extended groupings of time, such as the course of the seasons, lend themselves to a rhythmic interpretation since they can be foreshortened by thought, transformed into a conceptual series, and thus viewed as undulation.

Not only does man experience rhythmic patterns in the surrounding world, he carries within himself the continuous up and down of rhythmic waves: heartbeat and respiration. The experience of rhythm has often been connected with these organic functions[9] since they, in turn, seem to relate to rhythmic phenomena, such as walking, running, marching, and to certain rhythmic aspects of manual labor.[10]

Against this complex rhythmic background the specific rhythmic wave stands out; it is recognized as something concrete and orderly. The regular alternation of tension and relaxation characteristic of all undulating motion appears in the time arts as their "rhythmic" component. Whence the frequent identification of rhythm with regular recurrence in movement.

Rhythm as time transformed into durational patterns is a human concept and thus it depends on human awareness. Since the *"maximum filled duration* of which we can be both *distinctly and immediately* aware" seems to be about not more than a few seconds,[11] longer durations are subdivided by the listener according to inner tendencies which may largely depend upon physiological factors, such as heartbeat and breathing. In this manner, smaller durational subdivisions are empathetically supplied. A series of equidistant strokes of sound is grouped into accented rhythmic patterns.[12] When the time intervals involved are too large (extended) and thus not conducive to subdivision, the foreshortening thought transforms them into conceptual series, as we have said above.

If we take rhythm as the mode of appearance of time in music, then musical speed, the *tempo*, is the expressive mode of rhythm. Rhythm expresses time as a sequence which, as such, is structured and grouped, but not measured. Rhythmic order can be imagined without exact measurement; the powers of our memory foreshorten and extend. Yet, the moment a rhythmic sequence is expressed, is performed, sounds—it gains concrete dimension in time, it happens within a fixed period of time, at a certain rate of speed.

Tempo as a mode of rhythmic realization does not change the character of the rhythm. The rhythmic pattern remains the same whether its pace is sped up or retarded—since the proportion of the distance between one accent and the other remains unaltered. We have noted that rhythm experience and capacity to perceive sustained durations without empathetic subdivision are related. It follows that our interpretation of rhythm patterns also is related to the speed in which they are performed: a tempo which is too fast would lead to contractions of small rhythmic units; a tempo which is too slow would lead to their subdivision in the listener's experience. Thus it is essentially the tempo which determines the emotional effect of rhythms. Generally, slow movement is interpreted in terms of sadness, faster movement, in terms of joyfulness.[13] The same rhythmic passage may appear happily exuberant when performed at a fast tempo, but sorrowful when played slowly.

"Fast" and "slow," in this respect, do not necessarily directly correspond with the objective measurement of time (*e.g.*, by a metronome). They rather depend on the varying degrees of density in musical configuration. If there are a great number of configurations and if they follow each other in close sequence, the music appears lively; if the configurations are sparse and drawn out in long spans, the music appears to be slow. Our judgment of duration in music is similarly affected: the richer the musical configuration, the shorter appears the composition to us and vice versa.[14]

The musical work of art contains two aspects of tempo. One is firmly linked to its basic character and determines the composition as cheerful, moderate, slow (*i.e., allegro, moderato, lento*),[15] irrespective of the fact that it has no bearing on the rhythmic organization of the piece. This aspect is a functional part of construction, it controls the "meaning" of a movement, of a rhythmic passage. In this respect, however, the tempo designation reflects indirectly upon the rhythmic organization, because it rules out all patterns and groups of patterns which cannot be meaningfully interpreted at the designed rate of speed.

The second aspect of musical tempo is of a more accidental kind. It belongs to musical performance and provides both opportunity and freedom for the performer to deviate at times from the basic rate of speed. This entails temporary disturbance of the rhythmic order and of the prevailing tempo. The resulting feeling of tension has implications regarding the effect of the music; it is compensated by reversion to the original tempo.

Rhythm as structural grouping in time without which music cannot happen is an essential attribute of every musical work of art. As such it may be passively experienced. Meter, on the other hand, permits the active imposition of concepts of order, be they the composer's, or the performer's, or the listener's. Metric order can be part and parcel of the composition from its outset, as is the case with most contemporary Western works. Or it may be added in performance, as it happens when sixteenth-century music is subjected to the time and accent patterns of nineteenth-century meter. It also can be imposed by an ear accustomed to certain expectations and effects, as, for instance, is frequently the case when Western audiences listen to music of a non-metric kind.

Modern rhythmology interprets rhythm and meter in a number of rather divergent ways.[16] In part, this may be due to different realms of discourse. The psychologist is interested not so much in rhythmic structures as in the empathetic order imposed by the listener. For the musicologist who concentrates upon Western music and interprets its forms strictly in the rhythmic tradition of the classic and romantic periods, music is subject to a metric framework which music is forced to acknowledge while trying to overcome it. If, on the other hand, the musicologist develops his rhythmic categories with an eye on contemporary music and its forms which differ intrinsically from the metric ideals of classicism and romanticism, he will have to define rhythm in very broad terms—somewhat similar to those developed in philological and linguistic studies concerned with metric forms that are quantitatively indefinite, and intent upon a theory of "free rhythm" (*i.e.*, a theory of verses obliged neither to a definite meter nor to an order of rhyme and cadence).

The metric order of a musical composition is not necessarily self-generated. In vocal works it may be determined by the text. A closer examination of this situation reveals an essential difference between meter in language and meter in music. In music, the duration of the tones is precisely established, while in language the duration of the syllables remains indefinite. The metric order indicates which syllable is long and which one is short, but it does not indicate *how* long or *how* short. Music therefore enjoys a certain independence, in the sense that it can convert the quantities of language into qualities and vice versa.[17]

Relations similar to those obtaining between rhythm and meter appear also between the structure and the form of music. Music represents itself by way of structure (*Form*) and form (*Formschema*). The form indicates the overall design to which a composition must conform in order to be classed among the established formal types.[18] As the division of time over long periods is retrospectively interpreted in terms of rhythmic periodicity, musical structure may appear as rhythmic, musical form as metric. Structure in music is often effected by projecting immediately experienced rhythmic configurations into extended periods of time, while the overall musical form is outlined, as it were, by identifying its periods and sections with the metric order.

2 Rhythm and Meter in Music

Uniform duration can be interrupted by positive and negative impulses, intensification and weakening, and by accentuation.[19] In music this means that rhythm depends upon changes in intensity, pitch, harmonic function and timbre, and, of course, upon the change brought about by the absence of musical sound, the rest. To these factors of change affecting the single tone, we must add duration, which results from the relationship between one tone and the next, and, as such, obviously depends upon certain changes in the properties of the single tone.

While accents are created by changes in intensity, pitch, harmonic function or timbre, the rest creates silence: a specifically musical silence, filled with expectation of the new sonority that is to follow. Hugo Riemann referred to the rest as *"negativer Ton."*[20] Similarly, we may describe the rhythmic effect of the rest as negative accent. Such a negative accent is most obvious in metrical rhythms where the listener expects a regularly recurring beat. The inherent tension of the rest will depend on the kind of accent the preceding musical context seems to promise.

A sequence of accents creates a rhythmic chain which is C. Hoeweler's *"Verlauf."*[21] As a rule, the dominant system of metric order depends only on a sequence of accents of one and the same kind, be it now accents of intensity (*i.e.*, the chain of dynamic accents in Western music) or accents resulting from quantitative (durational) change (as in the music of Antiquity).[22] The metrical order of musical rhythm becomes especially clear when there are few or no conflicts between the various kinds of accents and their chains, and the dynamic accent falls on the tone which is also the highest in pitch and longest in duration (as is frequently the case in folk-dance melodies). Normally, the various accent sequences overlap. In music of metrical rhythm it is then the regular recurrence of identical (or related) events which increases the importance to the listener of one particular sequence rather than another and results eventually in habituation, so that the listener comes to accept this particular kind as the "natural order" of (his) music. We today, for instance, are easily tempted to transform a quantitative sequence ‒ ◡ ‒ ◡ ‒ ◡ into a qualitative one

♩ ♩ ♩ ♩ ♩ ♩

Intensities and durations, and the chains of changing intensities and changing durations, permit only a strictly rhythmic interpretation. Pitch sequences and sequences of harmonic progression involve orders of a different kind. Rhythm, harmony and melody usually are considered as musical components in their own right.

Frequently we associate with rhythm, and with arrangement in time in general, the concept of motion. This motion, which in our Western music is predominantly expressed by changes in intensity and duration, is strongly influenced by melody and harmony, although, actually, both melody and harmony are functions of the rhythmic progression in time.[23] In this respect, "rhythmic phenomena" can also be deduced from the "energetic properties of melody" (*melodische Energieverhältnisse*), as we note that *"melodischer Impuls nicht immer mit rhythmisch bedingtem Impuls zu identifizieren"* [ist].[24] We are able to distinguish, then, between music which is predominantly melodic, and types in which the harmonic or the rhythmic element prevails. The melodic character of a composition becomes more obvious as dynamic accents are restrained.

At the same time, rich melodic development requires an equally rich variety of durational detail, which is the natural result of the interrelation of accentuation by quantity (duration) and by quality (pitch). Compare, for instance, the following.

Examples 1 and 2: Variation XI and Variation XII in Mozart's Piano Sonata, K.284 (205b).

In our example, the tempo is also a determining factor. A slow, withheld pace, as a rule, invites rich melodic ornamentation of significant tones. This also enhances the effect of the rhythmic structure, but the melodic line predominates nevertheless. The rhythmic configurations are determined by the melody and not vice versa.

When motion and rhythmic drive prevail in a piece of music, a certain degree of melodic attrition may easily be the result. The pitch sequence is reduced to essential tones or fails to express a configuration of particular melodic significance.

Example 3: Béla Bartòk, "Ostinato" from Mikrokosmos Vol. 4, no. 146. Copyright 1940 by Hawkes & Son (London) Ltd, reprinted by permission of Boosey and Hawkes, Inc.

When the harmonic element is in the foreground, both melody and rhythm are frequently of reduced importance. Intense melodic lines and strong rhythmic development tend to distract from harmonic effects. In a composition which stresses sonorities, melody and metrical rhythm are frequently limited to bare essentials. (In music which uses non-metric, or altogether free rhythm, the harmonic element, if present at all, plays only a minor part.)

Example 4: Johannes Brahms, Symphony No. 1, Op. 68, 4th Movement.

Similarly, in cadences which primarily support the harmonic structure, larger note values are employed; sometimes, even the tempo is retarded.

Only seldom is an entire movement dominated by harmonic concerns, excepting, perhaps, the theme for a set of variations where the theme serves as a basis for subsequent rhythmic and melodic developments. Yet, within a vigorous rhythmic movement, the sudden bridling of musical motion by means of a section of primarily harmonic interest is of great significance. This effect is frequently used in Western music. It occupies an important place in the Renaissance theories regarding the effects of music.

3   *Historical Types of Rhythm*

We have described the rhythmic chain (*Verlauf*) in music as a sequence of accents. These accents, of course, are not uniform in character. We are able to differ between main accents and related accents of less importance. As a matter of fact, it is only through this interplay of main and subordinate accents that the motive, the basic formal unit, is established. The way in which the accents are connected, the way in which the motives follow each other—these are the determinant factors as far as any specific musical rhythm is concerned. It is impossible to list systematically categories of rhythmic expression. Moreover, our purpose is not to state "how it must be," but rather to grasp "what is" in order to find out what "might be"—in what manner the sequence in time is being expressed in music. In order to accomplish this, we must examine rhythmic formations as they occur in the history of music and test our conclusions on the works which represent the musical "reality."

One of the basic historical rhythmic types is the one which T. Georgiades and C. Sachs call *additive rhythm.*[25] Additive rhythm derives primarily from language. It is based upon prosodic models. The corresponding rhythmic motives reflect the characteristics of these models, such as patterns of long and short, or stressed and slack syllables which appear in music quantitatively determined (*e.g.*, expressed by a ratio of 1 : 2, etc.). The chain of motives, however, does not establish superordinate groupings or symmetric relationships. In this, additive rhythm follows the prosodic model, *i.e.*, the structure of language. Additive rhythm is not restricted to vocal music only. It may be found in purely instrumental compositions. Nevertheless (and ignoring instances in which it is patterned after vocal models or is actually a *cantus firmus* adaptation of such models, as in the organ compositions of the Renaissance and the Baroque)—instrumental music as such tends to symmetrical structures.

Since the prosodic model is frequently in metric form, its musical derivative is metrical rhythm. When prose texts are set to music, non-metric rhythm, lacking a regularly recurrent accent, may occur. In the Rumanian folk-song in Example 5, the rhythmic impulse is generated not so much by prosodic qualities of the text in general, as by the shout-like "Hâi," which seems to determine the character of the line. This shout-like quality finds musical expression in the drawn-out note values which in turn bring about the non-metric, recitativo-like character of the melody.

Example 5: *Hora lunga* from *Das Musikwerk, Europäischer Volksgesang,*
ed. W. Wiora, Cologne, Arno Volk Verlag, Sine anno, p. 15.

The typical situation involving additive rhythm, however, arises in connection with metric lines, particularly in the music of ancient Greece and
Rome, where music theory was perennially occupied with questions concerning rhythm and the interrelationship of language and music.[26] In the
music of antiquity, the basic rhythm pattern is the *pous*, the verse foot.
Musically, the *pous* represents the motive and not the measure.[27] Like
the motive, the *pous* comprises a determinate number of specific durations,
which are arranged in an equally specific manner in relation to *basis*
(*thesis*) and *arsis* (*i.e.*, the down-tread and lifting of the stamping foot of
the Greek chorus conductor, the *koryphaios*).[28] *Basis* and *arsis* are not
always of equal duration. According to their relationship in time, Greek
rhythmology differentiated between various rhythmic genera.[29] Here, the
difference between additive rhythm and modern metric concepts becomes
clear: in today's music, the various motives conform to an unchanging
meter and its regular beats; in the music of antiquity, each motive has
its own particular beat pattern and thus determines the meter. The motive, in turn, consists of smaller, indivisible durations, which the Greeks
called *chronoi protoi*. In the dactylic genus we can subdivide the longs
of the *spondaios* ‒ ‒ (♩ ♩), but not the shorts of the *prokeleusmatikos*
◡ ◡ ◡ ◡ (♫ ♫). The *chronoi protoi*, rather than *basis* and *arsis*,
are the primary time units of the verse foot. These smallest units are
added, the *pous* is additively established. But it is not regularly divisible.
Similarly, the feet are not combined into homologous groups. The lyrical
stanza often consists of verse feet belonging to different genera, as, for
instance, is the case in the Asclepiadean, the Alcaeic, or the Sapphic
stanzas. Antique rhythmology established rhythm by the addition of basic
time units into longs and shorts—into verse feet, which make up the poetic line. This additive character of the music of ancient Greece and Rome
is not contradicted by the fact that in its instrumental accompaniments of
vocal melodies, the (indivisible) short syllable could be ornamented and
melismatically elaborated. These embellishments did not represent subdivision. The listener did not "feel" their smaller note values as significant
independent units, but related them, as an entity as it were, to the syllable
of the text, which they embellished.

With the rise of complex polyphonic forms in the high Middle Ages and the resulting need for precise coordination in time, of all constituent voices, additive rhythmic techniques lost gradually in importance. The additive rhythm of antiquity survives in undiluted form only in folksong and, particularly, in the folksong of Greece.[30] Yet, additive formations continue to occur in Western music. Significantly enough, this always involves musical forms in which the text is an important element. Our example is the beginning of a madrigal by Luca Marenzio (1553–1599).

Example 6: Ahi, dispietata morte, from Luca Marenzio's *Madrigali a 4 voci* (1585). *Madrigalisti Italiani* I, ed. Lavinio Virgili, Rome, De Santis, 1952, p. 13.

In keeping with the rules of Italian sixteenth-century theory, Marenzio converts the qualitative accents of the text

$$´Ahi, \; dìspietáta \; mórte$$ into the

quantitative $$- \quad - \; \cup - \cup \quad - \cup$$

whence he derives the rhythmic pattern for the uppermost voice (generally the preferred voice in the Italian madrigal of the period). If ♩, as the smallest time value, is the "primary unit," the longer durations comprise:

Exclamation	*Ahi,*	strongly stressed	4 primary unit(s)	♩ + ♩ + ♩ + ♩
	di-	slightly stressed	2 " "	♩ + ♩
	-spie-	slack	1 " "	♩
	-ta-	strongly stressed	3 " "	♩ + ♩ + ♩
	-ta	slack	1 " "	♩
Cadential	*mor-*	strongly stressed	4 " "	♩ + ♩ + ♩ + ♩
retardation	*te*	slack	2 " "	♩ + ♩

Compared to antique lines, the rhythmic formation here is less consistent: four metric longs are expressed by no less than three different time values.

Yet, these different time values stem from the text, have been transferred to the music directly from the text. This holds true even for the cadence which coincides syntactically with the caesura of the verse line. If our madrigal were a monophonic composition, it would be a typical example of additive rhythm. But as we examine the remaining three voices, other determinants of the musical rhythm structure become evident. The harmonic progression of the cadence, for instance, necessitates a bass line, in which the syllable bearing the main accent in *dispietáta*, is reduced to ♩ —an obvious *barbarismo* (the term for incorrect text declamation adopted from antique terminology by the Italian theorists of the period). Furthermore, it is the regular alternation of up- and down-beats and the corresponding pattern of metric stress which causes the interchange of triple and duple time values in the inner parts:

dis- pie- ta- ta

instead of:

This points to the limits of purely additive formations in the Italian madrigal. It is the metric scaffolding of equal beats (and measures) which leads to ever new conflicts between the single parts, but which coordinates them into a meaningful polyphonic entity of rhythmic periods and sections beyond mere additive patterning.

Another form exhibiting additive rhythm is the *recitative*. In the last measures of the tenor *recitative* "Comfort ye" in Handel's *Messiah*, for example, the rhythmic relationship between text and music is even closer than in Marenzio's madrigal.[31]

Example 7: G. F. Handel, *Messiah*, I, 2 "Comfort ye."

make | straight in | the | de–sert | a | high-way | for our | God.

The accent pattern of the text finds corresponding musical accents expressed not only by metric stress, but also by the note values and the melodic choice. Melody and rhythm seem essentially determined by the text. Even the conventional harmonic structure places its strongest accent, the subdominant chord of A major, upon "high*way*," the most important word in the whole passage. There is hardly any conflict with the metric scheme of the music. Instances, such as "desert," where the notation may indicate a deviation from the text declamation, are easily negotiated by free rendition on the part of the singer. How little the exact note values mattered in performance is best indicated by the numerous *fermate* which were inserted later in this passage of Handel's score.[32] Even if they should indicate vocal embellishments, they cancel out completely the regular metric framework. The rhythmic formation is purely additive, save for the fact that patterning by dynamic accent has replaced the quantitative patterns of antiquity.

As additive rhythm develops in imitation of prosodic models, the dance (*i.e.*, the oldest species of instrumental music) favors another type of organization within time which is based upon parallelism and repetition and, in certain respects, leads back to the wave and our experience of undulating motion as the prototype of all rhythmic movement. Chains of identical figures transfer their structure upon subsequent chains; symmetric forms emerge in which rhythmic units are arranged in multiples.

Here, the connection with the dance explains the coincidence between rhythm (*i.e.*, the actual pattern of the chain) and meter (*i.e.*, the predetermined structure of the chain pattern). Deviations from this metric framework occur very seldom and usually indicate changes which engender new chain patterns in turn. Then, the original metric framework is simply being replaced, and the deviations now indicate a new metric framework.

A continuous chain of identical rhythmic figures implies an absence of tension. Exact repetition denies development; tension and relaxation cannot be expressed by such a chain alone. Only within symmetric forms, in the motive and in groups of motives, emerges such conative imagery.[33] The chain of identical rhythmic figures, uniform and without tension, may continue indefinitely. Only the interaction of harmonic and melodic accents is capable of bringing about the conclusion of such an endless chain by means of the cadence.[34] Characteristic harmonic or melodic accents arrest the motion, the note values grow larger, and the chain breaks off. In performance, of course, differentiations of various kinds emerge. The performer who adds embellishments and ornamental diminutions, who changes the note values and, by so doing, effects new groupings within the chain, who retards or accelerates the tempo—such a performer, doubtless, is able to raise the melodic and harmonic interest above the rhythm and thus deprive the motion of much of its impulse (as far as the listener is concerned).

Such a serial rhythm (*multiplikative Reihenrhythmik*) might have occurred as early as in the lost dance melodies of antiquity. It is manifestly present in the medieval *estampie*,[35] the first section of which is given below.

Example 8: Paris, Bibl. Nat. frc. 844, f. 104[v], "Quinte estampie real," after F. Gennrich, *Grundriss einer Formenlehre des mittelalterlichen Liedes*, Halle, Niemeyer, 1932, pp. 160 s.

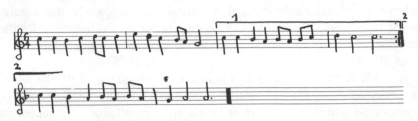

This dance melody is made up of several melodically independent chains, which are repeated with slight variations in the cadence, first ending on the third degree (half cadence), and the second time always ending on the *finalis* (full cadence). Thus the two segments become, retrospectively as it were, "halves" of a section which they establish as a symmetric entity. The subsequent chain repeats this rhythmic structure. The recurring steps and patterns of the dance require that the stress pattern remain unchanged. There are four accents in each rhythmic group (*i.e.*, a half or a third of a segment).[36] The distance in time from accent to accent remains the same. This is how serial rhythm differs from additive rhythm. It is not the smallest time unit which governs the rhythmic movement but the regular sequence of up-beat and down-beat, of a constant time interval which as such is invariable but can be subdivided as desired. It is this unchangeable stress pattern which determines the metric situation. Note values change from chain to chain, but the order of stressed and slack beats remains the same and maintains the same rhythmic structure for all chains.

In the modal notation of the twelfth and thirteenth centuries, serial rhythm appears as the basic structural principle also of vocal polyphony. A specific rhythmic unit, the *modus*, rigid in its arrangement of basic note values, determines the rhythmic continuity of an entire voice. These *modi* differ basically from the concept of meter in antiquity, although they too employ quantitative patterns and their nomenclature is derived from the usage of classic Greek and Latin rhythmology.[37] The *modi* are basically ternary units and only this one, single rhythmic genus in the antique sense, obtains throughout. The distance in time from accent to accent remains constant and thus permits the correlation of several simultaneously sounding voices, even though each voice may be in a different *modus* (as is sometimes the case). The long and short note values expressing the basic quantitative order, can, of course, be subdivided. As in the dance melody, the recurrent accent determines the rhythmic order which, in turn, supports the various durational patterns.

Example 9: Conductus Ave Maria. Third modus. Ms. Paris, Bibl. nat. fonds latin 15139 (olim St. Victor 813), fol. 259, after *Das Musikwerk. Die mittelalterliche Mehrstimmigkeit*, ed. H. Husmann, Cologne, Arno Volk Verlag, p. 32.

If we compare the rhythmic structures of baroque music with the modal rhythms of the twelfth and thirteenth centuries, we may find certain parallels inspired by the rhythms of the dance.[38] This holds true for the frequently regular steps of the *basso continuo*, as well as for the most characteristic type of motion in baroque music, the *ostinato* and its related forms, *chaconne*, *passacaglia*, and *folia*. Common to all these forms is the continuous repetition of a rhythmic-melodic phrase in the bass, and a group of correlated upper parts, some independent of the *ostinato* phrase, some motivically related to it.

The ever-repeated rhythmic chain of the baroque *ostinato* as such is without tension, without development. Nevertheless, it is here that we may clearly observe how tension and relaxation are supplied by *other* musical factors which are not subject to the "law of the chain"—factors such as harmonic progression, changes in sonorous density and in the texture of the polyphonic web. Another mode of rhythmic formation emerges: Tension and relaxation create periods and sections, determine the "form" of the music. This form lives within the general rhythmic flow which as dominant continuous motion tends to detract from the tensions it contains within. This explains the overall impression of uninterrupted motion which is so characteristic of baroque music.

This is particularly obvious in the *ostinato* and variation forms, where the impression of uniformity increases with the succinctness and finality of the repeated phrase. Yet it holds true also in the case of less defined, improvisatory musical forms.

Example 10: Johann Jakob Froberger, *Toccata* (DTOe X,2, 1903, p. 22).

The beginning of the *toccata* by Johann Jakob Froberger (1616–1667), in Example 10, shows serial rhythm structure in the continuous chain of eighth notes in the upper part. (The occasional sixteenth note subdivisions of *appoggiatura*-like, ornamental character, are hardly significant.) This chain does not break off when slower motion is introduced (measure 4). At once, the eighth note pace is taken up by an inner part and, later on, continued in the bass.[39] The serial character of this continuous motion in eighth notes is established by the strong dynamic accents which regularly fall on the first and the fifth eighth note in each measure. This effect is further supported by the melodic line and its syntactic points. The serial character, of course, is influenced by other musical factors. As the melody and the harmony become more independent, as the impression created by them becomes more autonomous—for instance, in slow passages—the serial character of the rhythm becomes more subdued. The continuity effect of regularly progressing motion is overpowered by the variety of melodic (and durational) as well as of harmonic effects.

The principle of serial rhythm is of considerable importance also in contemporary music. Serial structure occurs in Béla Bartók's Bulgarian rhythms, in jazz, in the variable rhythms of Boris Blacher's music, and, above all, in "serial music" which, in accordance with the rules of twelve-tone theory, subjects all musical aspects such as melodic progression, sequence of note values, volume and timbre to the laws of the "row."[40]

Serial rhythm appears also in the classic and romantic music of the eighteenth and nineteenth centuries. However, the rhythmic chain which, as such, is without delimitation, is now replaced by phrase and period, by the relationship of thesis and antithesis. The new motive emerges from the one that precedes it, according to the principles we observed before. The motives are not additive; they grow out of another. But now, the function of the second motive is *contrary* to that of the first one. The two are connected into one unit by an antithetic relationship. "Two-measure" groups, phrases, and periods are formed in this fashion.[41] This is clearly demonstrated by the beginning of Mozart's String Quintet in G minor, K. 516.

Example 11: W.A. Mozart, String Quintet, K. 516.

As in Example 10, regular eighth-note motion obtains which accounts for the striding, urgent character of the passage. However, the upper part dominates, and modifies the eighth-note motion. An ascending chord-line (motive 1) is countered by a step-wise descent (motive 2). Both motives are characteristically antithetic in structure. The same rhythmic motive without change in stress pattern, first is pointed upward in skip-wise motion, then downward in chromatic steps. But the up-swing of motive 1 is not canceled by this. The descent in motive 2 leads from g″, the climactic tone of the whole period, to a tone (e″), which is still higher than the highest tone (d″) in motive 1. A second phrase (motives 3 and 4) is needed for the return of the melodic line to its point of departure. The contrast between these two motives, however, does not lie in the change of melodic direction, but in a change of motion. The eighth-note pattern continues in motive 3, but, actually, serves only as embellishment of a descent in larger note values. In motive 4, motion in quarter notes has become a fact. The reduction to slower note values stresses the cadence which concludes the period. The contrast between motives 1–2 and 3–4 is again one of up-swing and tension vs. return and relaxation. This relaxation is of course rather incomplete—there is something left over, a remainder, as it were. The melody returns *almost* to its point of departure: the harmonic progression does not close with the tonic chord of G minor, but with the chord of the dominant. All this prepares for the second period, which, again, starts on the tonic. This second period relates to the first one not unlike the way the first "two-measure" group (motive 1–2) relates to the second (motive 3–4). In a similar manner, the larger formal units of classical music exhibit also a "dialectic" relationship, as for instance the relationship between principal and second theme, or between exposition, development, and recapitulation.

A characteristic difference between purely serial and antithetic rhythmic formations exists in their relationship to the meter. The antithetic formations also are predicated upon meter, perhaps more fundamentally so than purely serial structures, because the metric chain is germane to the principle of antithesis: stress follows slack; a weak accent is followed by a strong one; a weak and a strong measure form a two-measure group; a weak and a strong two-measure group, a phrase; a weak and a strong phrase, an eight-measure period, and so forth. With this metric framework of an entire composition goes however a certain freedom as far as the actual rhythmic content is concerned. The span of a motive does not always correspond to the time interval of one measure or half-measure, nor are the motives equal in length. In our example, for instance, motive 3 (seven eighth-note values) is one eighth-note value shorter than motives 1 and 2 (which each comprise eight eighth-note values), and motive 4 (nine eighth-note values) is one eighth-note value longer.

Just as additive rhythm is not limited to ancient music and purely serial rhythm is not limited to medieval music, examples of antithetic rhythm can be found in music other than that of classic and romantic periods. In a *canzonetta* by Giacomo Giovanni Gastoldi, dating from 1595, the beginning phrase is clearly antithetic in structure.

Example 12: G.G. Gastoldi, *Canzonette a tre voci, Libro II*, 1595, ed. G. Vecchi, Bologna, Edizioni di Quadrivium, 1959, p. 24.

An animated motive without definite linear direction is followed by one of descending motion in larger note values. The next two-measure group repeats this contrasting pattern with inverted directions. The descending tendency of motives 1 and 2 (d″—c″—b flat′—a′) is reversed (b flat′—c″—d″). Similarly, the harmonic progression which led in the first two-measure group from the "tonic" harmony (b flat—d) to the chord of the upper third (d—f sharp—a) begins in the second two-measure group on the chord of the lower third (g—b flat) and ends on the tonic harmony (b flat—d).

Yet there is one point in which the rhythmic structure of classic and romantic music differs basically from the music of any other period: only in classic and romantic music does the dialectic principle prevail in all aspects, so that rhythm and meter, structural content and overall formal framework, equally reflect the laws of this musical dialectic. Thus, we note in the Gastoldi *canzonetta* that the antithetic passages occur as if by accident and soon give way to other types of rhythmic formation.

4 *Conclusions*

In music, time is expressed through various types of motion. As melody, time is expressed in a sequence of different pitches and durations; as harmony, in the chord progressions; as "musical time," it is realized in the chain of dynamic and durational accents. The organic interconnection of these elements in time produces the primary unit in music, the motive, which represents also the primary complete expressive unit of the "musical becoming" as it unfolds in time, the first completed image of musical time.

The connection of the motives gives rise to different types of rhythmic formation. It is possible to connect motives according to the requirements of extra-musical models, such as text, dance patterns, or a piece of action which is to be illustrated by music (as, *e.g.*, the background music in plays, motion pictures, etc.). The number of these motives is without limit, save for the limits imposed by the extra-musical model. *The rhythmic chain in this case is additive.*

Then again, the form of one motive or, at least, its general character, may serve as the basis for all subsequent motives. A chain of homologous motives or of rhythmic figures of equal duration results. Now the number of rhythmic impulses is limited; it is determined by symmetric formations within the chain. Whenever a single motive is repeated, delimitation obtains only as far as the motive itself is concerned. The resulting chain of motives could continue indefinitely. As soon as the motives are interconnected and form symmetric groups of motives, the groups delimit each other. The number of motives now is determined by the symmetric groups. These motivic groups, which themselves form a chain of a higher order, so to speak, remain of course unlimited as far as their number is concerned. *The rhythmic chain formed by such motivic groups is serial.*

Finally, two related motives may exhibit dialectic relationships. The two motives thus delimit each other; they cannot be followed by another motive of the same character. Only as a two-motive group can they become a "dialectic unit" and thus a synthesis which may be followed by another antithesis. These complex units again are related to each other in the antithetic fashion of the original two motives. Thus, in a composition of consistently dialectic structure, each musical section is not only delimited by the preceding one, but it determines also the form of the section that follows next. *The serial groups which form the rhythmic chain are antithetic.*

We have described some types of rhythm as they occur in the musical literature of the West. These examples are not intended as an outline history of rhythm or its various forms,[42] nor as an attempt to identify various periods of music history with this or that type of rhythmic formation. Our purpose rather was to illustrate the different possibilities of rhythmic expression as they actually occur in the musical literature. There can be little doubt that a certain rhythmic type may be dominant in the music of one historical period—as additive chains in the music of ancient Greece and Rome, or antithetic serial formations in the classical style of the eighteenth century. On the whole, however, we see that the main types of musical rhythm can be traced throughout the course of music history.

Music is a temporal art, a specific realization within time. In the present paper, we attempted to show how strongly the musical rhythm, music's movement in time, affects all the elements of the art—melody and harmony, character and momentum of motion, sonority and dynamics— which make up that complex entity called musical form.

A Note on Rhythm and Time

As the behavior of every healthy child attests, the rhythm of a marching military band is a powerful device of attraction; on the adult level the interplay of rhythm and sympathetic induction is familiar to most of us, though perhaps not by that name.[1] The contagious nature of rhythmicity can be observed throughout the animal kingdom, and we may assert with very little reservation that every living thing that can move does also dance.[2] Writing about the problems of the scientific revolution and leisure, Prof. Cohen has remarked[3] that "The dancing teen-ager is drawn to a never-never land, to a 'once upon a time' which has no end, for all 'live happily ever after.' " One must not ascribe such feelings to teen-agers only, they are rather the community property of Homo sapiens. The subjective, human experience we are most likely to associate with the various manifestations of musical rhythm and often with rhythmic motion in general is a change in the apparent rate of the flow of time. The curious situation thus obtains that the most striking effect of the temporal arts on man is a feeling of transcending time. Some initial order may be brought into our thoughts about rhythm and time if we separate the sense of rhythm from the sense of time, and consider the latter as including the former but being more general in nature. Indeed, while regularly repetitive behavior is exhibited by all organisms, behavior which we consider a manifestation of a sense of time emerges only in humans and, as we shall see, only after a considerable period of training.

J. T. F.

Time Perception in Children

JEAN PIAGET[*]

Translated from the French by Betty B. Montgomery

The purpose of this essay is to show how child psychology can shed some light on the solution of certain epistemological problems of time. We begin by considering the two principal problems which underlie the study of temporal duration.

In classical mechanics, velocity appears as a relationship between distance and time. In the words of Descartes, space and time would be of a simple nature while velocity would be a relationship between them. On the contrary, in Einsteinian or relativistic mechanics time has become relative to velocity, while velocity became something absolute.

Here, then, is the first psychological problem: Is time expressed as a relationship, or as a simple and direct intuition?

The second problem concerns the well-known vicious circle which exists between time and velocity. Namely, we define velocity through the passage of time; but we can arrive at time, and measure time, only by accepting velocity. This vicious circle has been pointed out by the philosopher Le Roy, in one area, as well as by mathematicians and physicists such as Gustave Juvet in his *Structure des nouvelles théories physiques*.

The hypothesis which I should like to defend is that psychologically time depends on velocity, that time is a coordination of velocities, or, better yet, of movements with their speeds, even as space is a coordination of changes of place, that is to say of abstract motion made up of velocities.

This hypothesis seems to imply a sort of parallelism between time and space, but the parallel is not complete. This parallelism, nevertheless, is a

[*] Edited with notes by Emily B. Kirby.

202

classical one. According to Newton, for example, time and space are considered as two absolutes which make up part of the *Sensorium Dei*, absolutes which, according to Kant, have become a priori forms of sensation with parallelism between time and space. In the theory of relativity, time, in some way, is put on the same footing as space. But in spite of this blending, time and space are not completely isotropic.

In Relativity Theory, the first three purely spatial dimensions have as an attribute perfect reversibility, whereas time, to the extent that it is a physical unwinding, remains irreversible, demonstrating immediately that the parallelism does not go very far. From an epistemological viewpoint, one must say even more: space can be completely abstracted from its content in the measure of pure form and give way to a strictly deductive science of space, which would be pure geometry. By contrast, there is no pure chronometry; there is no science comparable to geometry in the field of time, precisely because time is a coordination of velocities and because when one speaks of velocity, one speaks of a physical entity. *Time* cannot be abstracted from its content as space can. *Temporal order*, in a sense, can be abstracted from its content, in which case, however, it becomes a simple order of succession. But *duration*, as we shall see, depends essentially upon velocities. Duration cannot be disassociated from its content psychologically or physically. From the point of view of psychology, Bergson's analyses of pure duration have amply shown the interdependence of time and its psychological content; similarly, from the physical point of view, time depends upon velocities.

In the first place, it is necessary to try to clarify the relationships between time and velocity in the area of physical time, where we define physical time as being that of processes external to the subject. It is here that it becomes useful to move to the field of child psychology and to ask: 1) how does the child arrive at the idea of velocity; 2) how does he acquire an idea of time, and 3) in the area of physical time, what are the genetic relationships between the concepts of velocity and time?

1   *The Child's Concept of Velocity*

About velocity, we ask first if the relationship between traversed space and time ($v = d/t$), a ratio which we can call the metric of speed, is really the first idea in the mind of the child. Or, differently stated, does velocity appear to be a concept of the relationship between distance and time, so that velocity is subordinated to duration as well as to space traversed?

It is nothing of the kind. We find that in children—and this concept is interesting from the epistemological viewpoint—intuition of velocity precedes the notion of measurability. This intuition is of a purely ordinal nature. In other words, it is based simply on the order of moving figures. The notion will naturally imply a comparison between two moving figures, and it will not be possible to apply it to the velocity of a single, isolated object. But, in the case where there are two moving figures, there will be an intuitive idea as to which has the greater velocity. This involves an estimate of velocity according to order of positions, which early in life a child will make by affirming that one moving figure which overtakes another is faster than the moving figure overtaken. In this case the idea of overtaking is independent of duration. In effect, the idea of overtaking simply implies both a spatial and a temporal order. If moving figure A happens to be behind figure B at time t_1 and then A is ahead of B at time t_2, the only ideas involved in the judging of overtaking are those of "behind" and "ahead of" in space, and "before" and "after" in time. These are simply ordinal ideas of rank and order which do not require the notions of distance traversed or of time elapsed. Nor do they evoke a notion of duration in terms of interval measurement. What is required is a notion of time sufficiently developed to comprehend an order of succession of events, and a spatial concept which recognizes that there is an order of position. The role of this concept of overtaking can be demonstrated by simple experiments which also illustrate that the young child has no notion of the measurability of speed.

For example, if one gives a child two tunnels of noticeably different lengths, he will immediately be convinced that one of the tunnels is longer than the other. One then puts two dolls moving on rods into the tunnels, while the child is able to watch both the entrances and exits of the tunnels. One asks whether one of the distances to be traveled is greater than the other, and all children point to the longer tunnel. The dolls are then made to enter the tunnels at the same time and emerge from them simultaneously. Young children maintain that the dolls moved at the same speed since they came out of the tunnels at the same time. These children, therefore, do not maintain their view concerning the inequality of the tunnels. After that, if the tunnels are removed and the experiment is repeated, the children say that one doll goes faster because it passes the other. In other words, since the overtaking of one doll by another is visible, there is a correct estimate because of the idea of order or rank; but while they have a situation involving only duration or distance, young children have no conception of speed.

This intuitive view of catching up with and passing takes place in four successive stages:

In the first stage, a very young child judges only on the basis of the event of arrival. Consequently, when there is a complete overtaking, there is a correct judgment; but when one moving figure does not completely overtake the other, even if this latter goes more slowly, the child declares the slower to be faster because it is the figure in front. The child considers only the point of arrival and ignores the point of departure.

In the second stage, the child takes into account both the events of arrival and those of departure. He keeps in mind the movement which he has perceived, and we can observe a rank-order procedure, this time based appropriately upon all of the elements of the situation.

In the third stage, we obtain an idea slightly superior to the notion of order described above. Here the child takes into account the distance between the moving figures, a distance which at first is large, then decreases, only to increase again after one figure overtakes the other. In the sense of Suppes, this process would be called a superordinal estimation.

Finally, when the child reaches the age of ten or eleven years, the idea of measurability appears. This is by no means a primary idea.

From the above we can conclude that there is an idea of velocity which does not depend upon duration; it is this idea which permits us to avoid the blending, and therefore the circular argument between velocity and time which was mentioned earlier.

2   *The Perception of Velocities*

Now, how does all this appear from the point of view of perception? Actually, alongside of those ideas which are the child's interpretational ideas and do not constitute perception, we can study the perception of velocity and discover that it follows approximately the same laws for children as for adults.

The perception of velocity has been studied by many researchers, particularly by Brown of the United States.[1] With reference to subjective or phenomenological distance and to subjective or phenomenological duration, he has tried to bring velocity into a metric model based on these subjective relationships.

We have undertaken a number of experiments on the perception of velocity without obtaining the same results as Brown. On the contrary, we have frequently observed, even with adults who have completed their psychological development and education, contradictions between their evaluation of velocity and of duration and distance. Our results suggest that from the point of view of perception adult behavior is based on the same laws of ordinal and superordinal judgments which have just been discussed above with respect to the pre-operational ideas of the child. The perception of velocity can be studied in three different situations.

The first situation: a comparison of the relative movements of two moving figures. In this instance everything can be explained by ordinal and superordinal relationships. In this situation, however, a systematic perceptual effect is observed. At the moment of an overtaking, a subjective effect of acceleration is reported in a great majority of cases. The moving figure which overtakes the other appears swifter; or even—and this amounts to the same thing—a subjective effect of deceleration occurs; the moving figure which has been passed appears to be slower. To put it another way, speaking in terms of perception, an ordinal effect based on overtaking appears again.

The second situation: In this case a single moving figure which traverses the visual field enters alone (since essentially here the question is one of visual perception). The eye movements take place freely, unrestricted by any device. In this situation, one might think that only a single source of motion is involved. However, in reality, there are two: the object which crosses the visual field and the eye movements. One can demonstrate that the perceptual effects observed in this case lead to the same conclusion that we noted above with regard to the two moving, external figures. Specifically, a well-known effect called the "apparent movement" effect[2] is observed. At the moment when the moving figure, which has not been seen previously, appears in the visual field and begins its journey, it gives the impression of being accelerated in relation to what its later speed will be: it appears faster at first, then it seems to slow down. This impression results from the fact that, at the instant of the "apparent movement" effect, the eye is directed to the point where the moving figure is going to appear and does not immediately catch up with the moving figure. Thus, there is a lag in the visual pursuit, and, consequently, an apparent acceleration of the moving figure. An explanation of this may be found in the effect of retinal persistence, or oscillations, which last until the instant when the eye has overtaken the moving figure. It is only then that the moving figure appears to have the constant speed which, actually, it had all along.

The third situation: This is the case where vision is steady because the subject's eye is fixated upon one point while the moving figure crosses the visual field. It is known that, even with the eye fixated, it is possible to evaluate the speed of a moving figure which crosses the visual field. The ordinal or superordinal interpretation remains valid because one can demonstrate that even in this situation there are two moving objects. It is a known retinal effect that speed seems greater when measured in the fovea than in the periphery of the visual field. The explanation of this phenomenon is as follows: in the fovea there is a concentration of cones, and the nerve endings are more numerous, and retinal persistence is greater than in the periphery of the visual field. Let us consider next the path of stimulation which crosses the retina and which simultaneously excites bundles of nerve fibers from the rods and cones (but successive bundles, because the moving figure stimulates them in succession). To put it another way, there is a greater spread of neural excitation in the fovea between the beginning of the excitation and the extinction of the excitation, which pro duces the effect of apparent acceleration. Therefore, we are again dealing with a comparison of two speeds, speeds compared with one another in an ordinal and superordinal manner without the necessity of mentioning either duration or space traversed. One speed is that of the passage at the beginning of the excitation, the other one that of the passage at the end of the excitation.

Consequently, in all of these situations, whether it is a question of preoperational childish notions, or whether the issue is elementary perception, one can discover a velocity psychologically independent of duration. This fact has been utilized elsewhere by a physicist and a mathematician, Abelé and Malvaux, in their suggestive book entitled *Vitesse et univers relativiste.* These authors attempt to construct a basis for Relativity Theory while escaping the vicious circle of velocity and time. To accomplish this they ask themselves (does this idea possess merit for physicists?), how are the notions of velocity and time developed psychologically? They followed our research in child psychology and found the ordinal notion of velocity useful; from this notion they constructed a generalized theory. They achieved this from a technical point of view by means of a machine which measures the movements of billiard balls, and from a mathematical point of view by combining according to a logarithmic law a psychological notion of velocity with the formalism of an Abelian group. Thus they derived an additive theorem of velocities and constructed a theory of velocities which does not use duration. The circular argument between velocity and time has thus been avoided.

3   *The Child's Concept of Time*

Now let us think about time itself, and ask ourselves if the same children whom we have used as subjects in our investigation of velocity developed notions of duration sufficiently simple to permit direct intuition, or whether their concepts of time and of duration depend on speed. It is the second hypothesis which is verified; one does not find that the young child has a concept of time which is radically independent of speed. The same conclusion is reached on the basis of perception.

The notion of time brings together two different entities which must be distinguished: one is the *succession* of events which furnishes us with temporal order. The other entity, the interval between separate events, furnishes us with *duration*, which we can fill in qualitatively (*A* within *B*, *B* within *C*, etc., without measurement) or with measured intervals.

First, let us examine the ordinal notions, for example, that of simultaneity. We put two dolls on the table in front of the child and make them leave from the same place, side by side, and then stop beside one another. At the moments when they start and stop we make an audible click as a signal. We then ask the children if the dolls started at the same time or not, and whether they stopped at the same time or not. If the two moving dolls have the same speed and if they leave from the same place and stop at the same place, the children have no difficulty in telling that the dolls started and stopped at the same time. If, however, the speeds are changed so that the two dolls have the same starting point and the same stopping time but one of them reaches a more distant point, the results are different. All the children agree that the dolls started at the same time, that the departures are simultaneous. However, they do not agree that the instants of arrival were simultaneous, for the moving dolls were not stopped at the same point. Then one asks:

"When this doll stops, is the other one still moving?"

"No, it isn't moving."

"Then they stopped at the same time?"

"No, they didn't stop at the same time, because that doll is ahead of this one."

What has been shown by this dialogue is that "at the same time" is a phrase having no meaning for the young child, in the situation described above. "At the same time" has meaning only in the situation where the two dolls stayed next to each other throughout the entire trip. For movements of different velocities, with different points of arrival in space, simultaneity does not yet have significance because the coordination of the time of one movement with the time of another movement presupposes a real understanding of the structure of time.

At about six years of age, while the child admits simultaneity, he does not yet conceive of the equality of synchronized durations. The child agrees that the two dolls have left at the same time and that they have stopped at the same time. But if one asks him whether one doll had moved for the same length of time as the other did, or for a shorter time than the other, or for a longer time, his reply is that one of them moved for a longer time because it has traveled further.

"But they stopped at the same time, didn't they?"

"Yes, they stopped at the same time; I gave the signal and they stopped."

"Then they moved for the same length of time?"

"No, that one moved longer because it traveled further."[3]

This experiment can be repeated with an inverted Y-shaped tube with a single control valve and with the two branches of the tube draining into two glasses. The two glasses may be of similar or different shapes. With a properly chosen inverted Y-shaped tube it is possible to regulate the levels of the liquid in the glasses so that these are equal for identical glasses but unequal for those of different shapes. We may then ask questions analogous to those we have asked before. In this experiment the question bearing on simultaneity ought to be clearer because a single valve controls the flow. Even if the child agrees (which is not always the case) that the water stops flowing into both glasses at the same time, he will still say that it must have run longer into the glass which finishes up having a higher liquid level.

Finally, we note that to make a comparison of movements through the same distances but at different velocities, one finds two distinct steps:

First, where the child replies "that goes faster and (consequently) takes more time"; and afterwards, the child says, "that one is faster and (consequently) takes less time."

The difference may be attributed to the fact that the first reply depends upon the result (faster is equated with farther and consequently with more time) and the second upon the process itself (faster means less time). The second case thus expresses some operations which bear on the transformation as such and not simply on the stationary result of configuration.

4   *The Perception of Time*

We will now examine the important question of what happens at the level of perception. Actually, in the domain of perception, it is necessary to exercise prudent control over the findings in the area of pre-operational ideas and to study the perceptual findings as well (regardless of the age of the subject, for adult perception is based on the same laws as perception by children). From the point of view of the perception of time, the three essential results are as follows:

First result: When one has a condition of movements of different velocities, one can demonstrate errors in the estimation of simultaneity. The majority of subjects estimate that the moving figure which is faster stopped earlier than the one which is slower even though the events can be shown to be simultaneous. This is a perceptual effect. It is interesting because of the parallel it permits us to establish with the responses of the child at the pre-operational level. Moreover, one can ask to what extent pre-operational thought is dominated by the dictates of perceptual configuration or of mental images.

Second result: This involves a comparison of two movements which last exactly the same time, but one of which is faster than the other. The movements are presented successively, in varying order. Comparative judgment is solicited after each pair of movements. Such estimates for short durations are readily made by adults and a systematic effect is observed. The moving figure which is faster creates a subjective impression of a motion of longer duration. Thus, here one finds on perceptual grounds the equivalent of childish judgments: faster equals more time. Fraisse,[4] who at first achieved differing results, repeated the experiment using a range sufficiently large from the point of view of the spatial field so as to permit appreciable difference in the perception of velocities. With a precision which characterizes his work, he often varied all those environmental conditions which play such a large part in perceptual measurement. The effect was then found.

Third result: We have made a similar evaluation of duration through the use of motions presented on films, so that the speeds of the motions could differ but their durations were synchronized. Under such conditions, duration is evaluated with the eyes fixated sometimes on the faster, at other times on the slower, movement. In the latter case, the majority of reactions are something of this order: faster equals less time. The third result differs from the second because here the attention is drawn to velocity, while previously it had been drawn only to the end result, the distance traveled through or the work accomplished. When two moving figures traverse the visual field, the subject must fix his eyes on one of them. However, he also sees the other one at the same time, thus observing a relationship which draws his attention to velocity. This is why, in the latter case, a reversal of perception is operating; and therefore, faster equals less time. The more rapid movement is judged to be that one which, from the standpoint of the estimated time span, is of shorter duration.

These different results concerning the evaluation of physical duration by the subject seem to indicate that physical time always appears as a relationship: as a distance traveled in relation to a given speed. One might write this $t = d/v$ (or even as the ratio of accomplished work to power, which comes to the same thing). And in the elementary stages of evolving childish notions, as in perceptions by successive comparisons, one of the terms of the ratio is always neglected. For example, only the numerator will be retained and time will be evaluated by distance moved through or by work accomplished; or if we concentrate on the other term of the ratio, then time appears as the inverse of velocity or power.

5   *Lived Time, or Psychological Time*

With psychological time, that is, lived time (*temps vécue* or *durée vécue*), we rediscover the same principles. The preliminary problem still remains the establishment of a relationship between psychological time and physical time. Is it possible to separate these two types of duration?

Two cases will be presented. If psychological time is without relation to physical events, then one can speak with assurance of psychological duration. But an experiment involving the absence of physical events would be very difficult to design for it should involve a dream or a daydream. On the other hand, it could be a question of time that had undoubtedly been lived, but lived in relation to ongoing physical events. In waiting, for example, along with lived time which seems long, there is a physical event: the time it takes the train to arrive, *etc.* It appears that, psychologically, the time which impresses the child the most at the outset is the time of the events which surround him. The child's attention is centered more on things going on around him than in his own introspection. Yet sooner or later the child will experience waiting for a longer or shorter time for the fulfillment of a desire. In fact we can now observe something intermediary between psychological time and physical time, because there is a relationship with external activity and not simply an introspection of internally experienced time.

Lived time, the psychological time, now appears to obey the same principles as time evaluated on physical grounds. Bergson, who has made pure duration (*durée pure*) the center of his reflections, has said, with reason, that time is an invention or it is nothing at all. What does this word "invention" mean? It means, among other things, that the feeling of duration cannot be separated from a course of action, in other words, from work accomplished; and, moreover, that work is no longer accomplished when there is no velocity to make things happen. Consequently, work is accomplished at a given rate of speed. One can then assume that psychological time is the time of work accomplished in relation to the speed of the activity in progress, or motor activity. Or, if I dare to use an idea from physics to characterize something which, in motor activity, depends upon the strength at our disposal as well as on the speed of the action itself, psychological time relates to power, for power = force $\times$ velocity.

The illusions of lived time can be interpreted in a very striking manner through use of the following schematic presentation. We are all well aware that an interesting task seems to cover a shorter period of time than a boring one. What is interesting, as has been shown by Edward Claparède and by Janet (*The Major Symptoms of Hysteria*), each in his own way, is a mobilization of the strength of the individual when he wholeheartedly attacks a task important to him. On the other hand boredom, disinterest, disassociation, can cause visible diminution of strength, or in other words, a shutting off of available energy.

Consequently, remembering the relationship according to which time = work $\div$ power, if one increases the power, the time seems to diminish. By contrast, during a period of expectation, the inverse occurs. External events do not happen at the rate we wish them to happen, or to restate this, the world outside us resists our activity. A displacement of space or time can also take place when we are actively engaged in performing a task. We all know, for example, that walking through heavy snow produces a different impression of duration from that arising from a walk on dry pavement. Time appears longer; it is because the work required in the former instance is considerably greater in relation to the same available forces.

The preceding analysis tends to make of psychological time a relationship resembling that of physical time, employing *mutatis mutandis*, the same formal relations. My colleague and friend Paul Fraisse thinks differently about lived time. He thinks it is possible to reduce lived time not to a relationship but to a direct experience, which is the experience of the number of changing events noticed by the subject. When the subject is aware of a large number of changing events in his activity, the time seems longer; while, as the number of changing events decreases, time appears to become shorter. In his scheme there would be only a single variable, and not two variables forming a relationship as in the hypothesis I suggested. I would be in complete agreement with Fraisse if he were to calculate the number of changing events with respect to a certain unit of objective duration. In such a case the number of changing events would be a frequency, and we would simply have a frequency representing velocity in psychological time *versus* motional velocity or speed of changes of place, such as frequently occurs in physical time. But Fraisse does not always mean this, and to demonstrate the role of the number of changing events, he has thought of all sorts of ingenious experiments. We have repeated one of them. The experiment was as follows: During a constant time span, the subject was presented, at two different rates, alternately, a certain number of pictures. At the slow tempo sixteen pictures were shown during four seconds, at the fast tempo thirty-two pictures were shown. Fraisse hypothesized, in the case of the thirty-two views, that the time would appear longer. He experimented with adults, asking them to estimate the time in seconds (we cannot do this with children), and found that among adults the two durations seemed equal. This can probably be explained by the fact that adults utilize some measuring units, because when it is a question of reassembling measurements of time, adults will have acquired this ability through repeated experience, at least in certain settings.

We have repeated this experiment of Fraisse's with children, using a distance measurement instead of a time measurement. We used rods of different lengths: a short rod meant a short time, a longer rod meant a longer time and the child simply had to make a choice between rods. Our collaborator, Marianne Meylan-Back, repeated this experiment on children using the equipment indicated. The results she found by this method are extremely suggestive for the problem which concerns us. Three stages appeared:

Among children until about eight years of age, whenever there are more events, or according to the terminology of Fraisse, "more changing events," the time seems longer. At this level of development, a higher frequency suggests a psychologically longer time.

But what is striking, and in my opinion decisive, is that at about eight years of age a kind of inversion appears. Between approximately eight and twelve years, the subjects reply that when the presentation speeds up it takes less time. To say that it speeds up is to say that it has a greater frequency. Therefore the question is one of frequency rate and not of movement speed.

In the third stage, one observes a balancing between the two responses. However, it is not a general stage of development, because one finds this compensation only among some subjects.

These results are informative from two points of view. First, they show that frequency rate develops in the same stages as displacement speed. This implies that, in the type of experiment described, frequency takes the place of velocity. As elsewhere, the reversal of the estimate of duration between the first and second stages is a result in agreement with our findings concerning speed of displacement. All of this seems to indicate that the number of changing events does play a role, because of a relationship between the objective unit of time and velocity, but not as an absolute factor of the number of changes.

With respect to velocity frequency, I shall add that we have been able to present examples such as the sound of a metronome at a fast or slow tempo, or flashes of light presented in a fast or slow rhythm. If we present in succession a high frequency and a low frequency, and then ask for a comparative judgment, we again find that the high frequency is perceived as taking a longer time. However, if we present accelerations and decelerations, in other words, if during the same perceptual experience there are modifications of speed, we can verify that for a slight acceleration the time seems longer than for a large acceleration. We again observe the reversal mentioned above and that it proceeds from the fact that the acceleration draws the attention of the subject to velocity, a phenomenon which is not produced when the subject simply perceives different frequencies which follow one another.[5]

6   *Temporal Operations*

Thus time appears as a relationship. This relationship gives way, as with all logical thought, to a progressive elaboration wherein we can distinguish a perceptual stage, a stage of pre-operational notions, and an operational stage.[6] Let us say a few words about the third level.

The temporal operations which make it possible for the child to develop logical concepts of time (at almost the same developmental stage when logical notions about velocity appear) can be reduced to three:

First operation: an ordering of events. It is easy to test whether or not the child distinguishes what comes before from what happens after, by asking him to reconstruct a sequence of successive events. Pictures which represent a sequence of events are given the child in random order. His task is to put them in proper order while discovering the temporal order of events.

The second operation must be the classification of durations. When many events succeed one another, not only do they take place in a certain order (for example: 1,2,3,4), but definite intervals separate them. A fundamental qualitative operation consists of classifying these intervals, that is, being able to understand that the time which passes between event 1 and event 2 is shorter than that elapsing between event 1 and event 3, or that between event 2 and event 3 the time elapsed is shorter than that between event 2 and event 4. To study this is a simple matter. One uses the flow of water from a pyriform glass into a cylindrical glass placed beneath the first one. The child will have to reconstruct the order of events by ordering a series of cards showing the levels of water at different stages of flow. In addition, we ask the child whether the time that has passed is the same, or greater, or less, than that between two other given stages. Thus the task is a problem of classification of the intervals posed before any measurement of time can take place. This is not simple. It supposes in effect that one is able to review the time sequence by memory. It supposes a sort of reversibility of temporal operations as opposed to the irreversibility of events. Some youngsters have said to me, "How do you expect me to answer! You ask me if it takes more time to go from there to there than from there to there. But now the water is already there and I can no longer tell. I would have to watch it go."[7] To put it another way, these children consent to judge perceptual time, but they cannot reconstruct the classifications of duration by logical operations, since logical operations necessarily assume reversibility.

Third operation: the measurement of time. However, before arriving at the realization of measurability it is necessary to coordinate order (succession) and classification of durations. Without this coordination, no temporal operations are possible. But this coordination is very difficult, as is apparent in the replies to some questions which appear to be transparently leading questions. With a seven-year-old child, for example, the following dialogue can take place:

"How old are you?"

"Seven years old."

"Do you have a friend who is older than you?"

"Yes, this one next to me is eight years old."

"Very good. Which of you was born first?"

"I don't know. I don't know when his birthday is."

"But come on, think a little. You told me that you are seven years old and that he is eight, now which of you was born first?"

"You'll have to ask his mother. I can't tell you."

Thus apparently it is difficult to coordinate the order of events on which a birth date depends with the classification of durations which determines age.

We return now to the third operation, which appears once the coordination we have just mentioned has been established. It is the synthesis of classification of durations and of succession. One finds here again, with reference to the genetic epistemology of time, what has been observed with

reference to the genetic epistemology of number:[8] a synthesis of the classification of sets and of seriation which generates the number series. In the
case that we are examining, the question is one of a synthesis of temporal
order and of classification of duration, with choice of an interval unit
which can be repeated successively for different intervals and which will
generate a time metric. But this metric is not, as was affirmed by Bergson,
a refined product and an artifact of science or of physics. For, spontaneous
temporal rhythm exists, like that used in ancient versification (Greek and
Latin verse) and in music (no matter how "popular" it is), with different
durations which are classified not simply in a haphazard and qualitative
manner but with a basic unit repeated time and again.

But these operations which permit the structuring of time from pre-
operational notions assume, with regard to the case of measurement, an
explicit coordination between duration and speed. In effect, when it is a
question of passing from this spontaneous rhythm, experienced simply,
such as that of music or verse, to a metrical technique something like that
of the hourglass or clock, one encounters the problem of "conservation of
velocity," because it is necessary to have a uniform speed in order to
measure time. Perhaps the question could be raised in these experiments
we have cited as to why we did not make reference sooner to a time-
measuring device. Would we not have facilitated the task of the children
by hooking up a large clock on the wall of the experimental room, or even
by putting an hourglass on the table which would have permitted them
to judge the equality or inequality of durations? We have tried these things
and the results were what we expected. At the pre-operational level, when
the coordination of order and interval has not yet been achieved, to look
at a clock or an hourglass doesn't help the child in any way to conclude
that the duration is the same for two different kinds of movement. The
child inevitably imagines that when he begins to run, the clock goes more
slowly than he does, or that when he walks, the clock goes faster. The
child does not have the slightest idea of the conservation of velocity which
underlies the concept of metric time and the sand itself seems to run independently according to the situation—either slower or faster. At this level
of development then, one does not encounter the possibility of a unit of
time transposable from the moment t_1 to the moment t_2. Such a measurement can be composed only by the synthesis of the two operations of serial
ordering and of classification, but in coordination with speeds. Consequently, not until the operational level is reached do we observe a fundamental relationship between time and velocity.

A Note on Psycho-Physical Isomorphism

In his Foreword to *Le Développement de la notion de temps chez l'enfant*,[1] Piaget refers to an exchange of views with Einstein. ". . . Einstein had asked him whether the subjective intuition of time is 'immediate or derived and whether it was integral with speed from the first or not.' With his attention thus oriented, Piaget considered the problem of time with particular regard to its relationship with speed. One might almost say that he sought situations in which the relationship time = distance/speed was apparent."[2] A detailed discussion of explanation in psychology and the problem of psycho-physical parallelism appeared recently in Piaget's *Traité de Psychologie Expérimentale*,[3] where he expresses hope in the possibilities of meaningful isomorphism between the relationships of time in experimental psychology and the description of time in physics.

Such apparently different fields of discipline as geometry and physics have become intimately connected by the emphasis of General Relativity Theory on the spatial, rather than on the temporal, aspects of phenomena. It is conceivable that through emphasizing some other selected aspect of the world—perhaps the temporal—physics and psychology may also become meaningfully connected. Except for the pioneering work of Piaget, there seems to exist, however, a paucity of scientifically acceptable data, and the question of any isomorphism between physics and psychology remains open to speculation. Thus, from experimental psychology we now turn to what might be called psycho-physically oriented speculation related to time and the philosophical questions of being and becoming.

J. T. F.

Time and Synchronicity in Analytic Psychology

M.-L. von FRANZ

Studies of the concept of time, entropy and information theory have resulted in some recent suggestions that topics in psychology be incorporated in the domain of inquiry of physics.[1] Earlier, C. G. Jung in his exploration of the unconscious and of the psychoid layers[2] of the human psyche also came upon certain problems where what we call "psyche" seems to fuse with what we usually call "matter" and where the applicability of the concept of time itself becomes questionable. It may, thus, be useful to give a brief report of his relatively secure, pertinent discoveries and list, as well, some open questions, reflections, doubts and proposed ways of research which might elucidate the obscure relation between psyche, matter and time.

1   The Unconscious Psyche and Its Structures

The greatest and most revolutionary finding of modern psychology is the discovery of the unconscious by Sigmund Freud, *i.e.*, that realm of the psyche which manifests itself relatively independently of the structure of our conscious mind. It is a well-known fact that an impulse, wish, thought, fantasy, or the like, can be temporarily unconscious (*i.e.*, forgotten). This fact would justify the use of an adjective "unconscious" but would not allow us to hypostatize a noun "the unconscious" as a self-subsisting entity. Further observation, however, has shown that unconscious or forgotten contents[3] often undergo specific transformations during their unconscious phase of existence. For example, they can become more emotionally important, automatic, stereotyped, or more mythological and archaic in form, or they can become associated with material which originally did not belong to them and thus they reappear in consciousness in a different condition. This fact forces us to conclude that the unconscious is a psychism sui generis[4] which seems to work differently from our conscious mind. Following William James, Jung and Pauli compared this concept with the concept of field in physics[5] and Niels Bohr pointed out that the relationship between conscious and unconscious is a complementary one.[6]

In the wakened state this psychism sui generis or the unconscious[7] in the human personality mainly manifests itself by interrupting the continuity of our conscious mental operations, *e.g.*, by the invasion of fantasies, speech mistakes, involuntary gestures, auditions, visions, or even hallucinations, and during sleep in the form of dreams.

In studying dream life as our major source of information about the unconscious area of the psyche, Jung found that a number of recurrent dream motives strikingly resemble mythological themes and motives as studied by the science of comparative religion.[8] Whenever such motives appear in the dream they also bring up or constellate strong dynamic and emotional reactions in the dreamer. They tend to produce passionate impulses such as sudden conversions, "holy" convictions, fanatical actions, obstinate thoughts, etc. In a positive sense, they can also bring about a passionate and creative pursuit of a theme. *Jung called the* (probably inborn) *unconscious, structural dispositions which cause the above-mentioned reactions "archetypes."*[9] Their products in dreams and semiconscious states, *i.e.*, the mythological fantasies, thoughts, symbolic actions, etc., he calls "archetypal."[10]

These archetypes actually coincide to a great extent with what zoologists call patterns of behavior, and the Jungian archetypes could, in a way, just as well be called patterns of human behavior.[11] There is, however, a difference in that the zoologist can only study these patterns in the outwardly carried out instinctual actions of an animal, while in our psychological studies of man we can also observe "from within" the accompanying subjective processes such as typical, recurring fantasies, representations, feelings and emotions.

The next most important discovery of depth psychology lies in the fact that these archetypal patterns of mental behavior contain not only an imaginative but also a definitely cognitive element which can influence the conscious mind in the form of sudden "flashes of insight"—as I will demonstrate later.[12] Jung calls the cognitive element of the archetype "luminosity" in order to distinguish it from the conscious daylight mind or ego-consciousness.[13]

The patterns of behavior in animals often include spatial as well as temporal elements in their radius of activity. Animals, for instance, have specific territories, migration routes, mating places, *i.e.*, spatial relations, whereas other patterns imply a sort of "knowledge" of future states of evolution, *i.e.*, temporal factors.[14] It is thus not surprising that Jung independently discovered similar facts while studying the archetypes or human patterns of psychological behavior.

2   Time in the Conscious and Time in the Unconscious

Physicists studying cybernetics have observed that what we call consciousness seems to consist of an intra-psychic flux or train of ideas,[15] which flows "parallel to" (or is even possibly explicable by) the "arrow" of time. While M. S. Watanabe[16] convincingly argues that this sense of time is a fact sui generis, others like Grünbaum tend to believe that entropy is the cause of this time sense in man.[17] O. Costa de Beauregard stresses that whatever its origin is, it runs parallel to the arrow of time in physics: "*le flux intérieur de la vie physique est lié de manière essentielle à un flux d'information entrante.*"[18] For the psychologist, this statement (that the train of ideas runs parallel with the arrow of time) is acceptable only as it applies to our conscious psyche, or to the concentrated "attention to life," as Bergson calls it.[19]

As soon, however, as we study the functioning of the unconscious psyche, the validity of this statement becomes questionable. It is generally admitted that people who have not succeeded in building up a concentrated state of consciousness or a sharply focused "attention to life," such as certain primitive groups or children, lack the ability of consciously "counting time." They rather live in a relatively and vaguely perceived stream where only a few emotionally important events are perceived and recorded, which then act as signposts in an indistinct flux of life.[20]

When the psychologist explores the phenomena which come immediately from the unconscious, such as visions, obsessive fantasies and above all dreams, the regularity, "speed" and distinct discernability of time become uncertain[21] or even questionable. It is as if in a dream one would rather experience a pictorially represented cluster of events perceived simultaneously, while the conscious mind when recording these dream events later would automatically put them into an order which we perceive as a logical time sequence. This seems to lend some support to the idea of a block universe with which the unconscious would be coextensive.[22] It seems that the conscious conception of time as an arrow and "flux" seems to become curiously relative (or possibly nonexistent) in the unconscious.[23]

Beyond the observation of dream material, this seems to be confirmed by the myths of all times if, as we said, such myths are admitted as archetypal in substance. In most mythologies and primitive religious systems there is the representation of a primordial time or an *illud tempus,* as Mircea Eliade calls it,[24] an "other" time beyond our actual time in which miraculous primordial mythological and symbolic events took place, or even still take place. This primordial time dimension has either now disappeared or continues to exist beside our usual time but can be reached only in exceptional unconscious mental conditions, such as in the dream state, coma, ecstasy, or intoxication. This primordial time is often described not as a time but as a creative *durée* in the sense of Bergson, or as an eternity meaning an actual extension of time without progress.[25] The Australian aborigines call this mythical time *aljira,* a word which also means "dream," for it is, according to them, the time-space continuum in which dream events occur. In other words, this mythical idea of a primordial time as the time dimension of dream events, visions, etc., seems to coincide with what modern depth psychology would now call the unconscious, for contact with this dimension is only possible by becoming unconscious.[26]

The testimonies concerning the relative "timelessness" of the unconscious do not come only from primitives, we find the same in all mystical experiences characterized by a feeling of oneness with the universe[27] and of timelessness.[28] This feeling of timelessness is inherent in the experience of the deeper layers of the unconscious, which Jung calls the *collective unconscious* because its structure is substantially the same in everyone. People who have had such experiences often claim that they can then also foreknow future events. The ordinary time experience with its subdivisions of past, present and future fades.[29] This is the "mystical" experience generally sought for in many Eastern religious movements. The feeling of oneness with all things is probably founded upon the cluster nature or contamination of unconscious contents.

We also find this same phenomenon of a timeless simultaneity in the appearance of unconscious products in the history of scientific discoveries. We may mention the famous example of Henri Poincaré[30] and a similar testimony from a letter of Karl-Friedrich Gauss to Olbers that he found the proof of an arithmetical theorem in this way: "Finally . . . I succeeded, not on account of my painful efforts, but by the Grace of God. Like a sudden flash of lightning the riddle happened to be solved. I, myself, cannot say what was the conducting thread which connected what I previously knew with what made my success possible."[31] Jacques H. Hadamard[32] and B. L. van der Waerden[33] have collected many more examples of such mathematical "illuminations" and Harold Rugg[34] of similar experiences in other fields of scientific activity. The always recurring and important factor for our theme is the simultaneity with which the complete solution is intuitively perceived and which can be checked later by discursive reasoning. This, in our opinion, is due to the fact that the solution presented itself as a sort of ordered Gestalt,[35] or, in Jungian terms, as an archetypal pattern or image and still carries with it some of the relative timelessness of the unconscious. These examples also demonstrate the cognitive element in archetypal patterns which was mentioned above.[36]

Similar evidence of a fading time sense in semiunconscious conditions can also be found in states of pathological overexcitement, intoxication and the disturbance provoked by prolonged isolation,[37] with their slowing of intellectual activity and accompanying intensification of dream life. All these factors seem to suggest that the time flux, as a subjective psychological experience, is bound to the functioning of our conscious mind but becomes relative (or possibly even nonexistent) in the unconscious. Thus Costa de Beauregard's suggestion to consider the unconscious as timeless and coextensive with the Minkowski-Einsteinian "block universe" appears to the psychologist a new and more precise formulation of the age-old archetypal idea of an "other time."

3   *Meaningful Coincidences of Consciously Perceived External Events with Archetypal Unconscious Processes*

In the course of observing the effects of activated archetypes or patterns of psychological reactions in his patients, Jung again and again encountered cases where rare outer chance events tended to coincide meaningfully with archetypal dream images. Such cases seem to occur only when an archetype is activated producing highly charged conscious or unconscious emotions. The following may serve as an illustration:[38] A woman with a very strong power complex, and a "devouring" attitude toward people, dreamt of seeing three tigers seated threateningly in front of her. Her analyst pointed out the meaning of the dream and through causal arguments tried to make her understand the devouring attitude which she thus displayed. Later in the day the patient and her friend, while strolling along Lake Zürich, noticed a crowd gazing at three tigers in a barn—most unusual inhabitants for a Swiss barn!

Taken separately, the causal background of the two events seems clear.

The power display in the woman probably brought about her tiger dream as a warning illustration of what was constellated in her. The three real tigers were in that barn because a circus was spending the night in town. But the highly improbable coincidence of the inner and outer three tigers in this woman's life seems to have no common cause, and therefore it inevitably struck her as "more than mere chance" and somehow as "meaningful."

At first sight there is no reason for concern about such events even if they prove highly impressive to the person who meets them, for the coincidence may be looked upon as mere chance. Jung, however, became impressed by the relative frequency of such occurrences[39] and named them synchronistic events. Synchronicity may be defined as simultaneity *plus connection by meaning* in contrast to synchronous events which only coincide in time. Before following up his argument further, we must recall that our laws of nature are statistical and thus valid only if applied to large aggregates. This implies that there must be a considerable number of exceptions which we call chance. Further, experimental approach postulates the necessity of repeatability and thus rules out a priori all unique and even rare events.[40]

We may now define chance coincidences as relating to the independence of the coinciding events from any common causal bond, and we can subdivide them in two categories, namely meaningful or meaningless coincidences.[41] A synchronistic event is thus a meaningful coincidence of an external event with an inner motive from dreams, fantasies or thoughts. It must concern two or more elements which cannot for an observer be connected causally, but only by their "meaning." What the factor "meaning" consists of transcends scientific comprehensibility for us at the moment, but the psychologist can show that it is conditioned by emotional and preconscious cognitive processes of a "Gestalt" or pattern nature. These processes sometimes seem already to have established an "order" or "system" before they come above the threshold of consciousness.[42]

They seem to depend on the activation of an archetypal pattern. In other words, the archetype sometimes preconsciously organizes our train of ideas.[43]

Meaningful coincidences are by no means rare and have therefore attracted not only Jung's but other people's attention, such as in more recent times, Paul Kammerer,[44] Wilhelm von Scholz,[45] W. Stekel,[46] J. W. Dunne[47] and several parapsychologists.[48] J. B. Rhine was the first to attack the problem with statistical methods,[49] whereas the others only collected more or less well-documented examples similar to the one I mentioned above.

4 *Philosophical Attempts to Formulate the Idea of Synchronicity*[50]

The idea that *all* events in nature are somehow connected by meaning (which is a purely anthropomorphic term) appears first in Western philosophy in the concept of the *Logos* of Heraclitus and in the idea of an *oulomeliē* (all harmony) or of a "sympathy" of all things in Hippocrates, Philo (25 B.C.–A.D. 42) and Theophrastus. This "sympathy" was conceived as the reason why similar external and internal events occur simultaneously even though the simultaneity lacks a causal basis. Philo and Theophrastus both postulate God as *prima causa* of this whole principle of correspondence in the universe. This idea survived in vague form throughout the Middle Ages and was still maintained by Pico della Mirandula, Agrippa of Nettesheim, Robert Fludd, and Paracelsus.[51] It also occurs in most alchemistic and astrological treatises of these times. The "cause" of this correspondence is for many of these authors either God or the World Soul. For instance, according to Agrippa this World Soul, being a spirit that penetrates all things, is responsible for the "fore-knowledge" in animals. Even Johannes Kepler still believed in an *anima terrae* that was responsible for similar phenomena.[52]

The only one who perceived of the emotional factor as being relevant in the occurrence of such phenomena was Albert the Great. He emphasized that such "magical occurrences," as he called them, can take place only when the right astrological constellation dominates and the person involved finds himself in an *excessive affect*.[53] This statement is interesting in so far as excessive affects are normally only constellated by an archetype and Jung observed the occurrence of synchronistic events only when an archetype was activated.[54] Albert's idea is not really his own but borrowed from Ibn Sina (Avicenna).[55]

None of these above-mentioned medieval authors, however, succeeded in freeing themselves from some vague idea of a magical causality which is characteristic of all primitive thinking. Only the great Arab philosopher, Ibn Arabi tried to give a purely acausal description of such events, though he still postulates the idea of God as prima causa. For him the whole universe is a theophany, created by an act of God's primordial imagination which goes on incessantly. This same divine imagination works within man and renews from moment to moment man's idea of the cosmos.[56] This is why certain real and intelligent inner imaginative activity (not a vain fantasy) in man can "click" with the outer facts. The Creation recurs from moment to moment in a pre-eternal and continuous movement by which the pre-eternal and post-eternal God manifests in all beings. There is no interval between these constant acts of creation because the moment of the disappearance of a thing is the same moment that its similar replica comes into existence. In its hidden form God remains in His *heccëitas*[57] but in His manifestation He assumes existential determination.[58] This happens by a continuous activation of the Divine Names.[59] The continuity lies in these Divine Names while on the side of the actual phenomena there is only an acausal connection of things. There are only similarities from moment to moment, but no actual empirical continuity and identity.[60] Whatever "view" man has of this Creation, is also created by the same God who shows Himself to man from within in His different Names or *ipsëities*.[61] This philosophical idea is, as far as I know, the most remarkable attempt in the Middle Ages to abolish the idea of causality, but it is still obliged to postulate a common prima causa of all things.

Inspired by Kant's reflections in *The Dream of a Spirit-Seer*, Schopenhauer has also made an attempt to account for synchronicity in his book *On the Apparent Design in the Fate of the Individual*. He pictured a double pattern of events like the lines of longitudes and latitudes. The longitudinal lines he imagined as representing the causal connections of events, the latitudinal lines as representing cross-connections of acausal meaningful coincidences. He thought that

both kinds of connections exist simultaneously, and the self-same event, although a link in two totally different chains, nevertheless falls into place in both, so that the fate of one individual invariably fits the fate of the other, and each is the hero of his own drama while simultaneously figuring in a drama foreign to him—this is something that surpasses our powers of comprehension, and can only be conceived as possible by virtue of the most wonderful pre-established harmony.[62]

Concerning the cross-connections, Schopenhauer considered the pole as being the origin in which all longitudinal lines converge. The pole is "the one subject of the great Dream of life," namely the transcendental Will or prima causa from which all causal chains radiate like meridian lines from the poles; and the parallels stand to one another in a meaningful relationship of simultaneity.

In Jung's view, however, we have no proof that there is only one prima causa and not a multiplicity of them. Moreover, the idea of an absolute determinism with which Schopenhauer still reckons has collapsed. Schopenhauer's postulate of synchronistic events being regular occurrences and the fact that he assigns a cause to them come from his wish to fit synchronistic phenomena into a deterministic idea of the world: but this goes far beyond what we can actually see or prove.[63]

Another remarkable attempt to integrate synchronicity into an image of reality is that of Gottfried Wilhelm Leibniz, who could be called the originator of the idea of psycho-physical parallelism. He thought of body and soul as two synchronized clocks. His monads or souls, as he put it, had "no windows," *i.e.*, no causal interconnections, but were "synchronized" in a "pre-established harmony."[64] Against this view Jung raises an objection which actually can and must be applied to all the above-mentioned attempts at explanation:[65]

The synchronicity principle thus becomes the absolute rule in all cases where an inner event occurs simultaneously with an outside one. As against this, however, it must be borne in mind that the synchronistic phenomena which can be verified empirically, far from constituting a rule, are so exceptional that most people doubt their existence. They certainly occur much more frequently in reality than one thinks or can prove, but we still do not know whether they occur so frequently and so regularly in any field of experience that we could speak of them as conforming to law. We only know that there must be an underlying principle which might possibly explain all such [related] phenomena.[66]

Instead of postulating a pre-established harmony of all things, Jung proposes to explain synchronistic events by the existence of a formal factor, namely by what he termed an *equivalence of meaning*. In his reflections he assumed that the same living reality expresses itself, but not causally, in the psychic as well as in the physical state. Jung called this hypothetical underlying "same reality" *unus mundus*.[67] The *unus mundus* is to be understood as an entity which consists of formal structures, systems or images, and of a knowledge which is prior to consciousness. He calls this knowledge: absolute knowledge because it is detached from consciousness.[68] We face here again, according to his hypothesis, a certain preconscious orderedness of some of our conscious representations.[69]

5 *Reflections on Synchronistic Events as Subspecies of a Possibly More General Principle of Acausal Orderedness*

According to Jung, synchronistic events should be taken as "just-so" facts, *i.e.*, as irreducible contingencies. Then he raises the question[70] of whether the principle of equivalence of psychical and physical processes, which according to his experience appear in synchronistic events, might not require some conceptual expansion. There exist two well-known examples of acausal orderedness, namely the discontinuities of physics and the individual mathematical properties of natural numbers.[71] Curiosities, like the tiger story described above, are, perhaps, only a rare subspecies of a more general principle of acausal orderedness or "of all equivalences of psychic and physical processes wherever the observer is capable of recognizing the *tertium comparationis*." Since Jung wrote this in 1952 it seems to me that the advance of natural science has done much to make this idea of a general acausal orderedness a feasible principle of explanation. Needham even goes so far as to say "perhaps we might characterize the only two components required for the understanding of the universe in terms of modern science as Organization[72] on the one hand and Energy on the other."[73]

In collaboration with Pauli, Jung proposed the following schematic representation of explanatory principles:

Quaternio of Concepts Required for the Explanation
of a Psycho-Physical Unitarian Reality

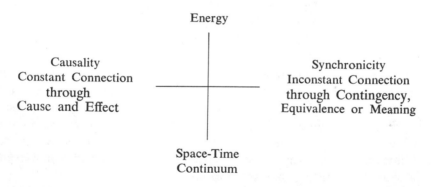

Energy

Causality
Constant Connection
through
Cause and Effect

Synchronicity
Inconstant Connection
through Contingency,
Equivalence or Meaning

Space-Time
Continuum

The principle of contingency or equivalence would be in no way replacing causality but would be seen as complementary to causal determination.[74]

In physics causality has become only a relatively valid principle and we now only look for "statistical laws with primary probabilities" as Pauli puts it.[75] One could also call these primary probabilities, lists of expectation values. In a similar way we could call the archetype a list of expectation values or "primary probability" for certain psychological (including mental) reactions. Just as expectation values in Quantum Theory cannot be measured to any desired arbitrary accuracy if, at the same time, we wish to measure their complementary parameters, so a unique manifestation in a list of psychological expectation values also eludes accurate observation if one insists on a sequence of temporal repetition.[76] Niels Bohr's idea that the conscious and the unconscious psyche are a complementary pair of entities refers to this problem. The archetype above all represents psychic probability,[77] but curiously enough sometimes it not only underlies psychic but also psycho-physical equivalences. Therefore Jung attributes a psychoid quality to the archetype or a certain "transgressiveness"[78] because it reaches out into areas of experience which we normally call material.

Normally and usually the archetype is only an introspectively recognizable form of an a priori psychic orderedness, but if an external synchronistic process associates itself with it, the external process also falls into the same basic pattern, i.e., it too is ordered. This form of orderedness differs from that of the mathematical properties of natural numbers or the discontinuities of psychics in that both occur regularly, whereas the forms of psychic orderedness when coinciding with outer events are acts of creation in time.[79] They represent creative acts within a pattern that itself exists from all eternity but repeats itself sporadically and is not derivable from any known antecedents.[80] Synchronistic events would thus be a relatively rare and special class of natural events while the contingent would have to be understood partly as a universal fact existing from all eternity and partly as the sum of countless individual acts of creation occurring in time.[81] These acts of creation are essentially unpredictable[82] and thus, at least seen from the point of view of depth psychology, they give time a specific and unpredictable turn[83] by interrupting the continuity of the time arrow. Psychology thus arrives from a different angle at the idea of a *creatio continua*—an idea which A. N. Whitehead and M. Čapek also postulate when they talk of a movement of nature[84] advancing creatively forward.

It has also lately been postulated that the physical concept of information is identical with a phenomenon of reversal of entropy.[85] The psychologist must add a few remarks here: It does not seem convincing to me that information is *eo ipso* identical with a *pouvoir d'organisation*[86] which undoes entropy. I can watch disorder without doing anything about it unless I am emotionally stimulated to put it into order,[87] and the emotional stimulus rising from within to establish order[88] comes, as the psychologist can prove, from preconsciously activated patterns of mental behavior, i.e., from the archetypes. It is only *there*, if anywhere, that we might discover instances or sources of negentropy.[89]

228

6   *The Problem of a Methodical Approach to Synchronicity*

Before discussing the problem of how to grasp the subspecies, the relatively rare and random phenomena of synchronistic events, it seems to me more important or rather more feasible to search first for a mathematical approach to the more general factor: the equivalence of meaning between a psychic and physical process, wherever the observer has the chance to recognize the *tertium comparationis*.[90] The question would then be: can we in some way quantify the psychological experience of "meaning"?[91]

As Jung noticed,[92] up till now psychologists can only approximately determine intensities of reactions. They do this with the "feeling function" which replaces the exact measurement in physics. But "the psychic intensities and their graduated differences point to quantitative processes. While psychological data are essentially qualitative, they also have a sort of latent physical energy." If these quantities could be measured, mass and velocity would be adequate concepts to use in describing the psyche insofar as it has any observable effects in time and space. It looks as if the most intimate connections of psychic processes with material phenomena belong to the realm of microphysics. But how can we get closer to this connection?

It seems to me that this would not be possible with the mathematical means we now have at our disposal, because our present use of numerical elements and their abstractions is purely quantitative and thus no more appropriate an instrument for exploring meaning than a black-and-white film would be for reproducing color. But this predominantly quantitative use of the natural numbers is a specialty of the Western mental outlook. In China, numbers were rarely used in this way but were seen above all as qualitative instruments of order. According to Granet[93] the Chinese did not use numbers as quantities but as polyvalent emblems or symbols which served to express the quality of certain clusters of facts and their intrinsic hierarchic orderedness. Numbers, in their view, possess a descriptive power and thus serve as an ordering fact for *"des ensembles concrets qu'ils paraissent spécifier par le seul fait qu'ils les situent dans le Temps et l'Espace"*[94] ("clusters of concrete objects, which they seem to qualify through positioning them in space and time"). In Chinese thought there is an equivalence between the essence of a thing and its position in space-time. This equivalence can only be expressed by number,[95] for numbers characterize arrangements rather than quantities,[96] but these qualitatively characterized arrangements (like our natural laws) also represent "regular connections of things."[97] Moreover, Chinese thought has been altogether more interested in synchronicity than in causality,[98] *i.e.*, they observed the coincidence of events in space and time realizing that they meant something more than chance. Namely they signify peculiar interdependence of objective events among themselves, as well as the subjective (psychic) state of the observer or observers.[99] Jung says,

> Just as causality describes the sequence of events, so synchronicity to the Chinese mind deals with the coincidence of events. The causal point of view tells us a dramatic story about how *D* came into existence. It took its origin from *C*, which existed before *D*, and *C* in its turn had a father *B*, etc. The synchronistic view, on the other hand, tries to produce an equally meaningful picture of coincidence. How does it happen that *A′ B′ C′ D′*, etc., appear all in the same moment and in the same place? It happens in the first place because the physical events *A′* and *B′* are of the same quality as the psychic events *C′* and *D′* and further because[100] *they are the exponents of one and the same momentary situation.* The situation is assumed to represent a legible or understandable picture.

Using binary numerals, but with the above-explained qualitative idea of numbers in mind, the Chinese invented the technique of the *I Ching* (Book of Changes), in order to grasp by means of numbers the equivalence of physical and psychic systems. What Costa de Beauregard says about the modern experimental physicist standing before the cosmos like a dice-player, whose empirical information comes ultimately from what his successive "throws of the dice" reveal to him about a universe whose laws are essentially contingent,[101] has already been observed by the Chinese, for they have long tried to gather information about synchronicity by just such a "throw of dice."

According to Wang-Fu-Chih (1619–92) the whole of existence is a continuum which is ordered in itself. It has no manifest appearance and thus cannot be observed immediately by sense perceptions, but its inherent dynamism manifests in images whose structure participates in that of the continuum. This idea of an irrepresentable continuum containing latent pictorial structures forms an exact parallel to Jung's concept of an *unus mundus.* The images in Wang-Fu-Chih's continuum form the sixty-four basic symbolic situations of the *I Ching.* Because they are ordered and therefore participate in the world of numbers, they can be grasped through a numerical procedure, which only works if handled truthfully.[102]

In no way do I want to suggest that we should adopt ideas or techniques bound into the specific cultural development of the East. What

I am suggesting is that our arithmetical concepts, especially our theory of numbers, might have to be expanded by studying the qualitative ordering aspect of number in collaboration with depth psychology before we can use any numerical method in comprehending synchronistic phenomena.

7   *Conclusions*

Synchronistic events are, according to Jung, meaningful coincidences of external and internal events. They seem to happen at random, and to be contingent to the principles of causal explanation. Possibly they are a subspecies of a more general principle of acausal orderedness. The only existing primitive attempts to grasp synchronicity are the divinatory methods of East and West. Most of them use numbers as the basis for understanding synchronistic events.

According to Jung[103] number is an instrument of our mind to bring order into the chaos of appearances, a given instrument for either creating order or apprehending an already existing regular arrangement or "orderedness." It is a more primitive mental element than concept. Psychologically we could define number as *an archetype of order which has become conscious*. Jung considers it an archetype because, as the arithmetical revelations mentioned in Section 2 demonstrate, the unconscious often uses number as an ordering factor much in the same way as consciousness does.[104] Thus numerical orders, like all other archetypal structures, can be pre-existent to consciousness and then they rather condition than are conditioned by it. Number forms an ideal *tertium comparationis* between what we usually call psyche and matter, for countable quantity is a characteristic of material phenomena *and* an irreducible *idée force* behind our mental mathematical reasoning. The latter consists of the "indisputability" which we experience when contemplating arrangements based on natural numbers.[105] Thus number is a basic element in our thought processes, on the one side, and, on the other, it appears as the objective "quantity" of material objects which seem to exist independently outside our psyche. If we do not want to agree with the Néopythagoreanism of D'Arcy Thompson, who attributes a world-creating faculty to number and thus sees it as the cause or causes of existence,[106] Jung proposes looking at the natural numbers rather as manifestations of the *unus mundus* or of that one reality which does not "cause" the coincidences of physical and psychic structures *but rather manifests in them*. This actually happens in every act of conscious realization[107] concerning external or internal facts. Thus, studying the "meaning" aspect of each individual natural number might best bring us closer to a comprehension of a more generally existing phenomenon of acausal orderedness, *i.e.*, synchronicity, and to representing it in numerical lists of expectation values.

The curiosities of the more special synchronistic events, however, seem to me for the moment essentially unpredictable. Apparently they point to a basic creative factor which manifests itself at random within the flux of time. Mythologically it has always been thought that these events also "undo" time, *i.e.*, represent instances of negentropy (or interrupt the mathematical density of time). I would not dare to assert this, however, because it does not seem to me that we have yet penetrated deep enough into the conditions which surround synchronistic events.[108] C. G. Jung's idea of synchronicity is only intended as a working hypothesis which might help to solve some of the riddles of meaningful non-causal coincidences without abandoning causal thinking.[109] Our time concept, however, would have to be revised if we are to include not only the conception of a countable-uncountable infinity but also the interruption of it by essentially unpredictable acts of a *creatio continua*—acts which do not happen completely at random but within a certain set of structural patterns which are in some way linked with the natural integers.

A Note on Synchronicity

Thornton Wilder's masterpiece, *The Bridge of San Luis Rey*, traces the inquiry of a Brother Sebastian into the lives of five people.[1] These five travelers, three men and two women, died together by coincidence as it seems, as an ancient Peruvian rope bridge plummeted under their weight into the gorge below.

The mind of Brother Sebastian conceived a meaningful connection between the five simultaneous deaths and his dream of "experiments that justify the ways of God to man." He realized, no doubt, that in the lives of these people as in his own life the law of causality could certainly be traced, but it must have appeared to him hopeless to apply causality to the coincidence of his apparent desire to witness "an act of God" and the fatal event which befell so many people. Brother Sebastian was an empiricist. He remarked, "If there were any plan in the universe at all, if there were any pattern in a human life, surely it could be discovered mysteriously latent in those lives so suddenly cut off. Either we live by accident and die by accident, or we live by plan and die by plan." The deaths which he witnessed constituted in his judgment a superb experiment, including proper control. "Here at last one could surprise His intentions in a pure state." In Jungian language, there was an archetype activated in the unconscious of Brother Sebastian. In any case, to the unfolding of the histories of these five people he dedicated his life. He tabulated his "proofs," but found the task more difficult than he had foreseen it, and "was driven to the use of minus terms." He compiled a book about the lives of these people; but the book was found heretical and his life was ended by the decision of his superiors. He "called twice upon St. Francis and leaning upon a flame he smiled and died."

Wilder's beautiful creation is an intuitive literary examination of the concept of synchronicity, for it is directed to an inquiry about meaningful, non-causal connections.[2] The occurrence of such events, while not frequent, is nevertheless common experience. They are interesting, for they exude a static, deterministic atmosphere and appear to "defy time." Systematic analysis of meaningful coincidences, as defined by Dr. von Franz, may lead to a better understanding of the functioning of the mind and, conceivably, might help in delimiting its domain. The basic plan, as I see it, is to provide a conceptual schemata suitable for the inclusion of the psychoid factor in our description and knowledge of nature. Not surprisingly, the resulting ideas of synchronicity suggest a philosophical view of life as much as a theory of the psyche.

J. T. F.

The Time Sense in Psychiatry

JOOST A. M. MEERLOO

In the observation of patients and in the study of medical concepts, the ideas of time sense, time awareness and time function are employed in a variety of meanings. A man who has just been saved from drowning may have experienced an apparent time span of eternity during the few seconds before he became unconscious. The events of his life may have passed through his mind almost instantaneously. His report on the temporal strangeness of his experience will differ from that of the schizophrenic patient who would tell us that he lives in an empty eternity without time. The first man would relate a reactive, eventful grasping of the history of his life as a defense reaction to his calamity; the second one would relate in symbolic metaphors his subjective feeling of total withdrawal from reality and his identification with the dead. My principal aim in this essay is to describe my encounter with the variety of meanings of time, such as these two examples, as used in clinical description.

When a clinician speaks of time sense, he uses unwittingly a cluster of different concepts and speaks in different symbolic systems. I know that my division of these concepts, while not altogether arbitrary is, nevertheless, tentative. The divisions also partially overlap. Yet, scientific research begins with description, division and organization of the data, and that is what I propose to do.

The clinician—especially the psychotherapist—who treats people in conflict with themselves or with their environment, experiences his patients' struggle with "time" in various ways. Words and concepts are used to describe inner experiences of which the patients are not usually aware, semantically and consciously, but which they nevertheless give as a very essential meaning for their suffering. What they attempt to communicate is related in some way to their time of existence on earth, to their views of the problem of life and death and to the impact of these views on their emotions, thoughts and behavior.

In psychotherapy, as well as in the prolonged, more systematic form of it called psychoanalysis, the sense of time is continually under scrutiny. Though it can exist as a psychological agnosia (denial of the sense of time), the patient in psychotherapy becomes more and more aware of his struggle with special aspects of the problem of time. Yet, what is simply called *the sense of time* is a combination and integration of numerous psychological and physiological functions forming a cluster of different but related elements.

Many divisions and distinctions could be made of the different aspects of time sense as they emerge, especially in the intensive emotional exchange during psychotherapy. While remembering the intrinsic danger of any such grouping, I elected for pragmatic reasons to divide this hierarchy of functions and their associated states of awareness into the seven categories used in this essay.

1   *The Patient's Concept of Physical Time*

The Newtonian idea of time represented perfect and universal ordering; quite conveniently time appeared as the awareness of such order. Cosmic time was defined as an entity by itself, irrespective of events. Contemporary physics seems to prefer the operational approach. We are permitted to speak about yesterday and tomorrow on a local scale. But for events or for bodies taking place at a distance in relative motion, the absolute quality of Newtonian time is replaced by an apparently self-consistent but rather complex concept of many "times." Worrying about the physical concept of time has increased among patients since the publication of my first studies on the time sense some thirty years ago. Science fiction, jet transportation and space flight have brought people in contact with theoretical problems of time. There is a detectable increase of formal interest in the concepts of Newtonian and Einsteinian time, even though most patients cannot fully comprehend the ideas involved. Usually we find that behind their theoretical ponderings other, more personal questions are hidden. In their questions on the physics of time they express only a great yearning for timelessness, for a nirvanic state without anxiety and fear, for a haven without the need to anticipate death and catastrophe.

2 *Evolutionary Time and Time-binding Processes*

History, that is, the passage of time and the succession of events, plays such enormous roles in the formation of human behavior and of the anatomical structure of man that a note on the relation of man's evolution to time is relevant to the study of time sense and therapy. Darwin's theory of evolution shook the belief that man has a biological place outside the animal kingdom; it also shook the medical ways of thinking by introducing systematic *historical thinking* into the study of life. Darwin perceived man's evolution as a precipitation of repeated challenges and responses in which the fittest individuals of a species survive. The evolution of man is thus understood as the history of such crises. In clinical psychology, too, we study events which are in continual, historical change, in retrogressive or progressive development. We investigate not only "permanent" rules of being but also unique occurrences, crises and coincidences. For a historical approach to the clinical concept of time sense, it is not important to ask what things "are" or how they can be reduced to other, simpler phenomena. What we wish to know is how man incidentally feels and thinks about them.

There have been several attempts to make us understand biological changes with time. For instance, the neuro-anatomists Constantin von Monakow[1] and Cornelius U. Ariens Kappers[2] worked out what they called the "time building principles" in organisms, especially as they relate to the central nervous system. According to them time and evolution are the eternal transformers of the anatomical aspects of living material. All functions of life are rooted in the primary instinct of life—a structural concept—a purposive striving, called *horme* by von Monakow and *entelechy* by other biologists and which I would like to call the inner piloting system. The horme has a general mnemic function in that it stores all received stimuli as engrams. We may call such a biological engram a *pre*memory.[3] Biological evolution gradually builds the conscious and unconscious individual memory by integration of manifold engrams. The individual receiving new stimuli transforms these perceptions into new impressions and adjusts his reaction pattern through the incorporation of new memories. Evolution and growth mean a continual addition of new reaction patterns to the organism as a result of new challenges and new responses.

The neuropathological concept of time-bound localization (chronogenous localization) describes a dynamic evolutionary development as the basis of localization of psychic functions in the brain. Every function of the organism has its own history, starting with the primitive function of the horme and developing into higher differentiated patterns of life. Every living entity reacts to an outside stimulus. Such reaction is always combined with a primitive "unconscious" feeling called by von Monakow the protopathy, or archaic emotion. On a higher level of awareness protopathy becomes emotion. The primary, unconscious selective function of the living instinct, the horme, either leads to attraction or rejection, either to adaptation or flight. Von Monakow calls this selective function the *protodiacrisis*, the archaic judgment. On a higher level of awareness, protodiacrisis gradually evolves into conscious judgment. The primary unconscious autoregulation and selective evaluation develops into highly differentiated moral conscience and moral judgment.

Similarly, the various functions of life gradually evolve from mere instinctual protofunctions into conscious awareness of functions. During this development a refinement and division of functions takes place, a differentiation that, however, remains related to the differential anatomical response of the central nervous system. That is why every response and function has its own, "in time" localized history. We may say that comparative anatomy studies the evolutionary history of new anatomical form adaptations to new challenges. Originally there are only the primary instinctual reactions leading to primary engrams. Later, the aggregate of all engrams determines the reaction of the living being at its position along its phylogenetic and ontogenetic history. We have to keep in mind that the different stimuli-responses, having changed various reaction patterns, at the same time remain specifically connected patterns. This time-bound correlation of response is also one of the tenets of psychoanalytic experience.

This course of events is, in short, what Ariens Kappers calls *neurobiotaxis*. Kappers searched for insight into the determining factors that gave order to the neurons in the nervous system. His study revealed that the architectonic influences in the nervous system are in accord with general psychological laws. He could show by his comparative neurological studies that time and history force the evolution of the central nervous system. Simultaneity of various stimuli and responses, of challenge and growth, link not only the psychological associations but also, originally, the nerve cells. The formation of nerve fibers is stimulated and linked by simultaneous occurrences. The result is an anatomical form picture, determined by the phylogenetic challenges of the environment on the living entity, and the chronogenous interrelated actions of cells and organs. The central nervous system is not a complex of localizations of functions merely existing beside each other, but it is a dynamically integrated expression of what took place during the phylogenetic and ontogenetic evolution of man. Thus the nervous system may be looked at as an interconnecting "time-binding machinery" in which the evolutionary history of challenges and the reactions to them are carefully filed. It is the plastic expression of man's biological history.[4] Through the modalities of medulla, mesencephalon, cerebellum, thalamus and cortex the brain developed into a historically determined complete unit. Wilder Penfield[5] calls it the built-in tape recorder of strips of time unlocking the past when one, for instance, stimulates the temporal lobes. Man, when expressing himself, makes use of a "kinetic melody," a symphony of functions, in which the whole organism is mobilized. According to the theory of neurobiotaxis, time is the "building principle," that means that in comparative anatomy a historical evolutionary figure can be traced in which past, present and even future can be recognized.

In every individual we can read his biological past. In his reactions he must partly obey the dictates of this past. The study of the growth of personality is a journey along the evolution of the race and the person. This study on the subjective time involvements of man is also such a journey. It is not only a trip along man's anatomical forms and functions and their mutual relations, but also along man's awareness of time, his awareness of becoming, of growth, of duration and man's subjective reaction to his involvement in time and history. In the part that is dedicated to time as a symbol I will come back to this.

We now turn to what is called man's "time-binding"[6] capacity, described by Korzybski as "Man's capacity to summarize, digest and appropriate the labors and experiences of the past."[7] He especially emphasized the subjective psychological aspect of this development as we can observe it in man's psycholinguistic behavior.

Another time-binding process is that of symbolization. For, just as the history of evolution is condensed and stored in the genes as bearers of genetic memory, the history of human thinking and communication is condensed and stored in human symbols.[8] The process of time-binding and condensing historical experience that first took place in the genes, determining man's heredity, had progressed beyond this biologically bound process and was assumed by man's cultural symbols. Psychoanalysis, too, depicts man as an incarnation of past times, as a product of many ancestors who transmitted their cultural endowment to their successors.

It began with man, a peculiar ambivalent animal who not only lives his own life instinctively but is capable of taking a distant view of his own life. Animals are thought to live only in a "now"; man is known to live in past, present and future. Man is a communicating animal who re-creates, for instance, the biological signals of pleasure and distress into persistent tokens through which he either expresses himself as an independent entity or hides his inner bearings. Man's highest expressive deed, the word, has become a polyphonous sign, an ambassador of his instinctual archaic needs. Through speech and words man learns to handle symbols and to condense reality into simple verbal signs. Man captures time and duration through his creative acts, his rhythms, words and cadences, his mathematical figures, poems and historical dreams. Thus the symbol as a historical communication became the inherited psychological gene and time-binding messenger of ancient traditional concepts.

3   Clinical Time

I would like to define *clinical time* as every person's unconscious and partially conscious concept of his personal life span between birth and death. The question relating to time, posed by many patients, expresses great anxiety for the unknown, for the mystery of death. For them time means mortality. As Heidegger expressed it: "We perceive time only because we know we have to die." For some people the fear and the mystery of the unknown can become so terrifying that they prefer sudden death by suicide rather than existence in restless anticipation.

What Aristotle called *athanatidzein*, the yearning for immortality, the yearning to go beyond man's confinement in time led, indeed, to many symbolic actions. In these man tries to reach "beyond time," to an unrepressed life in the Sabbath of Eternity. While so yearning, he can get lost in delusions such as erecting monuments that would forever make him remembered, so he believes, by future generations. Tyrants and dictators often suffer this "delusion of monuments." We also find this delusion in the act of murder, inspired by the magic thought that the souls of the murdered would become the murderer's servants and prolong his life. It is hard for man to admit that as a history-making creature, in Hegel's words, he pays the price of his individuality by being capable to contemplate death.[9]

Man's personal experience of near-death has a well-known influence on his awareness of time. Such increased awareness can be clinically observed, *e.g.*, in some of those who thought they would die and in a few of those who received electro shock therapy. The former may relive, as already mentioned in the introduction, their whole lives in a few seconds. This may be explained as a defense against the idea of departure from life; an inner voice says: "I do not really go into a state of nothingness, I see my whole life before me." The latter describe, as two of my patients have done, contrasting impressions. They can experience a general feeling of befogged mind with impairment of thinking, simultaneously with a feeling of appeasement to history and time.

Brown[10] describes very aptly how man's unconscious quest to repeat the ecstatic and traumatic moments of the past is in conflict with the flow of time. The repression of the now and of an insecure future is a strategy in the service of warding off death. Even in children we can observe the compulsion to repeat pleasure *ad nauseam* and into utter exhaustion. Childhood is the period of continually testing lengths of time and duration as well as space and perspective. Being together with bored elders makes time empty and lonely for the child and can be an early conditioning for later feelings of boredom and emptiness. The same expression of the restless quest for novelty may be found in older people. It is an evasion of *being*, under the guise of *continual becoming*.

Many a neurotic's obsession with time, such as being habitually too late, represents, besides an infantile obstructive defiance, a horror of growing up, the horror of departure and leaving the parental nest. So he must delay. His obsessive eternization of the one moment of the past intends to make permanent the archaic ties with mother. The fear of free time and leisure—of the mature combination of freedom and responsibility—belongs to the great fears of our era. That is why people must kill time.

The man who does creative work with his hands with the intervention of only simple tools takes part in a rhythm conditioned by nature. Carpenters and shoemakers of old were, in their trades, part of a world of rhythmic existence to which they subjected themselves in awe and satisfaction. The rhythm of work contributed an element of intimacy, largely missing in our present epoch. The use of sophisticated machinery destroyed this participation in world rhythm. It introduced a different tempo and an often frightening acceleration which continually affects the emotions. Wasting time in the age of technology became a great sin.[11]

The baby's first adaptation to environment takes place through a series of innate rhythmic movements which get usually quickly repressed in the established rituals of the nursery.[12] Rhythm is not only an initial biological necessity. It remains the basis of various cultural interactions such as marching, dancing, music and many other ritualistic expressions. In that part of my essay that describes the estimation of time span the importance of the confrontation between inner rhythm and outer rhythm will be emphasized.

"Clinical time" as defined before is the unconscious and partial conscious awareness of one's own individual life span as it is often expressed in the inadvertent question: "When is my time up?" There exist some people who know instinctively the moment of their death.[13] Predictions such as: "Doctor, you may do what you want, that day I will die nevertheless," remind me of the magic of the medicine men who are able to bring a spell of doom onto a person who violates a tabu. I have witnessed such a person die on the indicated day.

4　The Biological Sense of Time

We may define the biological sense of time (also called biological clock) as the awareness, usually subconscious, of cyclic functions in our organism. They can be, but are not always linked to astronomical cycles, such as the phase of the moon. The theories of Freud's friend and collaborator Fliess about the various cycles of activity in man have acquired renewed attention.[14]

Not only animals are dependent on cosmic time and show cyclic functions dependent on annual-moon-and-diurnal rhythm. In man we can also observe various comparable patterns.[15] There exist, for instance, diurnal cycles in body temperature, periodic diseases which are most prevalent in specific seasons, electric heart cycles and periodic brain-wave cycles. There are hormonal rhythms parallel to the moon cycle (*e.g.*, the menstrual cycle).[16] Recently, studies have been made relating the pineal gland to biorhythms. There exists a diurnal change in neurovegetative functions which in some people can lead to temporary disturbances when rapid travel by jet plane forces them to shift too quickly from a day pattern to a night pattern.[17]

We suspect that in some diseases the diencephalic centers of the brain, regulating adaptations to cycles of nature, do not function well. In post-encephalitic patients we can observe, for instance, that they can come out of the state called Parkinson rigor and become very lively again only during a few hours of the night; their diurnal cycle is inverted.

In general we may say that man possesses an inner clock probably re lated to the "world clock." In psychiatry we observe various periodic psychic phenomena dependent on the interactions between the private clock and "world clock." Some depressions reach their climax in the morning hours, others appear in specific seasons.

In some neurotic disturbances, emotion-laden unconscious memories are repeatedly re-experienced at specific times. We may speak here of a conditioned traumatic recall. In dream analysis this periodic time-bound reliving of the past can be observed. I have witnessed acute coronary reactions occurring on memorial dates.

5   *The Estimation of Time Span*

It is generally known that a certain span of time can seem of short or long duration depending on its emotional content. In darkness and waiting, or during anxious anticipation, minutes and hours can last much longer; in gaiety time flows too fast. Alcohol shortens time. Those who suffer from lack of sleep usually estimate nighttime as too long.

When in psychotherapy the patient has no resistance and can bring up important material, the hour usually goes much too fast for the patient as well as for the analyst. Grief can make the hour last much longer. In hypnotic experiments we can suggest a shortening of the time span or make the time go slower, and this distorts the patient's estimation of time span according to the familiar circumstances. The accelerated, awkward time before departure, or on long-distance telephone calls, is well known. In both cases you want to say much, but anticipation of the coming separation or the feeling that your time is limited takes over.

The evaluation of time span resulting from the confrontation and interaction between inner rhythms of the body and outer rhythms or schedules of the environment starts very early in life. Rhythm can be thought of as a direct experience of time span. A disturbed early interference with time adaptation may lead to later behavioral disturbances. A child conditioned to a confusing disarray of external actions will later lack in sense of time and estimate of time span; he will also lack adaptation to external rhythms. In some cases of schizophrenia the disturbance in early rhythmic confrontation and interaction between child and mother can be related to later disturbances in the sense of time.[18] The incorporation of external rhythms and other recurrent phenomena into the innate personal symphony of rhythms may be looked at as the first beginning of ego. That means that adjustment, adaptation and the incorporation of maternal schedules are accomplished by each individual in a personal way.[19]

6   *The Historical Sense of Time or Orientation in Time*

The historical or gnostic sense of time is a mnemic capacity to marshal along an imaginary time line, in their historical order, fragments of time and a succession of events through which one has lived. This comprehension of succession is the innate filing system of our own historical experiences. Yet, as Freud remarked, in the unconscious there is no fore and after.[20]

The processes of the system *Ucs* (= the unconscious) are timeless; *i.e.*, they are not ordered temporally, are not altered by the passage of time, in fact bear no relation to time at all.

Freud points out also that the myths of the race disregard an actual time element. The newborn child has no historical time. His sense of time is *tabula rasa* except for the racial information and genetic code provided by the genes. The concept of historical time appears only later when the transformation of innate adaptational responses into human conditioning and learning has taken place.[21] In those emotionally deprived early in life there usually is a lack of time orientation, since there is nearly no contact with any conditioning loving reality. One can conjecture that time awareness in the infant is synonymous with the child's confrontation with reality.[22]

Disturbed time orientation, however, does not have to relate to early emotional deprivation. Psychiatrists, for instance, are familiar with what is clinically often called "disorientation in time" or the Syndrome of Korsakow. In this ailment, because of organic brain disturbance or because of psychological causes, memories fall out and there is no comprehension of succession and duration. Disturbance in chronology and time-agnosia is often one of the first symptoms of an oncoming psychotic phase. A historical amnesia can also develop as a result of conflict and frustration. In such amnesia the voluntary memory for parts of life is repressed and the patient cannot localize events of his life along his own time line. In schizophrenics it is known that they either deny the existence of temporal sequences on the basis of their infantile delusions of omnipotence or deny that they lived before the onset of their psychosis. It is also known that sequential orientation can be experimentally disturbed by depriving a person of all sensory impressions.

The historical sense of time is an intricate function, for it is related to our sense of identity. Patients with compulsion neurosis, for instance, who have great difficulty in safeguarding their sense of identity because of their continual obedience to ancient conditioning commands, find that they rigidly have to obey a mechanical order in their historical sequences. They stick to a tyrannical inner schedule out of fear of real life. They become static historians as it were, without asking themselves if there was any meaningful relationship between events.

The autonomy of mnemic functions—the unobtrusive reliving and acting out of engrams from the past—can lead to confusion in the inner "filing system." *Cryptomnesia*, with its intricate psychiatric problems of delusion and hallucination, belongs to this study of orientation in time. Cryptomnesia means that a recollection of a hidden perception from the past is mistaken for an original experience now. A subliminal experience of years ago that never came to conscious awareness may appear as a strange, mystically acquired knowledge. This is often seen in hysterical patients who sometimes try to explain it as a memory from a former incarnation.

An inner feeling of both strangeness and familiarity is experienced in *déjà vu*, in which the person distinctly feels that he went through the same experience at some earlier time, but he is not sure what this experience was. *Déjà vu* often gives a feeling of living beyond time. It is in general explained by an actual experience triggering off an unconscious memory without the memory becoming conscious. This memory may be a forbidden infantile wish of which the person becomes subliminally aware again; it may be the recall of ancient knowledge related to the actual experience; it may also be related to a repressed dream. The symptom is most often seen in hysteria and epilepsy.

In brain disturbances there can be tremendous feelings of confusion and of being lost which result from the lack of ability to "file" daily events.

7   *The Sense of Continuity*

Man's search for multiple relationships in human aims and goals leads to a sense of continuity, a sense of being a link in a greater historical development. Time and history are experienced not as a juxtaposition of moments of time but as something integrated and meaningful on its own, lasting beyond man's personal life. Temporality and duration are the key elements in human self-awareness and existence. Past, present and future are experienced as related to each other, yet there is at the same time the awareness that man lives in different "time coordinates" and is himself the meeting point of various historical developments. *Panta rei*; man is part of the stream of becoming. Past, present and future have become one.

What is condensed above in a few words is of clinical importance, for so many beginning schizophrenics complain about this lack of awareness of continuity. In their personal universe of time they experience a sudden emptiness, a being alone in eternity. They are caught by the great urge to undo time, to rise beyond fate and death and causality. The various historical developments are not experienced as coordinated, time does not flow any more. This is their subjective way of announcing a withdrawal from reality. The schizophrenic catastrophe—the experience of breakdown of inner structure—is often explained by them as a downfall of the outer world, as aimlessness of existence and as the hell of timelessness. Timelessness here means: no future. They often have the experience of living in a timeless, archaic world without rhythm, without night and day.

A touching description of the inner experience of disrupted continuity may be found in Kafka's *Diaries*:

> This last week was like a total breakdown. . . . The clocks do not synchronize; the inner one chases in a devilish, or demoniac, or at any rate inhuman manner; the outer one goes haltingly at its usual pace.

The static sense of time orientation mentioned earlier meant mere succession, mental time or the dynamic sense of time means continuity and interrelation beyond mere succession. For example, there exist musical rhythms that divide time in small time spans. They give time a rigid cadence, a jumping, repetitious aspect. There are other rhythms that make time fluid and continuous. It is the secret of the creative musical composer to make time continuous and immense. In his symphonies he makes time an immensity without duration.

8   The Symbolization of Time

From the moment word and creative gesture were born, man has tried to catch and condense his subjective experience of time and duration in manifold symbols.[23] They are the signs and communications transmitting in a condensed way experiences from one generation to the other. Words are time-binders, in the sense of time-binding defined earlier. A word contains an evolution of various meanings attached to it in its etymological development and unconsciously reverberating in the final semantic token. Understanding word and symbol and their development is grasping human history in a nutshell.

The relation between language and the concept of time is related to our process of symbolization. The English word *time* has manifold meanings. It originates from a root *ti*—to stretch—and is in this way directly related to time span and duration. But *ti* is also related to *temps* and *tide*, thus indicating its relation with seasonal and atmospheric rhythms—like tempest and temperature. Another linguistic link is to tidy, tidiness and good tidings. Psychoanalytically, we are very much aware of the relation between time, cleanliness and orderliness.

Conscious awareness of time implies an awareness of self as mentioned earlier. Yet, the tool of such awareness is both given by and limited by the verbal habits of a language.[24] There are languages for instance that can only express the actual present.[25] Such languages force people to experience time as a series of immediately felt actions. Other languages have two past tenses, one for events that continue influencing the present and one for events that do not. Thus we find the difference between "dead" time and continuity in time superbly expressed.[26]

In psychoanalytic therapy the process of symbolization of time can be communicated and explored in a multitude of ways. In this section I will give a brief survey of these various symbolic communications from patient to therapist in which the patient depicts his own battle for identity and his own stand in history. The method of exploration we follow is free association and dream analysis as taught to us by Sigmund Freud.

The time as experienced in dreams is without localization in time line and without any direction. Past and present often are painted beside each other in a distorted symbolic way. Eternity and too shortness of duration alternate. Our organism is not even capable of estimation of the dream time. It is as if the observing inner pilot were stepping outside the time line when looking at dreams. He can view the past, present and future in one single moment. In dreams, "there is nothing corresponding to the idea of time . . . the Id knows no values, no good and evil, no morality" (Freud).[27] The unconscious has no conception of time as duration and continuity. Only analysis and interpretation make it into a continual process.

We may ask now: What are the most frequent symbols of time we encounter in the history of symbols and which are intuitively used by our patients?

Time eats itself away. This is the age-old image of the ouroboros, of the serpent or dragon swallowing its tail. This token means both death and eternity. In Chinese symbols it becomes the Pi, the round jade disk with a hole in its center, the symbol of heaven which through history transcends into the aura that crowns the heads of the saints. Later this symbol becomes the wheel of time. The mythical father Chronos devours his own children. One compulsive patient said it just the same way: "Time eats me, time is my enemy." In another way Father Chronos is depicted as an old man, the year that dies making place for the child, the New Year. In dreams time often represents the hostile father, the castrative old man with a scythe who cuts the cord and separates mother and child. In other dream images the time-serpent often appears as a kind of mermaid dragging her victims into oceanic oblivion. Sometimes time is symbolized as the devouring great mother, but sometimes also as the symbol of rejuvenation, the serpent perpetually shedding its skin.

Time's arrow symbolizes the Western world with its emphasis on progress and growth and with its view of time as an entity going forward in a straight line from the infinite past to the unknown future.

The concept of time's arrow shot into eternity symbolizes for many people the feeling of the irreversibility of fate and the hasty agitation of life that devours man minute after minute. Time is for them the moving river that drags us all to an unknown future. The illusion of a proceeding and passing time going from one now to another now is probably related to the awareness of an inner biological clock with its diurnal cycle of fresh awakening, followed by tiring till we need regeneration in a refreshing sleep. Our day experience is projected into our total life experience and time's arrow symbolizes the finality of our individual death.

Music, too, traps us in the idea of passing time; song and symphony are gone before we can have a second look at them.[28] Unwittingly people are reminded of the fleeting arrow of time.

Buddhism, with its strong belief in reincarnation and eternal repetition, sees time as a circle with no beginning or end, or as a spiral. Being here and now is irrelevant to Buddhist philosophy. Every being has its own cyclic time and its own space, its own time span experienced subjectively as its time of life.

Time as the silent surface of water. Especially in Hindu symbolism, time is portrayed as a placid, silent pool within which ripples come and go. The ripples are our temporary lives from which we must go down into the great, eternal Nirvana. Death is a stepping out of time, where time means the opposite aspect of timelessness and eternity; it means the way the unconscious conceives of its own existence as omnipotent.

Time as the two-faced god Janus. The Roman god Janus, the god of

248

gates and transitions, looks with one face into the past and with the other into the future. Later in history the same double function was attributed to St. Michael. The past can either be dead time or the time for which we must be responsible, or the time which we must make good in a future existence. The future signifies the great unknown, danger and fear, the expectation of the final judgment.

Time as submission to or revolt against paternal commands. Man's time concept, initially related to biological functions, became gradually more and more involved in parental relations as we have seen already in our explanation of the time span. It starts with the interaction between infantile rhythm and paternal scheme. For man, time and schedule symbolize the compelling and intruding external forces that offend his infantile feeling of magic omnipotence. For the so-called over-punctual anal character, being on time means being clean and obedient to the rule. It expresses a complicated, ambivalent relationship, the external submission to paternal commands despite internal rebellion. These people live in a kind of masochistic time: "I submit to your schedule but inwardly I prepare for revenge." Yet, these same characters may delay their submission, they may "kill time," symbolically kill the coercive forces, kill mother and father in a magic way. The strategy of delay is an ambivalent attack on those who command us.

"Waiting for somebody" has become in our society of schedule the symbol of either rejection or being less powerful. Hurt pride, depression and worry may fill the interval of waiting with dreary anticipation of revenge. Samuel Beckett's play *Waiting for Godot,* which depicts the tragedy of continual passive anticipation without any creative fulfillment, is a representation of such feelings. Primitive man did not and does not mind waiting; but in our culture waiting is often interpreted as an act of hostility. Here time, too, is experienced as the great censor of life, which keeps us caught in restrictions and coerced moral behavior. One of my patients expressed it sarcastically: "Time is civilization," by which she meant that strict scheduling and meticulousness were the great weapons used by adults to force the youngsters into submission and servility.

Time as an obsessive repetition compulsion. The most difficult struggle in therapy is with those patients who have accepted their initial conditioning training and experience as an automatic command that must be repeated ad infinitum. They have masochistically internalized their dependency and obedience and the schedules of their commanders. It is as if they ridicule the parent by overdoing with chiding punctuality what they are told to do. They have lost all trust in a self, all trust in growth and continuity; and with automatic conscientiousness they fight all individual conscience and morality. Habit is their law, spontaneity their enemy. They use the same pseudo-rebellion when they automatically deny every command, or automatically delay their response. Time is for them only the initial conditioner in the parental cage and they are continually searching for a new tyrant to make them miserable. Existing as automatons and robots they cannot live without commanding engineers.

This paradoxical defense against mortal time and wasting time may be found behind the technical "speedomania." Speed became the great killer of the technological age; the haste to nowhere is our escape from ourselves. Yet, man who learns to meditate knows that he cannot waste time.

Time is money. The proverb that tells us to use our time in a profitable way explains to us also the deep unconscious involvement with money and earthly possessions. Homo sapiens, the erect walking animal, chose the reversed way of adaptation to environment by shaping an environment that had to adapt to him. Yet, with his more cultivated forms of creation he follows the same instinctual magic strategy as animals do with their body products. Man externalizes himself in his creations; they are part of him and he cannot easily separate himself from them. Through his creations man tries to escape death and decay.

As said before, animals instinctually follow this magic strategy with their body products, encircling their field of action and possession with them. A dog who levels his paw sets out a frontier against a competing dog. Man has multiplied this instinctual reaction in a thousand ways. All that his hands and mind can shape is turned into one great defense against the emptiness of the world, against the emptiness of death. Hoarding and creating become for him an escape from death. "Time is money" means that time and money can buy defense against the finality of death. Time is something to be saved like money, avariciously. Money, genius and power are substitute symbols for time.

Pathological hoarding of everything is clinically most often experienced in the neuroses and psychoses of the aged as a huge mental barrier against the great, approaching separation of death.

Time as boredom. Boredom is another symbol of the great separation of death. It recalls, rightly or wrongly, the loneliness and deprivations of childhood when there was nobody to play with and to be loved by. As one patient said it: "The clock ticked my life away and nobody loved me." Many addicted patients experience time as empty space unless they can taste the artificial ecstasy of the drug. The emptiness of time is for them the relentless foe they must continually escape. A patient addicted to barbiturates expressed her ambivalent attitude toward her dead father as follows: "Time is relentless master, time is boring. The only thing you can do is to kill time with masturbation or with drinking or drugs." In our technical age, with more and more time available for leisure and with less emphasis on strong personalities, the problem of boredom and consequent addiction to drugs, games and amusements has acquired enormous proportions.

Time as the sexually desired parent. A patient was, in her dreams, fascinated by the sundial in a garden. The dial was staring at her, reproaching and seducing her. This distorted and symbolic language of the dream reminded her of her first sexual experience in one of the city parks. The lover had been a father figure to her; the clock symbolized the hidden wish for the real father. Grandfather clocks appear in many dreams as an old face with grasping hands and watching eyes. In modern initiation rites a watch is often given to the initiate as symbol of his sexual maturity. One of my patients who was eating watches in his dreams expressed thereby at the same time his Oedipal rivalry with his father. He had to devour this relentless taskmaster in order to possess his mother.

Time as the beloved. To lovers time symbolizes the moment of communion and shared emotion. This is their time in which they experience the beloved as more valuable than themselves. It is the time of which the poets sing. Even in popular songs time becomes the beloved: "I did not know what time was till I met you." Where there is mutual affection time is not empty any more. Giving time is universally experienced as an act of love. As often paraphrased: "From now on we create together a new eternity." Symbolically time has become the creative fertilizing power. Patients who cannot love complain about time as being empty and frigid. One girl, plagued by various compulsive activities, compared the empty time with her feeling of coldness during sexual intercourse. It symbolized to her the empty time with a tyrannical, seductive father who teasingly aroused her but never satisfied her. The ticking of the clock was to her always a warning signal for her guilty incestuous feelings.

Time as creation. The moment the patient can look at himself somewhat objectively, when he does not only live in his private time but can look at himself as part of various continuities, he can better accept the finality of his personal life. At that moment he feels himself becoming part of a greater creation. Though he does not lose his infantile sense of omnipotence, the act of being productive and creative together with other creatures gives him new satisfaction. Man in the act of creation is unaware of time because he becomes part of history and continuity. He gives his time and his attention, which means he gives himself to others.

Sometimes time is symbolized as the quintessence of loving creation, as an orgastic experience. "Those two minutes in which I really lived!" in the words of Dostoevsky, signify a condensation of affirmation, an ecstatic experience full of meaning to the person. In the final dreams of therapy time is usually experienced as the Creator, as the omnipotent father figure, as the life force that makes man.

Only man hopes and only man has a notion of time, and death. Though time remains for him a *memento mori*, creative man builds his own time.

Comments on Time and
the Uncanny

Manipulation of what we recognize as the proper temporal sequence of events is a well-known device of narration; alleged ability of traveling in time is employed with increasing frequency in literature and has formed the basis of such science-fiction classics as H. G. Wells's *The Time Machine*. In the performing arts even mediocre plots are apt to hold a large audience if they involve claims of non-inferential knowledge of the future. But the fascination of "time travel" is not recent. In every culture and age, persons credited with or accused of the gift of prophecy have been marked for special attention: they were honored and respected, or threatened and punished.[1] A definite attention, but one of lesser degree, has been paid to seers of the past.

My comments are prompted by observations such as the above, and derive from the following three assertions. (1) Non-inferential knowledge of the future, or the study of evidence which appears to support claims of such knowledge, tends to bring forth subjective responses dominated by a feeling of the "uncanny." (2) Non-inferential knowledge of the past can also appear strange and disturbing, but seldom with the same intensity as in condition (1). And (3), knowledge of distant, current happenings assumes strangeness only as an adumbration of events which, in the experience of the subject or of the observer are still in the future. We suggest that the feeling of the "uncanny" in these categories attests to the specific manner of growth of the sense of time in the child, that this growth sequence is identical with the steps of the historical development of the time sense in man, and, perhaps, even in organic evolution.

Our arguments will pertain to this vague but familiar reaction of strangeness, and not to the question whether the nature of time permits prediction of the future or divination of the past by extra-sensory perception. For discussions of the parapsychological study of precognition the reader must be referred elsewhere.[2]

Mankind's destiny, for Freud, as for St. Augustine, is "a departure from and an effort to regain, paradise; but in between these two terms man is at war with himself."[3] Psychoanalysts view this war as a struggle at the instinctual level between Eros and aggression, where these categories broadly correspond to the common terms of good and evil or love and hate. On this view, the life of man is an unceasing conflict between forces for the preservation and extension of life and forces attempting to destroy life. This inner struggle, however, can be turned outward as desire for power and achievement.

The Eros-aggression struggle is a uniquely human phenomenon; at the biological level, "life and death are in some sort of unity" but, "at the human level they are separated into conflicting opposites."[4] These oppositions can be and usually are repressed, that is, relegated for their operation to the unconscious and, therefore, they become the source of anxiety. It looks as if "the specifically human capacity for anxiety does reflect a revolt against death and individuality, or at least some deep disturbance in the organic unity of life and death."[5] Or, in Kierkegaard's words, "Time does not exist without unrest; it does not exist for dumb animals who are absolutely without anxiety."[6] In Dr. Meerloo's words: "though time remains for him a *memento mori*, creative man builds his own time."

In a paper first published in 1919, Freud asks the question, what kind of things appear uncanny?[7] Through a comparative word analysis of several languages and through a review of properties and persons thought to be uncanny, he concludes that "the 'uncanny' is that class of terrifying which leads back to something long known to us, once very familiar."[8] As one of his illustrations, he refers to the strange feeling of being lost or powerless in the face of certain repeated but unlikely events, such as coming upon the same unusual name many times in a row.[9] Freud argues that the uncanniness relates to the return of the mental content of the once familiar but subsequently repressed infantile repetition compulsion to the threshold of consciousness.

If, then, non-inferential knowledge of the future specifically, and knowledge of the past less emphatically, and knowledge of the present occasionally tend to appear uncanny, and if we take Freud's arguments seriously, to what process of repression may these feelings be referred? The answer suggests itself: early in the individual's life, separation of events into the three temporal categories was unknown, and time-ignorant existence was familiar. With the ensuing struggle for individuality, the ignorance of time was replaced by knowledge of time and the original way of looking at events became unfamiliar. If the degree of uncanniness is an indication, expectation constitutes the "deepest layer," corresponding to the earliest discovery of *a future*, followed by the discovery of *a past*, and finally by conscious knowledge of *a now*, which is probably the latest step in psychological evolution.[10]

"It is characteristic of the whole early development of the life of ideas," wrote William Stern,[11] "that they do not appear so much as memories pointing to something in the past, but expectations directed to the future. We meet here for the first time a general law of development: Reference to the future is grasped by the consciousness sooner than that to the past." Common experience seems to confirm this, for in the growing child verbalization of expectation usually precedes expressions of memory, and his exploration of the "now" appears the last.

I believe that to a time-ignorant unity of life and death, children, savages and saints can sometimes return, and perish if necessary without believing that death is a calamity.[12] Perhaps this behavior is an example of the capacity of regression in the service of the ego, a theory first introduced by Ernst Kris,[13] and further elaborated by R. Schafer.[14] Using Freud's distinction between the primary and secondary processes, Schafer remarks that "Thinking under the primary process tends to be unreflective, timeless and concrete; under the domination of the secondary process, thinking is reflective, shows time perspective, and uses abstract concepts corresponding to reality relations." The regression in the service of the ego facilitates the individual's access to the more primitive primary process; in our present context this means a return to a time-ignorant existence.

There are reasons to believe that the historical development of time sense in man followed in broad lines the sequence we have postulated for the individual's time sense. On this view, the primitive thought first expresses itself in the hopes and fears about the future as man attempts prophylactic actions against the dangers of future contingencies.[15] This stage is followed by his appreciation that the forward perspective can be viewed retrospectively.[16] Thus, we have a curious parallel in that the ontogenetic evolution of time sense has an analogy in the socio-cultural heredity of time and constitutes one of the "continuities in cultural evolution," to borrow Margaret Mead's phrase.

As we venture to ascend along the evolutionary complexity of life, it

255

appears that instinctual provisions for environmental changes anticipated for the distant future are observable in many organisms which give no hint of any but very short-range memory. They act as though their interest in the future is better established than their interest in the past. Indeed, experimental studies show that the capacity of delaying responses to eliciting stimuli is extremely limited below the level of man.[17] In primates vestiges of longer memory may be found, but long-time memory, and interest in the history of the community, are unique to man.

The psychological part of our argument relates to the concept of the unconscious. There can be little doubt, perhaps because it is so difficult to think of alternate propositions, that the unconscious itself is a result of organic evolution, even if the operational methods of this evolution are only ill understood, if at all. It would be difficult, writes Marie Bonaparte, "to conceive of a mind sufficiently independent to create out of nothing the forms under which the universe may be apprehended."[18] It is much more likely that "the human species has become attuned to the environment to which it belongs and has evolved in complete harmony with it."

One way to think of this is, perhaps, that evolution has provided us with a tool, or even a goad, which enables us to pursue our goals more efficiently than other living creatures on earth. But, if the superiority of such time-knowledge above the time-ignorance of earlier evolutionary stages is admitted, it should follow that any ontology of time, including one derived from studies of the results of experimental science, must include the acknowledgment of the temporal nature of reality rather than its negation. Otherwise a self-defeating condition arises for, as Dr. Needham has already concluded[19] on historical grounds, "any habit of decrying it [time] cannot be favorable to the natural sciences."

In general, whatever problems reality might present, the resolution of those problems will be more difficult through the rejection of a more advanced knowledge of the world in favor of the time-ignorant approach which, as we have tried to argue, represents a more primitive theory of reality both in the history of the individual and in the history of the community. Thoughts of time-ignorant existence may then be a haven to which we can occasionally return as individuals for repose in our struggle of becoming, but we dare not remain there or take it too seriously, as a matter of practice.

J. T. F.

Subjective Time

JOHN COHEN

An essay on time should begin at the beginning, if one can be found. Alas, this is difficult. For the essayist is denied the trick of the storyteller who, by a magical incantation "Once upon a time . . . ," transports his listeners to a world beyond time where his tale is set like a solitary isle in a shoreless sea. Let us be content to start then as natural historians and compile a short catalogue of the flora and fauna of temporal experience: the *apparent* duration of an interval of time or of the sequence of events, the feelings of what might be called "pastness" and "futureness." These are familiar features of everyday life, but we may see them in better perspective if we make the assumption of continuity at least in some aspects of the temporal experience of man and animal and in their adaptation to the rhythms and periodicities of nature. This does not mean that man's "sense of time" is superior to that of animals, any more than his auditory or olfactory powers are supreme in the animal kingdom, for it is conceivable that our reliance on watches and other artificial aids has led to an atrophy of our "sense of time";[1] and it is salutary to reflect that the minute beach crab (*Talitrus*), with barely 1 millimeter of nerve substance, is able to "calculate" the time of day, the course of an hour, as well as the angle formed by its bodily axis and the position of the sun.[2]

The assumption of continuity leads to the idea of an inner or biological clock. We owe the germ of this idea to Henri Piéron,[3] who suggested that if the speed of our organic processes were to be modified, for example, by varying our temperature, the result would be a proportionate increase or decrease in our mental time. The more general principle relating speed of reaction to changes in temperature had already been established by Van't Hoff, and Arrhenius had demonstrated that this principle was also true of biological processes. Piéron's suggestion inspired the dramatic experiments of his pupil, François,[4] four years later. In one situation, for example, he asked his subjects to tap a key at the rate of three times a second. He then raised their temperature by diathermy and found an increase in their tapping rate. Piéron went on to argue that just as clock time seems to pass slowly when temperature is high, so clock time would seem to pass rapidly when temperature is lowered. It is as if a feverish person would say: "Has an hour not passed yet?" and a freezing person: "Has an hour passed already?" Similar notions were independently arrived at by Hoagland[5] and he has successfully established a systematic relationship between temperature and "tempo" in a variety of situations. He supposed that there exists in the brain a chemical "pacemaker" which regulates the speed of its metabolism, and hence the rhythm and precision of human timekeeping.

In a lucid review of the biological status of such pacemakers or inner clocks, Thorpe[6] has discussed the recent evidence which bears on the origin of patterns of nervous discharge, patterns which are meaningfully distributed in time as well as space and which take their temporal cues from various sources, peripheral, proprioceptive or central. There is evidence for the existence of more than one localized pacemaker region within a single cell, and in the central nervous system rhythms may occur independently of external periodicities. Furthermore, inner clocks may be set to work with great reliability by an initial alternating stimulus or even by a single stimulus, provided it is powerful enough. Such triggering effects are readily linked with highly sensitive phases in the development of an organism, and they take us into the realm of imprinting or single-trial learning, which appears to characterize the early development of the human babe as well as the young of a variety of other species. Even more significant for the psychologist than the effect of imprinting is the successive *timing* of experiences in the life of the growing child.

The process of learning, during the developmental period, involves not only content and method ("what" and "how") but also discovering the optimal stages of maturation and receptivity ("when"). This holds equally for intellectual, emotional and social learning. In short, there is a correct "moment" in the life of the child when he is ready to learn a given lesson, whether it be to accommodate himself to his fellows in the classroom, to control his temper or to solve quadratic equations. The task of parent and teacher is to strike while the iron is hot, neither too early nor too late. This view assumes that, granted a preformed genetic code, the "moments" are determined by the rate of epigenetic decoding or unfolding of potentialities, the rate of decoding (or unfolding) itself being governed by factors within the individual as well as by factors in his social and material environment.[7]

1   *The Microstructure of Subjective Time*

The experimental study of the human sense of time to which investigators have devoted themselves during the past century may be regarded, in the light of this biological orientation, as an attempt to determine properties of the inner clock. With two notable exceptions—Piaget[8] at Geneva and Fraisse[9] at the Sorbonne—psychologists have been chiefly occupied with the precision with which people judge and compare brief intervals of time of the order of seconds. Such judgments and comparisons are bound to be mental constructions (or representations) superimposed on an organic appreciation of the passage of time, and governed by factors of a personal, social and cultural character.

I propose to give a brief sketch of this work. A great deal of effort has been invested in attempts to discover an interval to which people are optimally sensitive. Many would argue that such an interval lies in the so-called "indifference zone" from 0.6 to 0.8 second. Within this range estimation is said to be relatively precise, with a tendency to overestimate shorter intervals and to underestimate longer intervals. Fraisse,[10] among others, has suggested that the "indifference zone" corresponds to the duration of natural rhythms: walking, heartbeat, spontaneous tempo of tapping, the forming of mental associations and so on. It is only fair to point out, however, that others[11] take the view that the indifference zone may be nothing more than an experimental artifact.

Insofar as we are relatively more precise in judging seconds than minutes or hours, this may be because, in the former instance, we turn our attention to the interval itself, but in the latter, our minds wander and our judgments are then indirect, based on such cues as the number or kind of activities that have occupied the time. The frequency with which mind wandering takes place may be determined by an experiment[12] of the following type. Each member of an audience is provided with a bell push and instructed to press it whenever he becomes aware that his attention has *returned* to the lecturer. The bell pushes are so arranged that when any one is pressed a corresponding torch bulb lights up in an adjoining room, this being recorded on a moving film. Under such conditions people press the bell push, on the average, about four times during a forty-minute period, the rate being somewhat higher in a group listening to music; individual differences are striking, about 3 per cent "wander" every two minutes. The attention of normal children seems to return, on the average, about once a minute, the duration of the wanderings ranging from a momentary flicker too short to be timed to about four minutes; the average period of inattention of deaf children, who learn by lip reading, is about five seconds.

It must not be inferred from what I have said that all intervals longer than a few seconds are estimated in the same fashion. We may have different systems of estimation for intervals of increasing orders of magnitude —minutes, hours, weeks, months, years—and all these systems may be subject to exaggeration or contraction; but as we move from short intervals the operation of the inner clock becomes more and more obscured by reflective processes.

The measurement of the temporal span of attention[13] has intrigued many investigators. What is the interval of time over which a series of stimuli may extend and still be perceived or experienced as "unitary"? It has proved more difficult to discover the upper than the lower limits of such intervals, because of the ambiguity of the idea of a unitary or "momentary" experience. Upper limits ranging from about 2 to 12 or many more seconds have been indicated, and lower limits from 0.01 second for sound to 0.12 second for continuous light.[14] At the lower limit the subject has to distinguish the instantaneous from what seems to have some duration. This is no easy task, hence the attempt to determine the duration of a psychological moment on the basis of its information-bearing capacity. Thus in listening to a list of words, a value of 0.1 second appears to be the irreducible "moment," in the sense that the listener is unable to assimilate anything in briefer intervals.[15] An alternative device has been to measure the duration of an impression in terms of the critical rate at which a succession of sensory elements is experienced as a continuous flow. Here a value of about twenty per second has been indicated.[16]

Limits for temporal span are not to be confused with thresholds of separation and order, which have been established with some accuracy in auditory perception.[17] An interval of 2 milliseconds separating two sounds enables the subject to say that he hears two sounds, not one, but he needs an interval of about 20 msec. to decide the order in which the two sounds occur. In general our auditory is superior to our visual equipment in making fine temporal discriminations, which is not surprising since sounds are, so to speak, strung out in time while visual displays are spread out in space. Each sense, furthermore, has its distinctive powers of synchronization, retention and recall.

Instead of asking how brief an interval need be to separate two excitations, we can concentrate on the interval itself and ask: what is the minimum perceptible duration between two excitations? In the same modality of sense it ranges from 50 to 100 msec. for vision, and from 10 to 20 msec. for hearing as well as for touch. When the excitations relate to *dif-*

ferent modalities, the threshold rises from 30 to 60 msec. for hearing and touch, to 50 to 70 msec. for vision and touch, and to 80 to 100 msec. for vision and hearing. To this we may add that an "empty" interval delimited by auditory stimuli seems shorter the more intense the stimulus;[18] but if the interval is defined by a continuous tone, the effect is the other way round; the more intense sound will seem to last longer than the less intense sound.

Whatever the validity of these various thresholds, they do not reveal absolute properties of the external world. Against such an error of judgment Coleridge[19] warned us a century and a half ago. "The delicious melodies of Purcell or Cimarosa," he wrote, "might be disjointed stammerings to a hearer whose partition of time should be a thousand times subtler than ours," just as "the edge of a razor would become a saw to a finer visual sense."

2   Subjective Pastness

So far we have been concerned with what we may call the microstructure of subjective time. In the experience of "pastness," and "futureness," we move to another scale where the inner clock is no more than a subsidiary device in human timekeeping.

The capacity to recall earlier events in the order in which they occurred depends to some extent on the quality of pastness which we associate with the separate events, a quality which enables us to say how long it seems since the occurrence of any particular event. Estimates of elapsed time may be remarkably accurate even after several days spent without cues to clock time: two persons placed for forty and eighty-six hours respectively in a soundproof room without temporal cues guessed the time correctly at the end within about half an hour.[20] Alternatively we can study subjective temporal relationships to past events by asking subjects to mark off, on a given line, a length corresponding to the seeming lapse of time since some particular occasion: how long does it seem since yesterday's lunch? since last Christmas? since you left school? and so on. Linear representations of this kind follow a distinct pattern. Estimates of intervals up to about six months from "Now" show a relative contraction as they become more and more remote from the present, while estimates of intervals greater than one year do not show this contraction but are more or less in proportion: the length marked off for five years, for instance, is about five times that marked off for one year.

This contrast corresponds with the familiar fact that an event an hour or a day ago seems relatively far away by comparison with something that happened a week ago, and a month ago is not felt to be four times as remote as a week ago. Linear representations of intervals between six months and one year from "Now" may be either in proportion, or not in proportion, to the interval.[21] It is of interest to note in this connection that a line drawn to represent the apparent duration of an ordinary day is about 1.6 times the length of the line for a "short" day; and the same ratio characterizes the apparent duration of a "long" day compared with an ordinary day.

The subjective unit of elapsed time does not remain constant as we grow older. It is a commonplace that with advancing age, the calendar years seem progressively to shrink. In retrospect every year seems shorter than the year just completed, possibly as a result of the gradual slowing down of metabolic processes, and hence retardation in the rate of healing, with advancing age.

What I have said about linear representations of subjective time should not be taken to mean that such a form of representation comes naturally to everyone, as it did, apparently, to Barrow (Newton's teacher): "So shall we always represent time," he wrote, "by a straight line."[22] While many people spontaneously use spatial images to represent the subjective past and future, by no means do all use horizontal lines; some prefer ascending lines and others a closed figure.[23]

3   Subjective Futureness

"Futureness" deserves to be treated at some length because of its overriding significance in human life. A subjective future is supposed in all our activities. Without a tacit belief in a tomorrow nearly everything we do today would be pointless. Expectation, intention, anticipation, premonition and presentiment—all these have a forward reference in time. Our entire psychic life is permeated with the hope of things to come, which is the counterpart of nostalgia for the past. Implicit in all our actions are plans, however vague and inarticulate, for the future, and sometimes, as in saving and investment, this planning is deliberate. As we ascend the evolutionary scale the temporal horizon becomes more and more extended. This may be illustrated by the fact that in experiments on delayed reaction, the interval of delay may be increased at higher phylogenetic levels: the rat can sustain a delay of some 4 minutes, the cat, 17 hours, and the chimpanzee 48 hours. In man, the horizon may reach far beyond his own brief existence; from infancy onwards there is a growing capacity to relate what is happening at the moment to events foreshadowed in the more and more distant future.

This developmental change is no doubt related to the different conceptions of the future held by children of different ages. When asked to say what they mean by the future some five-year-olds reply with words to the effect that the "future has not happened, the past has happened." A characteristic reply at the age of nine or ten is that "the past comes before and the future follows the present." At the age of twelve or later, the replies refer to material differences: "there will be a better standard of living in the future than in the past," or, more commonly, to abstract features: "we have knowledge of the past but are ignorant or uncertain of the future," "the past, unlike the future, is done with and unalterable," "the past is stationary but we are always moving into the future," "we remember the past and think about it whereas the future is in our imagination."

Our orientation to the subjective future appears to have the character of a "gradient of tension": we become more and more vigilant as an expected event draws near in time. As the ticking of the clock records the passing of the hours and the fateful moment draws near our hearts beat faster, an experience well described in Adelbert von Chamisso's curious tale *The Shadowless Man*.

Now I remained with my eyes fixed on the hand of the clock, counting the seconds—the minutes—which struck me to the heart like daggers. I started at every sound—at last daylight appeared. The leaden hours passed on—morning—evening—night came. Hope was fast fading away as the hand advanced. It struck eleven—no one appeared—the last minutes—the first and last stroke of the twelfth hour died away. I sank back in my bed in an agony of weeping. In the morning I should, shadowless as I was, claim the hand of my beloved Minna. A heavy sleep towards daylight closed my eyes.[24]

If a sleeper has decided to wake up at a pre-appointed time, he becomes restless as the moment of waking approaches. Every examinee knows this feeling of mounting tension. So does a pregnant woman passively awaiting the birth of her child, a bridegroom the marriage ceremony and a prisoner his execution. The gradient may be demonstrated experimentally (as well as clinically) in animal and man alike. Pavlov's dogs, conditioned to be fed every thirty minutes, betrayed, by changes in breathing and salivation, that they "knew" when the next meal was due. They could fall asleep in the intervening period and awaken, after signs of restiveness, as the food was about to appear. In man we find that the ability to recall or recognize a task which has been started but not completed depends not so much on the amount done as on the amount that remains to be done: the less subjective time needed for completion, the easier the recall of a task regardless of how much time, within limits, has already been spent on it.[25]

The subjective future, so vital and indispensable a feature of human experience, appears to have no place in the *Weltanschauung* we owe to certain interpretations of natural science. Some philosophers assign a lower logical status to the idea of the future than to the idea of the past or present. The past and present, declares C. D. Broad,[26] are real, but not the future, which is a "nonentity"; so-called judgments about it are therefore not judgments at all, for they are neither true nor false. For the psychologist, however, the subjective future has the same reality as the past or present.

It is possible to obtain some indication of the way we conceive of the future, either in itself or by comparison with the past, by using the same type of linear representation described in relation to the study of the subjective past, though this is not a *measure* in any proper sense. Subjects are presented with a horizontal line of, say, ten inches, the extreme left of the line being marked "Now," and the extreme right, "End of Life." They are asked to imagine that this line represents the seeming duration of their future life, and they are asked questions of the type "How long does it seem to you from 'Now' to . . . ?" They then mark off a length of line which indicates their answer to the question. In one of several variations, the *midpoint* of the line is marked "Now," the left-hand side of the line referring to the past and extending until "Birth," and the right-hand side extending until the "End of Life." Some subjects are then asked to mark a point on the line to show how long it seems since their *last* birthday; others mark off a length to show how long it seems to their *next* birthday.

I should like to illustrate the results of such inquiries, in which several thousand subjects of all ages have taken part. The lengths of line marked off for different future intervals seem comparable only when they relate to one and the same unit of time: minutes, hours, days, weeks or years, but when the units differ, a different scale seems to be employed. This is not surprising when we think of the difficulty of reducing calendar to clock time.

When the line is marked "Now" at its midpoint a statistical adjustment has to be made to allow for the fact that subjects are not necessarily either at the *objective* or *subjective* midpoint of their lives. If we express median lengths of line for intervals of 0 to 3 months in the past as *unity*, we find the length for the same interval in the future is about 4. The same ratio of 1 : 4 also appears in comparing intervals of 3 to 6 months and 6 to 9 months in the past with similar periods in the future; but a year ahead is represented by a length only slightly, if at all, greater than a year in the past. It would thus seem that we represent the near future as a distorted mirror image of the just past.

Our subjective future may be said to reflect the subjective past in a wider sense, for our concern is not limited to empty durations of time. The further ahead we look the more our vision is shaped by fantasy, just as a Proustian magic transforms the recollection of our distant past.

We encounter the future with the aid of that vital organ we call hope. Now there is something about hope—what we might call its gambling quality—which makes it resemble the cyclical conception of time characterizing archaic cultures. The longer the sequence of a true gambler's failures, the stronger his conviction that he will triumph in the end. That is why he increases his stakes with every loss. It is natural, too, for hope to intensify as one's affairs deteriorate. Archaic cultures seem to have this quality of hope built into them in the form of a cyclical conception of historical time. The contemporary epoch, in archaic culture, is looked upon as decadent by comparison with preceding epochs; and even in the contemporary period, the moment progressively worsens. But a belief in this order of time is a sign of optimism, for the worsening situation is felt to carry a seed of regeneration. Hence the idea that the phase which heralds the millennium will be marked by chaos and the battle of Gog and Magog.

The essence of the cyclical notion is that a stage is reached at which the world comes to an end and then recommences afresh.[27] Notions of this kind certainly survive in modern societies which are supposed to have adopted a linear conception of historical time. For example, when we extend a wish for a "Happy New Year," we never attempt to stretch the wish for *two* years. We feel that this would not be valid, that the term of the wish expires at the end of the year, just as the validity of a railway ticket expires after a specified period. However, there is a tendency on the part of modern "progressive" societies to transpose the Golden Age from the primordial past to a future which will be achieved by not too long a series of Five-Year Plans.

4   *Sensorimotor Aspects of Subjective Time*

I referred at the outset to the manner in which our inner clocks may be influenced by temperature. Sensorimotor activity may lead to similar effects by accelerating or inhibiting cerebral processes, and hence affect the apparent duration of intervals of time;[28] the auditory background to a stimulus has been described as a "calibrating monitor" of this kind.[29] This aspect needs further discussion, which we can begin by considering the *tau*-effect.[30] If three points are marked on the subject's forearm and the interval of time between stimulating the second and third points is greater than that between the first and second, the subject declares that the distance between the second and third points is greater than that between the first and second though, in fact, it may be the same or even less. Estimates of the distances are thus influenced by the time intervals which demarcate them. Helson called this the *tau*-effect, and Geldreich[31] demonstrated a similar effect in visual experience.

A few years later, Abe[32] and Abbe[33] measured, also in the sphere of vision, the converse phenomenon, namely, the effect that varying the distance between two stimuli has on the judgment of the apparent duration of the intervals separating them. In our own laboratory the same effect has been independently established under different experimental conditions, which may be briefly described as follows.

The subject faces a repeated cycle of three flashes of light set horizontally or vertically in front of him, the ratio of the distances between the three flashes varying from 1 : 10 to 10 : 1, where the shorter distance is one foot. Each cycle of three flashes is repeated after an interval of 5/3 of the total cycle, which ranges from 0.6 to 6.4 seconds. The subject can control the timing of the middle flash and he is told to adjust it so that the interval of time between the first and second flash appears to him to be the same as that between the second and third. Under these conditions he makes the interval of time between the flashes spaced further apart *shorter* than the interval of time between the two flashes closer together, and the bigger the ratio of the two distances, the greater the difference between the two time intervals allotted to them. Thus if two *equal* time intervals are presented, delimited by flashes of light, one interval will seem to last longer as the relative distance between the corresponding flashes is greater. The magnitude of the effect varies with the direction of the flashes of light, being smallest in the upward direction, greatest in the downward, and intermediate horizontally.

If we replace the flashes of light by three different tones, we find an analogous effect, though weaker, possibly because the subject feels that he can safely ignore pitch, while he adjusts the middle tone to bisect the cycle of time. A kindred effect may be produced by employing two *continuous* tones to indicate the time intervals, instead of three brief tones. The effect of tone on duration is now more marked: the subject allots a shorter duration to the higher tone than to the lower one, and the greater the difference between the two tones the more striking is this effect. Thus apparent duration is demonstrably influenced by auditory as well as by spatial features in the pattern of stimulation.[34] We have called these influences on apparent duration *kappa*-effects.

We may now ask whether apparent duration is influenced by comparative distance when this is *passively* experienced. This question, we may note, was prompted in 1956, when space travel seemed to come within the bounds of the possible and a dispute took place on its effects on aging. The debate itself was inspired by an inference from Relativity Theory, first drawn by Langevin, that if a space traveler moved at a speed close to the speed of light, he would age less rapidly than the terrestrial friends that he had left behind; from the point of view of the latter, one year for the space traveler would be equivalent to a decade on earth.

Let us imagine that something of the nature of a *kappa*-effect holds for journeys in space. Our traveler would be subject to acceleration on embarking and to deceleration on landing, and we could suppose that his estimate of the duration of the parts of his journey which varied in velocity might be influenced by the corresponding distances through which he had traveled during the respective intervals; the time for the longer distance might seem disproportionately long and the time for the shorter distance disproportionately short, as compared to clocks on the spaceship. Similar effects, it could be argued, would appear in a man who flies blindfold from London to Paris in an hour and immediately continues the journey from there to Baghdad in another hour; the second lap of his journey might seem to him to last longer than the first.

In the event, it proved more practicable to come down to earth and study passengers in a vehicle on the road with respect to the apparent duration, distance and speed of their journeys. The windows of the vehicle were blacked out or the subject was blindfolded, in order to exclude visual cues to distance. The subject was told that he was to be taken on a journey during which a bell would ring. That was all. When the journey was over, he estimated its duration, distance and speed (*a*) before the bell and (*b*) after the bell. (In experiments now in progress, the driver, not the passenger, makes these estimates.)

The principal outcome is a demonstration of interdependence in the apparent duration, distance and speed of a journey such that if two parts of a journey take the *same* clock time, that part seems to last longer in which the distance and speed are greater. This we call the *kappa*-movement effect.

Two additional effects may be inferred, under the same experimental conditions: 1) the shorter distance is overestimated, and the longer distance is (if anything) underestimated. We may infer from this, together with the *tau*-effect, that if two parts of a journey are of *equal distance*, that part will seem greater which is traveled at a slower speed for a longer time; and 2) the slower speed is overestimated and the faster speed is, if anything, underestimated. From this, together with the *tau*-effect, we may infer that if two parts of a journey are traveled at the *same speed*, that part will seem faster in which the distance and time are shorter.

It is natural to inquire whether the judgments of apparent duration, distance and speed are mutually consistent. Are they interrelated in a manner analogous to the interrelations of the physical variables, time, distance and speed? The answer is in the affirmative so long as the passenger is traveling at uniform speed.[35] When, however, there is a change of speed, either by acceleration or deceleration, there is a disturbance in the "symmetry" of the relationships. Passengers then believe that they have been traveling for a longer time than is implied by their combined estimates of distance and speed. Their estimate of speed is excessive as judged by their combined judgments of duration and distance, but they underestimate distance as judged by their combined estimates of duration and speed.

A *tau*-movement effect[36] has also been demonstrated under conditions in which the subject walks in one half and runs in the other half of his journey: to many people the distance seems greater when walking than when running. However, this statement merely refers to the mean tendency and obscures the fact that the opposite effect occurs just as frequently, except that its magnitude is less. The individual to whom apparent distance increases as the associated time interval is greater seems conscious of the passage of time, but tends to ignore speed. The individual to whom apparent distance seems less seems very conscious of speed: he feels he has covered a shorter distance when walking because he moved more slowly than when running. The former is more and the latter less sensitive to the passage of time. Such individual differences may be related to variations in metabolism and tempo.

What is clear from these experiments is that people do not seem able to judge distances independently of the correlative time intervals. But the influence of the interval of time on the judgment of distance is not in the same direction for different individuals. So we must conclude that *kappa* and *tau* are statistical effects, unlike illusions to which everyone is subject, though not necessarily to the same degree.

5   *Disorders of the Inner Clock*

Like any other chronometric device, our inner clocks are subject to their own peculiar disorders and, in spite of the substantial concordance between private and public time, discrepancies may occur as the result of sheer inattention. Mach[37] noted that if a doctor focuses his attention on the patient's blood, it may seem to him to squirt out before the lancet enters the skin and, for similar reasons, the feebler of two stimuli presented simultaneously is usually perceived later. We also seem to serve as somewhat inefficient clocks when it is a question of comparing an interval that is passing with one that has elapsed. For example, if a subject is asked to read aloud, at the rate of one per second, a series of digits presented in random order, and subsequently estimates the duration of the elapsed reading period, the ratio "past time"/"present time" proves to be less than unity, where "present time" is measured by the number of digits read. The ratio declines with increase in the interval during which digits are read aloud and it reaches a stable level at intervals of about twenty seconds.[38]

Major discrepancies between private and public time appear in heightened emotion, in physical or mental illness, in hypnosis and under the influence of drugs. Normal life requires the capacity to recall experiences in a sequence corresponding, roughly at least, to the order in which they actually occurred. It requires in addition that our potential recollections should be reasonably accessible to consciousness. These potential recollections mean not only a perpetuation within us of representations of the past, but also a ceaseless interplay between such representations and the uninterrupted input of present information from the external world. Just as our past may be at the service of the present,[39] so the present may be remotely controlled by our past: in the words of Shelley, "Swift as a Thought by the snake Memory stung."

The time order in the recall of the past, which is an essential element in the sense of personal identity,[40] is disrupted in Korsakow's syndrome and in early senile dementia. In paresis, too, a patient may say that he has not grown older since his illness began.[41] By contrast, in "depersonalization," the immediate past may seem to the patient exceedingly remote, an effort similar to that found in other psychotics who are sure, after five minutes, that they have been kept waiting for six months.

Distortion of temporal judgment may be deliberately induced in a hypnotized individual.[42] He may be told, for instance, that ten minutes are to be allotted for a given task assigned to him, the actual time being only ten seconds; yet he is afterwards certain that he has been active for ten minutes. It is hardly necessary to say that the feeling of the hypnotized person does not in fact endow him with the capacity for performing a ten-minute task in ten seconds. Otherwise, Tristram Shandy, who was distressed by the thought that it took him two years to describe the first two days of his life, might have sought reassurance from a hypnotist. Temporal distortion under hypnosis is not more mysterious than other phenomena of hypnosis. It reveals that the hypnotist is able to make the subject accept beliefs uncritically even if they are at variance with objective circumstances.

Under the influence of mescalin, hashish and *cannabis indica*, the effect seen in fever is greatly enhanced.[43] De Quincey tells us that one night might seem to him "of a duration far beyond the limits of any human experience," and one of my own subjects under mescalin stated that "his sense of continuous time was lost." Walter de la Mare[44] has given us an excellent description of the effect of mescalin on his friend J. Redwood Anderson. Mr. Anderson felt his thoughts to be greatly accelerated but his movements as well as the events in his environment seemed much slower than usual, although to an observer his *actual* movements were feverishly quick. When he got up from his chair to open the door, he appeared to the observer to be moving rapidly, but he himself, by the time he had reached the middle of the room, had the feeling that it was years since he had risen from his seat and he could hardly recall what he had set out to do. In spite of all this, he was uncommonly precise in judging the passage of time. What looked like an enormous tapemeasure, marked in seconds, minutes and days, presented itself to his eyes. A pointer moved along this scale and all that he had to do when asked to tell the time was to glance at the pointer; the scale seemed to him as real as the trees in the garden.

It has long been suspected, at least since Pierre Janet drew attention to the fact, that there is an "unknown faculty" for judging the passage of time which characterizes somnambules as well as those under hypnosis. We can recognize the phenomena described in the older works as nothing more than a conditioning of the inner clock. J. M. Bramwell, T. W. Mitchell, J. R. L. Delboeuf, among others, adduced clinical evidence of the existence of these phenomena, which are sometimes encountered in persons with remarkably rapid powers of arithmetical computation. As Gunn[45] remarks, when Delboeuf instructed "J" to tweak the cook's nose in 1,600 minutes' time, he might just as well have said "at 9.10 A.M.," provided "J" could make the calculation quickly enough.

The special effects of hypnotism and somnambulism on the awareness of the passage of clock time must be distinguished from the alleged timelessness of the Freudian unconscious. "It is constantly being borne in upon me," wrote Freud, "that we have made far too little use in our theory of the indubitable fact that the repressed remains unaltered by the passage of time—this seems to offer us the possibility of an approach to some really profound truths. But I myself have made no further progress here."[46] It cannot be said that others have made much progress, although Marie Bonaparte[47] has attempted to dissect the concept of timelessness. She casts doubt on the suggestion that repressed psychic content remains unaltered by time, whatever we may think consciously. But the evidence she adduces is essentially intuitive, and her contention that "the unconscious has no knowledge of time" is hardly more helpful than her claim that "the unconscious does not perceive time."

In hypnosis, as under the influence of certain drugs, in dreams, in ecstatic states, the order "past→present→future" may cease to exist. What remains is an omnipresent in which there is neither a before nor an after. This internal "detemporalization" of experience is the hallmark of the writings of Proust, Joyce, Thomas Mann and others of this school, and it is in this sense that we have to understand the Freudian dictum of the timelessness of the unconscious. Ortega has remarked on the "sluggishness" of the narratives of Proust, Dostoevsky and Stendhal. In Proust especially, Ortega argues, the prolongation of the moment is carried to its limit. "So slowly does the action move that it seems more like a sequence of ecstatic stillnesses without progress or tension." There is no plot worthy of the name and no dramatic interest, only pure "motionless description": the river of time is frozen.[48]

A peculiar type of temporal derangement is encountered in those obsessives who cannot bear time's unalterable flow and behave as if indeed they could reverse it,[49] as in *déjà vu*. Some would like to do away with time, like Rousseau, who threw away his watch and thanked heaven that he would no longer need to know the time of day. Had he been a citizen of Butler's *Erewhon*, where the mere possession of a watch renders one liable to imprisonment, he would have felt totally free from anxiety.

Not far removed from such persons is the patient whom Ernest Jones[50] has described as suffering from a so-called "God-complex," marked by a feeling of omnipotence in relation to time. *His* time is valuable by comparison with that of others; *his* time is correct, and he alone is justified in being unpunctual. *His* recollection of past events is the only true one, just as *his* prediction of future events will alone be fulfilled.

Even a cursory review of the psychopathology of time would be incomplete without a reference to the manner in which so-called "experimental neurosis" may be induced by imposing temporal demands greater than an organism can bear. The reactions of animals under conditions of temporal constraint may help us to understand human disorders in the tightly time-bound cultures of our day. If Pavlov's dog was fed every half hour and one meal was missed, the dog would secrete two or three minutes before or after the half hour had elapsed. But if the animal was given a readiness signal by sounding a metronome half a minute before the half hour had elapsed, its "sense of time" became sharper and the secretion would not begin if the metronome were sounded too early by a minute or more. H. S. Liddell has found similar effects in sheep and goats, and established that it is the readiness signal which proves the last straw. A sheep in a Pavlov frame given an electric shock every six minutes becomes restless two or three minutes beforehand. But a ten-second readiness signal given before the six minutes proves too much for the poor sheep. This signal forces it to develop a keener "sense of time," but presently the stress becomes too great and the animal's behavior is totally disrupted.[51]

6   Social Aspects of Subjective Time

Apart from the pathology or deliberate disruption of the inner clock, there is yet another source of variation in temporal experience due to the value which people place upon their time. We all know the individual who cherishes and hoards every moment, but there are also people who not only enjoy wasting their own time but also take a special delight in dissipating other people's time into the bargain. Subjective time for such people is symbolic in the sense that the value placed upon it is displaced from some source of which they are unaware. Some theorists of the psychoanalytic school have suggested that hoarding time is an anal characteristic just as the dissipation of time signifies a promiscuous scattering of seed.

An individual who values his time wants to be in control of it, and this may influence his choice of a vocation. There are some who find it intolerable to be paced by a clock, to begin and stop work at a precise moment which they are in no position to choose. They prefer to be bound to the task itself. And there are differences in the extent to which temporal precision is demanded from diverse occupational groups. In general, the higher the status of an occupation, the less punctuality is expected.

We may note at this point the differences in the value and calibration of time among peoples at different levels of culture. Simple traditional societies punctuate the passage of time by regularly recurring events or tasks of social significance, such as milking, watering, or homecoming of cattle.[52] So temporal divisions may vary from tribe to tribe. The Aranda of Western Australia divide the day into twenty-five parts. The Tumerehá Indians reckon the year as ten months plus two further months during which the year is dead. The Cree Indians do not count the days when they cannot see the moon.[53] The Trobriand Islanders are said to lack temporal distinctions between verbs, and they speak neither of the past nor of history as we do. Past events, real and mythical alike, are included in the universal present or in a different kind of time, but not in a previous phase of present time.[54] In the Luapula Valley, the "sense of time" is twofold, depending on whether it is linked with a rigid personal history, or with an amorphous universal history. Time in personal history differs from that in universal history, and there are variations from one personal history to another. Furthermore, the history of any sub-clan is self-sufficient and cannot be compared with the time of any other sub-clan; all periods are defined by events which cover the periods.[55]

Industrialized societies need a finer measure than can be given by social events or by bodily rhythms such as the growth of hair or nails or the menstrual cycle. It is not enough to say: "I need a haircut, so it must be time to pay my rent." Hence the need for a calendar subdivided into equal units regardless of social or private rhythms.

But there is nothing absolute about our own conventional division of physical time, nor about the value we place upon it. The value we attach to our time reflects the socioeconomic pressures to which we are subject and our habits of zestful industriousness. Where indolence is the rule, it is hard to arouse a sense of urgency to induce workers to work at speed. "There is not an inhabitant in this island," wrote a onetime Governor of Ceylon, "that would not sit down and starve out the year under the shade of two or three coconut trees rather than increase his income and his comforts by his manual labor."[56] That was in 1802, but there are still vast areas in Africa and Asia where no value is placed, as in the West, on doing something quickly for its own sake. The mark of modern progress in technological societies is increase in speed. A device is "better" if it can act more quickly. The pride of industry and nations is founded on the making of machinery which operates at ever greater speeds. And in the Olympic games, although there is here a long tradition, we have a spectacle in which the peoples of the world compete with one another in performing some task a fraction of a second quicker than anyone else.

7   *Subjective Time in Myth and Art*

We have yet to consider subjective time in myth and art. Far removed from the simple beat of the inner clock and yet a vital form of subjective time is its representation in myth. Here time has no fixed metric, no uniform flow, but is apprehended rather as possessing inherent properties which are derived from actual events. In myth, time, like space, is demarcated into zones sacred or profane, lucky or unlucky, propitious or unpropitious. Hesiod's *Works and Days* provides a calendar of the months and days telling which are favorable and which unfavorable for different occupations. Each period is animated by a personal spirit and good and ill fortune are assimilated to the time when an event takes place. In the

Iliad, the quality of time has one value for the victor and one for the vanquished, and so acquires the character of fate or destiny. Past, present and future merge with and are sometimes identified with one another, the present moment being laden with the past and pregnant with the future. In attenuated form we perpetuate this feeling of time when, on a solemn anniversary, we recollect an event that happened in the distant past. The two-minute silence on Armistice Day is blended in our minds with the original moment in 1918 when World War I was brought to an end. But we do not, as in myth, credit a day with an event yet to come.

A species of this mythical time, which appears to be realized in ecstatic, oneiric and trance-like states, constitutes the frame of reference for an increasing number of contemporary films made by producers eager to free themselves from the shackles of clock and calendar time. The camera assists this extraordinary endeavor. Time is treated like space, and loses its irreversibility. The director can portray the past, present and future of his characters in any order he chooses, and he can vary this order as he pleases. What is more, he can compress within hours or minutes events which, in actuality, endured for years. In a word, time's arrow is abolished while transforming its normal scale.

A host of symbols of time have sprung from the imagination of poets and artists. Shakespeare alone "has implored, challenged, berated and conquered Time in more than a dozen sonnets. . . . He condenses and surpasses the speculations and emotions of many centuries":[57] "Wasteful time"; "Devouring time"; "Time's fickle glass"; "Time . . . delves the parallels in beauty's brow." We owe to antiquity two basic images: first, the Greek *kairos*, a decisive choice-point in human affairs, as represented in the figure of Opportunity, a man with wings on shoulders and heels, with scales precariously balanced on a knife's edge, together with the wheel of Fortune, and a forelock by which he could be seized. Representations of this sort made their appearance until the Middle Ages, when, in the image of Fortune, they blended the idea of equilibrium with that of transience. A second image, from Iranian sources, is *Aion*, a symbol of eternal and inexhaustible creativeness, associated with the cult of Mithra. Aion too is a winged figure, but with the head and claws of a lion wound round by a snake and holding a key in each hand. A variation on this theme is the winged Orphic god, *Phanes*, a youth surrounded by signs of the zodiac.

The antique images stood for boundless power and plenitude. Emblems of decrepitude and decay—the hourglass, the scythe or sickle, crutches—came much later, merely by error, so it has been suggested,[58] due to confusing *Chronos* with *Kronos*, oldest and most formidable of the gods and patron of agriculture. The mistaken identification was perpetuated by the neoplatonists who saw in *Kronos* a symbol of cosmic mind or *noos* and this was fused with *Chronos*, "father of all things." The sickle of *Kronos*, which could signify either agriculture or the castration of *Uranus*, is said to have been reinterpreted as a symbol of temporal corrosion; and the story that *Kronos* devoured his children came to be seen in the fourth and fifth centuries as Time consuming all things. This false syncretism is blamed for the notion of a hostile and capricious Time. Hence the portrayal of justice as a function of time, though this idea has an older birth: in the *Antiope* of Euripides, justice is "Time's Daughter," and in the fragment *Bellerophontes*, we meet the lines:

> For Time, who from no Father springs,
> applies
> His levell'd line, and shews man's foul
> misdeeds

Nevertheless, other qualities of subjective time found their way into medieval and Renaissance art. Thus Bronzino in the sixteenth century depicts Time as the Revealer, "unmasking falsehood and bringing truth to light," and his contemporary Reverdy treats Time and Chance as antagonists.

The experimenter has striven to identify the properties of the inner clock, and to trace the parallelism, such as it is, between private and public timekeeping. But the thought of time goes beyond this. It brings to mind the ideas of corrosion and decay, the knowledge of inexorable and irreversible aging and death. Hence man's efforts to arrest time, to cast off his chronological chains, and to build cities and monuments, pyramids and empires which can resist the teeth of time. Hence, too, his pursuit of a mirage of love which does not wither or fade with time, and his dream of a glory which is outside time. The more man reflects on time, the more his mortality weighs upon him, and the more he realizes that "all our yesterdays have lighted fools the way to dusty death."

A theory of time adequate to a "world picture" must encompass human experience as an integral part of nature. And in this experience is the bitter foreknowledge of one's own death, a confrontation with the certainty of dissolution. Death is a biological event which comes at the end of one's days, but the thought of death is our lifelong companion. This thought must find its place in the psychology of time which, as Proust affirmed, is as surely needed as a geometry of space.

PART III
TIME AND LIFE
Rhythm, Life, and the Earth

Introductory Note to Part III

According to Judaeo-Christian and Islamic traditions the creation of the world took place in six days and six nights. The writers of the Book of Genesis and of the Koran extrapolated their experiences of consecutive sunsets and sunrises to an epoch which, even to them, must have seemed distant. It is remarkable that they did so without any apparent doubt of the validity of the method of their description. In current cosmology the time scale of the universe, somewhat liberally called the age of the universe, is thought to be of the order of 4×10^{12} days. This figure is usually expressed in the more practical units of seconds or years; but the idea of the day is never too far distant from these units. This is not surprising, for the authors of the older as well as the newer theories were, perforce, submerged in an incessant variation of lightness and darkness and in a multitude of other periodicities, as were the lives of all plants and animals since their appearance on earth. Through "the immense journey," to use Loren Eisley's poignant phrase, as bio-substance evolved, it acquired the rhythmic nature of its environment.

The possession of a biological clock may be thought of as concomitant to a certain organic knowledge of time, but not necessarily to any time sense as we understand time sense in man. The very question whether talking about time means talking about clocks is open to discussion. For, if we believe in the absolute time of Isaac Newton, including its relativistic modification, then time is something which exists independent of the presence or absence of clocks of any sort. On the other hand, if we believe in the idea of relational time as put forth, for instance, by Leibniz and Mach, then events such as the operation of clocks can be considered to "make" time.

Be that as it may, the affinity of clocks to time is generally admitted. We should not hesitate, therefore, to begin our survey of time and life with a rapid statement of the experimental evidence for the biological clock, obtained mainly from studies of plant life.

> Here's flower for you!
> Hot lavender, mints, savory, marjoram;
> The marigold that goes to bed with the sun,
> And with him rises weeping. . . .
> —WINTER'S TALE

This is followed by a discussion of the factors which are believed to be responsible for rhythmic behavior, of the advantages such rhythmic behavior offers, and of the general role of rhythm in the ecology of the living.

> For winter is now past,
>   the rain is over and gone.
> The flowers have appeared in our land,
>   the time of pruning is come:
>   the voice of the turtle is heard in our land:
> The fig tree hath put forth her green figs:
>   the vines in flower yield their sweet smell.
> Arise, my love, my beautiful one, and come.
> —CANTICLE OF CANTICLES

Having established that biological clocks do seem to exist, we can then ask what the functional basis of such clocks might be. From short-term processes we will then turn to the longest process known in living matter, that is, to the temporal structure of evolution, and conclude Part III of *The Voices of Time* by a survey of ideas suggesting that time is a biological epiphenomenon.

J. T. F.

Experimental Evidence for the Biological Clock

KARL C. HAMNER

It has been known since earliest times that animals and plants show diurnal and seasonal rhythms, and that certain patterns of behavior could be anticipated or predicted with considerable accuracy. Only comparatively recently has it been shown that this behavior results not only from responses to physical conditions of the environment but involves an innate capacity to meter the passage of time. It was discovered in 1920, for example, that plants would flower at a particular time of year because they had some way of measuring the length of the day and, therefore, could adjust their physiological processes to the calendar day. Even before this, it was found that bees could be trained to feed at a particular place at a particular time of day. It has long been known that migrating birds had a sense of direction, and attempts to explain their ability to navigate have led to a great deal of speculation in past years. Very few of these early speculators proposed, however, what now appears to be the true explanation, namely, that birds have an internal clock enabling them to use the sun or the stars for orientation purposes.

It appears possible that the mechanism whereby the passage of time is ticked off is the same in every organism whether it be a simple unicellular plant or animal or a more complex organism such as the night-blooming cerius or man himself. While the nature and functioning of this mechanism is not understood, it is sufficiently recognizable to be given the name of "the biological clock." The present essay will deal with the discovery of the biological clock while Prof. Cloudsley-Thompson's will consider its implications in animal behavior.

The discovery of the time sense has come from three different fields of biological research: 1) the study of photoperiodism, or the responses of living organisms to the seasonal changes in the length of the day at latitudes some distance from the equator, 2) celestial orientation and time sense in insects, birds and other animals and 3) the study of endogenous rhythms or the diurnal behavior and activity of many living organisms.

1 *Photoperiodism*

In 1920 Garner and Allard[1] reported their discovery of the influence of the relative length of day and night on the flowering response of many plants. They called this response photoperiodism and classified all plants on the basis of their photoperiodic responses as 1) long-day plants, 2) short-day plants or 3) indeterminate plants. The plants of the first two groups exhibited what is called a "critical day length"; the long-day plants flowered in days longer than this initial day length while the short-day plants flowered in days shorter than this. The indeterminate plants did not exhibit a critical day length in their flowering responses. The discovery of photoperiodism by Garner and Allard was somewhat unique since, in all of the previous world's literature, it does not appear that anyone had so much as speculated on the possibility that the length of day was an environmental factor governing the seasonal behavior of plants and animals. Having discovered photoperiodism in plants they, of course, considered the possibility that migration of birds and the seasonal behavior of many animals might also be governed primarily by day length. This speculation was easily confirmed. A few of the seasonal behavior patterns attributed to day length include the autumnal coloration and subsequent abscission of leaves of deciduous trees, the seasonal behavior of insects, the formation of winter pelts in animals, the sexual behavior of many animals and many other phenomena too numerous to mention.

It is not surprising that the discovery of photoperiodism led to a great upsurge in related research. However, most of this research did not deal with the problem of how organisms measure the length of the day, but rather with the detailed methods of sensing daylight by the organisms. For example, plant research led to the discovery that leaves were the organs which perceived the photoperiodic stimulus. After being exposed to the appropriate day length, leaves transmit a message to the buds of the plant which cause them to form flowers instead of leafy shoots. Much of the subsequent plant research has dealt with attempts to understand the nature of this message, to extract the active principal from living plants and to identify it chemically. These efforts have not been completly successful to date. On the zoological side it was found that perception of the length of day could take place in animals which were blind or in which the optic nerve has been severed. A perception of day length, therefore, does not necessarily involve sight, but is believed to influence sensitivity to light of tissues which affect the endocrine system. Many animal studies have dealt with the endocrine system involved in the interpretation of the day length. Of course, many investigations on plants, animals and insects have involved simply the question of whether or not a particular response was photoperiodic.

Perhaps one of the most important steps in the development of our understanding of time measurement in photoperiodism was the discovery by Hamner and Bonner in 1938 that the effects of a long dark period could be nullified by a very brief light signal applied in the middle of that dark period.[2] A short day naturally contains a long dark period, but it was found that such a short day would be interpreted as a long day if the long dark period were interrupted for a very short time with but a few foot-candles of light. This discovery led many botanists to believe that the responses were simply to the length of the dark period. Another important discovery made at the same time was that the ordinary cockle-bur (*Xanthium pennsylvanicum*), a short-day plant, could be induced to flower by exposing it to a single short day. This plant, when maintained on long days (eighteen hours or more of light), does not form a single flower bud. But if such a plant is exposed to a single short day and returned to a long-day regime it subsequently flowers abundantly. Furthermore, if this plant is grown in continuous light it remains vegetative; but if, after it has produced several leaves, it is exposed to a single long dark period (signaling a short day) and returned to continuous light, it will form flower buds. This provided additional evidence suggesting that the length-of-day response was dependent upon the length of the dark period. Results such as these influenced the thinking of botanists for a number of years, since, in attempting to find the method whereby organisms measured the length of the day, they looked for timing processes which occurred in darkness. We might consider such a timing process as an hourglass type of clock in which a process proceeds in darkness until certain products are accumulated in sufficient amounts to induce a secondary process. Numerous theories have been developed involving possible reactions operating on the hourglass principle which could explain how organisms measured the length of the dark period with accuracy. One great difficulty, however, has been the fact that, whatever the possible mechanism, it must be such as to be only slightly affected by changes in temperature.

In recent years it has become apparent that plants and animals do not measure the length of the day on the hourglass principle, but rather that the length of the day is determined in relation to some kind of endogenous rhythm with alternate phases of approximately twelve hours' duration and that the two phases of the rhythm have differing sensitivities to light. This brings the time measurement of photoperiodism in line with other studies of time measurement. A more detailed discussion of photoperiodism will be postponed until a background has been provided by discussion of other timing phenomena.

2   Celestial Orientation and Time Sense

Work on bees illustrates beautifully the time sense of these insects and the application of such a time sense to their amazing ability of direction-finding. Man has long recognized the direction-finding ability of the bee, and, in fact, the term "beeline," meaning a straight line between two points on the earth's surface, was coined in recognition of the fact that a bee, when laden with food, flies directly from the source of the food to the hive.[3] The bees' ability to tell time has not been of such common knowledge, although over sixty years ago a Swiss doctor by the name of Forel made observations on this ability.[4] His observations have been amply confirmed and extended by recent research.

Bees can be trained not only to feed at a particular place at a particular time of day, but they can be trained to feed at two different places at two different times of day or even three different places at three different times of day. If the investigator places a circle of identical feeding trays some distance away from, but completely around a hive of bees, he may then perform the following experiment. Let us say he places food in the tray northwest of the hive at 10 o'clock each morning, in the tray east of the hive at 12 o'clock each day, and in the tray southwest of the hive at 4 o'clock each afternoon. After a few days, by making observations without placing any food on any of the trays, he may demonstrate that the bees have been trained to go in the correct direction at the right time of day. The bees come to the experimental tray at the correct time of day, and in such numbers that there can be no question that they are expecting food at the right place and time. Furthermore, if during the night the entire hive is moved to a new location with new terrain and landmarks, the bees still search in the direction and at the time of day to which they had been trained. In addition, if, while the bees are feeding at a tray, one covers the entire tray with its bees and moves the tray to a new position and then uncovers the bees, they will leave the feeding tray and try to find the hive in the direction that would have been expected from the old position. Under these circumstances the only obvious landmark the bees have for finding direction is the sun. If one covers feeding bees with a black box in the morning and releases them in the afternoon, the bees head directly toward the hive even though the sun has changed position in the meantime. Their time sense has enabled them to allow for the change in position of the sun during the intervening dark period. An ingenious experiment was performed with a hive of honeybees in which the bees were trained in New York to feed in a given direction. The entire hive with its bees was then transported by jet plane to California. On being released the bees did not head in the same geographical direction because their internal clock was still operating on New York time. It took the bees several days to adjust to the new local time.

The bees' known and fascinating ability to communicate with one another also involves employment of the biological clock. When a scout bee locates a group of nectar-laden flowers, it is of obvious advantage to the hive to know about this. Upon returning to the hive the scout goes into a "tail-wagging" dance during which she informs the other bees of the direction and distance of the flowers from the hive. Through observation ports in the hive it has been possible to study and interpret this dance. During the dance the bee indicates the location of the sun in relation to the source of nectar and, even after several days of confinement without being able to see the sun, the bee will still correctly plot the sun's position in its communication of the direction of the last known source of food. While it is clear that bees have an accurate sense of time and navigate to a rich food source using the sun as a compass, it is also true that they may use a prominent landmark to supplement their direction-finding.

3 Navigation of Birds

Man has long marveled at the ability of birds to fly great distances and arrive at a specific destination. Even young birds, in their first migratory journey, if captured and held in captivity for some time, would upon release head in the right direction even though their migratory companions had long since departed. Many hypotheses have been put forward to explain this ability of birds to know direction, but it was not until 1949 that Gustav Kramer performed his critical experiments and came forward with the explanation which is generally accepted today.[5] Kramer noticed that starlings held in a cage out of doors indicated by their behavior the direction in which they wished to migrate. He noticed that when the skies were heavily overcast, the birds failed to indicate this preferred direction. By shielding the birds from the direct rays of the sun and by using appropriate mirrors Kramer proved that the birds' sense of direction was dependent upon the apparent position of the sun in the sky. In other words, if the apparent position of the sun in the mirror was 90° from the actual position of the sun, the preferred direction of the birds was 90° from the proper direction of migration. In feeding experiments Kramer and his co-workers found that they could train the birds to feed in a given compass direction. They proved that the birds obtained their compass direction by sighting on the sun and that they compensated for the sun's movement during the day. Furthermore, they found that if they placed an artificial sun in a fixed position in relation to the cage, the birds changed their direction in feeding attempts throughout the day. This showed that in their direction-finding they were compensating for the presumed movement of the sun across the sky during the day.

When a starling has been trained to feed in a given direction in a large exercise cage, it can remember this direction even though confined for several months in a small cage. After training starlings for some time to feed in a given direction Kramer and his co-workers confined the birds in a small cage with an artificial light-dark regime which was approximately six hours out of phase with the natural day. When such birds were again placed under natural sunlight in an exercise cage, they sought food in a direction about 90° from the true direction. In other words, during the days of confinement their internal clock had been rephased and was six hours out of phase with local time. This suggested that the internal clock, by means of which the birds obtained direction from the sun, was maintained in phase with local time by the light-dark regime of the locality.

Kramer's work was followed by that of other investigators who showed that migrating birds could use the stars to obtain direction at night. By capturing migrating birds and placing them in a planetarium they found that when the appropriate star pattern was cast on the ceiling, the birds indicated their preferred direction of migration. On the other hand, if the star pattern on the ceiling was rotated, the indicated direction of migration also rotated. It was found that no particular star or small group of stars was involved in this direction-finding ability, but rather the pattern of the stars in the night sky. It was also found that the birds compensated for the movement of the stars across the sky during the course of the night. The direction-finding ability of migrating birds is a recent discovery and is today the subject of intensive research. Much more could be said about this fascinating subject, but it is sufficient here simply to point out that the ability to sense direction using either the sun or the stars depends upon an accurate awareness of the correct time of day, which in turn depends upon the biological clock.

Before leaving the subject of direction-finding one must say something about the homing instinct. It appears that the biological clock may also be involved in this particular phenomenon. For example, homing pigeons, when released, will usually head in the general direction of home. On the other hand, if, prior to experimentation, the birds are exposed to an artificial light-dark regime out of phase with local time, upon being released some distance from home they will usually head in the wrong direction. Nevertheless, in some strange fashion the birds usually reach home eventually. Furthermore, it has been reported that while young birds during their first migration, if displaced, fly in a course parallel to the migratory path, older birds, if captured during their second or third migration and displaced a considerable distance from the migratory path, will, upon release, fly directly to their destination. What is involved here is the possibility, certainly not yet proven, that these organisms possess the ability for true bi-coordinate navigation. If birds such as homing pigeons and turtles (which have also been reported to possess a homing instinct) have such an ability, they must have not only an extremely accurate biological clock, but they must also have an instinctive knowledge equivalent to that contained in the sea captain's tables and charts. The scientific proof or disproof of this ability is eagerly awaited.

That the direction-finding ability is innate is beautifully illustrated in the work with sand hoppers (*Amphipoda*) which inhabit beaches. If their beach becomes too dry, these creatures travel toward the ocean to find a moister habitat. Populations which inhabit a beach where the ocean is to the west would naturally be expected to travel west to reach the water. On the other hand, populations on a beach facing east might be expected, upon desiccation, to travel east to reach the water, and so they do. If these animals are taken to the laboratory, and eggs from each of the populations are hatched and raised under artificial conditions, the progeny from the west-facing beach will, upon desiccation, travel west provided, of course, they can see the sun. Similarly, progeny from the east-facing beach will travel east upon desiccation. These creatures, therefore, not only have a biological clock which enables them to use the sun for orientation purposes, but they also inherit this ability together with a knowledge of the direction in which they wish to travel. The pond skater (*Velia currens*) also exhibits an innate direction-finding ability although, in this particular case, the ability appears to be of no value. Pond skaters collected from various places will, when placed on dry land and exposed to the sun, head directly south, apparently having inherited the misinformation that water should always be south of dry land.

4   Circadian Rhythms

The discoveries of photoperiodism, of the time sense and of celestial orientation have shown that many organisms have an ability to measure with considerable accuracy the passage of time. However, the discovery of another phenomenon, circadian rhythms, has provided most of our information concerning the possible nature of the biological clock which is probably in operation in all these phenomena. This phenomenon was apparently first noted in 1729 by DeMarian, who made his observations on plants. Many plants show diurnal leaf movement, extending their leaves in the daytime and folding them at night. DeMarian noticed that this diurnal leaf movement continued for several days even though the plants were placed in constant darkness. Other scientists, including Pfeffer in 1873 and Darwin in 1880, confirmed DeMarian's results. However, most credit must go to Bünning, who has been studying this phenomenon for the past thirty or more years.[6] He confirmed DeMarian's results and made many additional observations. He found that under constant darkness, the oscillations of the leaves had a period of approximately, but not exactly, twenty-four hours. Since most of the other rhythms discussed below also have a periodicity that is not exactly of twenty-four hours, these rhythms are called "circadian" rhythms (*circa* meaning about and *diem* daily). Bünning found that the period of the oscillations of the leaves could be changed from twenty-four hours by altering the light-dark regime to cycles other than twenty-four hours. For instance, if he gave ten hours of light and ten hours of darkness in each cycle, the leaf oscillations changed to a twenty-hour period. Similarly a thirteen-hour light and a thirteen-hour darkness cycle changed the period of the oscillation to twenty-six hours. This adaptation of the rhythm to suit the light-dark regime is called *entrainment*, and the phenomenon is considered of importance in constructing theories to explain the mechanism of the clock. Bünning found, however, that when any artificial light-dark regime was terminated and the plants placed in complete darkness, they immediately reverted to a circadian, or approximately twenty-four-hour rhythm. He also found that when he planted seeds in complete darkness, the seedlings exhibited no rhythm until they had been exposed to a single period of light. This single exposure, followed by continuous darkness, was sufficient to induce the circadian rhythm. Bünning also found that plants in continuous darkness have the rhythm of the leaf movements rephased by a single exposure to light. Furthermore, and this was quite surprising, he found that temperature had very little effect upon the period of the rhythm, the frequencies of the oscillations being once every twenty-four hours either at low temperatures or at high temperatures (excluding very high or low extremes). The many experiments of Bünning with plants and other organisms are too numerous to discuss here. The important fact to know is that the plant apparently measures the passage of time even when placed in complete darkness, as is shown by the slow rhythmic oscillations of its leaves.

We now know that circadian rhythms are exhibited by almost all organisms. The activity of many animals is diurnal, some animals being active in the daytime and others at night. Nocturnal animals particularly lend themselves to study of circadian rhythms since they can easily be placed in constant darkness and the periodicity of their activity measured on exercise wheels or by other means. The deer mouse, the hamster, and the flying squirrel have been studied in this connection. If these animals are placed in a small cage with an exercise wheel they usually run on the exercise wheel when they first "get up," usually about sunset. If a recording device is fitted to the exercise wheel and the animals are placed in a closed room in complete darkness, they will still awaken each night and go to sleep in the early morning at about the same time. The periodicity of their activity closely approximates twenty-four hours and may persist under constant conditions for many days or weeks. Certain select animals of this nocturnal group show such regularity in their periodicity that one can predict to within about a minute when the animal will awaken and begin to exercise. Some individual animals have periods in their rhythm of more than twenty-four hours and some periods of less than twenty-four hours, while others may be almost exactly twenty-four hours in duration. Another nocturnal creature which has been studied intensively is the cockroach. Studies with this insect have shown that its rhythmic activity (persisting in complete darkness) is related to the rhythmic secretion of endocrines by nerve cells close to the brain.

It has been found that there are many circadian rhythms exhibited by all kinds of organisms, both plant and animal. The fruit fly (*Drosophila*) has long been used in genetic studies. When the organism is cultured on nutrient media for experimental purposes, the adult tends to emerge about 9.00 A.M. If the cultures are placed in complete darkness the flies still emerge at about the same time each day. Another circadian rhythm is found in the luminescence of a marine dinoflagellate (*Gonyaulax polyedra*). This is a microscopic organism which gives off light in the ocean at night and creates the phosphorescence that occurs whenever the sea water is disturbed. If a whole series of test tubes containing this organism are placed in complete darkness and one of the tubes shaken each hour, it is found that the organism luminesces or glows only when shaken at nighttime and not when shaken during the day. Certain fungi show characteristic rhythmic growth and spore formation. If a little piece of the fungus is placed in the middle of a large dish containing nutrient media and the dish kept in complete darkness for several days or weeks, a characteristic circular banding appears in the growth of the fungus, one band for each day's growth. Circadian rhythms have been found in such diverse phenomena as the loss of water by lemon cuttings, phototaxis, such as the response to light of Euglena, the mating activities of Paramecium, and the size of the pigment cells in fiddler crabs (*Uca* spp.).

We are naturally interested as to whether or not such rhythms are exhibited in man. One immediately thinks of the sleep rhythm, but it is difficult to prove whether or not this rhythm is truly endogenous. However, there are circadian rhythms in the blood eosinophil count, serum iron content, body temperature, heart rate, blood pressure, urine production, and excretion of phosphate and potassium. It has been shown that these are circadian rhythms by taking a small group of human subjects to the high latitude of Spitsbergen where the sun shines continuously during the summer and there is very little diurnal variation in light or temperature. The subjects were placed in comfortable bases in uninhabited country and they carried out routine activity. Each individual collected urine samples at regular intervals and the samples were measured as to volume and analyzed for sodium, potassium and chloride content. In one experiment the clocks of all of the individuals and all of the clocks at the bases were adjusted so as to complete a twenty-four-hour reading in twenty-one hours. On another occasion the clocks were adjusted to a twenty-seven-hour cycle. The people ate and slept and carried on their activities according to the clocks. Nevertheless, the secretion of urine and the secretion of the salts measured remained on a twenty-four-hour rhythm. This was true for most of the subjects, although there was some variation and some of the subjects showed some entrainment, adapting themselves to the new cycle range. In any event, it is clear that circadian rhythms may be exhibited by humans as well as by nearly all the other organisms.

5   *Photoperiodism and Circadian Rhythms*

It is clear from the above discussion that the ability of many organisms to use the position of the sun or the stars for orientation purposes depends upon a knowledge of the local time. The photoperiodic responses of many organisms demonstrate ability to measure the length of the day. Furthermore, it has been shown that some organisms may be trained to feed at a particular place at a particular time of day. In addition, most organisms which have been carefully studied exhibit some kind of circadian rhythm and it seems probable that circadian rhythms could be demonstrated in almost any living organism if the proper experiment were devised. It seems possible, however, that bacteria and, probably, viruses, may not exhibit circadian rhythms or possess a time sense comparable to that exhibited by the higher organisms. The question arises as to whether or not the same mechanism is used by all these organisms to meter the passage of time or if there are different mechanisms for accomplishing the same purpose. Most biologists are inclined to think that circadian rhythms are a direct manifestation of the biological clock. Circadian rhythms are clearly manifestations of rhythmic endogenous changes. The phase of this endogenous rhythm at any particular instance might serve to denote the time of day, while the uniform pulsation of the rhythm might serve to meter the passage of time. Such a clock must be temperature-compensated since the circadian rhythms are known to be affected only slightly by ambient temperature variations. Furthermore, such a clock, even though it runs slightly fast or slow, must be one which is synchronized with the light-dark regime of the locality. However, the presence of an endogenous rhythm which could serve as a clock does not necessarily mean that it *is* serving as a clock. The shadow of a post on the ground does not make a sundial until the appropriate clock positions are marked out on the ground so that the arrangement is used to indicate time of day.

Perhaps the most direct evidence indicating that the time sense is actually related to endogenous rhythms has been obtained from studies of photoperiodism. The nature of the evidence obtained can be illustrated conveniently by the work on Biloxi soybean.[7] This plant is a short-day plant and fails to flower if grown on long days (eighteen–twenty hours of light each day). On the other hand, if it has been growing on a long-day cycle and is transferred to a short-day one, it promptly develops flower buds. Furthermore, if it is given a few short days and then returned to long days, flower buds are developed and the number formed is in direct proportion to the number of short days received. Thus the number of flowers produced provides a quantitative assay of the effectiveness of an experimental treatment. The most effective short day is approximately eight hours of light and sixteen hours of darkness in each twenty-four-hour period. If a group of plants which has been growing on a long-day regime is divided into small lots, each of which is to receive an experimental treatment, then the following experiment may be performed (see diagram). Each lot of plants receives eight hours of light in each cycle of photoperiodic treatment, and the length of the cycle is varied by varying the length of the dark period associated with the eight-hour photoperiod. When this is done, the plants which receive twenty-four-, forty-eight- and seventy-two-hour cycles flower abundantly (it should be noted that the plants receiving the forty-eight-hour cycles receive 40 hours of darkness in each cycle and those receiving seventy-two-hour cycles receive sixty-four hours of darkness in each cycle). On the other hand, plants which receive sixteen-, thirty-six- and sixty-hour cycles either fail to flower or develop only a few floral buds. If a curve is drawn relating the amount of flowering to cycle length, an undulating curve showing a rhythmic response with periods of twenty-four hours is obtained. The plant exhibits an endogenous rhythm of sensitivity to cycle length with maximum flowering at cycle lengths of twenty-four hours or multiples thereof and minimum flowering at cycle lengths out of phase with these periods. To put it another way, with eight hours of light in each cycle, the plant interprets the treatment as being equivalent to short day if the cycle length is twenty-four hours or multiples thereof, and it interprets the cycle as long day if the cycle lengths lie between these values. It is clear, therefore, that the plant does not measure day length by measuring the absolute length either of the light period or the dark period, but it determines the length of the day according to the time at which it receives light in relation to its endogenous circadian rhythm.[8]

Another type of experiment also indicates that the photoperiodic response is dependent upon a rhythmic sensitivity to light. As has been mentioned, soybean plants exposed to seventy-two-hour cycles, each of which contains an eight-hour light period, flower abundantly. On the other hand, if the sixty-hour dark period of each of these cycles is interrupted at various points by a brief exposure to light, one finds a rhythmic response depending upon the time in the cycle of the interruption. There are three twelve-hour periods during the cycle when these light interruptions strongly inhibit flowering, namely from the twelfth to the twenty-fourth hour, from the thirty-sixth to the forty-eighth hour, and from the sixtieth to the seventy-second hour. During the alternate twelve-hour periods, *i.e.*, between the twenty-fourth and the thirty-sixth and the forty-eighth and the sixtieth hours, such light interruptions may stimulate flowering. We have, therefore, a rhythmic sensitivity to light with alternate twelve-hour periods of stimulation and inhibition by light.

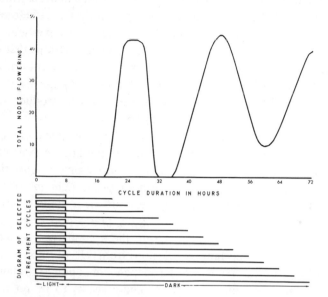

Summary response curve for Biloxi soybean of six representative experiments done in this laboratory. Plants were exposed to seven cycles, each cycle consisting of 8 hours of high-intensity light (1000-1500 ft-c) and associated dark periods of various lengths. One cycle for a few selected treatments are diagrammed below the graph for illustration. Total nodes flowering per 10 plants is plotted against cycle length. The standard error for high and low points of the curve was calculated. Standard error for flowering response at cycle durations of 24, 48, 60 and 72 hours was 0.15, 0.17, 0.45 and 0.25, respectively.

To further illustrate the involvement of an endogenous circadian rhythm in photoperiodic response, one may cite the following results. If soybean plants are exposed to seven consecutive short days (eight hours of light and sixteen hours of darkness) and returned to long day they will subsequently produce about 4.5 flowering nodes per plant. On the other hand, if seven short days are alternated with seven long days no flowers are produced. If we intervene between each of the seven short days a day of twelve hours of light and twelve hours of darkness there is no effect upon the flowering. However, if the intervening days are shorter than twelve hours, flowering is greatly stimulated, and if the days are longer than twelve hours flowering is inhibited. Complete inhibition of flowering occurs when the intervening day is longer than fourteen hours. Since the critical day length of Biloxi soybean is about fourteen hours, plants which have been growing on long day and are transferred to consecutive cycles of a twelve-hour day promptly flower. However, in the above discussions it was pointed out that a twelve-hour day is innocuous when used as an intervening day. Biloxi soybean, therefore, is a plant which fails to flower in long days because long days are actively inhibitory, and its critical day length is determined by the length of the light period beyond twelve hours, which is necessary to produce sufficient inhibition to prevent its natural tendency to flower.

The photoperiodic response of Biloxi soybean is dependent upon an endogenous circadian rhythm with alternate twelve-hour phases of differing sensitivity to light. During the first twelve-hour phase light stimulates flowering and during the second twelve-hour phase light inhibits flowering. Other short-day plants appear to respond in a similar fashion. In contrast to the short-day plants, the long-day plants appear to be stimulated to flower if they receive light during the second twelve-hour phase of each cycle. The important conclusion from all of this is that the photoperiodic response seems to be dependent upon an endogenous circadian rhythm of sensitivity to light. There seems to be no reason to believe that this endogenous rhythm is any different from the endogenous rhythms associated with the circadian behavior of the many organisms that have been studied in other connections. In the particular case of photoperiodism the organism is using the endogenous rhythm as a means of measuring the length of the day.

Clear-cut photoperiodic responses which can be arranged quantitatively are not as numerous as might be desired and experimentation clearly is difficult. Furthermore, only a few organisms are known to exhibit direction-finding ability or to perform a particular act at the same time every day. Consequently, most investigations on the possible mechanism of the biological clock have dealt with circadian rhythms.[9] These rhythms have almost always been demonstrated where organisms have been carefully studied to determine their presence. Consideration of the nature of the results of such experiments and the speculations concerning the possible mechanisms involved are presented in the next essay.

In anticipation of that discussion, it is worthwhile to consider some of the problems the investigator is faced with in attempting to determine the mechanism of the clock. In the first place the question has arisen as to whether or not the clock is truly endogenous. While circadian rhythms persist for many days or weeks in the laboratory under as nearly constant conditions as the experimenter is able to provide, some scientists claim that the organism may be receiving some kind of stimuli from the external environment which vary diurnally with the rotation of the earth. As will be seen, such evidence as is available seems to be contrary to this viewpoint.[10] If, as most biologists do, we accept the theory that the clock is truly endogenous, we are still faced with additional problems. Since unicellular organisms exhibit circadian rhythms and presumably possess a clock, does every cell of the larger multicellular organism also possess an individual clock? And if this is so, how are these clocks synchronized? A still more difficult question is how such a clock can be temperature compensated. If the running of the clock depends upon metabolic energy, as it must if it is truly endogenous, how is this energy supplied and why is it that treatments which affect the rate of metabolism do not affect the functioning of the clock? Is the oscillator involved of the relaxation or of the pendulum type or are there, perhaps, several oscillators involved which are coupled together in some fashion? These and many other problems must be dealt with in considering the nature of the biological clock.

Time Sense of Animals

J. L. CLOUDSLEY-THOMPSON

Rhythm is characteristic of all natural phenomena. Summer follows winter, new moon follows old, day follows night. "So do flux and reflux—the rhythm of change," wrote Thomas Hardy, "alternate and persist in everything under the sky." Recent research makes it increasingly evident that living organisms are no exception to this, for rhythms are apparent at all levels of biological organization. They can be seen in the lashing cilia of unicellular organisms, in the division of cells and nuclei, in locomotion, feeding and excretion. Groups of cells, tissues and organs are no less rhythmic, their periodicities being harmoniously synchronized within the living organism.

Some of the physiological rhythms of animals—for example, the spontaneous discharge of nerve cells and the beating of the heart—have a comparatively high frequency: others, such as menstrual and reproductive cycles, may have a periodicity extending over weeks or months. However, frequency does not provide a very good criterion for distinguishing between various types of rhythmic phenomena because different rhythms tend to influence one another.

In general, high-frequency cycles depend more upon the morphological and physiological characteristics of the individual, whereas long-term rhythms tend to be correlated with diurnal, lunar or seasonal cycles of the environment. But even here, the distinction is not absolute. Heart rate is clearly dependent upon the size of an animal, the surface–volume ratio, circulatory efficiency and so on—features which bear no particular relation to astronomical time. Nevertheless, it is well known that a twenty-four-hour periodicity is superimposed upon this and the rate of heart beat tends to decrease during sleep.

The physiological rhythm is never completely independent of environmental changes, although the relationship may not be immediately apparent. For example, the initiation of seasonal states of dormancy and the regulation of breeding cycles are related to the difference in the length of daylight as between summer and winter. This difference is also connected with leaf fall and the fruiting and flowering of plants, thus indirectly affecting herbivorous animals and, through them, the carnivores.

Indeed, if one considers the timing mechanisms of rhythmic and cyclical phenomena at all levels of biological organization, from rhythms within the cell to cyclical fluctuations in the numbers of animal populations, the only conclusion that can be drawn with any certainty is that, although periodicities are apparent throughout living matter, there does not appear to be any common ground between them. It is dangerous to lump together a number of unrelated phenomena under ill-defined names such as rhythmic or cellular activity—a practice as seductive as it is misleading.

Rhythmic phenomena are often subdivided into two main categories: "exogenous" rhythms, which are a direct response to physical changes of the environment and do not persist when conditions are kept constant; and "endogenous" rhythms, which continue, at any rate for a time, even when the environment is unchanging.

Endogenous rhythms are usually related to environmental changes although they are not necessarily a direct response to them. Thus, if cockroaches are subjected to twelve-hour periods of light and darkness, they are active mostly during the dark period, although they begin to stir from their rest and move about shortly *before* the light is extinguished. If the conditions of the experiment are changed and the cockroaches are subjected to continuous light or darkness, the twenty-four-hour rhythm of activity persists for some days, but eventually the rhythm is lost and the insects' activity is spread more evenly over the whole day.

If the periods of light and darkness are now doubled so that the "day" is forty-eight hours long, the cockroaches show the same twenty-four-hour rhythm of activity and rest as they did with the twenty-four-hour day. However, in this case the insects show an outburst of activity when the light is switched on as well as when it is extinguished. Observations of this kind have led to the idea of an innate endogenous rhythm which is synchronized by changes in environmental factors such as light, temperature and humidity. The rhythm is not a direct consequence of these changes: they act merely as "synchronizers" or "clues"—terms which correspond with the German *Zeitgeber* proposed by Professor J. Aschoff —in keeping the rhythm in step with the environment.[1]

It seems likely that, under natural conditions, several different synchronizing clues are at work at the same time, of which one is generally dominant in controlling the animal's periodicity. There may be competition between different clues in regulating an animal's rhythm, and changes in the animal's sensitivity and physiological state may bring about changes in the relative importance of different clues.

1   *Theories of Rhythm Causation*

Cosmic rhythms within which all living matter on Earth must exist are numerous. The earth's rotation relative to the sun, for instance, is responsible for the twenty-four-hour day. This rotation, relative to the moon, together with the moon's rotation about the earth, gives us the lunar day of 24 hours and 50 minutes from moonrise to moonrise. And the moon's arrival every 29.5 days at the same relative position between the earth and the sun marks what is called the synodical month. The earth, with its tilted axis, revolving around the sun every 365 days, 5 hours and 48 minutes, yields the year and its seasons. The daily and annual rhythms related to the sun are associated primarily with changes in light and temperature, while the lunar day and synodical month cause the tides of the ocean and changes in illumination at night.

Of the many types of biological rhythm known to exist, that which is related to the cycle of day and night is not only the most influential, but has been the most intensely studied. For this reason, and because the longer periodicities often operate through the influence of changing day length, it is with the rhythm of day and night that we shall be chiefly concerned in this essay. These diurnal rhythms are usually called "circadian" rhythms (as explained in the preceding essay) because their periods are not exactly twenty-four hours in duration.

When living organisms are removed from their normal changing environment and placed under laboratory conditions of constant light, temperature and humidity, their normal circadian rhythms of activity and rest, as well as various physiological rhythms, frequently persist for a considerable while. For this reason it is customary to regard endogenous biological rhythm as the manifestation of internal "chronometers" or "clocks." But the mode of operation of such biological clocks is not understood.

Man-made clocks are of two general types—those with intrinsic timing such as the hourglass, the pendulum and hairspring balance clocks, and extrinsic clocks like the sundial and electric clocks which depend upon an inflow of timing information. According to one theory, of which Professor Frank A. Brown, Jr. is the leading exponent, biological clocks are of the first kind, depending upon an inflow of exogenous or extrinsic factors.[2] The other theories, to which most biologists now subscribe, postulate the existence of intrinsic or endogenous clocks.

Exogenous factors. In addition to the obvious changes in light, temperature and humidity associated with the daily, annual, lunar day and synodical monthly rhythms mentioned above, there are changes in other forces, such as gravity, barometric pressure, high-energy radiation and magnetic and electric fields. According to Professor Brown, even when organisms are hermetically sealed under so-called constant conditions, they still derive rhythmic information from their environment.

Such information must be transmitted in an extremely subtle manner and depend upon some universal-time geophysical rhythm with simultaneous world-wide changes, such as changes in magnetic or electrostatic sensitivities.

Brown and his coworkers claim that organisms as little related as seaweed, potatoes, carrots and fiddler crabs (*Uca* spp.) continue to display both solar-day and lunar-day rhythms in their rate of oxygen consumption under constant conditions in the laboratory. In another experiment, oysters from New Haven, Connecticut, were shipped to Evanston, Illinois, and kept in a dark room. At first they continued to open their shells widest when it was high tide in New Haven waters; but after two weeks they were opening their shells widest at the moon's zenith and nadir positions with reference to Evanston. Thereafter they maintained this new schedule throughout a month of observation.

Brown's results, however, have been criticized for a number of reasons. In particular, it has been suggested that his statistics are invalid because a periodicity can be introduced into random data by the introduction of moving averages. On the other hand, it is not possible summarily to dismiss the immense number of significant correlations that Brown and his colleagues have amassed during the last years. For, even though no causal relationship has been established experimentally with any known periodic variable of the environment, if there is a statistical correlation between time series over a sufficiently long period it may well be assumed that there is some connection between them.

It has so far proved impossible to devise an experiment that will differentiate between an innate clock mechanism and one derived from extrinsic sources, since it is possible, so to say, to alter the hands of the clock relative to the position of the mechanism. Brown's theory is unassailable in terms of logic, and the critical experiment to prove or disprove it can probably be conducted only in outer space. However,

Professor K. C. Hamner[3] and his coworkers have recently shown that cockroaches, fruit flies, hamsters and various plants maintain their circadian rhythms even at the South Pole. Therefore, although it is still possible that some external periodic stimulus may regulate the biological clock, it cannot arise from any factor associated with the earth's rotation.

The answer to the problem may well be of vital significance to the possibility of future space travel. For, if biological clocks do depend for their accuracy upon the receipt of geophysical information, the removal of this might have dire effects upon the space traveler. Already it is known that stress symptoms result from the dissociation of rhythms from their ordinary relationship with one another, while epilepsy and various mental illnesses are clearly correlated with abnormal rhythms of the brain. Again, tumors have been induced in cockroaches by implanting the hormone-producing subesophageal ganglion from normal animals into insects whose natural activity rhythm has been artificially altered by reversing the natural light-dark cycle. It is doubtful that life could continue if all the clocks of the body were disorganized or stopped.

Imprinting. The imprint theory, which is associated with the name of Dr. W. H. Thorpe,[4] assumes that as animals and, presumably, plants develop from the fertilized egg they are initially arhythmic. But they then learn a twenty-four-hour rhythm from environmental conditioning or from the behavior of their parents by "imprinting," a type of learning which takes place very quickly and is almost irreversible. This conditioning is reinforced and made accurate by the environment as the animal develops, so that, when tested in a constant environment, the rhythm persists.

Although this theory has not yet been entirely disproved, the evidence in favor of inheritance is fast accumulating. Fruit flies have been bred for many generations in the dark and at constant temperature, and the rhythm which caused the insects always to emerge from the pupa or chrysalis at the same time of day persisted. Young rats raised in total darkness show circadian rhythms of rest and activity although their environment shows no diurnal changes. These observations were made by providing "running wheels" in which the rats exercised themselves when not in their sleeping quarters. The movement of these wheels was recorded and gave a measure of the distribution in time of the rats' activity.

It seemed just possible that the young rats had not inherited their rhythm, but had learned it through the influence of their mother, who had her own rhythm of activity which was reflected in the times at which she fed her young. Dr. G. E. Folk, Jr.,[5] was able to eliminate this possibility, however, by exchanging a "foster" mother with the real mother randomly at different times during the day and night.

Inherited Time Sense. The majority of biologists working in the field of biological rhythms now assume the presence of an inherited twenty-four-hour clock, and there is considerable evidence in favor of this hypothesis. For example, Professor J. Aschoff hatched and raised chickens in constant dim light and without any known periodic environmental factor. The chicks developed a rhythm having a frequency slightly less than twenty-four hours. In another experiment, different strains of mice were maintained under constant environmental conditions, some in light, others in darkness. Within each litter, the periodicities of individual mice were found to be very similar, but the frequencies of the rhythms of animals from different litters varied. These results were interpreted as proving the genetic origin of the natural period on the grounds that, if the rhythms were learned, there would have been an accumulative error which would have caused the third generation to have a frequency of other than 23.5 hours.

Much more evidence has been obtained from invertebrates, many of which show circadian rhythms which are apparently determined genetically. The clocks of all members of a population reared under constant conditions are not necessarily synchronized, especially if several generations are raised under these conditions. With fruit flies, however, Dr. H. Kalmus,[6] Dr. C. S. Pittendrigh and others have shown that the larvae can be set in the same phase by a flash of light as strong as a daybreak signal, so that subsequently adult insects emerge from their pupae at the same time of day. Moreover, mutant strains may show abnormal rhythms that are passed on to subsequent generations.

The behavior of insects and other small creatures in the dark is rather more difficult to study than is that of rats and mice. Their periods of activity and rest can, however, be investigated by means of "aktograph" apparatus. This consists essentially of a box pivoted like a seesaw in such a way that the smallest movement of the box causes it to tip, the movements of the box being recorded automatically.

When a field cricket (*Gryllus campestris*), for example, is placed in this type of apparatus, it can be shown to be active during the daytime, its rhythms being endogenous and independent of temperature. After weeks under constant conditions, the cricket's rhythm gradually dies away. However, it can be re-established by a single exposure to light or by a return to high temperature after a period at 5°C., suggesting that the twenty-four-hour period is inherent. But why does it tend to die away after long periods under constant conditions? I have suggested the following possible explanation: suppose that the animal's activity is controlled, not by one clock, but by a number of inherent cellular clocks kept in synchronization by an environmental factor like changes in light intensity. Deprived of this synchronization, the clocks will gradually go wrong and get out of step, just like a collection of wristwatches that are never reset by a time signal. When the cellular clocks get out of step, the animal loses its rhythm, but a single exposure to light or to the onset of rising temperature is sufficient to synchronize them. In fact, I think it is possible that exogenous rhythms, which do not persist under constant conditions, merely represent endogenous rhythms that rapidly get out of phase with the environment, and that there is no fundamental distinction between the two.

2   *Synchronization of Inherited "Clocks" by Environmental "Clues"*

Light is the chief environmental factor by which animal "clocks" are synchronized, but regular temperature changes can also be effective. Although, in general, the phases of the circadian rhythm are independent of temperature, temperature-determined rhythms have been shown to persist in cockroaches, spider beetles, rats, salamanders, and in the pupal emergence of fruit flies and flour moths. These temperature-synchronized rhythms are effective either in the dark or in constant light. Like the light-synchronized ones, they persist after the clue has been removed. Other factors which may serve to synchronize animal clocks under experimental conditions include barometric pressure and feeding rhythms, but their significance is usually slight compared with that of light. This is not really surprising, since daily changes in light intensity are more reliable synchronizing clues than temperature, humidity or barometric pressure, even though other factors may have greater ecological importance.

As I have already indicated, the biological clock is not a perfect timekeeper, unchanging in its regularity. Like mechanized clocks, animal clocks tend to go fast or slow if deprived of their synchronizing clues. This was first pointed out by Dr. M. S. Johnson,[7] who found that the nocturnal deer mouse (*Peromyscus leucopus*), has a remarkably persistent daily activity rhythm which goes on for up to eighteen months without synchronization from daylight. The rhythm persists in the dark and in continuous steady light, but in the latter case it is dependent on the brightness of the light. Johnson found that the constant light caused the mouse's internal clock to run slow, so that the daily active period became progressively later on successive days and could be shifted steadily round the clock with no tendency to be fixed at any particular time of the solar day or night. The amount by which the clock ran slow depended on the light intensity, but was constant for any given intensity.

Similar results have since then been obtained with white rats, which have a clearly defined rhythm of running activity. In continuous light the period of activity remains about twelve hours in length, but shows a regular, constant and definite change in the hour of starting from day to day, so that, over a number of days, the period of activity travels round the clock. The amount the clock loses again depends on the light intensity. In continuous light of intensity 5.1 foot-candles, the delay is as much as three to four hours a day.

So far, no method has been found of speeding up the biological clocks of these animals: the findings are consistent with the interpretation that light has an inhibiting action on their activity. The converse is true of other animals, such as stick insects, fireflies, fish and salamanders. I have found, for example, that the nocturnal tropical millipede (*Ophistreptus,* sp.) has a periodicity of 24.8 hours in constant light and only 23.0 hours in the dark. But other animals—day-active lizards, birds, bees and wolf spiders—show just the opposite effect: the period is shorter in constant light than in the dark.

In general, it seems that the activity rhythms of nocturnal animals are delayed by constant light, while those of day-active animals tend to be accelerated. The converse occurs in constant darkness. This shift in the daily rhythm allows the animals to keep pace with the seasons as the days lengthen and draw in. The duration of daylight in temperate regions varies greatly according to the time of year. In southern England, for example, it increases from about eight hours in winter to sixteen in summer. Obviously, if the animal is to maintain its regular daily activity it cannot synchronize to both dawn and dusk, since the period between them is variable. It now seems that most nocturnal animals tend to use dusk as synchronizer, while the day-active animals use dawn.

In the spring, as the days get longer, the intensity of light increases and the onset of the period of activity is delayed to compensate for the day length, so that the animal wakes progressively later. Similarly, the day-active animal has the frequency of its rhythm shortened, so that it gets up earlier as the spring days get longer. The converse is true in the autumn, as the days shorten and the nights begin progressively earlier.

It is difficult at first sight to understand how it can be possible to reconcile the acceleration or slowing of the frequency of rhythms under constant conditions with the exogenous factor theory of rhythm causation. According to Professor F. A. Brown, Jr., however, light and temperature have most impact during the sensitive period of the twenty-four-hour periodicity. When placed in constant conditions the rhythmic organism keeps resetting its sensitive period in a futile search for the night and day periods, a process known as "autophasing."

3   Physiological Mechanism

One of the most important characteristics of biological rhythms is that their frequency is, within wide limits, unaffected by temperature. This is to be expected because, if our clocks, for example, were to accelerate in warm weather and go slow in cold, they would be rather useless as timing devices. Consequently, temperature independence should not cause surprise when considered from a functional viewpoint. On the other hand, the mechanism is more difficult to explain in physiological terms. It is clear that the timing of biological rhythms cannot depend upon any simple metabolic process, or else their rate would be affected by temperature, for no single metabolic process has been found without this characteristic. At the same time, as we have seen, biological clocks are not insensitive to heat since temperature changes can act as synchronizing clues. And, although temperature does not alter the periodicity of the activity rhythm, it often influences the amount of activity that takes place. At high temperatures an animal may show greater maximum activity than at lower ones, although the rest period between bursts of activity is changed. Moreover, in all organisms so far investigated there appears to be a critical low temperature at which rhythms cease. This temperature may vary for different kinds of rhythm within the same organism. This is one of the main problems posed by biological rhythms: How does a physiological mechanism operate that responds to temperature and yet, in its most important functions, is temperature independent? This applies equally to "homoeothermic" or warm-blooded animals, and to "poikilo-therms" whose body temperature varies with that of the environment. Even mammals may display quite a wide range of body temperature. Furthermore, Dr. G. E. Folk, Jr., has shown that there is a circadian rhythm of shallow and deep hibernation in ground squirrels which tend always to awaken in the daytime, even under constant laboratory conditions, although the body temperature of the hibernating animals drops considerably.

It is possible to advance a number of hypotheses to account for the temperature independence of biological rhythms. It could be achieved as a result of the buffering action of numerous metabolic periodicities at each level of organization which are synchronized into an organic whole. There is, as yet, no direct evidence of enzymes whose activity increases at lower temperatures, but some oxidative enzymes of fish are known to show a decrease in activity after acclimatization to raised temperature. Presumably these enzymes also show an increase in activity when the temperature is lowered. Temperature independence could also be achieved by the coincidence of cold-stable enzymes with cold-labile inhibitors, which release more of the enzymes as the temperature is lowered.

Equally effective would be a chemical mechanism, part of which supplied a particular substance, while a second process destroyed it. If both processes were equally temperature-dependent, then the substance would accumulate at the same rate whatever the temperature. Alternately, if the rhythm-control process was one involving a change in entropy rather than a heat change, there would be but a slight temperature coefficient and the clock mechanism would tend to be independent of temperature. In the denaturization of most proteins, for example, there is a large change in entropy accompanied by correspondingly large heat changes of opposite sign, but if there were some protein in which such heat change were small, it could, perhaps, provide a mechanism for temperature independence.

Dr. C. S. Pittendrigh[8] has described an "endogenous self-sustaining oscillation" model, consisting of a complex system with constituent oscillatory processes. The control system is not a single temperature-independent process. However, the mutual synchronization of constituent oscillators would result in temperature independence over a limited range. Professor E. Bünning[9] also regards biological rhythms as relaxation oscillations because in the rhythms of both animals and plants there is a phase of several hours which cannot be delayed very much by chilling, while low-temperature treatment of the other phase causes a delay. The frequency of tension and relaxation are independent of temperature, but when cooled the oscillation drops to a lower energy level than before. Transients are a characteristic consequence of the disturbance of oscillation, and the aberrant effects observed during the initial adaptation of biological rhythms to temperature changes can be interpreted in this light. Of course, if biological clocks are shown to depend upon extrinsic factors (other than temperature variations), temperature independence no longer becomes a problem.

A considerable amount of research has been carried out in attempts to locate rhythmic control centers in animals. In the case of mammals, the most important results are those where rhythms continue after surgical operations, since the absence of rhythm in an operated animal is difficult to interpret. Endogenous rhythms have been shown to persist in rats after removal of many of the major organs of the body, including the stomach, adrenals, pituitary and thyroid glands, cortex, frontal tissue and corpus striatum of the brain. But if every cell is, or contains a clock, as I have suggested, these results are understandable.

Among invertebrates, clocks have been located in at least two instances. I have termed them "master chronometers" because, as I say, I believe most, if not all, cells either to be or to contain biological clocks for whose synchronization a control clock is seldom required. Consequently, I do not regard these master chronometers as being the only clocks in the animals that possess them, but merely as being rhythm control centers which regulate other cellular clocks. The first master clock was found by Professor G. P. Wells[10] to lie in the esophagus of the lug worm (*Arenicola marina*), or even in a slice of this structure in vitro. It controls the very regular activity of the animal which consists of three-minute bursts of feeding movements, whether or not food is present, and one-minute rests followed by locomotory movements every forty minutes. The rhythm of the esophagus can be likened to the activity of the mammalian heart. Unlike the heart, however, the rhythm from the esophagus is transferred through the nervous system to the entire animal, affecting every aspect of its physiology and behavior.

Another master chronometer has been located in a nerve center known as the subesophageal ganglion of the cockroach (*Periplaneta americana*). By an elegant experiment in which two cockroaches were joined together like Siamese twins, Dr. Janet Harker[11] has shown that a hormone is involved in the circadian rhythm. A cockroach which had previously been subjected to normal conditions of light and darkness was immobilized by the removal of its legs. It was then fixed, by means of a glass capillary tube, to the back of a mobile insect which had been kept in constant light throughout its life, so that it showed no rhythm of activity. The top cockroach then imparted to the lower one a rhythm of activity corresponding with its own, by means of a hormone passing through the glass tube. Further work showed that the hormone responsible is secreted by the subesophageal ganglion, under the influence of light entering the "ocelli" or simple eyes of the insect.

Temperature-independent cellular clocks have been demonstrated in a number of unicellular organisms. The most instructive example is provided by the work of Professor J. W. Hastings and Dr. B. M. Sweeney on the armored marine dinoflagellate, *Gonyaulax polyedra*.[12] This organism possesses a rhythm of induced luminescent flashing. This was stimulated by bubbling air through a suspension of cells at different times of day and measuring, in darkness, the amount of light emitted. When cultures were grown in alternating twelve-hour periods of light and darkness, the amount of phosphorescence produced during the dark period was forty to sixty times greater than during the light period.

Hastings and Sweeney have postulated that the clock system may operate by the formation of ribose nucleic acid in the nucleus whence it moves unidirectionally into the cytoplasm. Here specific enzymes formed in response to the ribose nucleic acid decrease the level of the substrates required for the synthesis of new ribose nucleic acid in the nucleus and thereby produce a temperature-compensated physiological rhythm. Since very little is known about the chemical events involved in clock systems, however, a great variety of explanations is possible.

4 Biological Advantages of a Time Sense

What, it may be asked, are the survival advantages of possessing an endogenous clock, to an organism normally inhabiting a fluctuating environment? The answer to this is complex. One advantage lies in the ability to maintain regularity in the temporary absence of environmental "clues," and another in the ability to pre-adapt to forthcoming changes. It no doubt benefits nocturnal insects to return to their daytime retreats before the daylight discloses them to predatory birds. Bees and other insects are able to visit flower crops at the time of nectar flow. No doubt all circadian rhythmical activities operate more smoothly and efficiently as a result of having an endogenous component.

It is now well established, as has been discussed in the essay by Professor Hamner, that some animals navigate by the sun, the moon or even the stars, orienting themselves relative to these at an angle which changes systematically with the rotation of the earth. Time-compensated astronomical navigation has been studied in birds, fishes, insects and other arthropods and is discussed elsewhere in this volume.

Yet another biological advantage of a time sense lies in the appreciation of seasonally changing lengths of daylight. It has long been known that the breeding seasons of most animals, like the flowering and leaf fall of plants, result from the interaction of two agencies: external environment and internal rhythm. In some species the environmental control of seasonal changes is all-important. In others, internal rhythm predominates, though never so completely as to make the organism quite free from environmental control. In either case the animal must be able to measure the length of daylight for this to act as a clue or *Zeitgeber* for the seasonal rhythm.

The final timing of migration and the breeding season of most species of animals, however, appears to be controlled not by a single factor such as photoperiodicity, but by a combination of external stimuli, including behavioral ones, so that the courtship displays or pre-migratory restlessness of one individual affect the physiological state of another.

The possession of a resting phase enables many organisms to persist in inconstant environments and regions that would otherwise be unfavorable for permanent habitation. The dormant state of "diapause" is usually characterized by temporary failure of growth and reproduction, by reduced metabolism and often by enhanced resistance to climatic factors, such as cold, heat and drought. It tends to occur in that state of the life cycle which is best adapted to resist the rigors of the climate (such as in the egg or in the pupal stages of insects, for example), and is usually initiated by photoperiod.

In insects and other arthropods the reaction is independent of intensity and total light energy, provided that the intensity exceeds a threshold value. In most cases, no doubt in correlation with their short life cycle, arthropods probably respond to the actual duration of light and dark as measured by this time sense, rather than to gradual changes in the lengths of daylight.

5   *Ecological Significance of Diurnal Rhythms of Activity*

It is not easy to evaluate the adaptive functions of circadian activity rhythms in animals for two reasons. First, the interaction of the various physiological and ecological components is usually extremely complex and, secondly, the environmental clues by which rhythms are synchronized may bear little relation to the ecological factors most significant to the animal. Thus, although light is the clue most commonly utilized, as we have seen, avoidance of high temperature, low humidity or the activities of predators may have greater ecological significance.

Diurnal movements are a characteristic feature of the behavior of free-swimming "planktonic" organisms. In general, most planktonic species avoid strong light, each showing a preference for a certain intensity. For this reason, few organisms are to be found in the surface layers of the sea during the hours of daylight. They are distributed at various depths according to their specific light responses. At dusk, they tend to swim upwards but, when all is dark and there is no light stimulus, they scatter. They migrate to the surface again at daybreak and later move downwards as the light strengthens. Doubtless the green plants, or phytoplankton, aggregate at the light intensity optimum for photosynthesis, and the reactions of the animals tend to maintain them in regions where they find a rich supply of food, but this is not the whole explanation.

Although light is the dominant factor controlling diurnal migration of plankton, temperature, salinity and aeration are also important. Moreover, it is clear from the results of tow-netting at different levels that all individuals in a population do not react in the same way to one particular set of conditions. Sir Alastair Hardy[13] suggests that vertical migration may have evolved because it gives the animal concerned a continual change of environment which would otherwise be unattainable for a passively drifting creature, and the resulting dispersal is biologically advantageous to the species. Water masses hardly ever move at the same speed at different depths, for the surface areas are nearly always traveling faster than the lower layers. So, although by swimming in a horizontal direction an animal will not get much change of environment in the sea, by moving upwards and downwards it can achieve an extensive degree of movement.

Life on land entails a number of problems for small animals which have a very large surface area in proportion to their mass. Many terrestrial invertebrates, such as worms, wood-lice, centipedes and other soil dwellers, avoid desiccation by remaining most of the time in a damp or humid environment, which they leave only at night when the temperature falls and the relative humidity of the atmosphere increases. Insects and arachnids, however, are comparatively independent of moist surroundings, because their integuments possess an impervious layer of wax which prevents desiccation. They are not, therefore, primarily nocturnal in habit. The exigencies of the physical environment play less heavily upon them and biotic factors such as competition and predation assume increasing importance.

Among larger animals, such as most vertebrates, biotic factors are of the utmost significance and the physical factors of the environment become relatively insignificant. Thus, many mammals can readily adapt themselves either to diurnal or nocturnal activity. For example, the African buffalo was very abundant until, in 1890, a terrible epidemic of rinderpest almost exterminated it in many places. Whereas previously the animals used to feed in herds in the open by day, the survivors retired to forests and dense swamps, feeding only at night. After a number of years, however, buffalos increased considerably and returned to their former diurnal behavior.

The timing of an animal's activities is important for a number of other reasons. There is an obvious biological advantage to a species in the synchronization of the activities of its members. Not only do the sexes have a better chance of mating, but the presence of large numbers at a given moment may be sufficient to satiate the appetites of predatory enemies at this vulnerable time. No doubt, too, the times of feeding of many carnivorous animals are related to the activity rhythms of their prey which, in turn, can avoid predators by adjusting their times of activity. As well as gaining protection from enemies, many animals escape competition for food by assuming a nocturnal activity period. Probably for this reason, primitive and otherwise less efficient animals tend to be nocturnal in habits.

Finally, the rhythms of parasitic animals appear, not unnaturally, to be synchronized with those of their hosts. For example, the parasites responsible for malaria, elephantiasis, and other insect-transmitted diseases circulate in the blood of the host at the time when the mosquito or other vector is most active. It seems likely that the rhythm of the parasite is synchronized by the physiological rhythm of the host, for if filiarial patients are made to sleep during the day and remain awake at night, the rhythm of the parasites becomes reversed in three or four days. But this again bears little direct relation to the ecological factor of greatest significance to the parasite, namely, the time of biting activity of its insect vector.

6   *Time Sense in Behavior*

It is reasonable to suppose, although not yet proved, that the central nervous system may act as a "master clock" in higher animals. The electroencephalograph has been cited as evidence of a cerebral cortical clock system since it has a simplified rhythm abstracted from the multitude of nerve cells, processes, channels and impulses of the nervous system. It is generally agreed that memory, foresight, judgment, intelligence, concentrated reasoning and so on become disordered when damage occurs to the cerebral cortex. The disorders in such processes, for all their apparent diversity, have one common underlying feature—the loss of timing. When memory fails, we find a defect in the recollection of time past. A failure of concentration is an inability to maintain mental activity ranged about the present, while lack of foresight, judgment and reasoning imply a failure of "forward memory" or prediction.

Throughout the life of its possessor the brain drives along, running through a sequence of activities. In lower animals these are largely instinctive, but in the higher mammals and man its activity is controlled by learning. A newborn child sleeps most of the time, its numerous brain cells firing in unison. Actions, at first random, develop into little sequences as growth proceeds. In response to any disturbing influence, the brain initiates sequences of action that tend to return it to its rhythmic pattern. If the first action fails to do this, other sequences are tried and the brain runs through all the rules it has learned, one after the other, matching the input with its various "engrams" until unison is somehow achieved. A normal person learns the "rules" of seeing by connecting some parts of the sensory input with motor acts that lead to satisfaction, such as naming and the fulfillment of communication. He at first learns to sweep his eyes along lines instead of in all directions at random. A man born blind, when he is given sight, has to learn to interpret what he sees. At first this is only a mass of colors, but gradually he learns to distinguish shapes though, to begin with, they can only be recognized in the same color and from the same angle.

Nerve impulses arriving at the central nervous system in a disorderly, that is, non-rhythmic way, produce sensations of pain and discomfort. Conversely, rhythmic stimuli are associated with pleasure and relaxation. The rhythmic nature of physiological activity is thus, in my view, ultimately responsible for behavior. And on an aesthetic plane we appreciate things that are essentially rhythmic—light and color, music and poetry—for the essence of beauty lies in harmony and rhythm.

Some Biochemical
Considerations of Time

HUDSON HOAGLAND

In 1933 I reported some experiments[1] indicating that the human time sense is basically dependent upon the velocity of oxidative metabolism in some of the cells of the brain. This essay reviews the evidence for this and considers, in the light of more recent findings, its implications in relation to our experience of duration and indeed to physiological time as mediated by nervous systems of animals other than man.

Before presenting this evidence it will be necessary to digress and discuss certain concepts of physical chemistry and enzymology that are essential to understanding the relation implied in the title of this communication. More specifically I wish to discuss the effect of temperature on chemical reactions and the way in which temperature, as an analytical tool, may shed light on the control by chemical pacemakers of physiological rhythms.

A chemical pacemaker may be defined as the slowest reaction in a sequence of linked reactions. The slowest step acts as a bottleneck limiting the over-all velocity of the chain of reactions from initial substrate to product. Elsewhere I have discussed the control of rhythms in nerve and muscle cells as determined by continuous metabolic processes.[2] These rhythms come under the category of relaxation oscillations in which some potential is built up to a critical value and is then discharged through resistances that are reduced during the discharge. Recovery involves again building up a critical potential against the re-established resistances with repetition of the discharge at the critical level of potential. Thus, for example, in the case of the beat of the denervated heart the slowest step in the metabolic recovery of excitability of the cells of the sinuauricular node limits the frequency of the heartbeat.

1   *Temperature as an Analytical Tool*

It is well known from elementary chemistry that heat speeds up chemical reactions; a rough rule of thumb is that reaction velocities are doubled or trebled for a 10° C. rise in temperature. A more formal description of this effect was given in 1880 by Arrhenius, who described an empirical equation relating chemical velocity to temperature. Over the years his equation has been given a firm theoretical basis by the physical chemists in terms of chemical kinetics.

The plot of the Arrhenius equation for experimental data is a straight line of negative slope when one plots the logarithm of the chemical velocity, also known as specific reaction rate, against the reciprocal of the absolute temperature. The equation describes a wide variety of simple bimolecular chemical reactions and it is possible to calculate directly from the slope of the line a constant of theoretical importance known to the physical chemist as the energy of activation. This constant is conventionally symbolized in the equation by the Greek letter mu (μ). It is equivalent to the amount of kinetic energy per mol which the molecules must acquire before they can react. More specifically it is the increment of energy, over and above the average energy of the molecules in the system, that is necessary for reaction to take place, and it is specific for the reactive valences. The magnitude of the mu-value can sometimes serve as a key to the nature of catalysts taking part in a reaction. The activation energy or mu-value cannot be measured directly; it can only be calculated from the slope of the line in the Arrhenius equation plot.

The Arrhenius equation may be written as follows

$$v = ze^{-\mu/RT} \tag{1}$$

where v is chemical velocity, z is a constant, e is the base of natural logarithms, T is the absolute temperature, R is the gas constant, and μ is the critical thermal increment or energy of activation, as defined above. In the literature of physical chemistry, the symbol E is used instead of μ for the energy of activation. To avoid the implication that mu-values of biological processes are necessarily identical with activation energies as used by physical chemists, Crozier[3] originally introduced the symbol μ and referred to it as the "temperature characteristic" rather than as the energy of activation. Work to be described below suggests that E and μ are identical.

Taking logarithms on both sides of equation (1) we obtain

$$\log_{10} v = c - \mu/2.3RT. \tag{2}$$

If experimental data conform to the Arrhenius equation, it is thus clear that a plot of log v vs. $1/T$ should yield a straight line with ordinate intercept at c and negative slope of $\mu/2.3RT$.

If frequency (f) of some physiological rhythm is directly proportional to the velocity of some underlying chemical pacemaker, we may write

$$f = kv = kze^{-\mu/RT} = ae^{-\mu/RT} \tag{3}$$

and again taking logarithms

$$\log f = c' - \mu/2.3RT. \tag{4}$$

Thus in comparing equations (2) and (4) it is clear that a plot of log f vs. $1/T$ would give the same value of μ as if we plotted log v vs. $1/T$ directly. In a plot of the logarithm of frequency vs. $1/T$ the intercept on the ordinate axis would be at c' instead of at c. The two plots would give parallel lines of the same negative slope equal to $\mu/2.3R$, or $\mu/4.6$, since R is 1.99, or about 2 calories per mol per degree.

The Arrhenius equation describes with precision a large number of homogeneous and heterogeneous reactions both in solution and in the gaseous state. The reader is referred in particular to Hinshelwood's monograph[4] for a discussion of the equation and the concept of the energy of activation. For our purposes we wish to see to what extent the equation may be used to describe physiological and biochemical processes with the view to examining mu-values as possible indicators of specific chemical pacemakers.

Enzymes and catalysts in general react with substrate molecules and in a number of instances the mu-value has been found to characterize the particular enzyme-substrate reaction. This is because catalysts promote chemical reactions that either would not begin at all or would proceed only if supplied with large amounts of energy available only at high temperatures. An enzyme as catalyst, by orienting appropriate bonds, enables a reaction to proceed in an orderly fashion at body temperature. It thus reduces the energy of activation for the reaction.

Figure 1 from a paper of Gould and Sizer[5] is an example showing how a variety of different substrates are oxidized by the dehydrogenase enzymes of *bacteria coli* in such a way as to yield identical mu-values of 19,400 calories. Despite the complexities of biological oxidations the appearance of identical figures for different substrates indicates that the same valence bond is the chemical pacemaker step in these reactions, and that over the entire temperature range the slowest step in the sequence of chemical events comprising the oxidation is the same one for all of the substrates.

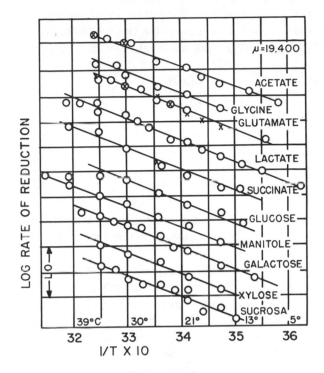

Figure 1. Log of rate of reduction (calculated from the time required for 75% reduction of methylene blue) in presence of various substrates by a suspension of *Excherichia coli* plotted against 1/T. The quantity μ in the Arrhenius equation = 19,400. Circles refer to the series of determinations made with a first bacterial preparation, crosses to determinations made with second separate bacterial preparations. For lactate, the crosses refer to determinations made with a preparation consisting of autolyzed cells. From Gould and Sizer (see note 5).

Just as the slowest worker in a chain of workers on a bench disassembling a piece of machinery sets the pace for the over-all operation, so the slowest enzyme-catalyzed link involved in the stepwise disassembling of a foodstuff molecule is the chemical pacemaker for the over-all processes releasing energy in the cells. Hadidian and I[6] demonstrated this some years ago in studies of the oxidation of succinic acid in the presence of succinic dehydrogenase and the cytochrome system. Succinic dehydrogenase removes hydrogen from succinic acid leaving fumaric acid. The hydrogen then combines with oxygen that has been activated by the cytochrome system to form water. In many animal cells, including those of the brain, this two-step linked reaction of succinic dehydrogenase and the cytochrome system is involved as part of the chain in oxidative metabolism.

The following paradigm shows the nature of the reactions.

$$
\begin{array}{c}
\text{COOH} \\
| \\
\text{CH}_2 \\
| \\
\text{CH}_2 \\
| \\
\text{COOH}
\end{array}
\xrightarrow[\text{dehydrogenase}]{\text{Succino}}
\begin{array}{c}
\text{COOH} \\
| \\
\text{CH} \\
\| \\
\text{CH} \\
| \\
\text{COOH}
\end{array}
+ 2\text{H} + \text{O}
\xrightarrow[\substack{\text{cytochrome-}\\\text{cytochrome}\\\text{oxidase}}]{}
\text{H}_2\text{O}
$$

By measuring oxygen consumption at different temperatures and plotting them according to the Arrhenius equation, a mu-value of 11,000 calories was obtained.

Now it happens that cyanide selectively poisons the cytochrome enzyme system but does not affect the succinic dehydrogenase enzyme, while selenite poisons this last enzyme but does not touch the cytochrome system. Too much of either of these poisons stops all oxygen uptake by knocking out one or the other of the enzymes. But lesser amounts of either poison can be used to change selectively the *relative* speed of one or the other of the two catalyzed steps. On adding increasing amounts of selenite we slowed oxygen uptake more and more and finally stopped it but without any change in the mu-value of 11,000 calories. In contrast to this, on adding cyanide we found, at a critical level, that the mu changed abruptly to 16,000 calories, and that increasing the cyanide slowed oxygen uptake progressively but we observed no further change in mu. However, if we added selenite to a reaction already slowed by poisoning with cyanide we found that the mu-value was shifted back to 11,000 calories from the cyanide-poisoned value of 16,000 calories. From these results we concluded that the mu for the succinic dehydrogenase link is 11,000 calories and that for the cytochrome system is 16,000 calories. This was in conformity with expectations since physical chemists had reported that the oxidation from ferrous to ferric iron gives a mu of 16,000 calories and the cytochrome system contains iron which is thus oxidized. Figure 2 from Hadidian and Hoagland shows a summary of these results: when the first step involving succinodehydrogenase is the pacemaker (being the slower of the two steps) 11,200 calories were obtained; when the second step involving the cytochrome system is pacemaker (being the slower of the two) 16,000 calories was the mu-value.

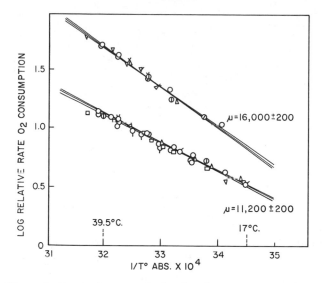

Figure 2. Composite massed plot for the enzyme experiments described in the text. Lines are drawn by eye through the data; the lines fanning out are for ± 200 calories around the means. From Hadidian and Hoagland (see note 6).

2   *Some Biological Rhythms*

Extensive studies of biological rhythms in cold-blooded animals conducted at different temperatures showed not only the reality of the Arrhenius equation relation but they gave mu-values that are grouped in certain well-defined modes. For example, the frequency of beating of the hearts of cockroaches as a function of temperature follows the equation and yields a mu-value of about 12,000 calories.[7] Figure 3, from Crozier, shows data on the frequency of chirping of crickets.[8] If one could calculate logarithmic and reciprocal temperature relations in one's head, the cricket could be a useful thermometer.

Crozier[9] in 1926 plotted a histogram of the distribution of mu-values for three hundred and sixty determinations involving frequencies of heartbeats, breathing movements, cilia beats, appendage movements and other physiological rhythms. The largest modes of occurring mu-values are at eight, eleven, twelve, sixteen and seventeen thousand calories.

In 1936 I made a histogram of cell and tissue respiratory studies of O_2 consumption and CO_2 production.[10] The same prominent modes were found as those Crozier had reported for the distribution of mu-values of

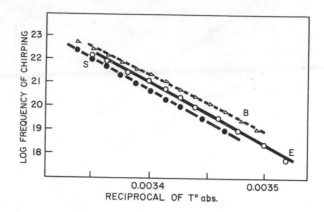

Figure 3. Data from three independent observers for the frequency of chirping of crickets as a function of temperature. From Crozier (see note 8).

physiological rhythms. These multimodal distributions were highly significant statistically.

Crozier interpreted these distributions to mean that in different organized tissue systems we have a limited number of enzymes, and that in some cases one and in other cases another will be the slow pacemaker step. If physiological rhythms are directly proportional to energy released by underlying respiratory processes, we would expect from the nature of the mathematical relations involved to find mu-values for the frequencies corresponding to those for the rate-limiting chemical pacemakers.

But this sort of thing is not confined to frogs, insects and other sub-mammalian forms. In my 1936 study I reported the alpha brain-wave rhythm in man as a function of his internal body temperature. Figure 4 shows how the alpha brain-wave frequency of a normal man is affected by temperature when one modifies his internal temperature by diathermy, *i.e.*, the passage of high-frequency alternating currents through the body. Figure 5 shows the Arrhenius plot of the data from this man yielding a mu-value of 8,000 calories.

Herbert Jasper[11] also obtained a mu-value in this range of 8,000 calories for the human alpha rhythm, and Koella and Ballin[12] studied brain-wave frequencies in cats as a function of temperature. Their data proved to be describable remarkably well by the Arrhenius equation, *i.e.*, a linear function was obtained when the reciprocal of the absolute temperature was

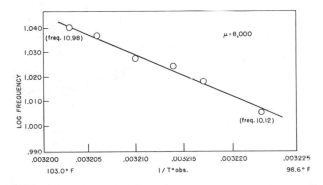

MEAN FREQUENCY FOR 80±SECONDS			°F
10.12	.055		99.2
10.42	.058		100.0
10.56	.064		100.6
10.67	.060		101.4
10.90	.040		102.0
10.98	.017		102.6

←— I SEC.—→ I 50 μr

Figure 4. Sample of brain waves from a normal human subject showing strong alpha rhythms. Mean frequencies for each temperature for 80 seconds of recording (*ca.* 800 cycles) with probable errors are shown at left. From Hoagland (see note 10).

LOG FREQUENCY

1.040 — (freq. 10.98) μ=8,000
1.030
1.020
1.010
1.000
.990

.003200 .003205 .003210 .003215 .003220 .003225
103.0° F   1 / T°obs.   98.6° F

(freq. 10.12)

Figure 5. Arrhenius equation plot of alpha brain-wave frequencies from the same subject whose brain waves are shown in Figure 4. From Hoagland (see note 10).

plotted against the logarithm of the dominant frequency. They obtained a mu-value of 8,000 calories for temperatures between 37° C. and 28° C.

Figure 6 shows data from a group of normal human subjects and two groups of general paretic patients whose temperatures were raised by diathermy. A mu-value of 8,000 calories may be seen for normal controls in the lower band of points. The middle band of points yields a mu of 11,000 calories and corresponds to data from clinically intermediately advanced general paretic patients. The upper band with a mu of 16,000 calories is from very advanced paretic patients. These three values correspond to three of our major peaks of activation energies described in the distribution curves referred to above. It would appear that the advancing

319

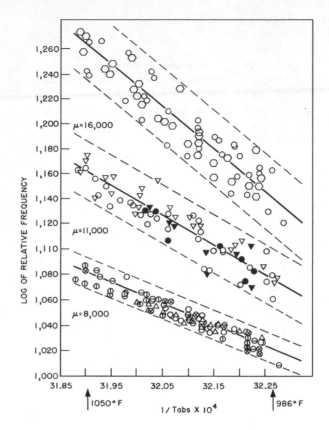

Figure 6. Composite plot of data showing a μ of 8,000 calories for three normal persons, a multiple-sclerosis patient and two very early general paretics. For two more advanced paretics $\mu = 11,000$ and for two *very* advanced paretics $\mu = 16,000$. From Hoagland (see note 10).

spirochete infection shifts the chemical pacemaker from one characterized by 8,000 calories, not yet identified with a known enzyme system, to one characterized by 11,000 calories, probably succinodehydrogenase, and finally to one characterized by 16,000 calories, the cytochrome system. In this connection it is interesting that there is well-known histological evidence of changes in the distribution of brain iron with the advancing infection. If such changes reduce the amount of available cytochrome iron this step should be slowed just as it is when we partially poisoned our enzyme systems in vitro with cyanide. In both cases we obtained a mu of 16,000 calories.

3 *Subjective Time, Internal Temperature, and Chemical Kinetics*

I should now like to return to a consideration of time. We all possess a private, internal appreciation of time. We can gauge within limits equal intervals of the experience we call duration. Even asleep, our time sense continues to function. It is possible for many persons to waken themselves at predesignated times to within a few minutes of accuracy after a night's sleep. However, our private time is not precise enough to serve the needs of society, and so man has measured time in terms of recurring happenings, he has constructed mechanical clocks, and learned to standardize his internal time sense with his widely accepted time conventions. Our public time is thus determined by relations of changing physical objects, and becomes meaningless except in terms of these relations.

Since clock time depends on relative motion, let us ask ourselves what is the source of motion determining our private or psychophysiological time? We judge time with our brain. To keep our brain cells functioning they must continually burn foodstuffs. This burning or oxidation is dependent upon motions of molecules, as are all chemical processes. A series of experiments has shown that our judgment of time depends on the speed of these processes.

If private, psychophysiological time is determined by chemical velocities, raising our internal body temperatures (as might occur in a fever) should speed the reactions, thus making more chemical change and hence more physiological time pass in a given interval of clock time than would normally be the case. If, let us say, two minutes of private, subjective time were thus to pass in one minute of clock time, we would think that time was dragging; on looking at the clock it would be slower than we think it should be. Lowering the internal body temperature, on the other hand, should have an opposite effect, making clock time seem to pass faster, since the reduced biochemical changes would make less time seem to have passed and our public clock would, in contrast, appear to run faster. In a fever, *other things being equal*, we should come early for our appointments, and at subnormal temperatures we should come late.

The first time I outlined these questions about time to myself was in 1932 when my wife fell ill with influenza and developed a temperature one afternoon of nearly 104° F. She had asked me to do an errand at the drugstore and, although I was gone for only twenty minutes, she insisted that I must have been away much longer. Since she is a patient lady, this immediately set me to thinking along the lines just indicated and then hurrying to find a stop watch. I then, without telling her why, asked my wife to count to sixty at a speed she believed to be one per second. As a trained musician she has a good sense of short intervals. She repeated this count some twenty-five times in the course of her illness, her speed of counting was measured with the stop watch, and her temperature was recorded each time. She unknowingly counted faster at higher than at lower temperatures. On the strength of this the experiment was repeated with several volunteer subjects in whom artificial fever was induced by diathermy. I later discovered some data published earlier by a French worker, François, on the frequency of tapping a prescribed rhythm at different body temperatures. All these data, both those from Paris and from my own experiments, showed the same result as regards judgment of time.[13]

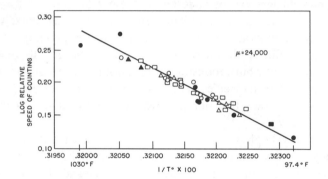

Figure 7. Temperature characteristic for the human time sense. Arrhenius plot for the effect of internal body temperature on the frequency of subjectively counting seconds and tapping prescribed rhythms for naive subjects. For discussion see text. From Hoagland (see note 1).

Figure 7 shows data related to the subjective human sense of time. The existence of a mu-value of 24,000 calories suggests that our sense of time depends upon the speed of a chemical step in some group of cells in the brain that act as a chemical clock. This clock cannot be the same as that regulating the alpha rhythm, for the mu-value of that clock is 8,000 calories in normal men. One might speculate that this chemical pacemaker involves oxidative metabolism in the cells of the reticular formation in the brain stem, since impulses from these cells to the cortex determine conscious wakefulness, and we "keep time" both awake and asleep.

With our physiological thermostats permitting normal *internal* temperature fluctuations of less than ½ ° F. around a mean of 98.6° F. (37.0° C.), the steady-state chemical events give us a linear, private time scale which we can standardize against our objective clocks. Biological oxidations, proceeding at a relatively constant speed, apparently furnish us with a basis for our uniformly flowing time scale.

Confirmatory evidence of this view of time comes from Heinz Von Foerster,[14] who develops equations describing data of learning and forgetting of nonsense syllables. He hypothesizes that memorization depends upon the deposition of impregnated units in the brain and their decay with forgetting. These might be protein molecules or substructures thereof. His equations enable him to predict a number of properties of learning and forgetting. This is especially interesting in view of the more recent remarkable work of Hydén and other molecular biologists indicating that engrams may well be specific protein molecules formed in neurones as a result of passages of nerve impulses.[15]

Von Foerster came upon a discussion of my time studies,[16] and he accordingly wrote me the following letter:[17]

"A few days ago I received the Transactions of the First Macy Conference on *Problems of Consciousness*, and I was especially thrilled with your presentation on 'The Chemistry of Time,' on grounds that you will see immediately. In 1949 I had the pleasure of talking in the Cybernetics Group of the Macy Foundation about an outline of a theory of the memory I developed a few years ago. You will find a sketch of these thoughts in the Transactions of the Sixth Conference [1949].

The main idea is to split up complex impressions into elementary impressions which are stored on quantized molecules (pseudo-isomeric change of protein molecules?). Due to the 'tunnel effect' the storage cannot last indefinitely.[18] These molecules must decay, with a decay-constant λ depending on the threshold E_z keeping the molecules in their higher quantum-state:

$$\lambda = \lambda_0 \, e^{-\frac{E_z}{kT}}$$

where T is the absolute temperature and k Boltzmann's constant. The decay-constant λ can be found from forgetting curves measured by nonsense-syllable experiments carried out at normal body temperature.

Antagonistic to this quantum-mechanical decay (forgetting), a physico-chemical process is working, which I called 'memorization K_0.' This memorization prevents the animal from forgetting completely already established memories. The physical dimension of K_0 is t^{-1} and can also be found by evaluating measured forgetting curves. K_0 is merely—to put it in your own words—'the speed of a chemical step in some group of cells . . . in the brain that acts as a chemical clock.' To make learning possible I showed that the ratio K/λ must be larger than unity. Since λ is a function of temperature, K must follow the same function to keep this ratio constant. Thus:

$$K = K_0 \, e^{-\frac{E_z}{kT}}.$$

Taking the logarithm on both sides one obtains an equation:

$$\log_{10}K = A - \frac{B}{T},$$

the Arrhenius equation for reaction-velocities, as you have shown in your presentation. This fact in itself already excited me very much and I hastened to compare the constants B in both our cases. In my case B is given by:

$$B_1 = \frac{1}{2.3} \frac{E_z}{k}$$

whereas in your case B is defined by:

$$B_2 = \frac{1}{2.3} \frac{\mu}{R}.$$

If my picture of these brain-processes should match with experimental data, found from an entirely different angle, these two quantities B_1 and B_2 should equalize,

$$\frac{1}{2.3} \frac{E_z}{k} = \frac{1}{2.3} \frac{\mu}{R}.$$

Since $R = kL$ ($L = 6.06 \cdot 10^{23}$), the equation above reduces to

$$\mu = E_z \cdot L.$$

Expressing E_z in electron volts—as I did—and expressing μ in calories—as you did—one has to convert the electron-volts into calories:

$$1 \; eV = 0.380 \cdot 10^{-19} \; \text{cal.}$$

and with $L = 6.06 \cdot 10^{23}$ one obtains:

$$\mu / \text{cal.} = 2.3 \cdot 10^4 \; E_z/eV.$$

I found from forgetting curves $E_z = 1.11 eV$ which gives a μ-value of:

$$\mu = 25,500 \; \text{calories}$$

a rather startling agreement with your value of 24,000 calories."

Only if consciousness depends upon awareness of duration, related to molecular kinetics determining memory traces, is it likely that agreement in calculations of this fundamental physical constant will result from such different approaches as those of Von Foerster's and mine. Fortuitous agreement in such a calculation seems most unlikely.

If our time sense depends basically on rates of biological oxidations in the brain, we would expect that a variety of factors other than temperature would modify it. Many experiments by psychologists confirm the view that filled time passes more rapidly than empty time as summarized by the adage that "the watched pot never boils." The literature also contains reports of effects of various drugs on the time sense. There is an especially interesting paper by Fischer, Griffin and Liss[19] on biological aspects of time in relation to model psychoses. In addition to describing their own experiments they give an excellent review of the problems of subjective time. In their experiments they modify physiological time by visual flicker stimulation and point out that relative changes between the outside stimulating frequency versus spontaneous receptor frequency can be brought about "from the inside" through metabolic changes in receptor areas of the central nervous system produced by psychotomimetic drugs, such as LSD, forms of Yoga discipline, and sensory deprivation, and they discuss these procedures in relation to the production of hallucinations.

In considering drug-induced changes of physiological time they compare excitability and depression in relation to certain classes of drugs. Excitation is associated with pyretogenic drugs of the amphetamine, mescaline, LSD type, and tranquilization with hypothermia-producing phenothiazine-type drugs. In relation to these drug effects, they write:

All these contrasting pairs of modi can be related to Hoagland's temperature characteristic of the human time sense. . . . Time contracts, the physiological clocks run faster, and events appear to pass slowly when body temperature is raised, whereas time expands, the physiological clocks run at a slower rate, and events appear to pass faster when body temperature is lowered. The same relation also exists for hyperthyroid (chronophage) and hypothyroid subjects respectively.

It is a common experience that time for the child seems to pass much more slowly than time for the adult. A year goes by rapidly for a man compared to his recall of childhood years. Seymour Kety[20] has reviewed available information, obtained by the nitrous-oxide technique, on over-all cerebral blood flow and oxygen consumption in man, and finds a distinct correlation of these functions with age. He reports a rapid fall in both circulation and oxygen consumption of the brain from childhood through adolescence followed by a more gradual but progressive decline through the remaining age span. Slowing of cerebral oxygen consumption with advancing years would, according to our considerations, make time appear to pass faster in old age, as indeed it does.

4   Temperature and Aging

It is well known that some living organisms can recover after being subjected to the very low temperatures of the liquefied gases such as nitrogen, hydrogen and helium. For such organisms at these very low temperatures, with metabolism virtually stopped, time, as a reciprocal of the rate of metabolism, must pass very rapidly as far as the organism is concerned. Since life in its concomitant aging processes is a series of chemical events proceeding irreversibly towards death, these temperatures must essentially arrest aging. Such organisms are, for all intents and purposes, projected forward into the future at these low temperatures. A stay of a century at −270° C., the temperature of liquid helium, with subsequent warming and recovery would, if it were possible to effect, be precisely equivalent to projecting the organism forward a century into the future—a sort of Wellsian time-machine idea.

Luyet and Gehenio[21] reviewed the subject of life and death at low temperatures. Some one hundred and twenty studies, beginning with those of Pictet in 1893, have demonstrated that a variety of small organisms can withstand the temperatures of the liquefied gases. These include some bacteria, protozoa, plant cells, and a few metazoans. The procedure in general involves rapid chilling and warming so as to vitrify the organisms and not permit their water to crystallize. Crystallization of water takes time and colloidal solutions with their large particles are relatively slow to crystallize. The crystallization range for protoplasmic systems, according to Luyet and Gehenio, extends for only some 30° to 40° below 0° C. Rapid passage through this range was regarded by Luyet and Gehenio as essential if organisms are to survive extremes of chilling. Moreover, the length of time spent at temperatures below the freezing or crystallization range usually has little effect on the per cent of revivable organisms. Organisms of more than a millimeter or so in size cannot survive this treatment since heat cannot be conducted from and to their interiors fast enough on chilling and on warming to prevent internal crystallization and death. The water content of the organisms also is important. In general, organisms with a low water content best withstand vitrification.

The germ plasm of mammals seemed to us to be of special interest in connection with vitrification since the storage of sperm would be of value in connection with the fertilization to produce offspring perhaps generations after the donor's death.

Luyet and Hodapp[22] vitrified and revived an appreciable number of frog sperm after removing some of their water content by plasmolysis with hypertonic sucrose solutions. They were, however, unable to revive any rat sperm. Shettles,[23] in the course of a study of the physiology of human sperm, had been able to vitrify and revive a few per cent of seminal human sperm immersed in capillary tubes in the liquefied gases and later rapidly warmed.

Hoagland and Pincus[24] reported studies of the revival of mammalian sperm after immersion in liquid nitrogen. A variety of procedures was used to test the motility of mammalian sperm after plunging them into liquid nitrogen at $-195°$ C. and later rapidly warming them to $35°$ C. by plunging them into a suitable balanced isotonic medium.

We found that human sperm suspended in fresh seminal fluid retained motility to a remarkable degree following immersion in liquid nitrogen and subsequent storage in dry ice at $-79°$ C. Samples in which 75 per cent of the sperm were alive before immersion yielded 67 per cent alive after sudden chilling followed by storage and rapid warming to $37°$ C. The same percentage of sperm survived whether they had been kept in the cold for minutes or for up to two months, when the experiment was terminated. Human sperm were remarkably viable under this treatment, compared to the sperm of other mammals. Investigation of rabbit sperm showed a few occasions in which less than 1 per cent were viable, and tests on several other animals were also ineffective in producing enough sperm to be used for determination of fertilizing capacity by artificial insemination.

In discussing these and related studies in 1943 I wrote:[25]

The problem of keeping sperm in suspended animation may be a very practical one for animal husbandry. The indefinite storage of the sperm of prize animals and their possible use to renew prize stock after it has degenerated by faulty breeding is at once apparent. A great horse like "Man O'War" could thus have immediate sons many generations after his death. It is even conceivable that at some time in the future we may systematically draw upon great geniuses of the past to father our human offspring. Possibly institutions which now store the memorabilia of our great departed may one day also store their vitrified sperm. Married couples today, when the husband is sterile, sometimes have children by sperm selected by their physician from a donor known to him but forever unknown to the couple. Social sanctions may ultimately extend this practice to illustrious men of the past not only for the fertile wives of sterile couples but for certain other women as well, especially since emotional revulsion might be reduced in the case of a non-contemporary donor long since dead. The historical test of genius would have operated so that part of each generation could be fathered by truly great sires.

While it still remains to be proved that vitrified sperm will keep indefinitely with fertilizing powers unimpaired, one is tempted with a variety of speculations. In a world rent by war and strife, the physically best are, in practice, selected for the most dangerous service. The possibility of the storage of vitrified and revivable human sperm available in the event of the donor's death may, at some future time, have considerable social significance.

This passage was written before the advent of atomic weapons, which adds to its cogency today, and Hermann J. Muller[26] has called attention to the long-term cumulative damage to our genes by human medical practices that are increasingly saving lives of people with genetic defects who reproduce and who in the past would have been eliminated by natural selection. (Unknown to me until recently, as early as 1935 Muller[27] had emphasized the desirability of changes in sex mores to improve mankind by the use of artificial insemination with stored sperm of genetically superior men.)

Polge, Smith, and Parkes,[28] while empirically testing a number of procedures to preserve mammalian sperm by freezing, discovered that fowl semen diluted with equal amounts of Ringer's solution containing 20%–40% glycerol enabled large numbers of the sperm to survive plunging into freezing mixtures at −79° C. The use of glycerol proved to be a breakthrough. It has resulted in the effective slow freezing and recovery on warming of both fowl and mammalian sperm. Sperm now can be kept in storage indefinitely and later used in artificial insemination. A book by Smith[29] on the biological effects of supercooling reviews the "States of the Art" in 1961. Fowl sperm frozen at −190° C. and kept for a year produced fertile eggs and normal chicks, and bull sperm stored at −79° C. for periods up to seven years and thawed had a high survival rate. Normal calves have been born from such frozen and revived sperm. Sperm frozen in 10% glycerol stored at −79° C. when warmed to 40° C. had a 75 per cent conception rate after a year. Today the storage of sperm for long periods and its artificial insemination of domestic animals are widely practiced in animal husbandry. Human sperm have also been frozen, stored and used in artificial insemination, with the birth of healthy, normal children.[30]

Techniques have been developed for the preservation by freezing of whole organs. A. S. Parkes in the foreword of Smith's book predicts that in the decade of the '60s "the preservation in deep freeze of whole organs for transplantation may become possible." Smith's book also describes the partial slow freezing and survival of small whole animals including hamsters.

We have discussed the use of temperature as an agent to modify certain aspects of physiological time. Birds and mammals in the course of evolution have developed remarkable thermostats to maintain constancy of temperature of their internal environments. These neuromuscular and neuroendocrine mechanisms have been extensively studied by physiologists for more than a century, and as Claude Bernard pointed out a long time ago, an important aspect of an organism's freedom is dependent upon the constancy with which it regulates its internal environment—that is to say, its homeostasis as later described by Walter Cannon. The constancy of the internal temperature is one aspect of this homeostasis making birds and mammals free to function over a wide range of external temperatures without disturbing important internal metabolic processes. Animals that do not control their internal body temperatures, such as the lower vertebrates—reptiles, amphibia, fishes—and invertebrates, are dependent upon the external temperature, *i.e.*, they are less free and tend to become sluggish and immobilized at low temperatures. It is interesting to consider that man's short-term subjective time scale may depend upon the constancy of his internal temperature. For so-called cold-blooded animals this would not hold. For them time would presumably pass slowly on warm days and rapidly on cold days for the reasons discussed in the early part of this paper dealing with the human time sense and its relation to temperature. Moreover, time would not appear to flow steadily in the linear sort of way familiar to us mammals. This is, of course, speculation, but aspects of the behavior of animals that do not regulate their internal temperatures compared to birds and mammals are consistent with this interpretation.

5 *Conclusions*

We have described studies of subjective time indicating that judgment of durations basically depends upon the rate of oxidative metabolism in the brain. We have used temperature as an analytical tool. Temperature affects many metabolic events and biological rhythms including the counting of seconds. Our analytical use of the Arrhenius equation indicates controlling mechanisms of metabolism and of their dependent rhythms in the form of the slow step in sequences of enzyme-controlled biochemical processes. The Boltzmann equation has been used by Von Foerster to describe in quantum mechanical terms the duration of memory traces following learning experience. A constant in this equation used for rates of decay of memory traces is found to have the same value for forgetting as that obtained by the Arrhenius equation for counting seconds as a function of temperature. Further evidence for metabolic chemical clocks is related to the fact that time passes for children more slowly than for adults, which appears to be related to continuing progressive decline of brain blood-flow and oxygen-consumption from infancy on over the years.

Biological time depends upon rates of metabolism. Lowering the temperature of an organism to the temperatures of the liquid gases stops its metabolism and its aging. If no irreversible damage is done and the organism recovers on warming, it is, for all practical purposes, projected into the future. We have summarized various studies of effects of low temperatures on living processes with particular reference to the storage of sperm for use in artificial insemination long after the donor's death.

Organic Evolution
and Time

HANS KALMUS

Organic evolution, like any other change, implies the most common current idea of time—something unidirectionally connecting an earlier situation *A* with a later situation *B*. Aristotle defines time as "the number (measure) of motion in respect of before and after."[1] And he continues, "time then is not movement, but that by which movement can be numerically estimated." Modes of thought which refuse the notion of time as Parmenides and Zeno did or which consider the time scale as equivalent to the three Cartesian coordinates may perhaps have their uses in some branches of physics but are quite unprofitable for the student of evolution.

The idea of evolution arose from the discovery of fossils many of which were obviously not relics of the contemporary flora or fauna and must have been left behind by plants and animals living in previous epochs. The two problems posed by these observations were, 1) how to date the individual fossils and 2) how to explain their occurrence, appearance and loss. In this essay we shall deal with both these problems, but mostly with the first.

It is again Aristotle who describes the idea—if not the methods—of dating, which is appropriate for the determination of geological age. He wrote: "Events have their places in time in a sense analogous to that in which any numbered group of things exist in number. And such things as these are embraced in number (that is in time), as things that have locality are embraced in their places."[2] In fact fossil faunas and floras are arranged in "horizons," which have bit by bit been assigned to one unique vertical sequence of sediments (strata), in which any stratum contains relics older than those in the stratum above and younger than those in the stratum below.

The stage on which organic evolution was enacted has been a gradually changing Earth—progressively resembling the Earth of today[3] on which days, months and years, perhaps of duration rather different from the corresponding units today, have been interacting in changing complexity. The dating of fossils in ordinary solar years is nevertheless quite legitimate when based on radioactive decay.[4]

Evolution being a macrophenomenon can be largely interpreted as a sequence of causes and effects, notions which again imply the everyday concept of time. And the nonrepeatability (irreversibility) of evolution can be deduced from its stochastic nature and is possibly indicative of a universal arrow of time.[5]

The idea of evolution has been applied to inorganic systems, to biogenesis and also to human history. However, while these processes do share certain features with organic evolution and may also interact with it, in this essay we shall be concerned only with the latter, that is, with certain demonstrable changes in time of plant and animal populations including man.

1 *Geological Time*

Succession of strata and fossils. Sediments and fossils were first arranged in a relative sequence although their absolute age was only studied later. Many abundant fossils, the *Leitfossile*, were found only in combination with certain others in a particular layer of sedimented rock. By the combination of stratigraphy[6] and study of fauna (rarely flora) the history of the Earth's surface and the history of life were laboriously pieced together to form the geological and paleontological record. First it was thought possible to accommodate this whole succession within a few thousand years perhaps compatible with the Biblical age of the Earth. Estimates, however, have gradually increased to the presently current values of thousands of millions of years. The age of fossil algae and fungi from Ontario has been given as between 1700 and 2000 million years.

Visualizing geological time. Such an enormous space of time is difficult to contemplate and several methods of doing so have been employed. One might for instance gauge it by extrapolating from the effects of observable geological agencies to consider how long it would take the glaciers and the rivers to destroy the Alps. This was Lyell's famous approach, which is not dissimilar to some charming medieval attempts at defining "eternity."

Another means of visualizing geological time is that of scale reduction.

As an example of the latter method, let us contract the whole existence of the world into a single year.

The world began on January 1st. Then conditions in which life could have arisen developed perhaps early in August. The oldest known fossils, life of the past imprinted in the rocks, were living things about mid-October, and life was abundant, much of it in the seas, by the end of the month. In mid-December, dinosaurs and other reptiles dominated the scene. Mammals, with hairy covering and suckling their young, only appeared in time for Christmas, and on New Year's Eve, at about five minutes to midnight, from amongst them stumbled man. Of these five minutes of man's existence, recorded history represents about the time the clock takes to strike twelve.[7]

In this illustration of relative time spans, twelve hundred years—the life span of a redwood tree—has been reduced to about one second.

Absolute estimates and their accuracy. Geochronology, the dating of events in the Earth's history, has made great strides in the last half-century,[8] though it must be said that the various methods employed—the study of sedimentation rates and of animal deposits, of climatic and astronomic cycles and finally the decomposition of radioactive substances—do not always lead to very consistent results.[9] Nevertheless the data thus obtained seem to agree within an order of magnitude and are sufficiently accurate to form a basis for evolutionary speculations. A few estimates of evolutionary rates are in fact based on time estimates of far greater accuracy.

An approximate idea of the antiquity and duration of the main geological periods can be gained from comparing the two scales on the right-hand side of the diagram on p. 333.

Changes of the lengths of day, month and year. The periods of the revolutions and rotations of Earth and moon may have changed in the past both relative to each other and when compared with some other "time constant" such as the half-life of a radioactive element. During historical time the succession of days, seasons and years has proceeded roughly at today's astronomical rates. A very slight retardation of the Earth's rotation has been inferred from the discrepancy between the hours of eclipses as reported by Babylonians and ancient Chinese and as calculated by astronomers,[10] and is tentatively ascribed to tidal friction. But some of the numerical data are questionable and the theory of the Earth's movements has been slightly modified so that we do not really know what has happened. Speculation from first principles has not yielded unequivocal results either. Be that as it may, it is fair to assume that in the temperate zones of the Earth day and night have alternated for hundreds of millions of years, and during many millions of years have had a period length of about twenty-

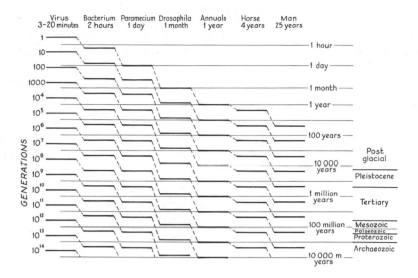

Number of generations characteristic of seven levels of organization (broken lines) which—had these levels remained constant—would have elapsed since a time stated in years (thin horizontal lines) or since a geological epoch. The number of years is entered on a logarithmic scale, beginning from the top with the "present," so that 1 cm. near the top indicates a much shorter span of time than 1 cm. near the bottom.

four hours. Direct evidence for this is found in patterns of diurnally changing growth in fossil insect cuticles, plant membranes, hairs, feathers and other striped structures. In addition, a universal feature of life on Earth, its circadian organization, may be taken as indirect evidence for the antiquity of the change between day and night.

The history and antiquity of the moon are at the present time once again in dispute and it would not be safe to speculate how far back the phases of the moon and the lunar tides have operated, though again it must have been a very long time. Nor is the relation between the lengths of a day and of a lunar month during the Earth's past at all clear.

The seasons, a result of the earth's skew axis and its revolution round the sun, have left their marks in fossil tree rings and similar growth patterns of shells and fish otoliths and scales, indicating that the year also is a venerable terrestrial phenomenon. Seasonal changes were thus persistent conditions in which much of life evolved. But a detailed discussion of possible changes in the axis of the Earth, with its attendant climatic sequels, is probably out of place in this essay nor would it be profitable to discuss possible changes in the length of the year.[11]

2 *Biological Time Scales*

The manifold processes which biologists study differ greatly in respect of the time which they take. Before discussing evolutionary time it is perhaps useful to consider various time scales and their mutual relations.

Metabolic rates. The fastest changes in biological systems are called *metabolic* (see the essays in this volume by H. Hoagland and R. Fischer). They concern such functions as are implied in the running of the cellular machinery of respiration, glycolysis and contraction. The rates of these processes are mostly determined by the interaction, transport and enzymatic change of comparatively small molecules like the carbohydrates, organic acids or salts. Restoration of the "steady state" of a cell after some small but measurable disturbance occurs after a time ranging from a small fraction of a second to perhaps a few minutes. The functions of our organs, including our brain, are based on these cell activities and involve periods of comparable durations. There is every reason to believe that the metabolic processes of the plants and animals of long ago were of the same nature as those of our contemporary life, including man, and that they proceeded at roughly the same rates.

Epigenetic rates. Epigenetic processes, which are implicated in the growth, development and aging not only of cells but of large organisms, are slower than the metabolic processes just described. Their rates are presumed to be determined by the synthesis, transport and interaction of such macromolecules as RNA and proteins. In microorganisms the synthesis of single macromolecules may be a matter of seconds, but in higher animals and plants usually many minutes and even hours are required before epigenetic responses become observable.[12] The life span of the higher organisms with all its successive changes is measured in weeks, in years and, occasionally, in centuries. Growth, development and aging occur only in conditions where the faster metabolic processes operate, and like the latter must be presumed to have been essentially similar in the past to what they are today.

Evolutionary rates. Evolutionary changes proceed at yet another, and usually much slower, rate. These rates can be measured in many generations, and may add up to millions of years.

Evolution presupposes the continuing and successful operation of metabolic as well as of epigenetic processes. Harmonious functioning and development impose severe limitations on the possibilities of change and also on its rate. Another constraint is the stability of the environment. Opposing these conservative influences are others which make evolutionary changes possible. The carriers of these changes, the molecules of DNA in the chromosomes, are subject to a multiplicity of influences—for instance, radiation-induced mutation, recombination, shifts in selection, and, at least in man, historical processes which do not operate at the lower levels of metabolism or epigenetics. A realistic consideration of organic evolution and time must take into account both the "conservative" and "progressive" influences. Less stringent constraints on evolutionary rates are discussed in Section 6.

3   *Measures of Evolutionary Changes*

Attempts at estimating evolutionary rates have been made by the practitioners of several disciplines working at different times and using greatly different criteria. This makes comparisons between various rates difficult and general statements concerning a single "rate of evolution" of little value. However, fairly precise estimates of several kinds of specifically described changes are meaningful, and hypotheses concerning their relations are possible. Three major approaches can be distinguished. We can measure: 1) the rate at which genera or species of a certain group of organisms arose during some geological epoch,[13] 2) the rate of change in some metrical characteristics of individuals from two populations forming a lineage,[14] 3) the rate of change in the frequencies of certain genes in populations sampled at different times.[15]

Attempts to relate the rates of morphological changes with the rates in gene frequencies underlying these changes have been rather tentative.[16]

Genus formation. Taxonomists dealing with living organisms are spending part of their time defining and describing the species of their chosen group. On the other hand many paleontologists are, for practical reasons, more concerned with the establishment of genera, the next higher taxa in the systematic hierarchy. Unfortunately, the species concept, and to an even higher degree, the ideas of what constitutes a genus, are highly subjective and controversial. When a genus is represented by one species only it is not clear what is a specific and what a generic character. A species may mean something quite different to a museum curator and to an experimental geneticist[17] and a genus may have a rather debatable rank in the various classes of animals and plants. Furthermore, these subjective entities can undergo considerable historical change. Haldane[18] has pointed out in this context that from the time of Linnaeus good systematists have considered the polecat (*Mustela putorius*) and the ferret (*Martes furo*) to represent not only different species but even different genera, while in fact

335

they give quite fertile hybrids and can therefore be accorded at most a sub-specific difference. Similarly, Dr. P. Leyhausen of the Max Planck Institut f. Verhaltungs-forschung (in a private communication) has assured me that he has good reasons for the belief that most of the "great" cat species would, when brought up together in captivity, not only mate but also produce offspring. Many dog breeds would by morphological criteria alone certainly be put into different species or possibly different genera. But they certainly are mutually fertile. On the other hand different genera of insects and of flowering plants are unlikely to produce fertile offspring. Unfortunately breeding experiments are impossible with fossils and estimates of speciation rates are of necessity based on the dubious and subjective criteria of what is a distinct species or genus. Nevertheless there is some persuasiveness in statements concerning increases in the pace with which new genera appear in certain groups during particular geological periods. The rate of appearance in the fossil records of new forms appears strikingly unequal over longer geological periods: "explosive phases"[19] of between 15 to 100 million years' duration following and being followed by much quieter phases. Possible examples of such rapid diversification are provided by the triassic ammonites, by several groups of mesozoic saurians, by the tertiary horses and elephants, some present-day insects and by many families of the carboniferous and post-tertiary terrestrial plants.

Estimates of speciation rates presuppose that two compared populations are indeed successive stages in one lineage. This, in view of the gaps in much of the fossil record, can be but rarely demonstrated. But some horse and ammonite material may indeed be continuous in time. Thus Simpson,[20] measuring some changes of teeth, has declared the average duration of a horse genus to be at about five and a half million years and Swinnerton[21] has found twenty million years as the average life span of a triassic ammonite species. The, since cambrian times (some 550 million years), unchanged morphology of lingula, a surviving representative of the lampshells (*brachiopoda*), indicates an even slower rate of evolution. The evolution of social insects—which since Darwin's time has posed a difficult problem,[22] is also worth considering in respect of its rate. Termites and ants have certainly existed for tens of millions of years, but the best documented single species probably is the honey bee. According to Zeuner (in an unpublished paper), among the forebears of *Apis mellifica* speciation has been reduced to almost zero, during millions, perhaps tens of millions of years, from the eocene to the oligocene, when considerable morphological changes occurred which were possibly concerned with socialization. Since that time thirty million years have elapsed, with hardly any morphological evolution.

It is tempting to associate high rates of evolution with diversification or change of a group's environment at the time of such developments. In general, then, organisms such as lingula living in a stable environment like the deep sea ought to change more slowly than those living in the more rapidly changing terrestrial or lacustrian habitats.

Metric changes: the "darwin." Changes in the dimensions or shapes of such solid organs as teeth, shells or bones which are preserved as fossils are accessible to biometrical methods. As Haldane[23] has pointed out, we can calculate the rate of change in the mean value of any measurement of such relics coming from populations which we believe to form a lineage, provided we can estimate the time between those populations. Such estimates, based on radioactive dating, are fairly accurate for the duration of the whole tertiary period so that estimated periods of ten million years would not be off by more than a quarter of a million years. Estimates of similar time spans during which mesozoic or paleozoic marine strata were deposited seem far less accurate and may be incorrect by a factor of two or more.

When using measurements for the purpose of estimating evolutionary rates it is necessary to bear in mind Haldane's considerations of a number of axiomatic and of empirical facts. First of all certain measures of structures from successive populations and some indices derived from them may differ considerably, while other measures and indices taken from the same fossils may differ less or not at all. From an evolutionary point of view the former are probably the more interesting, but one must realize that the choice of any particular measurement determines the rate of change observed. When comparing the evolutionary rates of different structures the absolute change of measurements must be transformed into percentages or expressed in terms of standard deviations.

While it is difficult to correlate rates of metrical change with speciation rates, one can usefully compare metrical or allometric changes of homologous structures during different periods and in related organisms. Linear measurements of structures not greatly subject to growth or reabsorption, for instance, of some adult mammalian teeth, the protoconchs of the bivalves or the inner chambers of ammonites, are the best for such studies of evolutionary rates, though occasionally areas or weights will be useful. On the other hand, when dealing with vertebrate bones which vary with the individual's age and condition, it is better to consider a measure of shape, which is fairly independent of these changes. For instance, a ratio

337

of two lengths like the cephalic index or an angle like the mandibular angle or some of the measures of allometric growth. Logarithmic, geometric or other transformations of some of these measurements greatly alter their percentage changes. Evolutionary rates of comparable structures in related organisms must be compared on the same scale.

Examples illustrating the points made above can be found in Haldane's 1949 paper but the details are too technical to be quoted here; only a few facts can be given. The paracones of unworn third molars in four tertiary horse lineages measured at varying periods covering several millions of years increased by several millimeters (to about double height corresponding to an average increase in the paracone height of 3.6% to 10.0% of the original value per million years, from Simpson's 1944 data).

Assuming these horses to have generation times of between two and eight years, the paracone heights in the slowest of these series would have a median evolutionary rate of 2×10^{-7} per generation. According to Haldane's estimation it would have taken the fastest evolving of these horse series about 1.5×10^{6} years or half a million generations to change its paracone height by one standard deviation amounting to a fraction of a millimeter. To establish such a difference one would have to measure very many teeth indeed.

The body length of the representatives of six suborders of mesozoic dinosaurs has been shown[24] to have increased manyfold during periods varying between twenty-two and ninety million years. In the fastest of these series, the sauropoda length changed in thirty-five million years from six to seventeen meters. The data show annual rates varying between 0.26×10^{-8} and 6.1×10^{-8} of the original value per year.

It would be manifestly absurd to compare directly changes in paracone length (millimeters) with changes in body length (meters). For such a comparison changes must be expressed as proportions of an original measurement. For this purpose Haldane has defined a unit of evolutionary metric changes as an increase or decrease in size by a factor of e ($= 2.71828$ ... the base of natural logarithm) per million years, roughly equivalent to an increase or decrease by $1/1000$ of the original length per 1000 years and has named this unit a "darwin." Basing the "darwin" on e does not imply any logarithmic law of evolution but is merely a mathematical convenience for the calculations involved in the study of rates. The definition of the "darwin" means that the logarithm of the character under consideration increases by one in a million years.

The horse rates mentioned above would then range around forty millidarwins, an order of magnitude which in the fossil record seems to have been rarely exceeded outside the primates. Rates of metric change below one millidarwin would be hard to measure and thus one can only estimate a limit for changes in lingula but cannot measure them. At the other end of the scale from lingula are some body measurements in the ancestors of man, the hominids, and among these measurements one indicative of rapid growth in cranial capacity; the length/height index. This probably has changed with an annual rate of about 3×10^{-7}. It took this measure 70,000 years or between 2500 and 7000 generations, to move through one standard deviation—which is twenty-five times faster than the changes in horse teeth and corresponds to around one darwin. Some metric changes in animals in the process of domestication and in plants bred for cultivation have been of the order of kilodarwins.

Changes in gene frequency. If one could breed from fossils one might try to assign the described metrical changes to allelic substitutions in one or several chromosomal loci. As this is impossible one can only think of genetical models which accommodate the observed facts. Again, following Haldane, we can develop such an idea. It is known that certain more or less harmless genes of the mouse (for instance, "brown") may in the homozygote increase some linear measurements (body length) by 1% to 2%. At a rate of 40 millidarwins—corresponding to the previously described dental changes of tertiary horses—a change of 1.5% effected by the (nearly) complete substitution of one allele by the other would take about 300,000 years—perhaps 600,000 generations. "If genes with such a large effect played a part in the evolution of the equidae, the substitution of such a gene for its allelomorph cannot have occurred more often than once per 300,000 years."[25]

Such a slow change could be accommodated by a mutation rate of the usual magnitude (10^{-6} mutations per locus per generation). However, we must assume that evolutionary change is the consequence of selection. If so, the differences in fitness of the three postulated genotypes must have been exceedingly small—far below the possibility of measurement. Metrical differences of 1% to 2% and more between the strains of many domestic plants and animals are well known, and are usually based on a multiplicity of small gene differences. It is thus very probable that evolutionary changes of similar magnitude will also be found to have been effected polygenically and only rarely by single gene substitution. The change in each locus would then have proceeded at a more rapid pace.

We have already mentioned that the rates of metrical changes which man has inflicted on the domestic plants and animals have occasionally exceeded corresponding natural evolutionary changes 10,000 times or more. Correspondingly rapid changes in gene frequency have also been described. As man is tightening his control over ever increasing areas of the biosphere, more and more plants and animals are being subjected to radically changed selective forces. Many species cannot adapt themselves to these environmental changes and become extinct, at first locally, while others for more or less fortuitous reasons continue to thrive apparently by means of individual adaptation. Others have been able to adapt by genetical changes.

An instructive example of a change in gene frequency following selection of known intensity is provided by the decline, during a century, of silver foxes in eastern Canada. Adults of *Vulpes fulva* are silver-colored when homozygous for an autosomal gene *R*, while the homozygotes for the wild type allele as well as the heterozygotes ("cross foxes") and combinations with other genes are not silver-colored and in fact are mostly red. Haldane[26] has analyzed the records of fox skins shot at four Moravian stations in Labrador and at other parts of eastern Canada. He found that the percentage of silver skins steadily declined, constituting between 11% to 15% of the skins collected during 1834–43, but only between 5% to 7% of those collected after 1924. He calculated that the average annual rate of red foxes killed was of the order of 10% or greater, but that 2% to 4% more of the silver foxes (that is 12% to 14% or more) were shot, presumably because their skins fetched a price four times higher. These rates correspond to an average annual decrease in frequency of the gene *R* of about 1.4%.

Rather similar high rates of gene substitution have occurred in several species of moths in those areas of Britain which have undergone industrialization and urbanization during the last one hundred and eighty years, resulting in the substitution of melanic mutants for formerly white ones.[27] This has also been observed on the continent of Europe and in the United States. Kettlewell[28] ascribes this change to bird predation, in that dark moths sitting on sooty surfaces are less conspicuous. He has indeed shown that under experimental conditions birds pick more light than dark moths of the species *Biston betularia* from a dark ("sooty") background and vice versa. In these experiments the frequency of the more conspicuous moths was sometimes halved in a single day, but natural selection was probably much less intense. A selection rate $s = 0.1$, that is, of 10% against the homozygotes (whites) would take forty-two generations to reduce the frequency of a recessive gene from 99% to 1%. Such rapid evolution ("tachytelic evolution,"[29]) is probably rare and may—apart from human intervention—only occur following such rather drastic changes as that from an aquatic to a terrestrial existence. Selection at a rate of $s = 0.01$ takes almost five thousand generations to reduce the frequency of an autosomal recessive gene from 99% to 1%, and it would take thousands or even millions of generations to eliminate it altogether. Such slow changes ("horotelic evolution"[30]) seem to be the most frequent.

Although speciation may be set in train by the establishment in separate populations of two different alleles, such a difference would be rarely discovered in fossil material. We can safely assume that species or taxa[31] which can be distinguished by the paleontologist will differ in many loci, perhaps in a hundred or even a thousand. Now these differences must have developed either by a succession of a great number of separate gene replacements, which in any case would take a very long time, possibly millions of generations, or they have been the result of a prolonged process of coadaptational changes occurring simultaneously at many loci. (This process is sometimes described as the accumulation of modifiers.) It might be thought that such simultaneous selection in many loci would be much quicker, but Haldane[32] has advanced good reasons why this is not so. Sometimes several genes must be substituted in the same individual before fitness is increased. Such individuals would be rare, and unless the substitutions are linked in an inversion they would separate during meiosis, so that the establishment of any such complex change would take very many generations. On the other hand clonal and asexual selection of complex gene combinations can be rapid.

Haldane has estimated that in horotelic evolution it takes on the average 300 generations for one gene substitution to become fully established. If two species differ at 1000 loci, it will take at least 300,000 generations to produce their interspecific difference. This would roughly correspond with the estimates of Zeuner[33] concerning some pleistocenic mammals (500,-000 years) and insects. Voles seem to have evolved more rapidly and other insects as well as carnivores[34] more slowly.

Summing up, there is thus a very great range in the rates of evolutionary change. At the extremes we have lingula, which has hardly changed during the 550 million years which have elapsed since the cambrian period. At the other extreme are the almost instantaneous changes of cyclomorphosis, periodic selection and polyploidy to be mentioned in Section 6.

The slow processes of the past are only dimly perceived from the fossil record and cannot be genetically interpreted; however, those changes which occurred almost while we were looking can be described in terms of gene frequencies, fitnesses and the other entities of dynamic population genetics. How far the great speed of these recent and often man-induced changes invalidates them as models for the mechanics of the very much slower natural evolutionary processes is a moot point.

4   *Years or Generations*

In his paper on the quantitative measurement of rates of evolution Haldane[35] has also discussed the relative merits of measuring evolutionary time either in years or in generations. If evolutionary rates depended on mutation rates, which Haldane considers doubtful, although it probably does apply to clonal selection in bacteria, to tissue cells in some plants and to certain parthenogenetic animals, the year would be as natural a unit as the generation. For mutation rates—whether spontaneous or radiation-induced—may equally be measured in years or in generations. If, however, as is more likely, evolution depends more on gene recombination and selection, then the length of a sexual generation is the more suitable unit of evolutionary time.

This measure has, however, considerable imperfections. The length of a generation is itself subject to evolutionary change and must have greatly increased in many phyla. The generation numbers in the diagram are consequently of a rather hypothetical nature. In many living species the length of a generation varies geographically; it is difficult to measure and, for fossil forms, it can only be guessed. Many insects and flowering plants propagate only once a year, while others, often closely related forms, produce two or three generations per year, and yet others take two or three years to mature. In species where several generations overlap one could in theory calculate a mean generation length from life and fertility tables. But he would be a bold man who would propose a definite figure for any such species, even for Western man. Thus only crude estimates of generation lengths can be made for most living organisms. For fossil forms one can, I think, follow Haldane's extrapolation and assume, for instance, that "for a small rodent the mean generation length was less than a year, for most ungulates more; for mammals not larger than a cow under ten years. . . . The longest mean interval between sexual generations is probably to be found in clonally propagated invertebrates such as corals, where it may possibly extend over centuries."[36] To these one may add some very long-lived trees. Taking all these uncertainties into account, it is nevertheless possible to calculate an approximate length of generation for a variety of organisms and to estimate the number of generations which have elapsed during a stated number of years or since a geological date, as was done in the diagram.

5   *Some Special Interactions between Time and Evolution*

Generation Times in Biologically Associated Species. The mutual genetic adaptation of two species living in some form of association is conditioned by the duration of a generation of each of these two species. This applies to symbiosis, parasitism, prey-predator relations, various forms of mimicry and other forms of interactions, of which we shall only consider a few. The life cycles of the associates are often synchronized in simple or complicated ways or alternatively other means of adaptation are employed to ensure survival of two participants of widely differing length of generation.

The great rate of multiplication of bacteriophages which divide once every few minutes makes them rather quickly exhaust the material of their more slowly growing bacterial hosts, leading to the latters' death (lysis). But for the occurrence of latent non-multiplying phages unchecked bacteriolysis would have long ago resulted in the extinction of all phage-prone bacteria together with the extinction of the bacteriophages. This course of events was presumably obviated by a mutation of the phage enabling it to suppress multiplication under certain conditions. These mutated phages survived and may have given rise in turn to genetic changes in the host bacteria.

Spontaneous as well as "induced" changes in bacterial and protozoan virulence which may compensate for high multiplication rates have been reported by epidemiologists and chemotherapists. It seems, for instance, that just before the First World War—many years before the advent of the sulfa drugs and of antibiotics—scarlet fever underwent a worldwide decrease in severity, presumably through a decrease in virulence on the part of the streptococcus responsible for the disease.

If we consider the treatment and prevention of bacterial and protozoan infections as a battle between these organisms and man we could say that the parasites—for instance, penicillin-resistant *staphylococcus aureus* or arsenic-resistant *trypanosomes*—are pitching their rapid evolutionary rates, and the resulting great numbers of individuals and mutants, against the skill of the modern doctor and the synthetic chemist. Very few, if any, of these battles have as yet been completely won by man.

Mutational responses defending the slower multiplying hosts against rapidly multiplying parasites have also occurred and may have been important in many different ways. It is, for instance, believed that the genes responsible for sickling, thalassaemia and the G6PD deficiency which occur in human populations exposed or formerly exposed to malaria are means of defense against the plasmodium parasites[37] and that the predominance of certain blood groups in different populations can be explained by their different effects in resisting such diseases as plague, cholera, typhus or syphilis.[38] But more mutations of this kind must have been rare, and their establishment by selection slow when compared with mutational and selective changes in the parasites. One is indeed obliged to explain the observed frequencies of these presumed, antiparasitic genes in man, not by the circumstances of the present, but by assumptions concerning a distant hypothetical past.[39]

Quicker reacting defense mechanisms have, however, evolved. Mammals and birds have developed immune mechanisms, which are independent of meiosis and seem especially designed to counteract the genetical flexibility of the parasites attacking them. While many details of the immunity mechanisms are still controversial,[40] it can be stated that the acquired specific immunity of human white cells, whatever its mechanics, arises at rates comparable to parasites' mutation rates.

Evolution of Timing Mechanisms. Generation lengths as well as timing mechanisms are significant factors in the evolution of systems of parasitism which include insect vectors. Interesting examples are the microfilariae, small nematodes, who at certain hours of the day swarm in the peripheral blood of man, dog or other mammals, while for the rest of the time they aggregate in the lung capillaries.[41] The particular hours of swarming of the different kinds of microfilariae at different localities appear to coincide with the hours of flight and biting activity of the local insect vectors, which carry the disease from host to host.

On some islands, where microfilariosis has been imported during the last hundred years or so, the filariae have "learned" to adapt their swarming hours to the local vector's activity patterns. This might have been achieved by varying the reaction time to a time cue, possibly the CO_2 concentration in the hosts' lung alveoli. It is tempting to ascribe the change to an initial single mutation.

Time as an Isolating Factor in Speciation. It is generally conceded that spatial isolation is an important factor[42] in speciation. Insofar as distance is implicated in this situation, spatial isolation could of course be equated with temporal isolation. However, time may also act as an isolating factor in many other ways. Differences in the hourly activities exhibited by the microfilariae may in the past have been the first steps in the splitting of a formerly common species. Differences in mating response to diurnal or lunar (tidal) rhythms may also have occurred, but these are not well documented. On the other hand it is known that some fish, for instance, certain species of charr (*Salvellinus*) and mullet (*Mullus surmuletus*) and cephalopods (*Sepia*) form separate populations, which aggregate and mate at different times either in the same or in different localities, though living in the same general area.

Considerable sexual isolation also occurs in multi-annual species such as in cockchafers and in several species of cicadas, which are characterized by regularly recurrent years of abundance, in themselves indicative of isolation. Biometrical studies may show whether in fact slight differences between these various strains have developed.

6   Time Available and Time Necessary

The extraordinary span of time during which life has existed on Earth, perhaps three thousand million years, is quite generally assumed to provide ample opportunity for any complexities of evolution one might think of. Abundant time is also thought to be indispensable for the operation of evolution. From at least one point of view both assumptions seem to me quite mistaken.

Permutation, Mass and Time. Let us consider the severe and indeed staggering limitations which time and mass impose on the realization of organic potentialities. Looking at evolution simply as a change of genes, we can discuss only the potentialities of one single species. How much time and how much mass would we need to produce, by the quickest means, all possible permutations of known mutants of the fruit fly (*Drosophila melanogaster*), one animal species among more than a million? Bridges and Brehme[43] list about 450 mutating loci in this species and it is conceivable that two or three technicians might just about keep a similar number of mutant stocks going in a laboratory. But what are the numbers of permutations of these mutants and what are the chances of ever producing all of them or even a large subclass of them? Let us assume that only two alleles would be represented at each of the 450 loci (in fact many polyalleles have been described), and that we want to have only one of every possible homozygous and heterozygous female. Their number would be 3^{450} or 10^{215}, though many of them would not be viable. One *Drosophila* fly roughly weighs one mg.; 10^9 of them weigh one ton. The mass of the Earth is about 6×10^{21} tons, the mass of the solar system about 2×10^{27} tons. Thus if one could transform all the Earth's mass into *Drosophila* females one could still only accommodate fewer than

345

$1/10^{185}$ of the enumerable types and if one could use the mass of the solar system, fewer than $1/10^{179}$. Thus the solar system is much too small to accommodate all permutations.

If one does not demand all the permutations to exist on one particular day, but explores whether they could have been produced during the 2000 to 3000 million years of organic evolution, one finds time also to have been far too short. Assuming that a generation of D. melanogaster lasts about a fortnight under laboratory conditions, this period is equivalent to between 40,000 and 60,000 million generations. If we assume that as before the entire Earth were made of female Drosophila and that these would change twice every month, this would still not quite accommodate $1/10^{173}$ of the enumerable permutations—an infinitesimally small fraction. Considering the enormous underestimates in the above assumptions, we can say that the discrepancy between potential and realized variability is quite staggering. Selection operates on the minutest fraction of the genotypes that might exist, and as was said before,[44] it is not the fittest permutations which are selected but only the best available at a particular time and place—a vastly different proposition. Evolution might thus be said to be short of time, and this may perhaps explain why it so often ends in failure. The above considerations are all based on rates of change which are vastly in excess even of the highest rates observed. Substituting these into our calculations would result in still more stringent constraints on the scope of evolutionary possibility and the rates of evolutionary changes.

The Minimum Time for an Evolutionary Change. Equally erroneous is the idea that evolution must in all circumstances be an exceedingly slow process, taking millions or at least thousands of years. In Section 4 of this essay we have already shown that evolutionary rates vary very greatly. Here only a few examples of quick changes will be given. Changes of shape (cyclomorphosis) on a vast scale can be observed every year in several lacustrian cladoceran crustaceans of the genera Daphnia and Bosmina,[45] and "periodic selection"[46] may completely change the make-up of a bacterial population even within days. In many flowering plants autopolyploidy, the duplication or higher multiplication of the chromosome number, has suddenly given rise to a new isolated "species," as happened in blackberries and many garden plants. During the last century hybridization, followed by polyploidy, produced a new grass Spartina townsendi, which now covers vast areas of Britain. New "species" may thus on occasion arise almost instantaneously.

7 *Evolution as a Stochastic Process*

Nonrepeatability (*irreversibility*). The directional character of evolution first obtruded itself naturally enough to paleontologists as soon as they tried to systematize their material. Based on numerous findings of his teacher, Dollo, published between 1890 and 1923, Abel[47] formulated his famous "law of evolutionary irreversibility." Unfortunately this inept term has been taken up by people having little biological insight and this has created considerable confusion. Dollo's thesis does not mean that a structure which has become useless after an environmental change cannot be lost—it obviously can, for instance, the fins or gills of animals after they have left the water; but it does mean that identical structures are not again acquired should the animals revert to the original environment, in our example, to the water. This could perhaps be better described as nonrepeatability rather than as irreversibility. Famous examples are certain littoral athecous turtles (soft shelled turtles of shallow seas) which are supposed to be descended from pelagic tortoises (tortoises of the high seas), themselves presumed to have had in turn littoral turtle ancestors. The charapac (shell) of the secondary littoral turtles is derived from quite different structures than the charapac of their primarily littoral ancestors. This fact shows that the loss of the original structures during the pelagic phase of this evolutionary sequence could not be reversed. Another instance is the teeth of fish-eating dolphins which resemble the teeth of fish, amphibia and reptiles. But the ontogenesis of these secondary "primitive" teeth clearly shows traces of complications explicable by their terrestrial ancestry. All this does not add up to a "law," but perhaps the modest claim could be defended that as yet no case of reappearance of a previously lost complex structure has been convincingly demonstrated.

However, some small scale reversals have been reported: Kurtèn,[48] for instance, describes the reappearance of two structures in living lynx both of which are absent in the presumed ancestors of the species and are missing in the other living felidae: about 10% of his lynx sample had a second lower molar tooth and 30% to 100% a characteristic complex structure on the lower carnassial tooth. Kurtèn mentions that lower second molars had, however, occurred as a very rare abnormality in other cats and one might also state that dental peculiarities of a similar kind are quite common in man. Comparable dental variation also occurs among the individuals of inbred mouse strains as well as between different mouse strains.[49] Thus, this sort of variation exists not only in successive populations (lineages) but also within populations or sub-populations of animals living at the same time.

Taking these findings as our basis we shall now attempt to develop a theory which will account for over-all nonrepeatability, but also allows some small-scale reversals. Before doing this a short quasi-historical discussion of "irreversibility" is, however, necessary.

Like the "logic of history," which "never repeats itself," evolutionary irreversibility has either been taken for granted[50] or it has been understood to be a consequence of the "arrow of time"[51]—which ensures that nothing can ever be the same again. While this sounds plausible and is probably correct when one deals with large parts of reality, it need not apply to certain specific features of a species and, accordingly, it can hardly be called a universally valid explanation.

Evolution and Thermodynamics. Attempts have been made to derive the directiveness of time in general from probabilistic considerations, in particular from the Second Law of Thermodynamics. The point of view adopted here is that for the directiveness of organic evolution we can accept a probabilistic explanation but not one derived from the Second Law. Briefly and crudely put, this thermodynamical derivation is argued as follows: Life is a complex chemical process. Chemical processes are characterized by an inescapable increase of entropy with time. Thus, identical states of energy cannot be repeated and evolution is irreversible.

To this argument many objections have been raised. Planck[52] has called it a "singularly unhappy attempt." Needham[53] pointed out that evolution, far from being a running-down process leading to "mixed-upness,"[54] looks more like a process designed at defeating entropy, in that it produces more and bigger systems of complexity and "improbability." A physical chemist, Clark,[55] has even described the Second Law as a "law of morpholysis" or death of forms. Most but not all biologists would subscribe to these judgments. Geneticists[56] for instance might consider evolution as a sorting out and gradual accumulation of coded information and certainly not as a mechanism for the dissipation of information. Recently Keyl[57] has demonstrated an increase with speciation of the number of chromosomal bands in the salivary chromosomes of *Chironomus*-midges. This band-duplication enables the animals during speciation to put more coded information onto the increased length of the DNA strands in the chromosomes. A general increase in the length of the chromosome complement and of the DNA strands contained in the chromosomes[58] is also apparent when one ascends from viruses via plants and lower animals to the vertebrates and man.

348

This seems to indicate the operation of some stochastic process, but not any loss of information. However, before pursuing this idea—a promising one in our opinion—it might be useful to consider the origin of the idea that organic evolution is a demonstration of the Second Law of Thermodynamics. This "thermodynamic heresy" appears to have arisen from mutual misunderstandings between physicists and biologists. Some physicists seem not to have realized what biology is. Eddington,[59] for instance, considered that "To restore the stone to its previous position we must supply extraneous energy which has the required amount of organisation." As Needham pointed out, Eddington did not realize that organization implies something above the physico-chemical level and that it is not synonymous with order.

Biologists for their part did not consider when and at what level thermodynamics is relevant for organisms. Lotka,[60] though very impressed by the power and formal beauty of the Second Law, nevertheless proposed more appropriate probabilistic models for evolution, but many others have been led astray. As a non-physicist one might argue against the application of the Second Law to evolution on two counts.

1) In a general way one could argue that the Second Law might not be as universally valid as is sometimes claimed, otherwise the physical world would not be what it is. To explain the present state of the universe one must assume either that before or during periods of "creation" entropy would decrease, or that it does not decrease now. It might not be unreasonable to assume that in the universe as a whole energy degeneration is being counteracted by creative processes of various kinds.

2) Even if one does accept that the whole universe forms a closed system and that therefore thermodynamics may apply to it, one could still maintain that biological systems are "privileged parts" of the whole where entropy decreases at the expense of other parts and where some canonical laws hold but not the Second Law. In any case, during the four thousand million years or so of the Earth's existence, there was enough radiant energy coming from the sun to maintain life. Evolution may or may not have increased the proportion of solar energy utilized in the processes of life, but there is no sign of any decrease of biological activity. Thus there is no reason to suppose that evolution has anything specific to do with an increase in entropy, probably quite the opposite. If for argument's sake we assume with Blum[61] that the Second Law holds even when we cannot make the necessary measurements to demonstrate its validity in a particular situation, as is the case with evolution, this is quite a different proposition from asserting that it has any *explanatory*

value for the processes of evolution or that it can account for its relative irreversibility.

Stochastic Models. There is, however, a different way in which statistical ideas can be used to explain the unidirectional character of evolution and which, in fact, has already been foreshadowed by Dollo himself. Needham quotes him as stating that,

in the last analysis it [*i.e.*, his law] is like other natural laws, a question of probability. Evolution is the summation of determined individual variations in a determined order. For it to be reversible there would have to be as many causes coming in the inverse sense, as those which brought about the individual variations [mutations as we should say today] which were a source of the prior transformations and their fixation. Such circumstances are too complex for us to suppose that they ever exist.

This essentially is also Lotka's[62] and Muller's[63] point of view.

Developments during the last decades in the theories of population genetics,[64] information theory,[65] and the stochastic processes[66] makes one hopeful that a competent mathematician may some day give a more precise form to Dollo's statement. He may also develop mathematical models which would permit a degree of reversibility corresponding to observed instances.

The probabilistic view of time is discussed by Satosi Watanabe later in this volume. Here we can only indicate the way in which the evolutionary aspect of this problem might be approached. If the genetical composition of a population at a particular time (t_1) is regarded as a store of discrete units of information, the genetical composition at a later time (t_2), for instance, one generation later, will be affected by mutation, selection and chance. If there is any genetic variability in the population, any particular outcome of those events will have a very small predictive probability (P_1) because there exist a very large number of alternatives. These are very much less numerous than the possible permutations but still enormous. A complete reversal at a yet later time (t_3) back to the original state at t_1 has a similar, usually even smaller, probability (P_2). Thus it is almost certain that at time t_3 yet a third situation will arise (with probability P_3) which is different from both the previous situations and, except in very special circumstances, this third situation is even more unlikely to revert at time t_4 to the state at t_1—the probability of this reversal transition being in the nature of a product of the previously stated small probabilities. A model of this kind clearly has the general properties of irreversibility and nonrepeatability.[67]

We shall now consider how it can be modified so that it exhibits some degree of reversibility. In the limiting case of clonal propagation and in the absence of mutation no genetic change occurs and the probability of the same situation (measured in gene frequencies) continuing is, at all times, unity. A single mutation in such a clone might easily be reversed. In sexual organisms, however, the situation is slightly more complex. For instance, one autosomal allele gaining in frequency at the expense of its alternative changes a population in one direction. But this might be reversed during a subsequent period, for instance in conditions of stable gene equilibrium[68] or under oscillating conditions.[69] Among a very large number of systems of this kind one should thus occasionally observe such a reversal, and that is, in fact, what may have happened to Kurtèn's lynxes and in other instances.[70]

That simple changes in gene frequencies are singly reversible and in fact characterize conservative genetic situations[71] is probably unimportant for organic evolution. As a rule it is the improbability of repeating long chains of events which results in the uniqueness of lineages. Certain ecological "types"—horses, elephants, aquatic mammals, trees, shrubs or climbers and many others—have, under similar environmental pressures evolved many times. But the resulting types always remained separate and demonstrably different in details of structure. Such diversity has also been shown experimentally, for instance, in sister colonies of *Drosophila pseudoobscura* derived from a common population; these usually diverge in some characteristics so that even small-scale evolution is not, as a rule, completely repeatable.[72]

8   *Conclusions*

The concept of organic evolution presupposes the idea of time as something unidirectional. Evolutionary events have been arranged in succession and dated by the methods of geochronology.

Evolutionary processes are slower than metabolic or epigenetic processes and can only proceed, while the others are at work.

Evolutionary change has been defined by the criteria of genus formation, by metrical changes or by changes in gene frequencies. Rather different values emerge when rates are calculated from the different approaches. When measuring different parts of the same individuals in the same lineages some change more rapidly than others. Thus the validity of the "darwin" as a general yardstick for evolutionary change is rather limited but it has its uses for comparative studies. There is no such thing as a "general rate of evolution."

As gene recombination is more important for evolution than gene mutation it is more reasonable to measure evolutionary rates in generations than in years.

The synchronization in respect of external rhythm of symbiotic organisms is of considerable evolutionary importance as is in a different way temporal isolation. Mass, time and to a lesser extent the "conservative forces" of genetics impose severe constraints on the potentialities and the speed of evolution.

Organic evolution can only very slightly and for a short way be repeated or reversed. Its irreversibility on a larger scale should not be deduced from the Second Law of Thermodynamics but from other stochastic models.

Comments Concerning
Evolution

As we transfer our attention from periods meaningful in the life of generations of men to periods comparable with the geological age of the earth, we find that the nature of our inquiry into the meaning of time changes. Instead of the psychologically and philosophically ambivalent creative-destructive time apparent in our lives, paleontology suggests a purposeful time of complexification, that is, essentially a process of creation. Earlier we have observed that in the case of the individual, time awareness relates to such matters as desire to control his environment, or to his fear of death and, at least from one point of view, to his psychological revolt against individuality. In man we were able to give meaning to ideas such as *purpose* and *personal identity* because we have assumed, though not stated, an analogy among all members of our race and because we have had at our command that wonderful tool of communication, language. But, how are we to inquire into the same problems of innumerable varieties and generations of creatures whose "inner life" by the analogy argument we cannot even guess and with whom no exchange of articulate thought is known or might, indeed, be possible? From being "insiders" who are in the company of the like, in the study of phylogenetic evolution we perforce become outsiders in the company of the unlike. This change unavoidably alters our methods of study. Individuality is replaced by the study of groups or kinds of organisms; individual purpose by the study of the restraining as well as of the generating schemes of life.

Organic evolution is not a smooth but rather a jerky progression involving, as it does, periods of rapid speciation followed by long spans of stagnation. It is precisely the eventful portions of this historical process, the surprisingly short periods in which species evolve, that constitute the most interesting single aspects of evolution from the point of view of time. A dramatic illustration is the very brief period of perhaps a million years in which the humanoid features of the brain have developed, as compared with the over two billion years in which life is believed to have existed on earth. In this connection purpose and plan in nature have often been discussed under many different names; what has seldom been emphasized is that directedness arguments tend to relate, by their very nature, to the impressive economy of evolutionary time rates. It seems that even the philosophy of evolution had to make allowances for this by calling for an indeterministic type of determinism which is "historical and not mechanistic," and which permits "multiple solutions and not only a unique outcome. It is therefore both nonrepetitive and nonpredictive."[1] The latitude of this view runs counter to those held earlier this century, that "the facts of biology encourage us throughout to pursue the reduction of biological to physical laws further and further with fresh hope."[2] Physico-chemical determinism, or just plain determinism had to be adjusted so as to accommodate accomplishments which appear to be unlikely, and account for a series of events which happened in a much shorter span of time than we might think they ought to have taken. But if so, perhaps the methods of speciation are such as to favor time economy.

As understood today, natural selection, the major tool of evolution, involves a type of feedback process[3] operating through certain agencies and constraints such as genetic variation, relation to environment, geographic separation of segments of population and relative success of reproduction. The existence of any temporal constraint in the feedback loop of natural selection could appear to be goal-directedness in the form of plan and purpose.

One philosophical objection which may immediately be raised against speculations involving directedness in evolution by the operation of some principle of time economy is that it introduces time *deus ex machina* to replace some extra-natural agency. This, however, is not necessarily so, as may be seen if reference to the intuitive content of the Minimal Principle of Physics is permitted for heuristic purposes in the context of the overwhelming complexity of life. A well known example of the Minimal Principle is that attributed to Fermat, also known as the Shortest Time Principle of geometrical optics. Fermat's Principle has been cited by some as an example of goal-directedness in nature. Such a philosophical point of view is rejected by most physicists, but not the physical content of the

Principle. In fact, it has been convincingly argued by Prof. Whitrow[4] through reference to the mathematical properties of certain physical relations that we cannot deduce goal-directedness in nature from the formulation of minimal laws, neither can we disprove the existence of such purpose as a consequence of our disability to do so.

The biochemical considerations of time suggest the necessity of homothermy for keeping the biological clocks of the individual in phase with other organic and inorganic clocks, and thus make ecological rhythms and social organization workable.[5] It would, therefore appear reasonable if evolution, through its many methods, would generally favor homothermic ability. In the view of Slobodkin,[6] homeostatic ability in general is, indeed, the only feature which is always maintained at a relatively high level in the evolutionary process, all other features of the organism being expandable. He speculates that "Evolution, in fact, is simply a consequence of the general homeostatic ability of organisms combined with the biochemical properties of genetic material." If this assertion truly reflects the techniques of evolution or at least an important segment thereof, it might be possible to assess with some rigor the influence of man's knowledge of time, and the effects of rudimentary time sense in those animals which manifest other than very short-term memory, via the necessity of homothermy, on homeostasis and finally on organic evolution.

Alternately, it is conceivable that our understanding of the temporal aspects of evolution may not become possible until after we have reached the point "at which the theory of evolution finally bursts through the bounds of natural science and becomes entirely an affirmation of man's ultimate aims."[7]

Leaving speculation for established science, we note that the biological advantages of time sense in the animal kingdom are well known, as was described by Prof. Cloudsley-Thompson. The role of time sense in the socio-cultural evolution of man is also quite evident.[8] Insofar as biological and cultural evolutions are interacting on each other, man's ability to learn from the past and prepare for the future must have also influenced his biological evolution. In this respect, Adolph Portmann remarked[9] that "In the case of man, natural evolution by means of the crossing-over of genes, selection and mutation can take place simultaneously with cultural evolution. But the cultural mode has come to be the more effective and dominant mode. The time is past when natural evolution could rival or even surpass the cultural evolution of mankind. Today already, the opposite appears to be the case."

In my view, our knowledge of time as an agency in natural selection pertaining to man himself is embodied in practical wisdom expressed, for instance, simply and powerfully in the Book of Ecclesiastes:

> To every thing there is a season, and a time for
> every purpose under the heaven;
> A time to be born, and a time to die; a time to plant
> and a time to pluck up that which is planted;
> A time to weep and a time to laugh; a time to mourn
> and a time to dance;
> A time to get, and a time to lose; a time to keep and
> a time to cast away;
> A time to rend, and a time to sew; a time to keep
> silence, and a time to speak;
> A time to love, and a time to hate; a time of war,
> and a time of peace.

J. T. F.

Biological Time

ROLAND FISCHER *

Between A.D. 725 and 1370 the Chinese constructed and operated elaborate astronomical clocks. Since then a voluminous literature has accumulated on the measurement of time. What it is that we measure, however, has been questioned by philosophers and is still an open question. St. Augustine's answer, from *The Confessions*, Book 11, is *"In te, anime meus, tempora metior"* ("It is in you, O my mind, that I measure time. I do not measure the things themselves whose passage produced the impress; it is the impress that I measure when I measure time. Thus either that is what time is, or I am not measuring time at all"). Augustine's interpretation has been considered for sixteen hundred years without substantial clarification of the basic nature of this cardinal feature of our existence: "We have a strong inclination to believe that the 'sense of time' of men and other animals, *i.e.*, the sense of differentiating between durations of varying length which have elapsed since a definite event, is based on nothing else than the strength of traces left by this event at various moments after its cessation. The weaker these traces the more remote in time the given event seems to be."[1] The main reason for the stagnation is inherent in human existence, of which time is such an integral part that objective analysis is difficult if not impossible.

The physicists Feynman, Leighton and Sands[2] state that "what we mean by 'right now' is still a mysterious thing." Efron[3] believes that the "now" is the moment of arrival of sensory data in the dominant temporal lobe. The peculiar slant in our contemporary understanding is that we believe that "the 'now' depends on the coordinate system."[4] This paragraph summarizes the progress we have made in clarifying the time concept of St. Augustine.

When a living organism is the "coordinate system," the concept of bio-

* This essay is dedicated to Dr. Paul Larose, Ottawa, with the author's affection and fondest regards.

logical time has to be introduced. Biological time, then, may be defined as a creature's awareness of its own duration and location within a restricted frequency range, in contradistinction to traditional concepts of time which attempt to be independent of life. What we commonly call "measuring time" is a physical process which, like the hands of the clock, measures space rather than time. Biological time, however, is the transformation of perceived sequences into durations. *"Du selber machst die Zeit, das Uhrwerk sind die Sinnen."**

In the first part of this essay, we will deal with time and space on biological organizational levels of increasing complexity—from the molecular to the organismic—and during states of excitation and tranquilization. We will examine the contraction of time or the expansion of space, that is, space-time "equivalence," observed on these levels and states, in the perceptual, psychological, physical and psychophysical contexts and we will stress biological perception as their common origin.

In the second part, life and time are treated as a unified process consisting of evolution, learning, perception and hallucinations and dreaming. These adaptive events are characterized by an increasingly efficient utilization of energy as well as an exponentially increasing time rate of change.

I would ask the reader to remember one sentence from Whitehead, already quoted in this volume by Prof. Benjamin, which was my *Leitmotif* when writing the present essay. "It is impossible to meditate on time and the mystery of the creative passage of nature without overwhelming emotion at the limitations of human intelligence."

1 *The Biological "Equivalence" of Space and Time*

Levels of Energy Metabolism. We have asserted that biological time is time within the "coordinate system" of the living organism. It is from this vantage point that we intend to view the various levels of our own organization to develop fully the concept of biological time. We shall start with the "equivalence" of space and time as it presents itself on various biological levels. Let us first turn to organismic life time and analyze the relations between the size of a metabolizing unit in space, its metabolic rate, and its duration in time.

On the *molecular* level of metabolism, the concentration of certain respiratory enzymes, such as cytochrome-*c*, cytochrome oxidase, etc., per unit of metabolically active mass ($kg^{3/4}$) is proportional to over-all energy exchange per kg—for rats, dogs, men, and cows.[5] The amount of mito-

---

* Angelus Silesius, *Gesammelte Werke* (Munich: Carl Hanser Verlag, 1949), p. 28.

chondria (that is, the tiny "powerhouses" within the cell) bears the same quantitative relation to total metabolism as to total body size, and it is likely that the relative amounts of these elements in any given tissue will prove to be the controlling factor in determining the regression of oxygen utilization on total body size of the species.[6]

In a broad and metabolic sense, *cellular* life span is closely correlated with the life span of certain cellular components or metabolites. Thus, the half-time of cytochrome-*c* in liver may be as representative of the turnover time of the liver cell as is the half-time of hemoglobin of that of the red blood cell.[7]

The mitochondrial *energy transduction* in aerobic organisms is based on citric acid cycle oxidations and fatty acid oxidation, coupling electron flow to synthesis of adenosine triphosphate (ATP), that is, oxidative phosphorylation. The components of the electron transport system are currently thought of as consisting of a series of physically interconnected lipoproteins with electron acceptors (coenzymes) tightly bound. The terminal catalyst appears to be *cytochrome oxidase* and may be the *rate-limiting factor*.

Other studies substantiate, although indirectly, the function of cyto-chrome oxidase as a rate-limiting factor. Chance's work[8] indicates that in the intact mitochondria, the cytochrome oxidase velocity constant is 4- to 23-fold less than that of the other reactions. Another evidence is that the absolute values of the total cytochrome oxidase activity in certain large as well as small animals were found to be very nearly the same as the maximal metabolism of intact animals.[9]

With cytochrome oxidase as a rate-limiting factor, increasing body size is known to be accompanied by decreasing respiratory intensity—in analogy, a small clock must have a faster swinging pendulum than a large one—*in vivo* and *in vitro*,[10] the duration of the metabolic process, life, can be conveniently expressed in terms of oxygen consumption (QO_2) per unit body surface. It becomes apparent that all animals, large or small, homeo-thermic or poikilothermic, burn the light of their lives with relative equality. Life, at least on the *organismic* level, is a democratic process: all of us must die, and the duration of our existence is the same—if measured in QO_2 per unit body surface. There is good evidence in the literature to support such a contention. Sacher,[11] for example, concludes that the mortality rate of different species and the rate of increase in mortality with age in the same species are positively correlated. This correlation forms a possible basis for the relative constancy in his life table for species as widely differing as the fruit fly, the mouse and man. In 1959, Sacher described a statistical analysis of the relation of life span to body weight for sixty-three species of mammals.[12] He established the quantitative dependence of the life spans of these species on body weights, brain weights and metabolic rates of adult representatives of the species. He regarded the life span as a physical dimension of a species on the same footing as linear or mass dimensions. In reformulating Rubner's theory of aging, he states that the life span of a species varies inversely as its basal metabolic rate. Fischer, Griffin and Liss[13] come to the same conclusion when comparing the above relation for men and women.

Now we may deal with the relation of time and space on the *perceptual* level. Percepts of space and time are related to metabolic rate since changes in the latter bring about concomitant perceptual changes. Physiological clocks run fast when metabolic rate is increased, while clock time is overestimated, subjects arrive early to appointments, time appears to pass more slowly. When the physiological clocks run slowly (corresponding to a decrease of metabolic rate), clock time is underestimated, subjects arrive late to their appointments, time flies by rapidly, the days seem to fly by "like magic."

Another manifestation of the relation between metabolic rate and time sense is exemplified by Lecomte du Noüy's experiments.[14] In these experiments, decreasing rate of wound-tissue regeneration and oxygen consumption for a constant surface area (space) is obtained with increasing age, that is, with decreasing metabolic rate (see Figure 28 in du Noüy). For example, du Noüy calculated the impression of "our passage" in time for a twenty- and fifty-year-old man to be four and six times faster respectively than for a five-year-old child.

It is particularly rewarding to study the relationship between time perception and metabolic rate of subjects during states of *excitation and tranquilization*, since the former state is connected with an increase in metabolic rate produced by hallucinogenic, pyretogenic (fever-producing) drugs such as mescaline, D-lysergic acid diethylamide (LSD), Psilocybin, etc., and the latter with a decreased metabolic rate produced by hypothermia-producing tranquilizers of the phenothiazine type. The relatively small, measurable increases in metabolic rate and temperature are only general systemic reflections of a specific localized (neuronal) excitatory process.[15] Psilocybin or LSD not only raises the body temperature and thus produces an overestimation of clock time, "time contraction,"[16] but also produces a simultaneous "expansion of space."[17] The latter phenomenon can be illustrated by the size of handwriting samples taken under standardized conditions before, during various phases of, and after the excitatory drug-produced hallucinatory experience.[18] Furthermore, a sequence of drawings executed during a hallucinatory experience induced by the ingestion of 100 μg LSD demonstrates the gradual expansion of space and its return to the initial level.[19] The administration of tranquilizers, on the contrary, results in a contraction of space, that is, a significant diminution in the size of the handwriting.[20] The size of drawings is also decreased if executed by patients under the influence of tranquilizing drugs.[21]

The upper part of Figure 1 shows the increase in handwriting space at

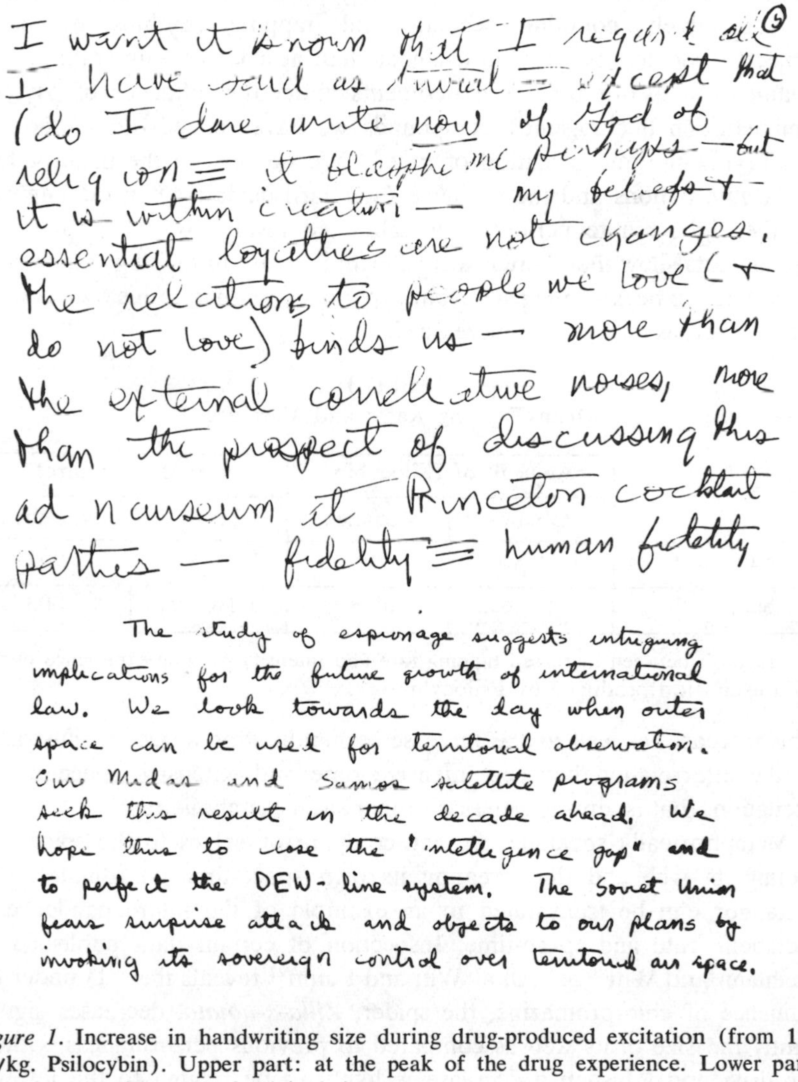

Figure 1. Increase in handwriting size during drug-produced excitation (from 115 μg/kg. Psilocybin). Upper part: at the peak of the drug experience. Lower part: control handwriting (⅗ of original size).

the peak of a Psilocybin experience, two and one-half hours after the ingestion of a mild dosage (115 μg of the drug per kg body weight). The lower part of the figure depicts the control handwriting of the same person one day later at the same hour of the day.

While nearby space, in this case, handwriting space, *extends* under the temperature and metabolic rate-increasing influence of psilocybin, time becomes *intensive*. It should be recalled that H. Hoagland's "Arrhenius plot"[22] depicts the effect of internal body temperature on the frequency of subjectively counting seconds and tapping rhythms by naïve subjects. One counts faster at a higher than at a lower temperature.[23] In continuation of our tapping experiments under the influence of psycho-tomimetic (hallucinogenic) compounds we have refined our method of quantifying the intensification of time.[24] We do present the mean values of the fluctuations and the variance (standard deviation) in tapping rate for the eight-minute periods. The tabulated results for two representa-tive subjects show that there is a significant increase in tapping rate in each subject at the peak of the psilocybin experience relative to his own control rate, as well as an absolute increase.[25]

TABLE I

Mean Tapping Rates and Variances

	At peak of Psilocybin		Next day (control)	
	Mean	Variance	Mean	Variance
Subject A	1.31/sec.	.001	0.42/sec.	.001
Subject B	2.21/sec.	.018	1.16/sec.	.003

The relation between increased tapping rate and intensity of time experience during mild excitation produced by Psilocybin.

This increase, as well as the increase in handwriting space, is illustrative of the interrelation between intensive time and extensive space during excitation, that is, during a period of increased metabolic rate.

Metaphorically speaking, we can compare ourselves to the spider ex-creting its web and thus creating its own space-time coordinates. Our metaphor can be transmuted to an example of the interdependence of metabolic rate and space-time. Inspection of certain data published by Heimann and Witt[26] as well as Witt and Baum[27] reveals that: 1) under the influence of chlorpromazine, the spider, *Zilla-x-notata*, decreases signifi-cantly the size of its web as compared to previous performances without the drug and 2) such a decrease is likely to be related to the lowered metabolic rate induced by the tranquilizer. This may be inferred from observation of the sizes of three webs built by one spider in June, July and November of the same year. Changes in the web size are clearly related to changes in metabolic rate during the growth and aging phases of the animal. Webs constructed during the former phase are larger than those constructed during the latter phase.

The following table represents a summary of changes in space-time coordinates under conditions of excitation and tranquilization produced by the sympathomimetic, pyretogenic hallucinogens such as mescaline, LSD, Psilocybin, etc., on the one hand, and the hypothermia producing phenothiazine type tranquilizers on the other.

Alterations in interpretations of space-time are not only concomitant with states of excitation and tranquilization induced by drugs, which raise or lower the metabolic rate. Other contrasting pairs of modi such as *youth* and *old age* also produce high and low metabolic rate respectively and thus concomitant changes in interpretation of space-time. The seeming contradiction that time passes more slowly for the young and faster for the old is resolved in the light of the above explication.

The Perceptual Context for the Conceptualization of Space-Time "Equivalence."

Space is a snapshot of time, and time is space in movement.—J. Piaget, *The Construction of Reality in the Child.*

If this is so, then space is the logic of things, and time is a relation of sequences and duration.—R. Wolf, "Ichstörung bei Wandlung des Zeiterlebens" in *Zeit in Nervenärztlicher Sicht.*

While our consciousness introduces succession into external things, inversely these things themselves externalize the successive movements of our inner duration in relation to one another.—H. Bergson, *Time and Free Will.*

We have dealt with systemic organismic changes, which are commonly referred to as occurring *within* the organism, and their relation to altered interpretation of space and time. The interpretation of space and time, however, also depends upon the intensity of the stimulating frequency, which is commonly referred to as external, from *without* the organism. The Pulfrich phenomenon[28] may be used as an impressive illustration of this statement: a steel ball with a diameter of three centimeters is suspended as a meter-long pendulum moving along a horizontal plane so that the observer sees the ball swinging from left to right. If the observer holds a smoked glass before only one eye, the horizontal movements of the pendu-

TABLE II

Changes in space-time brought about by (central) sympathetic stimulants and tranquilizers in humans

		Psychotomimetic central *stimulants*: LSD, mescaline Psilocybin, etc.	*Tranquilizers* of the phenothiazine type: chlorpromazine, etc.
Biological manifestations		Increased rate of oxidative phosphorylation	Uncoupling of oxidative phosphorylation[a]
		Body temperature raised[b]	Body temperature lowered
		Metabolic rate increased	Metabolic rate decreased
		Hyperglycemia[b] Mydriasis[b] Tachycardia[b] Piloerection[b]	Hypoglycemia[b] No Mydriasis No Tachycardia No Piloerection
		Pupillary dilation[c] Increase in frequency and amplitude of saccades in micronystagmoid eye movements[d]	Pupillary contraction[c]
		Speed of nervous conduction raised, 3 msec./1° C [e]	Speed of nervous conduction lowered, 3 msec./1° C [e]
		Tendency to raise critical flicker fusion value[f]	Lowering of critical flicker fusion value[g]
Chronological Time		Overestimated[h]	Underestimated
Perceptual Manifestations		Intensive; Contracts; "Arriving early to appointments"; Time appears to pass slowly[i]	Extensive; Expands; "Arriving late to appointments"; Time appears to pass by "like magic"[i]
		Increased handwriting rate[j] Nearby space (handwriting) enlarged[k]	Decreased handwriting rate[j] Nearby space (handwriting) diminished[l]
		Distant space diminished[m]	

[a] L. J. Abood and L. Romanchek, "The Chemical Constitution and Biochemical Effects of Psychotherapeutic and Structurally Related Agents," *Ann. N.Y. Acad. Sci.,* v. 66 (1957), p. 812; M. J. R. Dawkins, J. D. Judah and K. R. Rees, "Action of Chlorpromazine," *Biochem. J.,* v. 76 (1960), p. 200; R. Fischer and W. Zeman, "Neuronal Dye-sorption as a Histochemical Indicator of Nervous Activity," *Nature,* v. 183 (1959), p. 1337.

lum are transformed into elliptoid movements in space. The time difference created by the difference in stimulus intensity is transmuted into a
space difference. Moreover, placing the filter over the other eye reverses
the direction of the elliptoid motion.[29]

"Du siehst mein Sohn, zum Raum wird hier die Zeit."—R. Wagner,
Parsifal.

According to Efron,[30] the Pulfrich Illusion[31] develops because the temporal disparity in the perception of the moving image is interpreted
spatially, *i.e.*, when the neural messages from one eye are delayed due to
the reduced light intensity. In summary: when two stimuli (not only visual)
differ in intensity, they are relayed along the neural pathways with a differential delay into the hemispheric space of the observer; this delay is not
corrected by the central nervous system but interpreted as space.

Other phenomena also displaying the "equivalence" of space and time
are the so-called *tau* and *kappa* effects.[32] "It would seem from the two
phenomena that the spatial and temporal components of the space-time
events about which judgments are made are psychologically interdependent."[33]

Boynton has stated that space and time are also interchangeable in
communication devices, in the sense that either temporal or spatial resolution may be enhanced at the expense of the other in a system having a
given channel capacity.[34]

At this point we consider in retrospect the "equivalence" of space and
time on various levels of biological organization. Specifically, on the molecular, the cellular, the tissue and the organismic levels, during heightened
or lowered energy metabolism (*i.e.*, phases of excitation and tranquilization), and on the perceptual as well as on the communication levels.

The "equivalence" of space and time on the various levels bears a
striking resemblance to the "equivalence" of space and time in physics.

[b] A. Hofmann, "Psychotomimetic Substances," *Indiana J. of Pharmacy*, v. 25 (1963), p. 245.
[c] V. R. Carlson, "Individual Pupillary Reactions to Certain Centrally Acting Drugs in Man,"
J. of Pharmacol. and Experimental Therap., v. 121 (1957), p. 501.
[d] F. Hebbard and R. Fischer (see note 112).
[e] R. Magun (see note 29).
[f] Determined only for amphetamine by E. Simonson and J. Brozek, "Flicker Fusion Frequency,
Background and Applications," *Physiol. Rev.*, v. 32 (1952), p. 349.
[g] R. Hoehn-Saric, E. F. Bacon and M. Gross, "Effects of Chlorpromazine on Flicker Fusion,"
J. of Pharmacol, and Experimental Therap. v. 121 (1957), p. 501.
[h] H. Aronson and A. S. Silverstein, "The Influence of LSD-25 on Subjective Time," *A.M.A.
Arch. Gen. Psychiat.*, v. 1 (1959), p. 469; C. C. Bennett, *Drugs and Behavior*, eds. L. Uhr and
J. G. Miller (New York: Wiley, 1960), ch. 50, p. 676; R. Fischer, F. Griffin and L. Liss (see
note 13).
[i] Fischer, Griffin and Liss, *op. cit.*
[j] G. Grünewald and H. Mücher (see note 17).
[k] R. Fischer, "Selbstbeobachtungen . . ." (see General References).
[l] G. Grünewald (see note 20).
[m] E. Callaway III and S. V. Thompson (see note 18), also personal observation by this author
of subjects under Psilocybin.

The extension of time or the contraction of space (which in my view are exactly the same thing) can be expressed by

$$t' = t(1 - \frac{v^2}{c^2})^{-\frac{1}{2}}$$

The formula states that any process that takes t time as measured in one system will appear to take time t' when measured from another system moving relative to the former, while $v =$ the relative speed of the two systems and $c =$ the velocity of light.[35]

We encountered earlier the extension of time or the contraction of space when observing increasing body size paralleled by decreasing respiratory intensity and longer chronological duration (life span). In terms of energy metabolism (measured as oxygen consumption per surface unit), however, we have mentioned that animals, occupying large or small space, live equally long. There is an analogous relation on the level of (non-living) physical events: Events involving vast amounts of power and energy, such as the nuclear events culminating in an atomic explosion, occupy a large space or a short time, whereas events involving infinitely small amounts of energy, for example, mental activity, occupy very little space and are of very long duration. "If the duration of an event is very long, the energy may approach zero, a quite appropriate description of ghosts and materializations, if we choose to accept such phenomena."[36] "Mass and energy are inverse durations."[37] It appears that, using energy duration as a measuring unit, events occupying large or small space have equal existence.

As Minkowski said, "Space of itself, and time of itself will sink into mere shadows and only a kind of union between them shall survive."[38]

Could it be that the essential differences between physical space-time and biological space-time are due to the difference between their universal constants, which are, for the former, the speed of light at an assumed 300,000 km/sec., and, for the latter, the conduction velocity for impulses traveling along nerve fibers *varying* from 1 to 100 msec.[39] in inverse proportion to the size (diameter) of the fiber?[40] In addition to this, biological space-time is also greatly influenced by the intensity of the stimulus.

We are aware that we have committed the (not-too) original sin of comparing time as it is dealt with by the various disciplines, that is, on different levels of organized experience and that the concept of space-time does not carry exactly the same meaning in physics, biology, psychology, etc. These differences are slight, however, if compared with the one unifying origin of each space-time concept: its biological perception and conception, through sensory transduction or instrumentation. The differences in meaning suggest how far our specialized approaches remove the concept from the original perceptual context, that is, the frequency range of our spatio-temporal perception of simultaneities.[41]

The main rules of the biological game with special relevance to the perceptual context for the conceptualization of space-time "equivalence," may be summarized as follows:

1) Differences in intensity of stimuli are transformed by the nervous system to frequencies which are proportional to the logarithm of the intensity of the stimuli.[42]

2) Changes in intensity can be brought about by producing delay in relay of nervous impulses.

3) Therefore, the relationship between intensity of stimuli and intensity of stimulation can also be defined as the creation of spatio-temporal reality.[43]

The sophisticated experimental extension of the Pulfrich phenomenon by Efron[44] suggests that the conscious comparison of the time of occurrence of any two sensory stimuli requires the use of that hemisphere of the brain which is dominant for language functions. Efron presents evidence which indicates that those sensory messages (which carry information as to the time of occurrence) received by the non-dominant hemisphere are transferred to the dominant hemisphere by a pathway having a *delay* of 2 to 6 msec., the exact value depending mainly on stimulus intensity. The time necessary for transmission of a sensory message depends on the intensity of the stimulus. The transmission of visual stimuli is delayed about 10 msec. for each log unit reduction of intensity.[45] These delays differ in the different sense modalities, *e.g.*, conduction time is more rapid in the auditory system than in the visual.[46] Moreover, time delay and intensity differences produce changes in spatial localization in hearing ("directional hearing"),[47] vibratory sensations of the skin, taste sensation, olfaction and in the interaction of the latter two sense modalities.[48]

The relay of stimuli into the hemisphere which is dominant for language function is a necessary prerequisite for conscious thought formation. Verbalization, in the form of speech and writing, however, can be regarded as external manifestations of thought, that is, the coded expression of brain activity.[49]

"Thus, we can mean by 'sense perception' not just sensation and then perception, but a primitive process which includes all of what we call sensing, knowing, and attending."[50] Perception can also be defined, as by Werner,[51] as an analogous function (or process) to reasoning.

Holding up the Mirror of Sense Perception to the Image of Cognitive Generalization. The cognitive operation equivalent to our sense perception of simultaneity is called generalization. It is with the aid of generalization, that is, hypothetical simultaneities, that we embark upon the verification procedure. This verification is actually a sensory checking of the preceding cognitive generalization; both steps are essential features of the scientific method. If the generalization is verified empirically, within the accepted limits, a principle (law) has been established. Since, however, perfect verification is rarely obtained, the laws are framed as probability statements of events.

The verification procedure may thus be conceptualized as holding up the mirror of sense perception to the cognitive generalization: We demand that the mirror reflect the image.

An example of such a mirroring becomes apparent in Augenstine's[52] estimation of structural information content and constraints in proteins and English language samples. He compared Branson's[53] calculation of the frequency of occurrence of the twenty amino acids in twenty-six different proteins with the result of a similar analysis as to the frequency of occurrence of letters in paragraphs. In other words, letters were treated as amino acids and paragraphs as proteins. The results, then, show the distribution of the normalized frequencies for the letters and the amino acids to be almost identical.

Another example is implicit in the pharmacologist Clark's experiments.[54] He wondered how many different formulae could be used—with varying constants—to interpret the same set of figures, depicting relations between (chemical) concentration (or stimulus in intensity) and action (or sensory perception) if a variation of $\pm 5\%$ were allowed. Interestingly, he found Langmuir's formula expressing (molecular) adsorption equilibria and the well-known psychophysical Weber-Fechner Law, within limits, are strikingly similar exponential functions. Large portions of the two curves are even identical. It would seem that our laws which describe interaction on the molecular level are mirror images of those psychophysical laws which deal with the sensory representation of that interaction. Apparently, whether we look at chromosomes, cathedrals, or even into the mirror, we always look at our own brain.

Our previous reasoning and examples were brought forward to strengthen our hypothesis that the space-time "equivalence" *conceived* at each disciplinary level is contained in and is a variation of the main theme, *the perception of the beholder*. The subjectivity of any time, biological or physical, psychological or cosmological, is a consequence of this view; "but some times are more subjective than others."

2   *The Unity of Life and Time*

Concepts and Models. The peculiarly fundamental nature of time in relation to ourselves is evident as soon as we reflect that our judgments concerning time and events in time appear themselves to be "in" time, whereas our judgments concerning space do not appear in any obvious sense to be in space, as Whitrow has observed. Space seems to be presented to us all of a piece, he contends, whereas time comes to us only bit by bit. In order to study the temporal aspect of Nature effectively, men have strained their ingenuity to devise means whereby the peculiar characteristics of time are either ignored or distorted. Post-Renaissance mathematicians and physicists, for instance, have sought to explain it away in terms of the spatial.[55]

Gregory and Kohsen[56] believe that what we call time is something of a compromise between cosmological and microphysical events. Their point of departure is that according to Eddington the number of particles in the universe is 2.36216×10^{79}. This number, N, Eddington obtained from theoretical considerations, as $3 \times 2^{255} \times 136$. In round numbers N is of the order of 2^{264}, and consequently $\sqrt{N}$ is approximately 2^{132}.[57] "It follows that a universal frequency-scale of 132 octaves can, in principle, be set up but cannot be used for distinguishing 'past' from 'future.' " Such a scale has only many "presents" whose durations range from 1.4×10^9 years to 10^{-23} seconds. The former figure is the time taken for the expanding universe to double its linear dimensions everywhere, or for a light signal to traverse the Einstein universe. It represents the largest time unit in cosmology. The latter figure is the time light would take to cross a classical electron, the smallest amount of time in microphysics. The ratio of these two time durations is very nearly 2^{132}. The time building constructed by Gregory and Kohsen has approximately the shape of a five-dimensional pyramid; its base is composed of the smallest and most dense bricks.[58]

I do not intend to recapitulate here the sequence of events from the Ptolemaic through the Copernican, Newtonian and Einsteinian cosmological revolutions. It suffices to say that primitive mythological cosmology evolved into static classical systems, culminating in Newton's *Principia*. The present baroque cosmology is a natural outgrowth of Newton's mathematical stabilization of the universe.[59] It may also be regarded as an extension and refinement of the atomic concept of Democritus transformed under the influence of contemporary biological and sociological concepts, such as Darwinian evolution and our religious belief in progress.[60]

The Journey of the Photon. Contemporary concepts of man's place in the universe give rise to the construction of various models. One such model, that of Grebe,[61] interrelates the flow of energy throughout its stages of increasing organization and decreasing potential. It is a three-dimensional horn whose depth is determined by the wave length of the rhythm of oscillations, the time interval, or, broadly speaking, the period of the phenomena. It is millions of light years in length and depth. At the rim of the horn, time is measured in terms of the highest frequencies that reach 10^{25} cycles per second.

In discussing his cosmological model, Grebe asks the reader to follow the journey of a photon of light, which, having traveled millions of light years, being bent, twisted, diffracted, and recombined many times, having succeeded in dodging and hurdling the many specks of cosmic dust to which most of its companions fell victim, having even penetrated the atmosphere of the earth without being trapped by fog or smog, finally arrives at the leaf of a living plant. The photon has only one strike and, at that, has only about the same chance of making a hit as a big league batter has. That hit, if made with exact timing and placing, will convert the energy of the photon (transmitted at a frequency of 10^{15} cycles per second) to the CO_2 ball, converting it to carbohydrate and O_2 at only 10^{14} cycles per second. If it fails to hit, it merely warms the leaf at a frequency of 10^{13} cycles per second.

Szentgyörgyi[62] follows the photon journey further. According to him, that hit of the photon lifts an electron from an electron pair in the ground state to a higher energy level; then, as a rule, the electron drops back within a very short time to its ground energy level; however, a portion of the energy of the photon is trapped as chemical energy in the process. In order to do this efficiently, the photon must meet the electron with a specially built substance (mostly chlorophyll) and couple this substance to a system which converts the very labile electronic excitation energy into a more stable chemical potential, into chemical energy, that is, the energy of a system of electrons of a stable substance. Life (in the energetic sense) has shoved itself within a very short time to its ground level between the 10^{15} and 10^{14} cycle per second frequency ranges, and makes the electron drop back into its own machinery, after utilizing its energy.

It is the photosynthetic process which traps, stores, and channels the energy of photons. The basic energy cycle of life consists then of electrons boosted up by high energy photons gradually giving up their excess energy to drive the living machinery.

Other molecular events controlling the synthetic activities in cells are part of a negative feedback system maintaining cellular existence, *i.e.*, "stretching out" in space while enduring in time. Self-regulation is brought about through oscillations within a closed control loop—the genetic "memory" mechanism of a species—consisting of deoxyribonucleic acid, messenger ribonucleic acid, enzyme, metabolite and cellular regulators.

The "Stretching Out in Space" of Life. The phenomenon of life may be viewed as a process evolving to higher levels of complexity by utilizing chains of catalysts, or enzymes, making up interrelated metabolic cycles. Williams[63] states an important feature of the chains of catalysts, namely that they allow the use of space in dislocating chemical reactions. This stretching out in space—that is, the dispensing with the semi-permeable membrane and the substitution of a scheme for oxidative phosphorylation —is the story of the evolution of unicellular organisms into complex mammalian systems. The "clocks" which are postulated to account for certain periodicities in biological systems may be based, according to Spangler and Snell,[64] entirely on chemical kinetic systems,[65] without resort to diffusive or phase boundary effects.

There is a provocative line of speculation by Boyer[66] which points to the common mechanistic feature of such apparently diverse processes as oxidative phosphorylation and muscle contraction (stretching out in space). During both processes—oxidative phosphorylation as well as muscle contraction—structural changes in protein and accompanying electron transport are clearly indicated.

The periodicities or the biological clock phenomena may also be conceptualized as resulting from feedback of the end-products of biosynthetic pathways to interconnected chains of catalysts present in the mitochondria. Our metabolic suspension in time is a result of optimum homeostatic control under non-stressful conditions. Conditions of stress, however, result in a maximum power output—adenosine triphosphate (ATP) production —through release of homeostatic feedback inhibition. Biological time in the latter case is contracted on the molecular level whereas in the former it is expanded.

The Exponential Life Spiral of Time. If evolution is regarded as the stretching out of life in space, its beginning marks the birth of time. We may place the historical phenomenon in the form of an exponential time spiral of life between the 10^{15} and 10^{14} frequency ranges of Grebe's model of the universe. If we view life on earth historically, it may be regarded as a single event with evolution, learning, perception, and hallucinations and dreaming as the steps of this event, differing from each other only in their time rates of change (see Figure 2). Such a view is contrary to our habitual classification and description of systems and phenomena, which we relate to our own duration and frequency range (an ordinary human life has a duration of about 10^9 clock seconds).

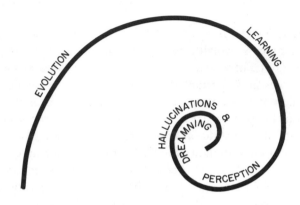

Figure 2. The Exponential (Logarithmic) Life Spiral of Time symbolizes the unity of evolution, learning, perception-hallucination and dreaming as adaptive events of increasingly efficient utilization of energy as well as increasingly rapid time rate of change. The origin of our spiral is the logarithmic spiral as first considered in 1638 by Descartes and then by Torricelli. John Bernoulli found all the "reproductive" properties of the "wonderful" spiral and wished to have the curve incised on his tomb: *Eadum mutata resurgo*; E. H. Lockwood, *A Book of Curves* (Cambridge: Cambridge University Press, 1963).

The relativity of our reference point can be demonstrated by taking a moving picture of a plant at one frame a minute and then speeding it up to thirty frames a second.[67] The plant will appear to behave like an animal, clearly perceiving stimuli and reacting to them. Why, then, do we call it unconscious? To organisms which react 1800 times as quickly as we react, we might appear to be unconscious. They would in fact be justified in calling us unconscious, since we would not normally be conscious of their behavior.

Our restricted range of consciousness can be demonstrated by exposing a man to a sequence of similar or nearly similar stimuli, such as a succession of frames on television. If the frequency of these stimuli is increased to about fifteen cycles per second, previously distinct images cease to be separated and a spatialization, or fusion, of events takes place.[68] The frequency ranges for this co-occurrence are varied in the various species as well as in the various sense modalities. Brecher,[69] for example, observed that four tactile stimulations per second with a rod on the snail's belly compels it to attempt to crawl upon a nonexisting coherent surface which it perceives as a simultaneity. This attempt is an illustration of the transduction of his simultaneity into our duration. In human terms, this could be expressed in a reformulation of Descartes' *cogito ergo sum*: I am aware of perceiving rhythmical sequences of simultaneities as metabolic rate variant durations; therefore I am. From that point of view, perception and learning are processes involved in the production of simultaneities; the cardinal difference between them is the speed with which we perceive simultaneities and the slowness with which we learn them.

This author's concept of life as a single event, exponentially receding and proceeding in time, is supported by scattered evidence in the literature. In the following paragraphs I would like to refer to evidence of parallelism between evolution and learning as well as the similarity between learning and perception.

Pringle[70] has examined *evolution and learning*, using the concept of "complexity" and analyzing it in terms of information theory. In his study, "complexity" and "order" are used to mean the number of independent parameters needed to define systems fully in space and time. He shows that synchronization of oscillators can lead, in a population of oscillators, to an "evolutionary" increase of complexity of rhythm, in a manner analogous to the increase of structural complexity which occurs in organic evolution. His model is consistent with the known features of the physiology of the central nervous system of animals, and it is capable of producing the increase of complexity found in the process of learning. The oscillators in his model consist of circular chains or "closed loops" of neurons, and it is suggested that such closed loops may generate oscillatory wave forms more closely resembling the electrical waves found in the cerebral cortex than the impulses characteristic of peripheral axons. The coupling required to produce synchronization of these oscillators is provided by the sharing of neurons.

Pringle then analyzes the six types of learning as classified by Thorpe[71] and shows them to be related to the properties of his model.

That *learning and perception* follow the same laws is well documented.

In 1935, Peterson suggested that perceptions are "learned reactions" acquired by the usual methods of "trial and error" learning,[72] and since the translation into English of Senden's *Raum-und Gestaltauffasung bei operierten Blindgeborenen* in 1960, there is ample material to answer the famous Molyneux question.[73] Ever since Molyneux posed his question in 1692, cases of persons born blind who later acquire sight have attracted close interest in view of their recognized importance for the theory of our ideas of space, time, perception and learning. More material has appeared recently on the perception of space and shape in the congenitally blind before and after a corrective operation. The accumulated evidence leads me to agree with Senden that subjective visual space is a form of intuition given along with the sensory content in every act of vision and that objective space is not a primary datum, but is only secondarily acquired from subjective space by grasping the positional relations of visual objects to one another within visual space. Since all parts of visual space cannot be objectified to the same degree, they do not all have the same objectively spatial character. Furthermore, the idea of shape is not due to a dynamic effect of the stimulus on the visual organ, nor to an alleged passively reproduced transmission of the stimulus occurring simultaneously at a purely physiological level; it is the outcome of a process of conscious interpretation in time. The development of the fully formed idea of shape involves a series of transitional forms as intermediate stages which are liable to vary between individuals since it is the individual who himself creates them.

J. Z. Young also agrees with Senden[74] when he says, "It is clear that into the brain of the child or young animal a great deal of information must be placed before it is possible to conduct what seems to the adult to be the simple operation of seeing."[75]

Even the body schema which gives rise to the "amputation phantom," that is, the perception of sensations from a lost part of the body, is basically a postural schema and must be learned. In a group of children who had undergone amputation of congenitally malformed extremities it could be shown that phantoms occur only if there has been some sensory and/or motor function before the loss of the limb, and that phantoms are rarely reported if the amputation is performed before the age of four years.[76]

Time, like space, is constructed little by little, according to Piaget.[77] The elaboration of a subjective consciousness of time in relation to personal activity, a universe without permanent objects, is followed by an elaboration of the temporal field which requires the development of images and language.

The importance of learning in connection with sex perception was studied by Money,[78] who found that hermaphrodites and persons with sexual deformities who are born psychosexually undifferentiated acquire a gender role and identity with learned experience. Once the psychosexual identity has appeared, it is indelible and imprinted; it is as influential as if it had been present at birth.

Another illustration of our contention that learning and perception differ only in time rate of change is Ostfeld's experiment[79] with subjects who had had relatively normal vision for the first several years of life but had become progressively blind. They reported that they could still "see" in dreams for from five to ten years after the onset of blindness, but that this ability slowly disappeared thereafter. The same blind people, when given 75 micrograms of D-lysergic acid diethylamide (LSD) orally, reported a diminished frequency of complex visual imagery as compared to normals and appreciably more auditory, tactile, olfactory and gustatory hallucinations. Unfortunately, we know of no data on drug-induced visual hallucinations with congenitally blind subjects.[80] We favor the view that learning and perception are analogous processes and that perception is to a large extent the end result of a learning process. It is only fair, however, to state the difficulty "that whatever is found can be interpreted without reference to the innate versus learning controversy."[81]

Prior to closing this section, another aspect of the Molyneux problem is also worth considering. In Furlong's formulation[82] the problem is: "Granted that a man has forgotten what sense experience is like and has been accustomed only to the experience in imagination, that is, he has the effective use of none of his senses; how will sense experience strike him when it is restored to him?" In what way will he re-learn the distinction between waking and dreaming experience?

The ability to distinguish between waking and dreaming would lie in the way in which percepts are related to the rest of the individual's experience. The difference in the elapsing time necessary for making a correct decision could be used to "diagnose" normality. The longer the time needed to decide whether one is awake or dreaming, the stronger the probability of a disturbance in reality testing.

After having discussed evolution and learning as well as learning and perception, we may proceed to further clarify the meaning of the life spiral in Figure 2 and discuss each of its four steps separately.

Evolution. The value for the rate of evolution obtained by Palmer[83] was 1 per cent per 10,000 years, corresponding to the value of a unit suggested by Haldane[84] as an appropriate one for measuring evolutionary rates. The unit is a "darwin," as discussed by H. Kalmus in his essay on evolution in this volume. This is the same rate as 1 per cent in 10,000 years, which seems to be about the value for man. The rate is extremely rapid compared with that for most animals which have existed on the earth for a much longer time than the million years during which man has been developing. Specifically, using the figures for *Pithecanthropus erectus* and modern man, with a time interval of 400,000 years, the evolutionary rate works out to be 1.2 darwins.[85] This indicates that, according to measurements of his skulls and jaws, man is evolving physically at an extremely rapid rate. Such a rapid evolutionary rate is characteristic of a new species. The rate begins to slow up after 50 million years, as judged from ancestral lines of fossils, and then dies down gradually as the species tends toward extinction. Whether or not this process applies to man is a matter for speculation. A genus can cease to be, either because all its members have died without issue, or because all its surviving descendents have changed so much as to be classified as members of one or more new genera. Simpson[86] concludes that all mammalian genera are relatively short-lived; the average life of a genus of carnivores is about 6.8 million years. Thus, in Haldane's words,[87] "A mammal must evolve or perish."

The only way for a complex system to persist is to evolve, or, in other words, to further increase its complexity. It appears that such an evolution proceeds at an ever increasing time rate of change not only on the biological level but also on the social and on the technological levels.

Learning and other slow-motion perceptual phenomena. Both learning and its symmetrical counterpart forgetting are exponential processes.[88] The quantum mechanical theory of memory of Von Foerster *et al.* implies the irreversibility of the process. As a spider excretes his web, we excrete our dimensions of existence, space-time, to catch as much world in it as we can. It is through this unidirectional web of time coupled to our metabolic rate that we learn about our lives, organic development, or the mathematical-physical world of Plank, Bohr, and Heisenberg. Strangely, our concepts of the world evolve, as does our time, at an exponential rate, as in growth or aging.[89]

Even such phenomena as banditry in Manchoukuo, ganging in Chicago, or the world's total wars over a span of more than a century can be conceptualized in form of a linear dependence of the logarithm of incidence against the magnitude of the deadly quarrels.[90] (In other words, the number of victims is proportional to the sizes of the aggressive groups.) Logarithmic-normal distributions such as the above come about if the increments accrue proportionally to the absolute value of the variable. Thus it is not surprising that sizes of cities are approximately logarithmic-normally distributed if we make the assumption that both increments and decrements of population will be proportional to the populations and if we wait long enough for the steady state to be established.[91] This may be an example of how perceived simultaneities are expressed as observed statistical trends and promoted to theoretical constructs called concepts.

Perception. That perception is an exponential process has been amply documented through the years from Fechner's discovery in 1860 to Stevens' proposed modification of it.[92] In order to send quantitative information to other distant points, the nerve cells cannot increase the intensity of the impulses; they only increase their frequency. Katz, and before him, Adrian,[93] found that this frequency is proportional to the logarithm of the "intensity" of the stimulus. The parallelism between this law and Fechner's Law treating the relationship between intensity of stimulus and intensity of stimulation is quite remarkable.[94] Fechner's Law is based on the observation that stimulus intensity must always be altered by a certain amount in order to produce a just-noticeable change of sensory magnitude; therefore, sensory magnitude can change only in quantum fashion. It was also assumed that these quantum steps of sensation were equal in magnitude throughout the whole sensory range until von Békésy in 1960[95] formulated the limitations which had to be imposed upon this latter assumption, namely, that it holds true only in the region of intensity where attention does not change. We also know now that Stevens' Power Law describes sensation magnitude *before* adaptation to a test stimulus, whereas the Fechnerian Discriminability Law describes it *after* adaptation.[96] Unfortunately there has as yet been no attempt at an operational synthesis of Fechner's Log Law with Stevens' Log Log Law,[97] although Treisman's paper[98] is a good departure.

It is our perception which compels us to perceive other processes as exponential. This can be demonstrated, for instance, with molecular weights of enzymes which can be arranged in geometric series,[99] with the flow in the proximal part of the mammalian nephron,[100] with loudness discrimination[101] and many others. Janisch[102] has compiled a vast number of exponential processes in his book. The special problems arising in mathematical similarity analysis may help to clarify the issues involved

377

in this philosophical question. Stahl[103] discussed methods of similarity analysis such as the Buckingham Pi theorem, which states that if a physical system can be properly described by a certain set of dimensional parameters, it may also be fully described by a lesser number of non-dimensional (similarity) parameters. The Buckingham method places major emphasis on the dimensional content of the variables, such as viscosity (M/LT), force (ML/T^2), etc. (Mass, Length, Time). Stahl pointed out, however, that M, L, T, etc., constitute a special algebraic system which is technically an Abelian group; they do not stand for ordinary real numbers, but only for numbers used in a special way—for multiplication or division, *not* addition or subtraction. Therefore, in my opinion, the special mathematical treatment of Abel *et al.* is merely an expression of the quantal nature of all biological responses and percepts. The philosophical question, whether our way of perceiving is the essential determinant of the exponential nature of all percepts, can be partially answered in stating that the chemical reactions perceived in the process of perception already contain the response (or product) as a logarithmic function of the stimulus (reactant).

Hallucinations and Dreaming. The title of this section is meant to include phenomena which in comparison to evolution, learning and perception may be characterized by an apparently more efficient use of energy as well as faster time rate of change.[104] Two essential points are to be re-emphasized prior to the formal treatment of the subject matter. The one refers to the only reality, namely, that which is contained in our experience; while the other defines the boundaries of a person not by the surface of his skin, but by the limits of his experience. These limits may extend from the distance of a few Angströms of the chromosome as viewed through a microscope to millions of light years when perceiving certain nebulae through a telescope.[105]

The phenomena to be dealt with are accelerated reproductions and rearrangements of structured past experience: hallucinations, dreams and hypnotic phenomena. Characteristic of them is the inverse relationship between the intensity of "outside" stimuli and "inside" anticipation (attention-expectation, vigilance) necessary for their occurrence; in agreement with Broadbent and Gregory,[106] who concluded that diversion of attention away from a stimulus produces an effect resembling a reduction in the stimulus intensity.

The accelerated reproduction of structured past experience amply documented by Cooper and Erickson[107] has recently been subjected to quantification by Weitzenhoffer.[108] The experimenter hypnotized his subjects so as to include a clock in their hallucinations and asked them to report the time in their hallucinatory world. The results obtained from "good time-distorters" reveal a continuous and uniform dilation or contraction of the time continuum. Interestingly the resulting curves resemble learning curves, a feature quite similar to the already observed similarity between the exponentially experienced processes of perception and learning.

Different types of eye movements accompany dreaming and perception of a "real" object, such as a beating pendulum, than the simulating, imagining and hallucinating conditions of hypnosis.[109] Specifically, involuntary pursuit eye movements occur in the absence of a "real" moving visual stimulus, suggesting that the necessary and sufficient prerequisite for the development of pursuit eye movements is not a retinal image but a *cerebral image*.

A few words should be said about the recent discovery of a correspondence between eye movement and reported dream content.[110] Investigations on normal human subjects confirm that dreaming is a periodic phenomenon associated with the so-called paradoxical stage of sleep accompanied by three-per-second rapid eye movements of a "saw-tooth" wave pattern.[111] The relation of this phenomenon to our analogous (unpublished) findings of an increased frequency and amplitude of saccades (three per second) and "saw-tooth" wave patterns, obtained when studying physiological nystagmus in a wakeful healthy subject after the administration of alcohol, is yet to be determined.[112]

Attention, as measured by reaction time, is also a periodic phenomenon related to EEG alpha frequency.[113] In measuring the reaction time of the human eye to visual stimuli it was found that there are preferential equidistant periods—(reaction) times—after which eye movement starts.[114]

Anticipation. What is anticipation (attention-expectation)? St. Augustine in *The Confessions*, Book 11, believed that "There is in a mind the expectation of things to come." "It is not then future time, that is long, for as yet it is not: but a long future, is a long expectation of the future." Anticipation is then the experiencing of our temporal succession in perceptual space.

The inverse relationship between intensity of stimulus and anticipation mentioned earlier is not only exemplified by dreams, hallucinations and hypnotic phenomena, but also includes the placebo phenomenon. It is known that an increase in anticipation such as is present during stressful situations, for example, surgical trauma, intractable pain, greatly increases the efficacy of placebos, *i.e.*, inert substances given instead of an active drug.[115]

Extreme examples of the inverse relationship between stimulus intensity and anticipation are phenomena of hypnotic suggestion characterized by intense anticipation and no measurable stimulation; some of these are experiences produced by mescaline, LSD, etc., the curse, Voodoo death and mental suggestion at a distance.

There is no overstatement in Hollister's succinct review[116] evaluating experiences produced by moderate doses of mescaline, LSD and Psilocybin: these drugs are as potent as hypnotic induction itself and "provide an experimental situation in which the results can be almost foreordained."

Elliott[117] described the curse as a wish in a magico-religious context; it is a wish, expressed in words, that evil may befall a certain person. Hence, it is a verbalized anticipation in imagination. The fulfillment of the curse is its development into the vow and the prayer, or into the ordinance and ultimately into the law built around its sanction.

The "Voodoo death" is a phenomenon occurring not only in Haiti[118] and Africa,[119] but also in contemporary Western society. Mathis[120] presents us with a case in which severe and eventually fatal asthma developed in a previously healthy adult male following his mother's prophecy of "dire results" if he went counter to her wishes. A similar "curse" of another domineering mother is described by Jilek with a similarly fatal outcome for the submissive son.

Vasiliev's recently translated report[121] contains a section by Platonov describing demonstrations of hypnotic suggestion at a distance. These were carried out in the presence of the participants of the All-Russian Congress of Psychoneurologists including its president, V. M. Bechterev, in December, 1924. (The results of these demonstrations do not fit into the already existing belief systems of the majority of our scientific community. This is, of course, insufficient reason for their dismissal.) One of the experiments consisted of suggestions at a distance; the sender reported: "I closed my eyes and mentally represented to myself the image of the sleeping subject (who was in a closed chamber)." The subject apparently immediately stopped talking and fell asleep. "Awakening was effected in the same way, *i.e.*, by my representing to myself the subject waking up from her sleep." This experiment was repeated three times with the same success. The point we should like to emphasize regarding these demonstrations is that the sender stated that, ". . . when I tried to influence the subject by means of a mental command, such as 'go to sleep!' or 'sleep!' such orders invariably remained without result. But when I visualized the image of the sleeping subject (or the subject awake) the effect was always positive."

In attempting to order these experiences, it may be stated that the described actions at a distance—curse, Voodoo death, and the experiments described by Vasiliev—appear to be analogous to and extensions of phantom limb pain, an experience of action at a short distance, even though they differ in that phantom limb pain occurs within the perceptual and physical space of the same person whereas, in the other examples, the phenomena are transmitted within the perceptual space of two persons with final manifestation in the physical space of the receiver (medium). In both cases, a more time-like event has been transformed into a more space-like one.

> *Je suis ma mère et mon enfant*
> *Dans chaque point de l'éternel*
> —Paul Éluard,
> *Poésie ininterrompus*

3 *Conclusions*

Our point of departure was the assertion that biological time is time within the "coordinate" system of the living organism. We have followed the creation and "equivalence" of space-time as an elaboration of a system of relations on the various levels of energy metabolism. We have also viewed the space-time "equivalence" within the framework of various disciplines and attempted ultimately to relate our perceptual and conceptual constructions of the world to the original biological context.

A unitary view of time and life has been postulated. A photon was followed through its journey from the sun until its energy was trapped by the photosynthetic process on earth making possible the appearance of life, and its "stretching out in space" through the reactions of oxidative phosphorylation.

The continuation of our "stretching out in space" appears to us as an irreversible transformation process, *evolution*, proceeding at an exponentially progressing time rate of change in the direction of increasing entropy. The exponential "life spiral of time" includes other adaptive events of increasingly efficient utilization of energy: *learning, perception, dreaming, hallucinations* and related hypnotic phenomena. These events of increasingly rapid time rate of change appear to counterbalance, at least locally, the trend toward ever increasing over-all thermodynamic disorder.

Molecular systems, polymolecular systems, cells, tissues, organs, and organisms—that is, associations of increasing complexity—become possible only at temperatures much lower than that of the sun.[122] These highly organized, more and more complex systems can code and decode more and more information and need for their maintenance less and less energy. The amount of energy needed by this author to contemplate with enthusiasm the possibility of an evolution toward order is infinitely small.

There is a whole array of phenomena which are approximately on the energy level of contemplation; they are, however, more succinctly defined by the lack of measurable stimulation and concomitant excessive anticipation necessary for their occurrence. These are the hallucinatory and hypnotic phenomena spanning the range from placebo reactions, curse, "Voodoo" death, and phantom limb pain to suggestion at a distance.

We reiterate after Efron[123] that the mysterious "now" is the interception of the stimulus by the speech dominant hemisphere. As true sons of our time reflecting the *Zeitgeist*, we have to conclude that existence from now to now involves the fusion of sequences into duration; hence biological time *is* creation endlessly aborted, travestied, corrected; a creation which, as Sartre[124] has shown, continually demands the retouching of the present and of nothingness.

It is my pleasure to acknowledge the support of the Ohio State University Development Fund. I am also grateful to the late Doctor C. C. L. Gregory and his wife, Anita Kohsen, M.S., Church Crookham, Hamshire, England, as well as Doctor John Gander, Columbus, Ohio, for their exceptionally original and stimulating discussions and to Miss Carolyn Kelley, B.A. and Miss Marsha Rockey, B.A. for their creative editorial assistance.

PART IV
TIME AND MATTER
Clocks, Man, and the Universe

"THE GREATLY ELABORATED INCENSE SEAL" illustrated overleaf was a timepiece featured in the *Hsiang Ch'eng*, a work on aromatics and incense which was popular in Medieval China. It was later included in the *Hsin Tsuan Hsiang-P'u*, or "Newly Compiled Handbook of Aromatics (Incense)," from which the illustration has been reproduced. Incense made from a variety of aromatic powders according to prescribed recipes was placed into the grooves of this incense seal, which was carved in hard wood, and lit at one end of the continuous path. It is probable that the beginning of the path was at the center of the seal, and that the incense burned for a period of approximately 12 hours. Such incense seals were particularly useful during times of drought, when the community water clocks became inoperative. The length of the entire path formed by the grooves is believed to be about 240 inches or 20 feet. According to a writing dated October 1329 A.D. and signed by a "Retired Gentleman of the Central Studio," who cannot be otherwise identified, this seal was presented in the form of a diagram to Tsou Hsiang-hun, a native of Yü-chang Province, an official of the Prefecture of Yü-li. He was a lover of literature, particularly proficient on "The Book of Changes" and held in high esteem by the Court.

Reproduction through courtesy of Silvio A. Bedini, from his "The Scent of Time," *Transactions of the American Philosophical Society*, new ser., v. 53, pt. 5 (August, 1963), p. 11.

Introductory Note to Part IV—
The Artisan and Time

Clockmakers, since the days of sundials and time sticks, have helped us answer the question, "what time is it?" They have seldom been concerned with the nature of time. They had nimble fingers and pragmatic minds. They pitted their prowess against the raw materials of their profession (wood and iron, mainly) and set their imaginations to the task of producing devices which could turn, move and occasionally make noises like a living thing. They wanted reliability so that their machines would work even when "the cold North wind blows," and in spite of "the Motion of the Ship, the Variations of Heat and Cold, Wet and Dry, and the Differences in Gravity in Different Latitudes." Their striving for precision workmanship and their desire to satisfy the aesthetic aspirations of their fellow-men resulted, at least in the West, in a delightful assortment of timepieces, from animal clocks to inclined plane clocks.

The history of clockmaking shows a continuous search for a device which can convincingly image the rhythmic activity of nature and man. The beginnings of modern clocks date to those (various) epochs in Europe and China when linear and cumulative time indicators (such as fire and incense clocks and clepsydrae) and devices which depended on starlight were replaced by contrivances exhibiting periodicity. One can detect a progression from Su Sung's astronomical clock to Essen's cesium clock in the improvements of the resonant member.

Timepieces are expected to produce cyclically recurrent indications whether by sound, sight or smell; they must also satisfy the beholder's mental image of what a clock ought to be. Interestingly, the second condition appears to be more fundamental. Recurrent indications alone are usually not accepted as time signals, unless backed by convincing arguments as to their "regularity" with respect to another, already accepted time standard. It is possible, however, to think of radioactive[1] or thermodynamic[2] clocks, for their operations can be shown to agree with our interpretation of nature and understanding of time, even though such devices would not operate in cyclically recurrent stages. The primacy of lawful uniformity which we require of our clocks, over simple repetitive indications, suggests that time involves some type of a relation between man and the universe, the clock being only a useful link in the chain.

By meeting the technical challenges of their trade, the horologers paved the way for the trained artisan who was necessary to make the laboratory apparatus of natural science and produce, later, the tools of industry. Together they created the clock, the instrument which was to become the most important tool of the industrial revolution, a weapon we might say, and which remains an indispensable companion to the men of all conceivable brave new worlds. While the importance of clocks increased, the concept of time itself became, conditio sine qua non, the fundamental variable of the physical sciences. The idea of time, the device named clock, and the broad mental construct called science have become inseparably bound because predictability is essential for the formulation of scientific laws.

If clocks, man and the universe are so closely related, it is not surprising to learn that people have suspected a hidden "sympathy" between mechanical clocks and the universe. Mechanical timepieces have sometimes been compared with the Newtonian universe, and quite meaningfully so. The clock was not only an image but also a miniature model of the Great Everything: stars, man and all. It connected the journey of the planets with the daily routine on earth.

In the Newtonian universe gravitational forces kept the planets in their orbits. Forces, energy and Laplacian determinism controlled the motion of heavenly bodies which were tearing through endless space. And the same forces, energy and determinism moved and controlled the clock.

The Newtonian world has been modified by two imposing worlds of contemporary physics: Relativity Theory and Quantum Theory. In the former, the motion of the planets is not described through the action of forces but through the properties of space-time trajectories which determine the orbits of planets and of falling apples. For a given set of initial conditions these trajectories are, so to speak, the easiest and therefore the only paths. Energy has become synonymous with matter, and matter the link between space and time. In the latter, Laplacian determinism in particle motion turned out to be irrelevant.

Our clocks have also changed. Of course, they still have the resonant member, but many have no perceivably moving parts. The resonating elements are aggregates of absorbers and emitters whose operation is accounted for by statistical laws which make deterministic knowledge of the particles' whereabouts unnecessary. Atomic clocks are quiet; they "wink" rather than "talk" and are accurate beyond the dreams of early clockmakers. Yet, all their excellence is not sufficient to make one of them run at the same rate as a like clock whose motion associates it with a different "time horizon," to borrow Dr. Whitrow's phrase, that is, with a different relation to the universe as a whole. And so the universe, whose spatial extent is thought to be described as finite but unbounded, makes its presence felt in the clock. It seems that the "sympathy" of the clock to the universe still exists. Perhaps what we are saying is that man is the link between them.

We are now equipped with timepieces which function not only when "the cold North wind blows," but even under conditions where there is no earthly North and no earthly wind. We also possess that wonderful "mixture of clear logic and unwritten superstition," the scientific method. With these tools at hand, in this fourth part of *The Voices of Time,* we shall follow the problems of the man with the clock, with a dedication from Pope's *Essay on Man*:

> Go, wond'rous creature, mount where Science guides,
> Go, measure earth, weigh air, and state the tides;
> Instruct the planets in what orbs to run,
> Correct old Time, and regulate the Sun.

J. T. F.

Timekeepers—
An Historical Sketch

H. ALAN LLOYD

1   The Beginnings

Primitive man must have noticed the natural division of time into periods
of light and darkness and into seasons of the year, but the first written
record describing man-made subdivision of the year and the day dates back
to the Sumerians at about 3500 B.C.[1] Their ideal year was 12 months, each
containing 30 days. In turn, each day was divided into 12 *danna* and each
danna into 30 *ges*. One *danna* was therefore roughly equal to 2 modern
hours, and 1 *ges* to 4 modern minutes. A tablet from the first half of the
third millennium B.C. mentions *danna* as a unit of length, presumably derived from the unit of time by the same name. Since, roughly speaking, the
sun travels some 180° of arc in 12 modern hours, it travels 1° of arc
in 4 modern minutes, that is, it travels 1/360th of the circumference of a
circle in 1 *ges*. It is not surprising, then, that *ges* also means "1" or,
"unity," thus doubling up as a measure of time and as a measure of angle.
Working backwards now, the division of time into 360 *ges* during the
period of one complete day suggests the division of the circle in 360°,
and vice versa. Historically, the division of the day and the division of
the circle are intertwined, time being measured by the apparent movement of the celestial sphere.

Some fifteen hundred years later, around 2000 B.C., the Egyptians
measured time by *shadow clocks*.* This shadow clock is a T-shaped con-
trivance lying flat on the ground, with the cross of the T raised and
placed in an East-West direction. A four-thousand-year-old document

---

* Definition of horological terms may be awkward, for they represent components of dynamic devices which must be seen in operation for complete understanding. Accordingly, in this essay, I shall attempt to give definitions by a combination
of words and drawings, asking the reader to attempt to visualize the operation of
the clock elements. I shall *italicize* terms unique to horology, the names of historically important *clocks*, and *concepts* useful in the study of clockmaking.

says, possibly about an eclipse and with reference to a shadow clock that, at one time "one cannot tell when it is midday, one cannot record the shadows."[2]

The earliest devices which measured time by employing some process other than the motion of shadows were based on the regular outflow or inflow of water, sand or oil, or on the even combustion of a commodity, such as oil or candles.

Water clocks or *clepsydrae* were employed in Egypt from *ca.* 1600 B.C. for use at night. A well-known example is the water clock of the Temple of Karnak, now in the Cairo Museum. This clock is the *outflow* variety, with its inner surface graduated for 12 hours of the night with variations for seasonal differences. The outflow orifice of these clocks was usually decorated with an effigy of Thoth, the god of the night hours, in the form of a cynocephalous baboon. In the *inflow* clock a controlled stream of water fell or dripped into a container. A measuring rod was fixed to a float with a pointer moving along a scale set above the clock. Simple outflow clocks were used in classical times in the law courts of both Greece and Rome for the limitation of the lengths of speeches. The Saxons used a bowl with a hole in the bottom which sank in a prescribed time when placed in a container of liquid. They also used graduated candles protected from drafts by horn shields and burned oil through a regulated wick from a graduated container.

Water was also used as a motive force for clocks with mechanical movements. Recently remnants of such a water clock have been found, showing that they existed in China in the eighth and eleventh centuries A.D. A thirteenth-century source[3] describes a *mercury clock* consisting of a drum divided internally with perforated radial plates. A weighted rope was wound on the drum so as to cause it to revolve against the resistance of the slowly varying mercury level. The rotations were recorded on a dial.

Water-driven clocks were used in monastic life to enable the sexton to call the monks for the various chapters at the correct hour. When, in 1198, a fire was discovered in the Abbey of Bury St. Edmunds in Suffolk, England, they recorded that "the young men among us ran to get water, some to the well and some to the clock, while yet others, with the utmost difficulty, succeeded in extinguishing the fire with their cowls."[4]

2   *The Escapement and Early Mechanical Devices*

An early attempt to make a weight-driven clock appears in the commentary of Robert the Englishman, in the *Sphere of Sacroboso*, which was written in 1271. "Nor is it possible for any clock to follow the judgment of astronomy with complete accuracy. Yet clockmakers are trying to make a wheel, which will make one complete revolution for every one of the equinoctial circle, but they cannot quite complete their work. But, if they could, it would be a really accurate clock and worth more than an astrolabe or other astronomical instrument for reckoning hours, if one knew how to do this according to the method aforesaid. The method of making such a clock would be this, that a man make a disc of uniform weight in every part in so far as could be possibly done. Then a lead weight be hung from the axis of that wheel so that it would complete one revolution from sunrise to sunrise. . . ."

Robert was not successful. Some means had to be found to check the running down of a weight-driven clock; some arrangement that would stop the clock's running down and then release a train (a series) of wheels, produce some indication and again permit the clock to start running down and then stop it again, and so on in sequence. The name of the genius who found the answer by inventing what we now call *escapement* is unknown. But it is generally accepted that the invention took place in Western Europe some time about the end of the thirteenth century.

The first escapement, represented in Figure 1 by the verge escapement, was the only one for nearly four hundred years.[5]

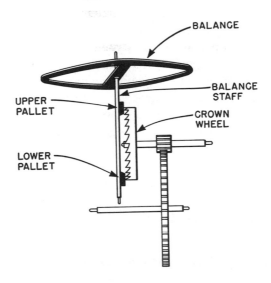

Figure 1. The principle of the escapement shown by one of its early embodiments, the verge escapement. Sketch from the author's private collection.

The saw-toothed *crown wheel* tends to rotate by reason of a driving force but is held in check by the *upper pallet* on the *balance staff*. As the balance oscillates, the tooth of this pallet releases a tooth of the crown wheel, the wheel starts to turn and is again locked, now by the *lower pallet*. The direction of oscillation is reversed and the lower pallet now frees the crown wheel, which the upper pallet then locks and so on.

Another early form of oscillator was the *foliot*, a crossbar instead of a ring. The ends of the crossbar were notched and small adjustable weights were hung on them as a secondary means of regulation. The primary means was by adjusting the driving weight.

Some of the earliest clocks using the escapement principle were made in Italy in the early 1300's. These clocks, like the water clock mentioned earlier, were for monastic use. They would strike the bell for the sexton's guidance, one blow each hour. Regulated hour striking was achieved through the construction of a *locking plate*, a disk with slots on its circumference which permitted an arm, raised by the striking train, to fall as often as the numbers of strokes needed to denote the hour. These early clocks are not believed to have had dials; in any case, the public would not have been able to read them. It is thought that the *clock dial* was invented by Jacopo Dondi of Choggia, Italy, in 1344. He was awarded the title of "Del Orologio" (The Horologist). His epitaph reads, in part: "Gracious Reader, advised from afar from the top of a high tower how you tell the time and the hours, though their number changes, recognize my invention. . . ."[6]

The earliest refinements and complications introduced into clocks were the indications of astronomical information: first, indications of the moon and its phases as well as its relative positions to the sun; secondly, the zodiacal indications. Later, dials were introduced showing the planetary motions. We call clocks showing astronomical events *astronomical clocks*; we shall return to their history at the end of Section 5 of this essay. The earliest of these that is known is an astronomical clock by Giovanni Dondi, a son of the Jacopo Dondi mentioned earlier. In 1364, after sixteen years of work, he completed his clock which showed mean time, sidereal or star time, and the time of the rising and setting of the sun. This is the first clock known to have been provided with a train to convert sidereal to mean time. Dondi's clock also showed the motions of the five planets then known and the motion of the moon. He took into account the slight eccentricity of the moon's orbit and incorporated a perpetual calendar for the movable feasts of the Church—feats not equaled by any other clockmaker for five hundred years after Dondi's death. He left a full description and complete drawings of his clock. An entirely new reproduction of his clock based on these drawings has been reconstructed under my direction and is now in the Smithsonian Institute in Washington, D. C.

The *first Strasbourg Clock* (1354) is reputed to have had astronomical indications, but we have no details. All that remains of this clock is the cock which surmounted the assembly. The wrought-iron cock reminded the listeners of St. Peter's denial of Christ. The cock spread and flapped its wings, opened its beak and crowed. It was used again in the *second Strasbourg Clock*, in 1574. The cock is now in the Strasbourg Museum.

The oldest surviving mechanical clock is that in the Salisbury Cathedral in England. The first record of it appears in the Salisbury Archives in 1386. Who the maker was we do not know, but possibly it was someone brought in from the Continent for the purpose of building the clock. The next oldest clock (1389) is the one in Rouen, France; this is the earliest surviving clock that had a quarter strike. Next comes the one in Wells Cathedral, England (1392).

Up to the middle of the fifteenth century all mechanical clocks were weight-driven. About the middle of the fifteenth century the idea was conceived of having a coiled spring for motive force. There is a portrait in the Fine Arts Museum in Antwerp, dating back to about 1450, which suggests that the use of coiled springs was known, for it shows a clock without any weights attached. It must soon have become apparent, however, that a spring drive cannot give the same constancy of driving force that a weight can. A new device, called the *fusee*, was invented.[7] The principle of the fusee is that of a lever equalizing the pull on the clock train as the spring unwinds. Another device for equalizing the declining force of a spring is the *stackfreed*,[8] found mostly in South German clocks and later in South German watches. A pinion is fixed to the arbor around which the spring is coiled; this takes into a wheel, which is not entirely cut, permits the pinion to make not more than about three turns, then the clock stops. These first three runs drive the clock with the most nearly even force. A spring bearing against an eccentric cam fixed into the wheel exercises an additional equalizing effect.

The verge escapement, although a great improvement over water clocks, left much to be desired, as the clockmakers of the day realized. Jobst Burgi, a Swiss, invented an escapement in which the two pallets were carried on separate arbors and the depth of entry into the teeth of the escapement was separately regulated for each. This escapement, known as the *cross-beat*, was a great improvement over ordinary verge; but within about sixty years after its introduction (in 1657, specifically) the pendulum as part of a clockwork was invented by Christian Huygens, the Dutch physicist, and the cross-beat was quickly forgotten.

Burgi was also the inventor of what is known as the *remontoire*, that is, any instrumentation which "winds up" the clock frequently so as to maintain the driving force equally and maintains the drive close to its peak power. In his first remontoire, small weights were enclosed in a box which descended on a rack once in twenty-four hours. At the bottom the box set off a spring that raised it up to the top of the rack. The spring only needed to be wound once in three months.

3   *The Pendulum*

In this historical sketch we are now approaching the middle of the seventeenth century, the epoch which saw a revolution in the building of timekeepers: the invention of the use of the pendulum. The principle of the *isochronism* of a pendulum, which Galileo discovered, is that the period of pendular swing is almost independent of the pendular amplitude. Sketches of apparatus fitted with pendulums appeared as early as 1480–1490 in the drawings of Leonardo da Vinci, but they are believed to have been used only for providing a rocking motion. Their motion had to be manually maintained and they were never adopted for the *mechanical counting* of the number of oscillations.

Galileo, besides discovering the isochronism of the pendulum, also invented an entirely original form of escapement, but it seems that he never actually succeeded in constructing a working clock by combining the two elements. Shortly before his death, however, he did communicate his secret to his son Vicenzo, who died in 1649, also before having completed the clock. It is thought that the clock was later completed for the Duke of Tuscany by Philip Treffler of Augsburg, Germany.

Before the invention of the pendulum there existed no man-made time standard. The rate of a clock was controlled by the amount of weight applied or by the torsion of the spring. The pendulum offered a device of fixed period, depending within limitations upon its length but not on the driving force. Christian Huygens studied the theory of the pendulum and concluded that a truly isochronous pendulum should have its bob swinging along a cycloidal curve[9] and not in the arc of a circle. The cycloidal curve is slightly more U-shaped than the arc of a circle. Huygens then designed a pendulum suspended on a silk chord with a verge escapement having an amplitude of about 40° in which the orbit of the bob was deflected from that of a circular arc to that of an arc of a cycloid. This did provide the correction but introduced more errors than it corrected, and after his death it soon ceased to be used. In fact, it was superseded by the brilliant invention by William Clement of the *anchor escapement* shown in Figure 2.

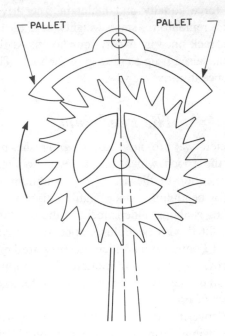

Figure 2. The anchor escapement of William Clement.

In the anchor escapement the faces of the pallets of the escapement are in the same plane as those of the teeth of the escape wheel, thus enabling the pallets to effect clearance within a much smaller arc than was necessary for the verge escapement. Within a small arc, say 3° or 4°, the two curves, a circular arc and the arc of a cycloid, coincide for all practical purposes and the pendulum becomes sufficiently isochronous. The anchor escapement is an invention only second in importance to that of the pendulum itself, for it forms the basis of practically all subsequent escapement designs up to the present day.

The narrow amplitude of the anchor escapement made it possible to encase a one-second pendulum (about 39″ long) and the *long case*, or *grandfather clock*, was born. Enclosing a clock with a verge escapement and a 39″ (seconds) pendulum with a 40° amplitude was not very practical.

394

The greatly increased accuracy of the pendulum beating the units of time (seconds, specifically) led to experiments with longer pendulums, beating one and a half or two seconds. Their use, however, was transitory except in the case of *turret clocks* (large public clocks) where heavy pendulum rods can provide stability. Thomas Tompion, the "Father of English Watchmaking," constructed two 14′ two-second pendulums suspended above the movement for the (then) new observatory at Greenwich, England. Their amplitude of oscillation is so small that it requires close attention to notice it, as may be ascertained by observing them in the Octagon Room of the original Greenwich Observatory. Although Huygens was well acquainted with the details of the anchor escapement, he never used it. This is surprising if we assume that he must have realized that this was a straightforward and practical solution for the problem of using the cycloidal curve he himself suggested.

About 1715 George Graham improved timekeeping with his *dead beat* escapement, an improved form of anchor escapement. This was the stand-and escapement for use in Observatories at Greenwich and elsewhere for over two hundred years.

4 *The Balance Spring and Temperature Compensation of Clocks*

The improvement in timekeeping in clocks now achieved with the anchor escapement spurred the watchmakers to find an equivalent improvement, since this escapement was not suitable for adaptation to watches. The dishonest watchmaker would fit his watch with a double-ended *dumbbell foliot* and verge escapement and would hide the upper half under a cock so that the lower half appeared to swing as a pendulum. In 1675 Christian Huygens published the details of his *balance spring*. This is a device, in a watch, which gave control over the main spring's driving power in a way analogous to the control the pendulum has over the driving power of the clock. Robert Hooke subsequently claimed priority, and the question of inventorship has never been satisfactorily cleared up. The invention of the balance spring was followed by George Graham's invention of the *cylinder escapement* (a tiny hollowed-out balance shaft) for watches. The combination of cylinder escapement and balance spring was widely used during the eighteenth and early nineteenth century. The number of escapements invented is now well over three hundred. Practically all, however, are but variants of the principle of the anchor escapement.

While in the industrialized parts of Europe progress was being made in the fabrication of more accurate clocks, the Bavarian peasantry started making wooden clocks as a home trade during the long winter months. These were forerunners of the *Black Forest Cuckoo Clocks* (which first appeared in 1740), and of the Eli Terry wooden clocks of Plymouth (now Terryville), Connecticut, of the early nineteenth century.

As the accuracy of timekeepers improved, some second order effects in the motion of pendulums and balance wheels became evident. Such effects were, *e.g.*, those caused by temperature and pressure variations. In about 1726 James and John Harrison, two English country carpenters who made wooden clocks, investigated the relative heat expansion of steel and brass. Based on their studies, they constructed a bimetallic *gridiron pendulum* compensated for heat expansion. George Graham had a different approach. He placed mercury in a glass jar and used it as the bob of the pendulum. With increasing heat the mercury and the pendulum rod expanded, inversely, the center of gravity of the two and thus the effective length of the pendulum tended to stay constant.

5   *Mechanical and Electromechanical Timekeepers from the Eighteenth Century to the Present*

The search for a reliable means of ascertaining a ship's longitude at sea had long engaged the attention of maritime nations. Christian Huygens ineffectually tried to produce a pendulum clock for use at sea, as did Henry Sully of Paris.

In 1714 the English Government offered a prize of £20,000 for the construction of a clock that would determine longitude within 34 miles or 30 minutes of arc, that is, would keep time with an error of not more than 2 minutes in 42 days. The brothers Harrison set themselves to the task by first trying to eliminate all friction. They bushed the pivot holes in their wooden clock with *lignum vitae*, an oily wood, and they developed what came to be called the *grasshopper escapement*, because of its very light action. Their first, somewhat cumbersome, and very heavy machine (72 pounds) had only a limited success. Two more "machines" were made by John Harrison along similar lines, but they were never tried at sea. He then changed his ideas completely and produced what may be described as a large watch. He reverted to the verge escapement with specially designed pallets, using a special seven-and-a-half-second remontoire and a special bimetallic temperature-compensation device. This watch, after much delay, won for him the first prize. Simplified versions of the Harrison watch designed by Arnold and Earnshaw are still in use today.

In 1770 Thomas Mudge[10] constructed his first *lever escapement*, which had the advantage that for a good part of the swing of the balance the escape wheel is free of the pallets.

The forerunner of our contemporary *Atmos clock*, self-winding through changes of temperature, is the clock James Cox produced in 1765 which was self-wound by change in atmospheric pressure. His clock was wound on either the rise or fall of pressure, whereas the Atmos clock winds only when the temperature falls.

Another second-order effect whose importance slowly became appreciated is the fact that the variation in the impulse given to pendulums varies their period, even though slightly. To counteract this variation, many forms of *constant force escapements* have been devised. The first one on record is that by Alexander Cumming in 1766. He had a small detent connected with each pallet. The detent unlocked in turn one of two small weights which then fell from a fixed height onto arms projecting from the pendulum rod, supplying the pendulum with impulses of constant magnitude. This principle was later developed into the *three-legged gravity escapement* by E. B. Denison, the later Lord Grimthorpe. He is the designer of the Big Ben in London's Westminster Tower.

In the early nineteenth century, owing to the lack in the United States of rolling mills of sufficient precision, good spring steel was not obtainable there so that there were no American-made portable clocks. To overcome this, Joseph Ives of Brooklyn, New York, invented the *waggon spring* clock. He connected the ends of the laminated spring by a cord or gut to the winding drum of the clock. His idea, however, also seems to have had a forerunner: there is a seventeenth-century clock in the Horological Museum at Le Locle, Switzerland, which employs this principle.

In 1830, Aaron D. Crane, another American, patented a clock in which the thin strip-pendulum spring hung vertically and was used as a torsion pendulum. This is a *year clock*, which takes advantage of the fact that the power necessary to drive a torsion pendulum is much less than the power necessary for an ordinary pendulum. In this case the United States was about a century ahead of Europe.

In 1843 Alexander Bain, Scottish psychologist and author, produced what is probably the first electric clock. He sank carbon and zinc plates into the earth, the ground moisture providing the electrolyte. The pendulum was electromechanically driven. The bob was a coil which swung between two permanent magnets. The clock was not particularly accurate, but it was a good beginning.

George Airy's clock of 1873 (he was then the Astronomer-Royal) embodied compensation for barometric pressure variations. A float on the surface of the mercury at the open end of a siphon barometer is coupled to a horseshoe magnet placed underneath the pendulum bob, to which two permanent bar magnets have been fixed on either side. Variations in mercury level were then used to vary the magnet-bob distance so as to correct for variations of the period caused by barometric pressure changes. This clock was used to determine *Greenwich Mean Time*, the prime basis for standard time through the world from 1873 until 1922, that is until Shortt's clock was installed, as we shall discuss shortly.

In 1899 Sigmund Riefler, a German, introduced his clock for observatory use. In his device the pendulum was suspended from a broad, thick spring which was ground thin at the point of bending, and the impulse was given by flexing the whole pendulum suspension.

In 1895 Frank Hope-Jones introduced his *synchronome remontoire*, which was designed to secure reliable electric contact for pulsing distant clocks, but without interfering with the main timekeeping function of the clock.

It has been known since the studies of Huygens that an ideal pendulum would be one which swings freely, has no work to do and receives its driving impulse at the dead point. In 1898 James Rudd of Croydon, England, built the first such *free pendulum clock*. He suspended his pendulum without an escapement above an auxiliary or *slave movement* from which it received its impulse. If the slave clock was too fast or too slow, its rate could be adjusted between succeeding swings of its pendulum.

The next approach to the free pendulum was that of Shortt of Exeter, England, in 1920. Shortt, like Riefler, suspended his pendulum on a strong spring ground, thin at the bending point. He used the termination of the impulse to synchronize a slave pendulum to which the energy-consuming operations were transferred. The slave had a permanent losing rate on the master. When sufficient loss had accumulated a buffer spring action speeded up the slave pendulum, but the difference never got larger than 1/240th of a second. This clock had a remarkable success and was installed in most leading observatories of the world and widely used until the advent of quartz crystal clocks, atomic clocks and electromagnetic devices in general.

In the 1950's the firm of Riefler brought out another clock in which a gravity impulse-driven pendulum is freely suspended on a knife edge. The pendulum carries a diaphragm with a pinhole. A pencil beam is focused on the diaphragm and as the pinhole permits the light beam to pass, the beam activates a photocell. The clock's signal is, then, the electric signal of the photocell which has thus been obtained without loading the pendulum.

Briefly we now return to the history of astronomical clocks. Following Dondi's early lead many such clocks have been built representing increasing accuracy and complication of mechanism. They, however, demonstrate only the same known principles. An exception to this is Jean Baptiste Schwilgue's *third Strasbourg* clock. In this, completed in 1842, he provided for the mechanical recording of the movable feasts of the Church, as Dondi had done in 1364.

But Schwilgue had now to contend with the Gregorian Calendar. Under the *Julian Calendar* leap year occurred every four years and had 365.25 days. This is 11 minutes and 14 seconds longer than the *astronomical year*. As we shall see in the essay by Professor Clemence, by 1582 this small difference had accrued to a total of 10 days. To correct this Pope Gregory XIII introduced further modification, still in use today, and known as the Gregorian Calendar. It is interesting that the Gregorian Calendar was not adopted in England until 1752, by which time the error was eleven days. It was adopted at various dates by other European countries. The Greek Orthodox Church did not accept it until 1924 and Turkey until 1927.

6 Divisions of the Day

The method of recording the hours of the day varied considerably over the ages. The Sumerians, referred to in Section 1, had a day of 12 hours starting at sunset, the Babylonians a day of 24 hours also starting at sunset. The Babylonian astronomers used a day of 24 equal hours, believed to have started at midnight. The Egyptians had a day of 26 hours, starting at sunrise. Monastic time was measured by 12 equal periods of daylight and 12 of darkness. These temporal hours, therefore, varied with the seasons. This system was also used in Japan, where it persisted until the end of the nineteenth century. The Italians counted 24 hours from sunset and referred to the 2-times-12 system as *French hours.* On the Continent of Europe the reckoning of 24 hours was some-times referred to as *great hours,* and the 2-times-12 system as *small hours.* The *Nuremberg hours,* abandoned early in the nineteenth century, con-sisted of 16 hours of light and 8 hours of darkness at the summer solstice and of the converse at the winter solstice. Public notices ordained when an hour should be transferred from one day division to the other.

In this brief summary I have sketched some of the important events in the history of timekeepers. By necessity the sketch is incomplete and the reader must be referred to the general references for this essay. Figure 3 shows the progress of the accuracy of timekeepers from the eighth century to our day.

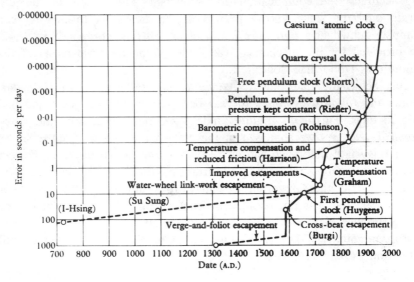

Figure 3. Chart showing the rise in accuracy of the mechanical clock through the centuries (amplified by J. Needham, with his approval, from the original of F. A. B. Ward, "How Timekeeping Mechanisms became Accurate," *The Chartered Mechanical Engineer,* v. 8 (1961), p. 604, after consultation with J. H. Combridge and H. von Bertele). Clement's anchor escapement, discussed in the text, would appear be-tween Huygens' first pendulum clock and the improved escapement of Graham, known as the dead beat Escapement (1715).

Time Measurement for Scientific Use

G. M. CLEMENCE

1 *The Physical Universe of Space and Time*

The physical universe is comprised of all events occurring in space and time which can be observed, or inferred from observations, and regarding which all (or nearly all) interested persons are agreed. Any phenomenon that can be observed by sight, hearing, touch, odor or taste is such an event. Events may be of a simple, ordinary kind, such as meeting an acquaintance on the street, or seeing a robin on the lawn, or they may be inferred by very complex reasoning with the aid of the most delicate instruments, such as the explosion of a star in the year 1006. Supposed events such as apparitions of flying saucers are excluded from the physical universe, since there is hardly any measure of agreement regarding their existence.

Also excluded from the physical universe is the universe of ideas. Although an idea, when it exists in the mind of a person, is certainly an event of a sort, considered as a concrete universal it hardly exists in space and time; wisdom, for example, cannot be uniquely associated with a particular place or a particular time, and hence it is not an event of the physical universe.

The universe of art, if indeed it can be properly distinguished from the universe of ideas, is also excluded from the physical universe. Whether a piece of music can be said to exist in space and time, when it is not being performed, has been a question for philosophical debate. In my opinion, music certainly has not the same sort of existence as a star; it has no definite place in space, and probably not in time.

Any physical event can be fixed by means of three numbers for space, and one for time. These four numbers are called *coordinates*. The three coordinates of position, or space, are of considerable variety, for they may involve an infinite number of points of origin from which the measurements are supposed to be made. In many cases location of the origin is implicitly understood and not stated. Sometimes only two or even only one coordinate suffices. But in principle, a specification of position always requires three numbers, distances being reckoned in three directions from a definite origin.

Whereas space has three *dimensions*, time has only one. But, as with space, the specification of an instant of time always requires an origin from which the time is measured. On the other hand, with the most common method of specifying time, by giving the year, month, day and hour of day, the origin of time never need be mentioned, because it is implied by the statement itself. It may be objected that to give the year, month, day and hour of day requires at least four numbers, whereas time, if limited to one dimension, should be expressed as a single number. We, of course, might have expressed all times in seconds and fractions thereof, but to do so would be cumbersome, and the advantages of indicating the season of the year and the time of day would be lost. It is purely by custom and for the sake of convenience that we use several numbers instead of one to express a time.

The word *time* is used in English in two distinct senses. One meaning is a duration; another is a specified instant of time. To avoid confusion between the two, a specified instant of time is called an epoch.

2   The Calendar

Unlike the units of length, which have been arbitrarily established by decree or acts of legislative bodies, the basic units of time are natural ones.[1] The two most important ones are the day and the year. The day is the period of recurrence of daylight and darkness, the year the period of recurrence of the seasons. These two periods are incommensurable, the year consisting of 365.242199. . . days. Some complications of the calendar follow, since it is generally desired to have the same month and day occur perpetually at the same season. Also it becomes necessary to distinguish between the year already mentioned—which is named the *tropical year*—and the *calendar year*, since it would not be practicable for the calendar year to contain a fractional part of a day. Under the present calendar, established for the Western world by Pope Gregory XIII in 1582, ordinary calendar years consist of 365 days, while years divisible by 4, designated as leap years, consist of 366 days, the extra day being added after the twenty-eighth of February. An exception is made for centennial years, which are leap years only if divisible by 400. With these rules the average length of the calendar year works out to be 365.2425 days precisely, the discrepancy with the tropical year amounting to one day after some 3000 years.[2]

Although the Gregorian calendar was established for use in the Western church, it has spread throughout the world, at least for commercial purposes; in some countries, like India, various other religious calendars exist as well. And in nearly every country there are holidays which are observed by only part of the population, but these are not generally counted as separate calendars.

In 1923 the orthodox oriental churches adopted a modified Gregorian calendar in which century years are leap years only when division of the century number by 9 leaves a remainder of either 2 or 6. Under this rule, their calendar will differ from the Gregorian by one day at the year 2800, but will agree with it again at 2900.

The month originated as a third natural unit of time: the recurrence of the phases of the moon, after which it is named. But no attempt is any longer made to adjust the calendar month to the synodic (phase-of-moon) month. Such adjustments would have to be so frequent as to be disagreeable, and the synodic month is no longer as important in everyday affairs as it was before the days of artificial illumination. So the calendar month is now to be regarded as a purely conventional unit of time.

The week of seven days is not based upon any natural unit of time, but is a purely arbitrary unit first established for religious purposes, and dating back to antiquity.[3] Some religious sects, especially some which celebrate the seventh day as their sabbath, are concerned about preservation of the unbroken sequence of days of the week, and the question has often been asked whether this sequence has ever been interrupted. The answer is that the sequence has been unbroken since the days received the names by which they are now commonly known, under the Roman Empire. In particular, the Gregorian reform of the calendar entailed no break in the sequence of weekdays. But information is lacking about which day has been reckoned as the seventh one at various times and in various places; for example, it is not known whether our Saturday coincides with the ancient Jewish seventh day. To settle the matter it would be necessary to consult records showing corresponding weekdays in different calendars, which were made in ancient times. No such records have so far been discovered.

The subdivision of the day into twenty-four hours, the hour into sixty minutes, and the minute into sixty seconds is also purely arbitrary, the hour, minute, and second not corresponding to any natural units of time. The number 24 results from the ancient custom of dividing days and nights each into 12 parts, the number 12 having a mystical significance, being a so-called perfect number, perhaps because it is the smallest number divisible by 2, 3 and 4. The number 60 has the same properties, and also is divisible by 5 and 6. But the precise history of the subdivision of the day is lost in antiquity.

Our method of reckoning time, then, is neither very simple nor very elegant, being a historical development over many centuries. Perhaps the most remarkable thing about it is its universal use. To a visitor from another planet it would certainly seem whimsical and obscure. Yet it is among the most firmly fixed of all our customs, partly no doubt because no official procedure exists for changing it. Historically, the calendar has been the charge of the Church, because in ancient times the priests were the only persons with sufficient knowledge for coping with the subject. Now, with the waning of influence by the churches over secular affairs, it is doubtful whether any would undertake a reform of the calendar. National legislatures and executives, while having the necessary authority, would be reluctant to act unilaterally, especially since a measure of confusion, if not worse, would certainly result. Committees of the United Nations have repeatedly declined to place the subject on their agenda, asserting that they had more urgent business to attend to. Notwithstanding these cogent considerations, a large number of different plans have been devised for calendar reform and promoted through popular news media and privately printed pamphlets. One plan would provide for thirteen months of twenty-eight days each, with the extra day (or two days in leap years) not having any weekday name; in this fashion every month would begin on Sunday and end on Saturday. All plans for reform aim at greater regularity of some kind, while admitting greater irregularities of other kinds.

3   Clocks and Watches

The measurement of time is basically an operation of counting units and subdividing them. In the case of the years and days no special instruments are necessary, nor are any commonly used. A wall calendar is a device for assisting the counting of days, but it is not itself a counter. In practice the count of years, months and days is kept with the aid of the numerous records that are made in everyday life. A man in isolation very quickly loses the correct count of days unless he keeps a record. The shorter units, on the other hand, such as hours, minutes and seconds, are counted with actual counting devices such as clocks and watches. At any instant a clock or watch, if running correctly, shows the count of hours and minutes (and seconds) that have expired since the preceding beginning of the day (or half-day).

An ordinary nonelectric clock, as discussed in its many embodiments by H. A. Lloyd in the preceding essay, consists of three essential parts: a source of power to turn the hands, an escapement to release the power at the proper rate, and the time indicator consisting of dial and hands. It follows that the quality of timekeeping depends on the precision of manufacture of the various parts. The best pendulum clocks, found in astronomical observatories, keep time to better than a tenth of a second per day, if maintained in a vacuum under constant temperature.

A nonelectric watch is built on the principle of the clock, but a balance wheel is substituted for the pendulum, as also discussed by Mr. Lloyd. Such watches, if well made and adjusted, keep time to a few seconds per day. It is interesting that accurate, theoretical work as far as I know has never been done on watch movements. Like the relaxation mechanism of piano keys, it seems to have developed chiefly through the skill of the artisan.

In recent years, watches driven by miniature electric batteries instead of mainsprings have appeared; they never have to be wound, but the battery does have to be replaced perhaps once a year. Some have more or less conventional escapements, while others have a tuning fork substituted for the balance wheel. They may keep time as accurately as a few seconds per week.

Ordinary electric clocks are controlled by synchronous motors which keep the clock continuously in phase with the master clock located at the powerhouse. The master clock, in its turn, can usually be relied upon to a few seconds, since it is frequently set correctly with reference to astronomical time.

When even greater precision is required, as for instance in the national astronomical observatories of those commercial nations which provide standard time for precise scientific applications, none of the devices mentioned above proves to be satisfactory. Instead, recourse is had to quartz-crystal clocks. A crystal of quartz, suitably ground and polished, exhibits piezoelectric properties. If it is set in mechanical vibration, an alternating difference of electrical potential is produced between its two principal surfaces. The vibration, whose period may be analytically related to the dimensions of the crystal, may be used to control the frequency of an alternating current which, in turn, drives the hands of a clock. Such clocks, if maintained at constant temperature and pressure, keep time with an accuracy of a few milliseconds per day.

4   *Fundamental Units of Time*

Any repetitive phenomenon whatever, the recurrences of which can be counted, is a measure of time. For scientific purposes some phenomena are preferable to others. On measurement of time with the highest possible precision, we place the following requirements. The indications should be accessible to all who need them; they should be continuous, that is, not subject to interruption; and they should be invariable. From the earliest times until about a decade ago the rotation of the earth has been the fundamental measure of time for all purposes, scientific and nonscientific. The rotation of the earth is most conveniently measured by observations of stars at an astronomical observatory, and by comparing such observations with a clock that is regulated to run as nearly consistently with them as is feasible. The clock is never actually set to time; instead, a daily record is kept of its performance, which serves the purpose just as well. The principle of the observations is as follows.

The stars are sufficiently distant to permit, in a first approximation, the assumption that a line from the center of the earth to a particular star has an invariable direction. As the earth rotates, this line traces out a parallel of latitude on its surface. We may imagine an observer to be stationed on this parallel. Then, once each day, he will see the star directly overhead. The time interval between two successive passages of the star overhead will be equal to that required for one rotation of the earth. With a suitable telescope and clock an observer can time such a passage of a star with a precision of about 0.006 second. He can increase the precision considerably by observing many stars and averaging the results. He can also take into account the very slow motions of the stars themselves, and of various motions of the earth's axis of rotation. These various parameters have been determined by centuries of observations. Such an observer is able to count and subdivide rotations of the earth with an accuracy of a millisecond. Of course the actual techniques are very much more sophisticated than is described here; I am speaking only of the basic principles.[4]

The measurement of time is inseparably connected with the determination of an invariable direction in space. Whether an invariable direction can in fact be identified, or even defined, is a question that has been much discussed by theoreticians, who usually express it in other words. As conventionally stated, it is a question of measuring absolute rotation, or equivalently of establishing an inertial frame of reference. While the experts are not yet completely agreed, the weight of opinion holds that to find an inertial frame of reference is theoretically impossible. The actual measurement of time is, however, not a theoretical subject but an experimental one; all experiments are subject to a margin of error, and on the practical side we proceed as follows. We imagine as many nearly inertial frames as possible; we compare them with one another, and if they are in agreement within the experimental errors, then we may call them inertial.[5]

So far three systems have been proposed: the system of the stars, the solar system, and the system of external galaxies.

The system of stars is in motion. Each star, as well as the sun and planets, is moving in its own orbit around the center of the galaxy, at a speed of some tens of kilometers per second. The orbits are so large that years of observations are required in order to detect the motions at all; even now, after some two centuries of accurate observations, the relative motions of only a few thousand stars in our own neighborhood have been measured with high precision. By analyzing these relative motions, we derive such information as we can about the rotation of the entire system of stars; the accuracy of the result, as judged by the internal discrepancies, is about a tenth of a second of arc per century, which is then the precision with which the inertial frame is established.

The orbits of the planets are slowly rotating, each in its own plane, and the speed of the different rotations may be precisely calculated by gravitational theory, and also observed relative to the stars. When the calculated motions are subtracted from the observed ones, we find a residual rotation for the stellar system that agrees with the previous result within half a second of arc per century. Thus, within the observational errors mentioned, the stellar system and the solar system provide identical frames of reference, which in practice we consider to be inertial.

The system of galaxies may be considered as immediately inertial, because the galaxies are so far away that their motions across the line of sight are negligible for many centuries, even if they should be moving in this direction as fast as they appear to be receding from us (up to half the speed of light). It is not yet known whether this frame of reference agrees with the other two; observations for the purpose were commenced only a decade ago, and must be continued for several more decades in order to reach the desired precision.

5   *Zone Time*

We may require that each observer should mark the beginning of his day when the vernal equinox is either directly overhead, or bears due north or south, as he happens to be south or north of the equator. Astronomical clocks regulated on this principle are called sidereal clocks, and they are said to keep sidereal time. Sidereal time would not, however, be convenient for everyday use, because as the earth revolves around the sun, the sun appears to move continuously among the stars. The rate of motion is such that during approximately 366 sidereal days the sun rises and sets 365 times. Accordingly, our ordinary clocks, and many astronomical ones, are regulated to run slow on sidereal time, the amount being 3 minutes 56.555 seconds per day. These differences add up to precisely one day in a tropical year, and keep the clocks in step with the rising and setting of the sun, on the average. Furthermore, ordinary clocks are set so that noon by the clock occurs near the middle of each period of daylight. Now, since when it is daylight on one side of the earth it is night on the other side, it follows that a traveler going around the earth in an eastward direction must continually advance his watch. It would be inconvenient for travelers to readjust watches and clocks at every moment, and so, by international convention, it is done by the hour. The earth is divided into twenty-four time zones, which are not exactly equal in area but are adjusted to follow natural and political boundaries of various sorts. Within each zone all clocks keep the same time, while in the adjacent eastward zone they are one hour faster, and in the adjacent westward zone one hour slower. Thus, at any one instant of time, all clocks on the earth indicate the same minute and second, while each hour of the twenty-four is indicated by some group of clocks.[6]

This system of time, which is called *zone time*, is very simple and convenient, and would be difficult to improve upon, but unfortunately it is easily disrupted by whim and local fiat as exemplified, for instance, by the practice of Daylight Saving Time. According to this arrangement, certain geographic areas elect to keep the time of the next eastward zone during some of the summer months each year, causing some confusion.

The traveler proceeding eastward around the world advances his watch by one hour twenty-four times during his journey. Thus another convention is required, so that he will not be one day out of step upon his return. The International Date Line was established, running approximately north and south through the Pacific Ocean, avoiding inhabited islands. A traveler crossing this line from west to east sets his calendar back by one day, thus enjoying the same day of the week twice in succession; in traveling westward he advances the date by one day, thus missing one day altogether if he happens to cross the date line at midnight.

For some scientific applications, and for commercial activities involving a large part of the earth, it is desirable to have a single world-wide time. The one selected is the zone time of the meridian of Greenwich, England, from which longitudes are measured. This time is named Universal Time (UT), and sometimes is called Greenwich Mean Time (GMT). Universal Time is never altered, and hence it is suitable for designating an epoch without ambiguity.

Since the earth moves around the sun in an ellipse, not a circle, its orbital motion is not uniform, but is about 6 per cent faster in January than in July. Since the rate of rotation of the earth is almost constant, it follows that *with respect to the sun* it is variable, and the time shown by a sundial differs in general from the time shown by our clocks. Sundial time is called apparent solar time and it differs from zone time in two distinct ways. First, there is a constant difference depending on the location of the dial within the time zone, which vanishes in the middle of the zone, and increases to about half an hour at the two boundaries with the adjacent zones. The exact amount of this difference may be found as follows: find from a map the difference of longitude between the dial and the standard meridian for your zone (which is an exact multiple of 15°); the clock will differ from the dial by four minutes for each degree of longitude difference, the dial being fast on the clock if it is east of the standard meridian and slow if west. Second, superimposed upon this constant discrepancy is a variable one, which vanishes on April 15, June 14, September 1 and December 25 of each year, and reaches maximum values of −14 minutes in February, +3 minutes in May, −6 minutes in July and +16 minutes in November, the positive values indicating that the dial is fast on the clock. This variable discrepancy is called the equation of time. Some sundials are provided with so-called *analemmas*, from which the equation of time may be read off and added or subtracted from the indication of the dial.

6 *Ephemeris Time*

Although time based upon the rotation of the earth (that is, a count of days and their subdivisions) is still indispensable for many purposes, including navigation at sea and surveying on land, it is not suitable for precise scientific applications, for it was discovered about thirty years ago that it is not invariable. The discovery was made by H. Spencer Jones,[7] by comparing the rotational time (which is still the time indicated by our ordinary clocks and watches) with the time required for the moon to revolve around the earth, the time required for the earth, Mercury, and Venus to revolve around the sun, and the time required for the satellites of Jupiter to revolve around Jupiter. All of these different motions are somewhat irregular, but the irregularities can be calculated very accurately and allowed for. It was found that all of these various time standards agreed in indicating that the rotation of the earth was irregular, making our clocks during the past century at times some seconds fast and at other times some seconds slow. The causes of these irregularities in the earth's rotation are not very well understood, but they are suspected to relate to some sort of motion in the liquid interior of the earth.

It was necessary to define a better standard of time, and in 1956 a new definition of the second was officially adopted. Formerly the second had been defined as 1/86400 of the day. Now it was defined as 1/31556925.9747 of the tropical year.[8] The new measure of time is named *ephemeris* time, from ἐφημερίς meaning a calendar or tabulation, and implying an almanac that shows the positions of celestial objects corresponding to various times which are listed at the left side of the table. Such a tabulation (which was under the author's direction from 1945 to 1958) is *The American Ephemeris and Nautical Almanac*, an annual British-American joint publication issued in the United States by the U.S. Naval Observatory about one year in advance of its effective date. It contains positions of the sun, moon, and the planets for every day of the year and much related information.

Ephemeris time appears to satisfy all scientific requirements in principle, but it is subject to one drawback: it is known with full precision only some years in arrear. Several years are required to accumulate enough astronomical observations so that by combining them an epoch can be established with full precision, and this epoch is necessarily always some years earlier than the latest observations. While this drawback is unlikely to be fatal to most experiments, it is a nuisance to those who are impatient. Fortunately, during the past twenty years an entirely new kind of time standard has appeared; the so-called atomic clock.

7   Atomic Time

It was discovered that hyperfine transitions and inversion frequencies in the spectrum of certain elements depend almost entirely on the substance and only very weakly on its environment. Such transitions, therefore, are well suited to form the primary reference source of frequency standards. Inasmuch as these frequencies are very high, they are not used directly but are made to regulate an electric clock. Such a clock, when controlled by suitable resonances, is referred to as an atomic clock. The epoch of such clocks must be set, however, by means of astronomical observations.

The most commonly used such "clock" or time standard is called the cesium clock[9] because it is the splitting of the spectral lines due to the magnetic moment of the cesium nucleus which is utilized in them. The specific method used involves a continuous beam of excited cesium atoms which is permitted to traverse an evacuated space wherein they return to their unexcited states emitting photons of the desired reference frequency. These frequencies are compared daily with that of a quartz crystal, and a record is kept of the performance of the crystal, which serves as a clock. Although the clock is not actually indicating the correct time, the error is known, and is easily allowed for.

Other substances and other techniques may also be used to obtain frequency standards. Hydrogen enjoys some popularity as do masers, lasers and rubidium gas cells.

In accordance with the theory of relativity,[10] atomic clocks do not keep uniform time, but depend on the gravitational field where they happen to be placed. Their running also varies with the latitude of the place, the height above sea level, the distance of the earth from the sun, the distance of the moon, etc. But all of these variations can be calculated and allowed for, thus arriving at a suitable scale of time.

8   Analogies between Space and Time

It was remarked earlier that any event may be fixed in space and time by stating the values of its four coordinates, three coordinates being required to fix its position, and one to fix its epoch. The epoch is the temporal coordinate that corresponds to the three spatial coordinates. Coordinates of the same sort may be added or subtracted. Thus, the difference between two positions is a distance, or interval of space, and the difference between two epochs is a duration, or interval of time.

In dimensional analysis we may mix units of distance and of time together in various ways, obtaining new sorts of physical entities. For example, we may think of velocity as being the ratio of a distance to a duration, and acceleration as the ratio of velocity change to duration. We might even continue the series indefinitely in this way, but we have no special word for expressing the ratio of acceleration change to duration.

We could invent other units, such as seconds per foot, square feet per minute, etc., and we sometimes do so for special applications. The basic entities in this class are distance, velocity and acceleration.

Another class of entities is based exclusively upon time intervals, the notion of distances not being needed. A frequency, for instance, is the ratio of one unit of time to another and is, therefore, a pure number. When we say that there are 365.2422 days per year we are stating a frequency. When we say that the frequency of a particular alternating current is sixty cycles per second, we are stating the ratio of two units of time, the cycle and the second.

Just as we extended the notion of velocity to obtain acceleration, we may extend the notion of frequency to obtain a new entity, expressible as a definite number of seconds per day per year, for example. We have no special word for this entity, but by analogy with acceleration we may call it acceleration of time scales, or time-acceleration for short. An example of time-acceleration occurs when a clock gains two seconds the first day after it is wound, four seconds the second day, six seconds the third day, etc. In this case it is said that the clock is accelerating two seconds per day per day.

A question of basic physical importance is whether the time scales found in nature are accelerated upon one another or not. Note that time-acceleration is not absolute; it makes no sense to say that a clock is accelerated or it is not, but only that one clock is or is not accelerated upon another. We have already seen that the time scale furnished by the rotation of the earth is variable, which is equivalent to saying that it is sometimes accelerated and sometimes decelerated with respect to ephemeris time. But what of the scales of ephemeris and atomic time? (We assume, of course, that the indications of the atomic clock have been corrected for the local variations already mentioned.) There is no a priori reason for assuming them to be identical, and it is entirely possible that one of them may be continuously accelerated upon the other. Indeed, it is thought by some theoretical physicists, notably by Dirac and Jordan,[11] that such an acceleration does in fact exist, only it is too small to have yet been measured. Dirac's basic postulate was that the constant of gravitation is diminishing with the passage of time, at the rate of about a part in 10^{10} per year. If this is the case, then it is possible to imagine another scale of time, accelerated with respect to the first, such that in this new scale the constant of gravitation would remain truly constant. One of the two time scales would have a logarithmic relation to the other, such that the age of the universe, if it is ten thousand million years in one time scale, may be infinite in the other one. The implications for philosophy and theology, as well as for theoretical physics, would be profound.

Whether one of these scales of time is identical with atomic time and the other with ephemeris time is an open question.

Eventually the question whether atomic time is accelerated on ephemeris time will be settled by experiment, by comparing the two time scales. The duration of the experiment will have to depend on how great the acceleration is expected to be. If it is as great as the theoreticians suppose, about 0.003 second per year per year, it will be possible to detect it by the year 1980, or even sooner if the techniques of observation should be improved. The experiment consists simply in observing the motions of the planets in the atomic time scale; if the two time scales are different, the planets will eventually drift away from their calculated positions by perceptible amounts.

In summary of this section, we find that the analogies between space and time involve such elements as position and epoch, velocity and frequency, and acceleration and time-acceleration.

9   *Relativity and the Measurement of Time*

Relativity is discussed in other essays in this volume. Therefore, I shall concentrate only on those principles of the theory which relate to the problem of time measurement for scientific use.

Special Relativity Theory treats of measurements of space and time made by two observers in uniform relative translation along a straight line. General Relativity treats of measurements of space and time made by an accelerated observer, that is, an observer moving with constantly changing velocity. General Relativity is meaningful in terms of everyday experience, because we are continually accelerated by the gravitational field of the sun, and of the other planets.

In the case of Special Relativity we speak of experiments that cannot actually be performed; nevertheless, it is worthwhile to discuss the subject, because the actual accelerations are so weak that they are hardly perceptible, and it is somewhat simpler than General Relativity.

According to Special Relativity, if two observers are in relative, uniform, rectilinear motion, and equipped with identical clocks and yardsticks, each will measure the other's clock as running slower than his own, and each will measure the other's yardstick as shorter than his own, provided that the yardsticks are aligned along the direction of the relative motion. Neither distances nor durations are conserved under different conditions of uniform motion; there is, however, a relation between the two that is con-served: a so-called invariant relation. Now the statements just made are not meaningful in themselves; they only become so if we *specify precisely how the measurements of time and space are to be made.* As to the clocks, we assume that the observers may signal to each other by means of instantaneous flashes of light. Then, if observer A signals observer B from a mirror at B and returns to A, and A observes the reading of his clock the second time, and if A averages his two readings, and if B has read his clock when the signal arrived there, then, if B and A arrive at the same numerical result, the two clocks are said to be synchronized. If similar signals are sent at a later time, and if one clock has gained or lost on the other, that fact can be easily ascertained. Similarly, B can send a signal, which reflected from A, and thereby ascertain the state of the clock at A, from time to time. Then A and B will each conclude that the other's clock is running more slowly than his own.

Objection is sometimes made to Special Relativity in saying that it pre-dicts only appearances, not reality. Such an objection, however, begs the question. The only practicable way to make measurements is first to decide how they are to be made, and then to accept the results at their face value; otherwise, confusion would reign supreme.

Whereas Special Relativity treats of observers in uniform relative mo-tion, General Relativity treats of accelerated motion, which always takes place when an observer is freely falling in a gravitational field. An impor-tant deduction of General Relativity as concerns time measurement is that a clock at rest runs more slowly in a strong gravitational field than in a weaker one by a factor of $1 + GM/c^2r$, where G is the Newtonian gravi-tational constant, M the mass of the body, r the distance between the point in question and the center of mass of M, and c is the speed of light. This effect has been tested experimentally by moving an atomic clock from the basement of a building to one of the upper stories, and then back again, and the deduction is amply verified.

It should be clear even from this brief review that the precise measure-ment of time, although it is a highly specialized occupation engaging the activity of only a few hundred persons, is basic to our knowledge and understanding of all physical and chemical processes taking place on or in the earth and outside of it.

Comments—
Relativistic Dialectics

We have arrived at an understanding of how to measure time. We learned that a local clock, that is one stationary with respect to the experimenter, is thought to be running at the right rate and showing the correct time if we find our ideas of what the clock should show with respect to the motion and configuration of celestial bodies consistent with our ideas of the best—one could almost say simplest—theory of the motion of these bodies. Time measurement turned out to be a "democracy of regular motion," in which the majority is judged to be correct.

Relativity is primarily a theory of the macroscopic world; also, as Clemence has noted elsewhere,[1] "a clock must be defined as a mechanism for measuring time that is continually synchronized as nearly as may be with ephemeris time," which is equivalent to identifying ephemeris time with the coordinate time t of relativity. Therefore, in these introductory thoughts we will be considering relativity in its relation to astronomical rather than atomic time.

All astronomical time standards refer back to circular motions with respect to an absolutely non-rotating framework. It follows that we cannot measure time more accurately than we are able to ascertain the non-rotation of that "non-rotating" coordinate system.[2] The desirability of referring to the detectable distant masses of the universe suggests the relational view of time and hints at the possible usefulness of Mach's Principle. This process of recourse to the universe might not be necessary were we able to give a theoretical construction to a clock; this, however, as Einstein pointed out in 1923, we cannot do.[3]

Nevertheless, there is general agreement on what we mean by here-and-now, even though the "now" is the most complex and controversial of the three temporal categories.[4] Assuming its understanding in an unproblematic way, we may inquire what exactly we are to understand by there-and-now and, in general, by time at a distance. What precisely we may admit as a measure of time at a distance so as to obtain logically acceptable laws of physics is the fundamental proposition of Special Relativity Theory.

The Special Theory does not, however, disclose much about the over-all design or structure of the universe; it remains, therefore, a theory of local interest. General Relativity extends our horizons by proposing a theory of gravitation and by searching for intrinsic restraints on a universe held together by gravitational forces. Inasmuch as the operational determination of astronomical time relates to the idea of absolute rotation, and since analysis of absolute rotation is in the domain of General Relativity Theory, it is to the cosmological content of that theory that we might have to look for suggestions about the nature of time—at least as long as we are unable to give a theoretical construction to a clock.

In the following articles relativity is approached three times, representing three different interpretations of some aspects of this rich source of physical knowledge. While considering the question of what the Theory of Relativity can or cannot contribute to the problem of time, it may be useful to keep in mind two warnings. One is that "mathematical structures and immediate experience are initially independent entities, and the rules of correspondence establish a precise relation between them."[5] The other is that "Geometry (G) predicates nothing about the relations of real things, but only geometry together with the purport (P) of the physical laws can do so. Using symbols we may say that only the sum $(G) + (P)$ is subject to the control of experience. Thus (G) may be chosen arbitrarily, and also parts of (P); all these laws are conventions. All that is necessary to avoid contradictions is to choose the remainder of (P) so that (G) and the whole of (P) are together in accord with experience."[6]

We will find it useful to direct the dialectics of Relativity Theory about time to at least two problems. One problem is the finding of the rule of correspondence between time as it is generally understood right here, and the same entity as it would be inferred by others elsewhere. Second, we may inquire whether the relativistic view of time, as it follows from the purpose and geometry of the theory, says anything about the existence (or absence) of an incalculable element in the universe.

J. T. F.

Time in Relativity Theory: Arguments for a Philosophy of Being

OLIVIER COSTA de BEAUREGARD

Translated and adapted from the French by David Park

The fundamental discovery of special relativity theory is that experimental facts admit a joint definition of time and length measures which entails a physical equivalence[1] between them. The present essay will review the historical development of this idea and its consequences for physics, and give a speculative discussion of the profound influence that it is likely to have on our philosophical views of the world.

1   The Relativity Principle of Classical Dynamics

The relativity principle of classical dynamics was a true predecessor of the special relativity principle of Einstein, and it was already related in many respects to fundamental aspects of the time problem.

The formulation of this pre-Einsteinian version of relativity lies between the absolute space principle postulated by Newton, and what may be called the relative motion principle of classical kinematics. According to Newton's absolute space principle[2] there must exist an absolute spatial reference frame relative to which all movements can be thought of as taking place. This idea turned out later to be metaphysical in character, *i.e.*, deprived of operational support. By stating this principle Newton gave a sort of formal status to common sense feeling; it may be that the postulate had its motivational root in the common experience of living on solid ground.

In complete contrast with the absolute space principle, the relative

417

motion principle of classical kinematics seems at first sight to be experimentally established. This new principle emerged during the development of classical kinematics; it follows directly from Euclidean geometry and Newton's principle of an absolute time.[3] It states that *any* two solid reference frames, in whatever relative motion (translational acceleration, rotation, or arbitrary motion), are kinematically equivalent for the description of movements.

The relativity principle later discovered in classical dynamics is neither of these two, but lies, so to speak, between them. On the one hand, there is nothing in dynamics to substantiate the idea of an absolute reference frame but, on the other hand, dynamics gives a precise way to characterize absolute accelerations or rotations[4] (which classical kinematics cannot do). Thus, according to dynamics, the class of fundamental reference frames of space is neither as restricted as Newton supposed it to be, nor as broad as the purely kinematical relative motion principle would have it. One deduces in classical dynamics that the class under consideration is restricted to solid reference frames all in uniform relative translation with respect to each other; experimentation then allows the full characterization of these so-called Galilean frames.[5]

The simplest operational characterization of the class of Galilean frames refers to the inertial motion of a point particle. But, as Thomson and Tait have stressed in their famous *Natural Philosophy*,[6] this implies a simultaneous operational characterization of what may well be called a Galilean time scale t, for it is obvious that a point motion which is rectilinear and uniform, when referred to any Galilean frame and to a Galilean time scale t, will generally not remain so if referred to a non-Galilean frame and/or to a non-Galilean time scale $\tau = F(t)$. Therefore (and this is an important point for our purpose) *it turns out that there is a very close connection between the appropriate physical definition of a time scale and the Galilean relativity principle.*[7]

Natural clocks, that is, clocks evidently displaying Galilean time, may be any kind of inertial motion or, more generally, any motion implying the fundamental Newtonian formula of dynamics; such motions form the physical basis of mechanical clocks of either astronomical or laboratory size.

Now we must discuss the question of the measurability of time. My point of view will perhaps become clear if we briefly review the analogous development of a quantitative scale of temperatures. In the theory of heat it is found that temperatures are rendered *measurable* through the introduction of Kelvin's thermodynamic scale, or at least through the laws, due to Boyle and Gay-Lussac, for the compressibility and dilation of perfect gases. The point is that, before these definitions, the additivity of two temperature intervals could not be validly defined. But the discovery of the *universal laws* of perfect gases and/or of phenomenological thermodynamics created a new situation, permitting a universal relation of temperatures to other physical quantities; namely, to volume and pressure ($PV = nRT$) by use of a "perfect gas" thermometer, or, even better, to the mean value of the kinetic energy of (monatomic) gas molecules through the Maxwell-Boltzmann formula $\frac{1}{2}mv^2 = \frac{3}{2}kT$. The presence of the universal constant R or $k = R/N$ (N, Avogadro's Number) in the perfect gas formulas is significant, as it expresses a *physical equivalence* between temperatures and pressure-volume products or kinetic energies.

A very similar situation occurs in the time problem. Before the discovery, by Galileo and by Newton, of the universal laws of inertia and of inertial response to forces, the physical status of clocks was quite similar to that of thermometers before Lord Kelvin: there was no possible guarantee that a unique and valid definition of a time scale could be extracted from the performance of, say, sand or water clocks, or even from astronomical clocks.[8]

So, the new *universal law* (contrasted to the previous multiplicity of uncongruent physical clocks) may well be taken as Newton's formula for a point particle, $F = m \frac{d^2x}{dt^2}$. In this formula there is of course a universal constant present, but this constant is traditionally taken as equal to one with the dimension *zero* through our appropriate joint definition of the units of force, mass, space and time.[9] While the additivity of space intervals (in Euclidean geometry), of forces (through arguments based on statics) and masses (quantity of matter[10]) is taken as obvious, the additivity of Galilean time intervals as expressed in Newton's formula is established through the universal character of this formula.

An alternative statement is that the universal constant implicit in the Galileo-Newton formula establishes the physical equivalence between forces and mass-acceleration products—an equivalence which is directly experienced in the form of "inertial forces."

Our conclusion is that the Galileo-Newtonian universal laws of inertia have rendered time "measurable" in very much the same way that the universal laws of thermodynamics have rendered temperature measurable.

The profound significance of this remark is to be found in Einstein's and Minkowski's special theory of relativity.

2 The Relativity Problem in Classical Optics and Electrodynamics

When it had become clear that neither classical kinematics nor classical dynamics were able to define by themselves an absolute reference frame, it was hoped that studies in some other branch of physics would circumvent this apparent failure. In this regard kinematical optics, *i.e.*, the optics of moving systems, seemed at first quite promising. Indeed, the nineteenth-century physicists believed that the Huygens-Young-Fresnel optical waves were propagated in some appropriate medium which they named the "luminiferous ether"; and this hypothetical medium seemed likely to take the place of Newton's hypothetical absolute space. For example, according to classical kinematics, the spherical waves emitted at velocity c by a point source at rest in the ether would be expected to have velocities ranging between $c + v$ and $c - v$ in a reference frame moving with velocity v ($v < c$) relative to the ether. So began the long history of the physical connection between kinematics and optics, the conclusion of which was to be Einstein's remodeling of kinematics after the requirements of electromagnetic theory.

In 1818, Arago proposed to detect the earth's "absolute motion" by measuring the refraction of starlight by a prism. This was a turning point in the history of physics, though it is clear today that Arago's way of questioning nature was not the most unambiguous one; Angström later improved the Arago test by using a source, a receiver, and a prism all at rest in the laboratory, so that no problem of a relative motion between the source and receiver was implied.

Nature's answer to Arago's question was negative: the observed refraction was the same as if the source, the receiver and the prism were all at rest relative to the ether. This came as an intellectual shock. Fresnel's answer to the riddle, known as the "ether drag postulate," was extremely remarkable: the formula was so adjusted that the effects of velocity v relative to the ether were eliminated up to the second order in $\beta = v/c$.[11] Thus the problem of finding a second-order effect was implicitly raised. When Veltmann[12] and Potier[13] had produced a theorem showing that due to Fresnel's formula the absence of first-order effects is absolutely general, the problem of finding a second-order effect was explicitly raised and this of course was the prologue to the famous Michelson-Morley experiment.

Before we come to this experiment, however, some more thinking on the state of affairs resulting from the work of Arago, Fresnel, Veltmann and Potier will yield profound insight into the relativity problem as viewed from its optical side. Our discussion will approach relativistic kinematics in a post facto way which is perhaps unfamiliar to some readers but which allows us to stress the most essential aspects of the subject.

It must be noted first that Fresnel's answer to Arago's result was metaphysical in that its wording still implied the notions of an ether and an "ether wind," while its formula was precisely built so as to eliminate (in the first order) all observable effects of the ether wind. A quite parallel situation arose later, but this time in the second order, with the "contraction hypothesis" that was Fitzgerald's and Lorentz's answer to Michelson's negative result.

Moreover, Potier made it clear that the Fresnel formula expresses a purely kinematical law of universal character,[14] namely, a composition law between three relative velocities: light vs. refracting medium, refracting medium vs. laboratory frame, and light vs. laboratory frame. This feature of the first-order Fresnel formula closely parallels that of the second-order Fitzgerald-Lorentz formula.

In 1908, von Laue showed that the Fresnel formula is merely a special case of the relativistic velocity composition law. The reciprocal step was taken in 1952, when Abelé and Malvaux[15] showed that if the Fresnel formula (in Potier's form) is postulated as the infinitesimal composition law of a group, the Einstein-Minkowski kinematics can be deduced.[16]

The group concept, of course, has been historically,[17] and is still essentially, one of the foundation stones of relativistic kinematics. But, in the seventies, the concept was hardly available to physicists; so the whole story had to be re-enacted in a strikingly parallel fashion, in the case of the second-order ether-wind effect.

In 1878, Michelson and Morley applied their interferometer to the problem of finding the supposed second-order effect of the ether wind. Once more no such effect appeared. Once more theoreticians formulated an ad hoc hypothesis: the Fitzgerald-Lorentz hypothesis, implying a universal formula of longitudinal contraction of material bodies under the ether wind. Once more there was something "metaphysical" in the discourse, the postulated ether wind and absolute frame of reference having no experimental counterparts. And once more the proposed formula was of a universal character, and purely kinematical in its nature.

In the meantime, two important concepts had come to maturity, whose association constitutes the key of the problem. On the one hand, various thinkers, among them Mascart[18] and Poincaré,[19] had become convinced that the relativity principle of dynamics is in fact the universal relativity principle, valid in all branches of physics. They concluded that the law of preferential equivalence of all Galilean frames is in fact a kinematical law which is also valid in optics, electrodynamics, etc. On the other hand, continuous group theory had emerged as a doctrine.[20] In essence, relativistic kinematics follows as a consequence of the application of group theory to optics or electrodynamics.[21]

It was the young Einstein[22] who showed that the unobservability of ether effects leads to a new joint definition of the length and time measures according to which the velocity of light is found to be the same in all reference frames. These measures are precisely those implied in the Lorentz transformation formulas connecting inertial frames[23]

$$x' = \frac{x - vt}{\sqrt{1 - \beta^2}}, \; y' = y, \; z' = z, \; t' = \frac{t - vx/c^2}{\sqrt{1 - \beta^2}}$$

($\beta \equiv v/c$) which read in reciprocal form

$$x = \frac{x' + vt'}{\sqrt{1 - \beta^2}}, \; y = y', \; z = z', \; t = \frac{t' + vx'/c^2}{\sqrt{1 - \beta^2}}.$$

The main difference between the new Lorentz group and the corresponding classical Galileo group

$$x' = x - vt, \; y' = y, \; z' = z, \; t' = t,$$

$$x = x' + vt', \; y = y', \; z = z', \; t = t',$$

is that the transformation affects not only the spatial variables, but also the time variable t; thus, *a "proper" time t is attached to each inertial frame* or, in other words, *an inertial frame is not merely a spatial reference frame (as in the Galilean case) but also a temporal reference frame.* To emphasize this important difference, inertial frames are called Lorentzian rather than Galilean in the relativistic kinematics. It is well known, and obvious, that the limiting form of the Lorentz formulas when one lets $c \to \infty$ is the Galileo formulas; it may thus be said of the new kinematics what is written of the New Testament: that "it does not destroy, but it fulfills the Old."

By setting t or $t' = 0$, one verifies easily that the Fitzgerald-Lorentz contraction is built into the Lorentz formulas; the point is that, in the Einsteinian presentation, this contraction is reciprocal. Each of two

Lorentzian observers finds that the yardstick carried by the other seems shorter than his own.[24] This conclusion seemed highly paradoxical in its day; but there is in it no more mystery than in the well-known Euclidean fact of mutual foreshortening of distant objects known to be of equal size. A somewhat similar situation relates to time measurements, but here the situation is especially interesting because of an apparent paradox which has no analogue in the contraction of length. In picturesque form, the paradox considers a pair of twins, one of whom is an astronaut and undertakes a long space voyage while his brother, an administrator, stays home. Since the astronaut's time scale is contracted by the motion, it is expected that the astronaut should be biologically younger than his brother on his return. On the other hand, if there is no such thing as absolute motion, one might at first think that the astronaut could be taken as the reference point, with respect to which the administrator travels and returns. Then by the same argument as before, the administrator should be the younger when the journey is over. This "twin paradox" has been much discussed, but its resolution is basically simple. The doctrine of relative motion applies only to uniform motion in a straight line, and the astronaut is distinguished from his brother by the acceleration he undergoes. The astronaut is finally the younger of the two.

This closes our survey of the problem of kinematical optics. Before we start discussing Minkowski's remarkable interpretation of Einstein's theory, it will be useful to explain how, according to Duhem's and Poincaré's epistemological views,[25] Michelson's experiment "allows and suggests" the new relativistic joint definition of length and time measurements.

3 *Operational Commentary on the Results of the Michelson-Morley Experiment*

In this section we shall show that there exists a close connection between four classes of experiments: 1) experiments of the Michelson-Morley type; 2) optical measurements of length;[26] 3) Hertzian measurements of time[27] and 4) measurements of the speed of light. These considerations suggest that electromagnetic waves were truly predestined to furnish the scales for distance and time. They also permit us to understand without mathematics that to use these optical or Hertzian scales makes the speed of light in vacuo an invariant by definition.

1) Michelson's interferometer is essentially an optical scale arranged

so as to measure changes of length by detecting differences in the number of standing light waves along two rigid scales. It follows that the most direct interpretation of the negative result of Michelson and Morley is that the number of wave lengths of light emitted by a monochromatic source at rest with respect to a rigid scale and spread out along the scale is independent of the orientation of the scale. In 1887, when it was first firmly established, this result seemed highly paradoxical. But ever since the development of quantum mechanics in 1925–26, a positive result of the Michelson-Morley experiment would have seemed equally paradoxical. In fact, quantum mechanics describes any solid body as a standing de Broglie wave of complex structure,[28] and since waves of light are considered in quantum mechanics as a special case of matter waves, there is clearly no reason to suppose that the two kinds of waves should exhibit different kinematical behaviors.[29]

2) Michelson's negative result is required by the theory and practice of the optical measurement of lengths. If it were possible to detect the so-called "ether wind," any comparison between a rigid scale and an optical wave length would have to be preceded by a determination of the direction and velocity of the wind.[30]

3) Suppose on the other hand that one decided to use the period of a monochromatic optical radiation as a time scale. This raises no kinematic problem analogous to that just discussed. Now the problem is a dynamical one, for we must know whether the new time scale is (in the non-relativistic limit) identical with that furnished by a body in uniform motion. Quantum mechanics again gives an affirmative answer: it is well known that the inertial motion of a particle is unambiguously described through a monochromatic plane wave[31] whose mathematical description is manifestly covariant in character.

We have thus explained how Michelson's negative result permits and suggests that we adopt at the same time the wave length and the period of a monochromatic optical wave as our standards of length and time.[32] Relativity thus legitimizes the situation that exists, and we can understand that it is the forms of the equations of d'Alembert and Klein-Gordon,[33] together with the Lorentz group under which they are invariant, that renders optical wave lengths and periods the natural measures of space and time. An optical wave is chosen rather than a matter wave because of the properties implied by the simpler equation of d'Alembert.

4) But to adopt the wave length and period of an optical wave as our standards of length and time is ipso facto to declare that c is an absolute

constant, a coefficient of equivalence between space and time. This is because of the exact relation

$$\lambda = c\tau$$

between the wave length λ and the period τ.

Finally, we may remark that a number of the most modern determinations of c follow the above conceptual scheme very closely. In the microwave cavity measurements of Essen[34] and of Hansen and Bol,[35] the spatial dimensions of the wave are determined by measurements of the cavity (some of them optical), while the period is compared with astronomical time. In the band-spectrum method of Plyler *et al*[36] and Rank *et al*,[37] c was determined by measuring separately the periods and wave lengths of the same molecular spectral lines.

At its first appearance, relativity seemed to mark a victory of optics and electromagnetism over mechanics. This was because neither kinematics nor dynamics had up to this time recognized the importance of the constant c. By now, not only the kinematics given us by relativity but also the dynamics given us by de Broglie, Heisenberg and their followers have assimilated into the physics of waves the constant c in an essential way. Today it would be possible to deduce all of relativistic kinematics, not from electrodynamics and optics via d'Alembert's equation, but more generally from the properties of matter described by the Klein-Gordon equation.[38] All is therefore once more in traditional order, with the quantum theory of the electromagnetic field a special case of a more general theory of mechanics which is the theory of quantized fields.

4 *Space-Time "Equivalence" and Minkowski's Four-dimensional Geometry*

The relativistic "equivalence" between space and time is strongly suggested by the elementary expression for a field component of a light wave

$$\phi = \cos 2\pi(\frac{t}{\tau} - \frac{x}{\lambda})$$

where τ and λ are the wave's period and length, or by the form of d'Alembert's equation, of which ϕ is a solution. But the exact nature of the "equivalence" is expressed only by the Lorentz formulas which transform the spatial coordinates x, y, z, and the time coordinate t so as to leave invariant the quadratic form

$$s^2 = x^2 + y^2 + z^2 - c^2t^2.$$

It was Minkowski who established relativistic kinematics in a canonical form by showing that the Lorentz transformations can be regarded as rotations of a set of four Cartesian axes in a four-dimensional space-time with pseudo-Euclidean metric.[39] To verify this, we need only express the parameter β of the Lorentz formulas in terms of a new "angle" θ by

$$\beta = \tanh \theta$$

whence

$$\frac{\beta}{\sqrt{1 - \beta^2}} = \sinh \theta, \qquad \frac{1}{\sqrt{1 - \beta^2}} = \cosh \theta,$$

and the Lorentz formulas become

$$x' = x \cosh \theta - ct \sinh \theta, \qquad x = x' \cosh \theta + ct' \sinh \theta,$$

$$ct' = ct \cosh \theta - x \sinh \theta, \qquad ct = ct' \cosh \theta + x' \sinh \theta,$$

$$y' = y, \quad z' = z.$$

These are indeed the formulas for a "hyperbolic rotation" in the plane whose coordinates are $x^1 = x$, $x^4 = ct$; this is a transformation which leaves invariant the "hyperbolic distance" $\sqrt{x^2 - (ct)^2}$ together with the coordinates $x^2 = y$, $x^3 = z$. The relation between the original and the transformed axes is shown in Figure 1.

One can go even further and force the geometry of space-time to be formally Euclidean by taking the fourth coordinate to be imaginary,

$$x^4 = ict, \ (x^4)^2 = - (ct)^2 \ (i = \sqrt{-1}).$$

The Lorentz formulas can now be written as

$$x' = x \cos i\theta + ict \sin i\theta, \qquad x = x' \cos i\theta - ict' \sin i\theta.$$

$$ict' = ict \cos i\theta - x \sin i\theta, \qquad ict = ict' \cos i\theta + x' \sin i\theta.$$

These are in the form of the ordinary expressions for the rigid rotation of a pair of axes, except that the coordinates are x and ict and the angle of rotation is expressed as $i\theta$.

Since the metric is pseudo-Euclidean, the cone defined by $s^2 = 0$ divides the directions of space-time into three classes (Figure 2): the

426

exterior of the cone, $s^2 > 0$, and the two interior regions with $s^2 < 0$. Since these classes are distinguished by the value of the invariant s^2, no Lorentz transformation can take a vector from one class into another. The directions corresponding to the three classes are respectively called space-like ($s^2 > 0$), future time-like, and past time-like, both of the latter with $s^2 < 0$. Because the numbers x, y, z, and t are always real, it follows that the axes of x, y, and z are space-like while the t-axis is time-like. If we require the Lorentz transformations to be continuous,[40] all positive time axes will point into the future half cone, however they may

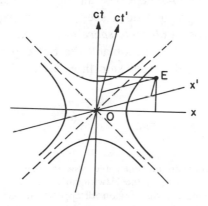

Figure 1. The axes x, representing a spatial direction, and ct, representing time measured in units compatible with those of x, can be used to represent the event E in a certain observer's space and time. For another observer moving relatively to the first, the coordinates of the same event are measured by the inclined axes x' and ct'. The hyperbolas are lines, or in general surfaces, which are described identically by both observers.

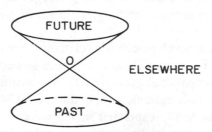

Figure 2. The double cone, shown with time dimension vertical and spatial dimensions horizontal to it, is generated by signals traveling inwards to and outwards from O with the speed of light. It divides space-time into three regions labeled *future*, *past* and *elsewhere*, which would be labeled identically by any observer, traveling with any uniform velocity, who coincided momentarily with O.

be transformed; the so-called "orthochronous Lorentz group" is thus defined.

We may note incidentally that expressed in ordinary units of space and time, the value of c is very large. The cone is thus very flat, and the *elsewhere* region is crushed between the *past* and *future* regions. Classical kinematics is obtained in the limit $c \to \infty$, which annihilates the *elsewhere* region (Figure 3). It is important to realize that the difference between classical and relativistic kinematics is not in the use of a four-dimensional space-time, but rather in the metric character first ascribed to this space by Minkowski.[41] To use a two-dimensional analogy, one may say that classical kinematics can be graphed on a sheet of ruled paper in which the coordinate axes are labeled *absolute space* and *absolute time*, while relativistic kinematics is inscribed on a blank sheet of paper, each point provided with a compass-card divided into three sectors labeled *past*, *future* and *elsewhere*.

Figure 3. Measured in conventional units, the cone in Figure 2 is very flat and the *elsewhere* region is very narrow. In the Newtonian limit, in which the finiteness of the speed of light is ignored, this region disappears altogether and space-time consists only of the future and the past, separated by the present instant.

We have seen earlier that the essential new idea involved in the transition from pre-relativistic to relativistic mechanics is that the equivalence of Galilean (now Lorentzian) frames of reference is a law not only of dynamics but of kinematics as well. In Minkowski's geometry the interpretation of this point is very clear: it is the privileged equivalence of strictly Cartesian coordinate systems for the description of phenomena, just as in Euclidean space.

Very early in the development of Relativity Theory, Minkowski and von Laue succeeded in writing electromagnetism as well as dynamics in the language of four-dimensional geometry. In formulating wave mechanics, de Broglie followed their example and, in the years 1948–49, the quantum theory of fields was finally expressed in the same way.[42] It is a goal to which all parts of fundamental physics must aspire, for example, classical statistical mechanics.[43]

The theory of gravity poses a special problem. Though there have been, and are still, many attempts at a Minkowskian theory of gravity, the most successful theory is undoubtedly Einstein's General Theory of Relativity

(1915), which is expressed in terms of a Riemannian geometry in four dimensions. Here the geometry of Minkowski is only a locally tangent approximation, as a tangent plane may locally approximate a curved surface in the geometry of ordinary experience.

In any theory whose mode of expression is geometric—and we have seen that relativity theories are of this kind—a very important concept is that of covariance. What is objective in a geometric theory is what is defined independently of the way in which coordinates are assigned to the space, for example, points, lines, surfaces, etc., together with the figures formed from these elements. An important class of geometric "objects" is formed by the vectors and the tensors that can be defined at each point. We owe to Minkowski the idea of expressing all the fundamental laws of physics in the form of relations covariant in space-time, that is, tensorial relations which take the same form in all systems of coordinates. Physics is thus related to the intrinsic geometry of space-time. Looking back over history, we can see that every time someone has succeeded in expressing the problems of a physical theory in relativistically covariant terms, it has led to advances in both fundamental understanding and technical skill; the most recent example is in the quantum theory of fields.[44] For this reason, the concepts of physical objectivity and of covariance have become nearly synonymous in Relativity Theory; that is, in a strictly relativistic discussion only those objects and relations which can be expressed covariantly can be considered as objective properties of the world.

As concerns philosophy and general culture, perhaps the most important consequences of the requirement of covariance concern the relation between past and future. In Newtonian kinematics the separation between past and future was objective, in the sense that it was determined by a single instant of universal time, the present. This is no longer true in relativistic kinematics: the separation of space-time at each point of space and instant of time is not a dichotomy but a trichotomy (past, future, elsewhere). Therefore there can no longer be any objective and essential (that is, not arbitrary) division of space-time between "events which have already occurred" and "events which have not yet occurred." There is inherent in this fact a small philosophical revolution.

Before Relativity Theory, many philosophers were inclined to consider matter as occupying a certain region of space but, in respect to time, as being concentrated in an instant without extension; this view was compatible with the kinematics of Galileo and Newton, but is incompatible with that of Relativity Theory. If matter has spatial extension, it follows (in virtue of the trichotomy mentioned above) that it has also extension in time.[45] This is why first Minkowski,[46] then Einstein,[47] Weyl,[48] Fantappiè,[49] Feynman,[50] and many others have imagined space-time and its material contents as spread out in four dimensions. For those authors, of whom I am one, who take seriously the requirement of covariance, relativity is a theory in which everything is "written" and where change is only relative to the perceptual mode of living beings. Humans and other living creatures, for reasons which one can try to explain,[51] are compelled to explore little by little the content of the fourth dimension, as each one traverses, without stopping or turning back, a time-like trajectory in space-time.

We can, of course, easily imagine the past experience of an individual as being diagrammed in space-time along a time-axis extending backwards into the past. In our view, the future is to be adjoined to the diagram in a corresponding fashion. There are writers who affirm that the future contains elements which are undetermined by the past and the present, and that the future light cone in the diagram must therefore be left blank. The answer is that there *is* a future. Nature "will take" one of the alternatives open to her, and it is this that we must imagine inscribed, even though we do not know what it "will be."

5   *Causality and Free Will: A Relativistic Formulation of the Problem*

When Magellan's companions reappeared from the East after having set out towards the West, a group of interwoven problems concerning the sphericity of the earth and the heliocentric theory of Aristarchus and Copernicus ceased to be academic; they now concerned everyone. Currently, thousands of technicians in many branches of physics and engineering use the findings of relativity in their work. Perhaps the day is not distant when school children must learn to think in terms of Minkowski diagrams. This will be no harder for them than it is to accept that the earth is a spinning ball, small in the general scheme of things, slowly revolving around a gigantic sun. The only difficulty is to learn that the everyday phenomena of astronomy, even though they do not at once suggest to our minds a heliocentric system, can nevertheless easily be explained in this way. We learn to do this as children under the guarantee that the heliocentric system is in harmony with the general principles of nature (*i.e.*, the laws of mechanics) while the geocentric system is not.

We encounter the same kind of difficulty in trying to adjust our intuitive views of time and space to the relativistic view, but the same guarantee encourages us to make the attempt.

Perhaps the most obvious points at which some effort must be made to harmonize relativity and intuition are the apparently one-way flow of time and the related problem of free will. We shall briefly discuss both.

The most conspicuous signs of the unidirectionality of time can be traced to our participation in the general evolution of the universe. The universe (at least the region of it known to us) consists of a markedly unstable arrangement of matter and energy evolving irreversibly in the direction of equilibrium. This subject is discussed elsewhere in this volume, and we may say that it is reasonably well understood. This has interesting consequences in the domain of elementary processes, such as the emission and absorption of radiation, governed by equations which are relativistically invariant and yet in which the law of cause and effect seems to demand that a distinction be made between the past and future directions of time.

The nature of the question is illustrated by a simple example. Imagine a quiet pond with leaves in the water around the edge. Into the pond at the point P a stick is dipped and withdrawn. Waves travel outward from P towards the edge of the pond and when they arrive, move the leaves in the water. This is an example of cause and effect, chosen to put in evidence the wave mechanism by which the dipping of the stick causes the leaves to move. It is obvious in it that the effects follow the cause in time. Now, formally speaking, the entire process is reversible. If it were reversed, the leaves would move, a circular wave would detach itself from the bank and converge towards P, at the right moment the stick would dip in and out and the pond would then be silent. In this process the roles of cause and effect are reversed, and it is possible to show that the entropy of the universe would *decrease* as a result of it.

In this example we see the interconnection of three tendencies in nature: that waves move outward from a disturbance and not inward towards it, that effects follow their causes and do not precede them, and that natural processes increase entropy and do not decrease it. In the first and second of these, nature seems not to use the symmetry given it by relativistic laws: the third shows why this is so; even elementary processes, or at least almost all that we would wish to discuss, are involved, in a probabilistic fashion, with the general increase of entropy. And since the process occurring in the pond is analogous to such processes as the radiation of energy by atoms and the interaction of charged particles via the electromagnetic field, we see that the irreversible evolution of the universe as a whole imposes, *via* probability, its dissymmetry even on the microscopic scale.[52]

We come now to the question of determinism versus free will. It is an exceedingly complex one, especially because it cannot be analyzed without a detailed use of quantum mechanics.

The subject concerns relativity because if a conscious being is able to make free decisions, these decisions influence the future and not the past and once again a dissymmetry exists. Further, it seems to be urgently implied that the future half of the Minkowski diagram cannot be filled in, even in principle, because decisions upon which it depends have not yet been made. To this we have answered above that *something* is going to happen and that is what is written down. This does not answer the question of determinism, however, and we must proceed to make a basic distinction.

There are at least two different ways in which we can imagine events to be inscribed on a Minkowski diagram: as a mathematical description in terms of the psi-function of the entire universe, and as a simple record of the sense impressions of one observer. We have a free choice. Theoretical physics imagines a universe which evolves subject to precisely causal laws except for processes in which conscious beings intervene. Such interventions are governed by essential indeterminacies whose result is that our successive sense impressions cannot in general be put in exact causal relation with each other. In this sense, causality is often said to have disappeared from physics and, in this sense, a four-dimensional diagram of an observer's sensory experience will not consist of causally related events. One of these two modes of description is causal, but it is not causally related to sensory experience. The other is a direct transcription of sensory experience, but it is not causal.

It should not be thought that the two possibilities represent different schools of thought or that they conflict with each other. Instead, they are in Bohr's[53] sense complementary ways of describing the same thing.

Just as the freedom of the will is an experiential category of our psychic life, causality may be considered as a mode of perception by which we reduce our sense impressions to order. At the same time, however, we are concerned in both cases with idealizations whose natural limitations are open to investigations and which depend on one another in the sense that the feeling of volition and the demand for causality are equally indispensable elements in the relation between subject and object which forms the core of the problem of knowledge.[54]

Each view, the deterministic one and the one that emphasizes freedom of choice, has its appropriate mode of expression. In the *Heisenberg picture* of quantum field theory the psi-function, assumed to contain the completest possible specification of the universe, is independent of time. The equations governing the various interacting fields are formally the same as they would be in classical physics. The psi-function in any representation denotes which possible states are occupied and which are not. In the

Heisenberg picture the occupation numbers of the states never change: nothing happens. This picture is a reformulation in quantum terms of the classical and deterministic conception of nature in which everything is written once for all.

In the *interaction picture*, the various fields are described by equations which omit all interactions. The psi-function, on the other hand, reflects the effects of the interactions and evolves continuously in time. It represents the changing face of Nature as we know it, while the Heisenberg picture may be said to represent it in God's view. The crucially important point is that the two pictures are complementary descriptions of the same thing. A mathematical transformation enables us to pass from one to the other at will and to know that the physical content of the two pictures is exactly the same. This is an interesting example of how two viewpoints, which from a non-mathematical point of view may appear irreconcilable, may in fact be merely complementary. In one sense they are mutually exclusive, but both are necessary to our understanding of the Universe and its relation to man.

These, of course, are the author's own speculations on a difficult subject which has not yet been settled in a way generally accepted by all thinkers. They are given as an example of the perplexing problems which are raised as soon as the mind-and-body relations are thought of not "in space at a given instant," which would not be relativistically covariant, but in space-time.[55]

6   *Conclusions*

We may say that Special Relativity Theory has shed much light upon old riddles in kinematical optics and in innumerable other problems of contemporary physics. In this respect, Minkowski's theory of space-time has provided a powerful tool of far-ranging validity; perhaps future generations will say that Minkowski was a second Euclid, for he has made a theory of geometry marvelously describing the physical world as we know it today.

Relativity theory has drastically changed the concepts we use to describe the physical world, that is, in a broad sense, our cosmological view of things. This is a new Copernican or Magellanic revolution, the consequences of which are almost unexplored. Of these consequences we have tentatively submitted an example in the last section—a few other ones being also very interesting to consider even if they cannot be settled *hic et nunc*.

Time in Relativity Theory: Arguments for a Philosophy of Becoming

MILIČ ČAPEK

On April 6, 1922, at the meeting of the French Philosophical Society, Emile Meyerson, one of the outstanding philosophers of science at that time, asked Albert Einstein a point-blank question: Is spatialization of time, *i.e.*, the tendency to regard time, "the fourth dimension," as not being essentially different from the spatial dimension, a legitimate interpretation of Minkowski's fusion of space and time? Meyerson was obviously prompted to ask this question by the fact that the above-mentioned static interpretation of space-time was—and, as we shall see, still is—present not only in numerous popular or semi-popular presentations of the relativity theory, but in serious scientific and philosophical treatises as well. For our purposes it will suffice to give only two illustrations from the period prior to 1922; some more recent examples will be given later.

With Minkowski space and time became particular aspects of a single four-dimensional concept; the distinction between them as separate modes of correlating and ordering phenomena is lost, and the motion of a point in time is represented as a *stationary curve in four-dimensional space*. Now if all motional phenomena are looked at from this point of view they become *timeless phenomena* in *four-dimensional space*. The whole history of a physical system is laid out as a *changeless whole*.[1]

The second illustration is perhaps even more characteristic:

There is thus far an intrinsic similarity, a kind of coordinateness, between space and time, or as the Time Traveller, in a wonderful anticipation of Mr.

Wells, puts it: "There is no difference between Time and Space except that our consciousness moves along it."[2]

And in the footnote on the same page the author added:

It is interesting that even the terms used by Minkowski to express these ideas, as "Three-dimensional geometry becoming a chapter of the four-dimensional physics," are anticipated in Mr. Wells's fantastic novel. Here is another example (*Time-Machine*, Tauchnitz, ed., p. 14) illustrative of what is now called a world-tube: "For instance, here is a portrait (or, say a statue) of a man at eight years old, another at fifteen, another at seventeen, another at twenty-one and so on. All these are evidently sections, as it were, three-dimensional representations of his Four-Dimensional Being which is a fixed and unalterable thing." Thus Mr. Wells seems to perceive clearly the absoluteness, as it were, of the world tube and the relativity of its various sections.

This is explicit enough. All the ingredients of the static interpretation of space-time, as it still lingers in the minds of some physicists and philosophers, are contained in these two quotations.

In asking Einstein, Meyerson did not hide his own negative attitude toward the spatializing interpretation of relativity. His argument consisted of five logically related parts. The first three are merely three different aspects of one argument. He pointed out first that the privileged character of the temporal dimension is preserved in Einstein's cylindrical model of the universe in which, unlike the spatial dimensions, time is uncurved and unidirectional. He then recalled Einstein's own words that "we cannot send wire messages into the past." Thirdly, he pointed out that the law of entropy which "guarantees" the irreversibility of time remained intact within the relativistic framework. Meyerson's fourth argument was very probably inspired by Bergson: the spatialization of time in the relativity theory is, according to him, merely the last manifestation of the perennial tendency to treat time in a space-like fashion. The fifth argument should probably have been put in the first place since, as we shall see, it underlies the rejection of the backward-moving causal actions. Meyerson recalled Hermann Weyl's proposal to speak of three-plus-one dimensions of space-time rather than of four dimensions, since in Minkowski's formula for the spatiotemporal interval the time variable is preceded by an algebraic sign different from that of the three spatial variables. Thus the heterogeneity of space and time is reflected even in the mathematical symbolism of the relativity theory.[3]

What is surprising in Einstein's answer to Meyerson is not so much his apparently complete agreement with him—Meyerson, after all, quoted

Einstein himself to support his criticism of the spatialization of time—but the very briefness of his reply: "It is certain that in the four-dimensional continuum all dimensions are not equivalent."[4]

Today, in the light of Einstein's later utterances, it is clear that the very briefness of his reply was due to the fact that he was not especially interested in this question. Only thus can we explain the vacillations and ambiguities in his attitude toward this particular problem. These ambiguities were documented by Meyerson himself when he discussed the same problem in a more systematic way in his book *La Deduction relativiste* in 1928.[5] While he recalled again Einstein's statement that "we cannot send wire messages into the past," he also quoted another of Einstein's utterances according to which "the becoming in the three-dimensional space is somehow converted into a being in the world of four dimensions." One can hardly have a more radical formulation of static interpretation! Yet, Einstein's response to Meyerson's book was enthusiastically positive. He not only praised it as "one of the most remarkable books written about the relativity theory from the standpoint of epistemology," but he also explicitly agreed with its central thesis, that is, with his rejection of the spatializing interpretation of the world of Minkowski.[6]

This, however, was not Einstein's last word on this subject. In 1949 Kurt Gödel wrote a vigorous defense of the static interpretation of space-time.[7] According to him, the relativization of simultaneity destroys the objectivity of the time lapse and thus substantiates "the view of those philosophers who, like Parmenides, Kant and modern idealists consider change as an illusion or an appearance due to our special mode of perception." As an additional argument against the objectivity of time lapse Gödel adduced the mathematical possibility of certain cosmological models in which it would be possible "to travel into any region of the past, present, and future, and back again, exactly as it is possible in other worlds to travel to distant parts of space." Such a trip could be described by a world line similar to the F-$(H$-$N)$-G-F line of the diagram in this essay. Gödel even makes an estimate of the quantity of fuel and the velocity of a rocket ship needed to make such a fantastic trip. Such a rocket would be a realization of the time-machine of H. G. Wells and Kurt Gödel would subscribe to Silberstein's view that the famous British fiction writer anticipated relativistic physics.

We would expect that Einstein, who two decades before endorsed Meyerson's criticism of the static interpretation without reservations, would have rejected unequivocally such an extreme form of spatialization of time. But the very opposite happened: Einstein's comment on Gödel's essay was distinctly, though cautiously, sympathetic.[8] Did Einstein then forget his previous view that "we cannot send wire messages into the past"? Is not Gödel's hypothetical rocket merely an oversize form of signal traveling into the past? Such doubts are hardly fair if we read Einstein's comment carefully. Einstein indeed modified his view in the following way: it is impossible to send wire messages to the past on the macroscopic scale; but this is not necessarily true for microscopic phenomena which seemed to be reversible. Not only this; if we concede with Gödel the possibility of the closed world lines on a huge megacosmic scale, says Einstein, then the relation of succession itself becomes relativized; for on a circular world line it is a matter of convention to say that *A* precedes *B* rather than vice versa. In other words, Einstein as late as 1949 considered the possibility that the irreversible time is confined to what Reichenbach called "the world of the middle dimensions" while it may be absent both on the cosmic scale and on the microphysical level. It is true that he added cautiously: "It will be interesting to weigh whether these (*i.e.*, cosmological solutions) are not to be excluded on physical grounds."[9] Despite this reservation it is clear that Einstein was closer to the spatializing interpretation in 1949 than in 1928.

Einstein was not alone in his vacillations on this point. Herrmann Weyl, Sir James Jeans, Hans Reichenbach and others shifted their views on this subject, sometimes even within one and the same book. A more consistently negative attitude toward the static interpretation was shown by Paul Langevin and, contrary to what Meyerson claimed, by Arthur S. Eddington;[10] and among philosophers by Bergson and Whitehead. It is true that the attitude of the latter two was mostly inspired by their general philosophical outlook, even though the effort to grasp the concrete physical meaning of the relativistic formulae was not lacking in either of them. Despite all criticisms the spatializing interpretation still lingers, though more in the minds of philosophers than in those of physicists. Besides the relatively recent essay by Gödel (1949), there was Professor Donald Williams' article with the challenging title "The Myth of Passage" (1951). Even more recently, Professor Willard Quine claimed that the discovery of the principle of relativity "leaves no reasonable alternative to treating time as space-like."[11] Among contemporary philosophers of science two most vigorous defenders of the becomingless view of space-time are Olivier Costa de Beauregard and Adolf Grünbaum. The former speaks of matter as "displayed statically in space-time." (*"statiquement deployée dans l'espace-temps"*), while the latter says explicitly that "coming into being is only coming into awareness."[12] Thus the opinion is still divided—

sometimes divided within one and the same mind. This shows clearly how complex and difficult the problem of correct interpretation of the relativistic fusion of space and time still is.

1   The Alleged Argument for the Static
Interpretation of Space-Time

The crucial issue which we face is as follows: are there any cogent reasons for the static interpretation of space-time or is the very opposite true? In other words, does an attentive analysis of the conceptual structure of the relativity theory support the becomingless view or does it suggest the very opposite?

The most frequent and superficially most plausible argument in favor of the becomingless view is based on the claim that the relativization of simultaneity definitely destroys the objectivity of temporal order. A pair of events appearing simultaneous in one frame of reference is no longer simultaneous in other inertial systems. Even worse, some events succeeding each other in one system may appear in a reversed order in another appropriately chosen system. Since there is no privileged frame of reference which would impart a mark of objectivity on any of these systems, what objective status can succession and becoming still retain? This is a standard argument and thus it is hardly surprising that we can find it in Kurt Gödel's essay to which we referred above:

The argument runs as follows: Change becomes possible only through the lapse of time. The existence of an objective lapse of time, however, means (or at least, is equivalent to the fact) that reality consists of an infinity of layers of "now" which come into existence successively. But, if simultaneity is something relative in the sense just explained, reality cannot be split into such layers in an objectively determined way. Each observer has his own set of "nows," and none of these various systems of layers can claim the prerogative of representing the objective lapse of time.[13]

Similarly Costa de Beauregard:

In Newtonian kinematics the separation between past and future was objective, in the sense that it was determined by a single instant of universal time, the present. This is no longer true in relativistic kinematics: the separation of space-time at each point of space and instant of time is not a dichotomy but a trichotomy (past, future, elsewhere). Therefore there can no longer be any objective and essential (that is, not arbitrary) division of space-time between "events which have already occurred" and "events which have not yet occurred. . . . This is why first Minkowski, then Einstein, Weyl, Fantappiè, Feynman, and many others have imagined space-time and its material contents as spread out in four dimensions. For those authors, of whom I am one, who take seriously the requirement of covariance, relativity is a theory in which everything is "written" and where change is only relative to the perceptual mode of living beings.[14]

We have to consider carefully what is correct and what is questionable in these passages. Gödel and Costa de Beauregard correctly pointed out that while the classical Newtonian space-time possessed a stratified structure in the sense that it was regarded as a continuous succession of three-dimensional strata, each of which represented a particular cosmic "now" or "present," the relativistic space-time does not yield to such stratification. No common series of such cosmic "nows" exist for different observers; the observers in different inertial frames split the four-dimensional continuum along different instantaneous "cleavage planes." Each such cleavage plane is a substratum of the events simultaneous for the corresponding observer, but—unlike in the Newtonian space-time—none of them possesses a privileged, objective character. This is the meaning of the relativity of simultaneity. But, contrary to what Gödel and Costa de Beauregard believe, from the relativization of simultaneity it does not follow that the lapse of time and change lose their objective status. Gödel's conclusion would have been correct if lapse of time or duration were completely synonymous with the classical even-flowing Newtonian time consisting of the succession of the world-wide instants. This had been accepted tacitly through the whole classical period in the same way that space and Euclidean space were regarded as synonymous. The fact that some critics of relativity in defending the objective status of universal time really defended the time of Newton merely added to this confusion.

What Gödel and modern neo-Eleatics do not consider at all is the possibility that the Newtonian time may be only a special case of the far broader concept of time or temporality in general in the same sense that the Euclidean space is a specific instance of space or spatiality in general. If we admit this possibility, then the negation of the Newtonian time entails an elimination of temporality and change in general as little as the giving up of the Euclidean geometry destroys the possibility of any geometry. Similarly, the present revision of classical determinism means merely a widening, not an abandoning of causation in general; despite the fears of some conservative philosophers, the probabilistic universe is not an irrational chaos, even though its rationality is of far broader kind than the restricted form of rationality characterizing the Newton-Laplacean determinism.[15]

2   Consequences of the Constancy of the World Interval

Let us now consider in detail the argument that the relativization of simultaneity implies without qualification a relativization of succession and thus destroys forever the objective status of "lapse of time." Its plausibility is undeniable: for if there is no objective "Now" unambiguously separating the past from the future, what objective status can succession still claim? In other words, if succession itself is relative, depending on the choice of our frame of reference, it cannot constitute an objective feature of reality. The last conclusion follows unquestionably from its premise; unfortunately (or rather, fortunately!) the premise itself is not correct. For it is simply not true that simultaneity and, in particular, succession of events are purely and without qualification relative. In making such claim we would be guilty of completely disregarding certain mathematical implications of Minkowski's formula for the constancy of the world interval. This formula follows from the Lorentz transformation and it shows in a condensed way the differences between classical and relativistic mechanics. In the former the spatial distance and the temporal interval separating two events E_1 and E_2 are *separately* invariant for each inertial frame

$$s = \text{const.}, \quad t_2 - t_1 = \text{const.}$$

(where $s = \sqrt{(x_2 - x_1)^2 + (y_2 - y_1)^2 + (z_2 - z_1)^2}$,

with $x_1, y_1, z_1, t_1, x_2, y_2, z_2, t_2$ being the spatial and temporal coordinates of E_1 and E_2 respectively). In Minkowski's space-time the constancy does not belong to the spatial distance and the temporal interval separately, but only to the quantity called "world interval," which is defined in the following way

$$I = s^2 - c^2 (t_2 - t_1)^2 = \text{const.} \quad (c = 3 \times 10^{10} \text{ cm/sec.}).$$

We can then distinguish three distinct groups of relations between two events according to whether the world interval is positive, zero or negative.

$$I > 0, \ I = 0, \ I < 0.$$

Each group should be considered separately.

a) When $I > 0$, then $s^2 > c^2 (t_2 - t_1)^2$; in other words, the spatial separation between the events E_1 and E_2 is greater than their separation in time multiplied by the velocity of the fastest causal action, *i.e.*, the velocity of electromagnetic radiation. This means that no causal interaction can take place between such events; they are not only causally unconnected, but even unconnectible, that is, intrinsically mutually independent.[16] Since the interval should retain its positive sign in *all* inertial frames of reference, and since this sign remains unaffected when $t_1 = t_2$ or when the temporal interval $t_2 - t_1$ changes its sign, we can see the possibility that the events E_1, E_2, succeeding each other within one group of systems, will appear simultaneous in another group, and will appear in *a reversed order* in still other systems. In other words, the simultaneity and succession of causally unrelated events is fully and without qualification relative. But this statement is restricted to the specific case just considered.

b) $I = 0$, or $s^2 = c^2 (t_2 - t_1)^2$. Since the spatial distance is equal to the separation in time multiplied by the velocity c, it is clear that this is the case of a photon or more generally of any quantum of radiation, in two successive "positions." It is obvious that in this case the interval $t_2 - t_1$ can never become zero unless the spatial distance itself would vanish at the same time; but in that case the events E_1, E_2 would merge. In other words, each photon at every instant is simultaneous with itself. This statement can be generalized: every world point—or rather world event—is simultaneous with itself in every frame of reference. (As we shall see, this is not as trivial as it sounds.) But as long as the spatial distance does not vanish, the corresponding time interval does not vanish either. In other words, two events of this kind, successive in one frame of reference, must never appear simultaneous in any other system; a fortiori, they can never appear in a reversed order. This is only natural; for two successive positions of a photon or, to use the undulatory language, two successive states of the vibratory electromagnetic disturbance, are simple instances of causally related events; the reversion of their temporal order would be equivalent to the reversion of their causal order. This would mean that what appears as a cause of a certain event would appear as an effect of the same event in another system!

Such a case was possible in classical physics; when Flammarion imagined an observer moving away from the earth with a velocity greater than that of light and seeing the earthly history reversed so that "Waterloo would precede Austerlitz,"[17] it was "science fiction" which, nevertheless, was compatible with the principles of Newtonian physics. Moreover, it did not contradict the unidirectional character of causal relations because the reversion mentioned above was only apparent. For physicists of the last century believed with Newton that it was possible, at least in principle, to distinguish the real temporal (and causal) order from the merely spurious or apparent one. The distortions of temporal and causal perspective produced by some relative motions, *e.g.*, by the motion of Flammarion's observer with respect to light, disappear in the only true perspective of the privileged frame of reference—absolute motionless space. But in the relativity theory the situation would be far more serious: because of the absence of any privileged frame of reference there is no way—in the special relativity at least—to differentiate between "apparent" and "real" order, and thus a reversion of causal order due to an appropriate change of the system would result in most serious discrepancies and causal anomalies. Fortunately, Flammarion's fantasy is excluded by the very principle of constant velocity of light according to which no material body can attain the velocity equal to that of electromagnetic radiation. For this reason, the succession of two states of the electromagnetic disturbance in the void can never degenerate into an apparent simultaneity in any other system; nor can it ever appear in reversed order. As the consideration of the third case will show, this is true generally of any couple of causally related events.

c) When $I < 0$, then $s^2 < c^2 (t_2 - t_1)^2$; the spatial distance is then smaller than the product of the separation in time and the velocity of light (and gravitation). This is the case of two events whose causal links are propagated with the velocity smaller than c; in other words, the connecting causal links are not the world lines of photons, but those of "material particles." Since the interval I must retain its negative sign in all frames of reference, the temporal interval of the events E_1, E_2 cannot vanish in any system: otherwise for $t_2 = t_1$, $s^2 < 0$, *i.e.*, their spatial distance would become imaginary. In other words, the succession of the causally related events can never degenerate into simultaneity in any other system; a fortiori, it can never be reversed. This is clearly a generalization of the result obtained in (b) and it can be summarized as follows: the succession of causally related events, whether they are joined by the world lines of photons or by those of material particles, is a topological invariant independent of our choice of system of reference. Or more concretely: *the world lines of any kind are irreversible.* Although this important conclusion was pointed out explicitly by Paul Langevin as early as in 1911,[18] it was frequently overlooked not only by the authors of popular or semi-popular expositions of relativity, but sometimes also by serious thinkers.[19] This was undoubtedly due to the fact that the absence of the metrical invariance of temporal intervals obscured the aforesaid topological invariants and that the case (a) was not distinguished from the case (b) and (c).

442

Before formulating our general conclusion, let us consider the conditions under which the spatial distance can vanish in the three aforementioned instances $I > 0, I = 0, I < 0$. We already saw that in the case (b) the spatial separation can disappear only if the corresponding time interval vanishes: $s = 0$, when and only when $t_2 - t_1 = 0$. This is the case of two events merging into one; every photon coincides spatiotemporally with itself, and, more generally, every event coincides spatiotemporally with itself and no change of the frame of reference could produce its dislocation into two different events. (We shall see the full significance of this apparent truism later.) In the case (a) the possibility of the spatial separation becoming zero in any system is excluded since $s^2 = 0$ would imply $c^2 (t_2 - t_1)^2 < 0$, *i.e.*, the time interval separating the events E_1 and E_2 would become imaginary. On the other hand, the spatial distance can vanish by an appropriate choice of the system in the case (c); the condition $s = 0$ determines the frame of reference in which a particle appears motionless, and E_1 and E_2 are two successive events "at the same place." But "the sameness of place" is completely relativized and has lost its original Newtonian connotation of a motionless part of the motionless space. Every moving particle can by an appropriate change of the frame of reference be converted into a motionless one, and no frame of reference has a privileged character: this is the meaning of the relativistic denial of the motionless Newtonian space. The only exceptions are quanta of radiation; they cannot be made motionless by any change of the standpoint; they are essentially and under all conditions in motion in all the systems. This is implied by the principle of constant velocity of light, and, more specifically, by the relativistic formula for the addition of velocities.

3   *The Dynamic Character of Time-Space*

Our conclusions are then as follows:

a) The succession of causally related events is preserved in *all* frames of reference. In other words, the irreversibility of the world lines, which are constituted by causal successions of events, is a *topological invariant*.

b) The succession of causally unrelated events is completely relativized.

c) Equally fully relativized is the simultaneity of all events with an apparently trivial exception of the simultaneity of each event with itself. The last part of this statement can be expressed in the following way: *absolute coincidences*, that is coincidences both in space and time, are as much topologically invariant as the temporal order of causally related events.

The propositions (b) and (c) are not logically independent. The relativization of the succession of causally unrelated events and the relativity of simultaneity of distant events are two related consequences of the fact that *the temporal order of all causally unrelated events remains undetermined*. A pair of such events can appear in certain temporal order in some systems, in a reversed order in other systems, and finally simultaneous in the third category of systems which constitute, so to speak, a boundary case separating the first two groups of frames of reference. No events can be judged simultaneous unless they are causally unrelated. Only if there were instantaneous causal connections in nature would the simultaneity of causally (in this case instantaneously) related events be possible.

This, indeed, was the case of Newtonian mechanics, and it is certainly not accidental that Galileo's transformation is obtained when we substitute an infinite value for the velocity c in the Lorentz transformation. Infinite velocity means instantaneous interaction. It is true that classical physics knew since Olaf Roemer's discovery in 1675 the finite velocity of light which in the nineteenth century was found equal to the velocity of electromagnetic waves; but it remained completely unaware of the limiting character of this velocity. No upper limit was imposed on the range of possible velocities, that is, on the speed of causal interactions. Thus for a considerable time the velocity of gravitation was believed to be infinite. Laplace still believed that it was at least 50,000,000 times larger than that of light.[20]

In truth, the assumed existence of the Newtonian space, spread instantaneously and orthogonally with respect to the "axis of time," was an embodiment of instantaneous connections; every geometrical distance in such space can in virtue of its instantaneous character be regarded as a world line of a point moving with instantaneous velocity. This network of instantaneous geometrical relations, constituting "space at an instant," was at the same time an objective substrate of absolutely simultaneous events. When we say, for instance: "Sirius is eight light years from the earth," it has in classical physics the following meaning: 1) that there is an instantaneous space at this particular moment in which the events both on the earth and on Sirius are located; 2) that because of its finite velocity the luminous message which I perceive now left Sirius eight years ago. It is clear that the difference between "Now" and "Seen now" was fully recognized by classical physics; but although the objective "Now" was by definition unperceivable, it was in principle inferable and calculable on the basis of the classical theorem for the addition of velocities which was applied to the relative motion of the luminous source and the observer.

The belief in the distinction between "Now" and "Seen now" was due to the fact that classical physics—unlike the general theory of relativity today—accepted the distinction between static geometrical space and its changing physical content. "Now," that is, absolute simultaneity, belonged to the former; "Seen now," that is, the perceived, spurious simultaneity, belonged to the latter. It was this distinction which inspired the search for an absolute frame of reference which would be the substrate of the objective simultaneity. It is sufficiently known how this search, carried on by the experiments of Michelson, Morley, Trouton, Noble, Tomaschek, and Chase, ended in the failure which inspired the most comprehensive and revolutionary revision of the traditional concepts of space and time. The profound and far-reaching meaning of this revision is still not always fully understood now, more than a half century after the formulation of the special theory of relativity.

Thus we read frequently, and not only in semi-popular treatises, that the simultaneity of distant events, absolute for Newton, "was made relative by Einstein." To use such a language is highly misleading. It suggests almost inevitably that behind the inherent relativity of the human frames of reference there lies hidden the true absolute simultaneity, the absolute "Now," even if it may remain forever inaccessible to our knowledge. It is far more accurate to say that the simultaneity of distant events was *eliminated* instead of being merely relativized. What objective status could possibly exist for an entity which is unobservable by definition, and which is an inferential construct different in different frames of reference, none of which possesses a privileged character? It is thus not sufficient to join the adjective "relative" to the noun "simultaneity"; the noun itself should be dropped because of its lurking ontological connotation. Einstein himself did not hesitate to do it:

There is no such thing as simultaneity of distant events; consequently there is also no such thing as immediate action at a distance in the sense of Newtonian mechanics.[21]

This correlation between simultaneity of distant events and the network of instantaneous connections can be expressed in a far more explicit way. The class of objectively simultaneous events constitutes the space of classical physics at a certain instant. Conversely, any instantaneous three-dimensional cut across the four-dimensional world process contains the events objectively simultaneous at that instant. Thus the simultaneity of distant events implies their juxtaposition and vice versa. This is what Newton had in mind when he claimed that "every indivisible moment of duration is everywhere."[22] The cosmic "Now," in virtue of its universality, is instantaneously spread everywhere; this is the meaning of the classical correlation of absolute simultaneity and absolute space.

But such instantaneous three-dimensional cuts, admissible in the physics of Newton and Laplace, are excluded by the physics of relativity. Contrary to Newton's belief, there is no moment of time which is present everywhere. This lack of correlation between Now and Everywhere was expressed by various thinkers in different ways. In Eddington's words, there are no "world-wide instants";[23] according to Whitehead, there is not such a thing as "nature at an instant";[24] or, as A. A. Robb said, "there is no identity of instants at different places at all"; in other words, "the present instant, properly speaking, does not extend beyond here."[25] *Since there is no absolute space correlated with each instant of time, there is no absolute juxtaposition which would serve as a substratum of absolutely simultaneous events.* But while there is no juxtaposition of events which would be a juxtaposition for all frames of reference, *there are certain types of succession which remain such in all systems.* As we have seen, these types of succession are represented by causal chains, that is, by the world lines of material and luminous "particles." Unlike spatial juxtaposition, the irreversibility of the world lines has an *absolute* significance, independent of the conventional choice of the system of reference. We can hardly have a more convincing illustration of the dynamic character of space-time.

We may anticipate the following objection: what about the relativization of the succession of the causally unrelated events? Is it not as fatal to the ontological status of time as the relativization of juxtaposition is to the ontological status of space? Not speaking of the fact that the succession of causally related events still remains invariant, we must not forget that the relativization of the simultaneity of distant events and the relativization of the succession of causally unrelated events entail each other (see above). If we substitute with Einstein the term "elimination" for that of "relativization," it becomes clear that the succession of causally unrelated events is as much devoid of concrete physical meaning as the simultaneity of remote events. Nothing in nature corresponds to either of them. To continue to refer to them as something "real, though relative" betrays the pre-relativistic modes of thought. Such expressions result from an incongruous overlapping of two incompatible languages, the Newtonian and relativistic; it is the resistance of our Newtonian subconscious which prevents us from saying boldly and consistently that simultaneity of distant events as well as the succession of causally independent events simply does not exist.

For there are only two types of relations in the relativistic universe: that of successive causal connections and that of contemporary causal independence. Since the universe consists of the dynamical network of the irreversible causal lines, their irreversibility which remains absolute in the relativity theory is conferred to the universe as a whole. Needless to say, it is not the irreversibility of the Newtonian time. The world process according to Newton consisted of the irreversible series of the world-wide instants, that is, of the Now-Everywhere planes; and we have seen that no such cleavage planes are admissible in the relativistic universe. We have seen that a three-dimensional space, at any moment, is an arbitrary instantaneous cut in the four-dimensional process and that such artificial cuts were superseded by the four-dimensional regions of causal independence ("elsewhere" of Eddington, "co-presence" of Whitehead) which separates the front cone of causal future from the rear cone of the causal past. But this does or at least should make clear two important points. First, the impossibility of three-dimensional instantaneous cuts radically transforms the classical concept of space; space now is incorporated into the four-dimensional world process in which the classical space of Newton is a mere artificial instantaneous cross-section. Second, the fact that the past and the future are now more effectively separated than in classical physics certainly does not weaken the objective status of succession. Thus, all these evidences point to one important conclusion: the relativistic union of space with time is far more appropriately characterized as a *dynamization of space* rather than a spatialization of time.

4   *Impossibility of the Backward-flowing Time and of the Self-intersecting Causal Lines*

The impossibility of the backward-flowing time follows directly from the irreversibility of the causal lines and is embodied graphically in the relativistic time-space diagram. The world lines emanating from any Here-Now event must be contained in the causal front cone of Absolute Future. They can never radiate into the region of "Elsewhere," which is forbidden to them by virtue of the limiting character of the velocity of light. It means that their angle with the local time axis can never be 90°;

this would imply the existence of infinite velocities and the resulting flattening of the frontward causal cone. With the concomitant flattening of the rearward causal cone the region of Elsewhere would be squeezed out of existence. This would be a return to classical physics which indeed admitted the existence of the world lines orthogonal to the time axis; in truth, every distance in the Newtonian space belonged to this category. The whole of instantaneous classical space may be regarded as an infinitely dense network of such orthogonal world lines. The backward-running local time would require that the corresponding world line would be bent by an angle greater than 90°; in other words, the corresponding causal front cone would be turned backwards like an upturned umbrella. Such a case was impossible even in classical physics; in the physics of relativity, for which the past and the future are even more effectively separated by the region of Elsewhere, this is, so to speak, doubly impossible. No world line starting from Here-Now can ever reach the region of Elsewhere; a fortiori none can be bent backwards to reach the rearward causal cone of the past. In the three-dimensional time-space diagram by which we symbolize the relations in the four-dimensional world process, the space angle of the frontward cone can never attain the value of 2π; a fortiori, it can never surpass it. This definitely excludes all Wellsian fantasies about the travelers visiting their own past or the past of their own ancestors.

The impossibility of the backward-bent world lines clearly entails the impossibility of any line recrossing its past course. Such a case would be in conflict with another essential idea of relativity: the absolute character of spatiotemporal coincidences. We have seen that every world event coincides spatiotemporally with itself; consequently, it is simultaneous with itself. This is not as silly a truism as it sounds, especially when we formulate it in the following way: each event is simultaneous with itself and only with itself; or, in A. A. Robb's formulation, "an instant cannot be in two places at once."[26] This evidently was not accepted by the physics of Newton, according to which every instant of time was present through the whole of space; each Now was everywhere. Nor is it accepted by those who, like Gödel, accept the possibility of self-intersecting world lines. In the latter case there would be some events which, besides being simultaneous with themselves, would also be simultaneous with other instants in time! In other words, a certain event, corresponding to a single point in which the corresponding world line recrosses itself, would be simultaneous with a remote future instant. In such a case we would be clearly on the brink of magic; indeed, some

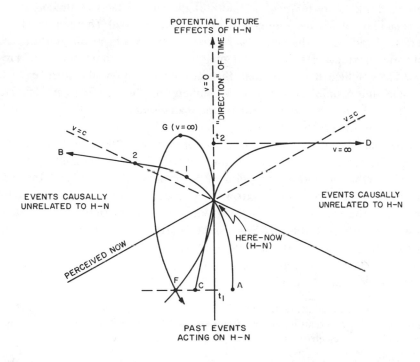

POTENTIAL FUTURE
EFFECTS OF H–N

EVENTS CAUSALLY
UNRELATED TO H–N

EVENTS CAUSALLY
UNRELATED TO H–N

HERE–NOW
(H–N)

PERCEIVED NOW

PAST EVENTS
ACTING ON H–N

THREE IMPOSSIBLE WORLD LINES

The diagram above, which is an elaboration of Figure 1 in Costa de Beauregard's essay, depicts three world lines, that is, four-dimensional orbits, whose existence is excluded by the limiting character of the velocity of light. They represent bodies moving in *H-N* (Here-Now) with the admissible velocity $v < c$; but all of them would acquire eventually a velocity $v > c$. A body moving along the world line *A-(H-N)-B* would acquire it beyond the point event 1; at the point 2 it would overtake the photon emitted from *H-N* and would enter the Elsewhere region (See Costa de Beauregard, Figure 2). This would mean that an observer in the Elsewhere region, contemporary in the relativistic sense with *H-N*, would perceive the signal from an event future with respect to *H-N*. The world line *C-(H-N)-D* is the world line of a body, or of a signal, moving eventually with infinite velocity. It would be equivalent to the realization of the Newtonian instantaneous space at time t_2. The third trajectory could be called a "Gödel line," after the distinguished mathematician Gödel, who adduced the possibility of certain cosmological models in which such travel may be formally permissible. Moving along this line would require that a body reach infinite velocity at point *G*, would turn backwards and after crossing the region of events causally unrelated to *H-N*, that is the Elsewhere region, it would enter the region of Absolute Past. Absolute, that is, with respect to *H-N* and all events causally subsequent to it. Leaving point event *G*, the body would eventually cross itself at time t_1. The point of intersection, *F*, would represent an event both successive to and simultaneous with itself. In causal terms this would be an event affected by its own future effects.

449

serious thinkers are tempted to interpret the alleged facts of precognition by a similar retroactive action of the future on the past! There is no place here to dwell on all the causal anomalies which would result and on the intrinsic discrepancies of the language in which similar situations are described; suffice it to say that the retroactive action of future events remains impossible ʾas long as we adhere to the requirement of the relativity theory that no causal action can escape from the frontward causal cone of Absolute Future.

One recent argument for the idea of backward-flowing time is its alleged usefulness in removing certain kinds of causal anomalies in the so-called creation and annihilation of microphysical "particles." Instead of saying that a pair of electrons was created, one positive and one negative—the positive one a moment later being converted into radiation —we can interpret this process as the world line of a single electron moving in a zigzag way through space-time. Hans Reichenbach, who mentions this "equivalent description" not without sympathy, is fully aware of its limited usefulness:

The anomalies of creating from nothing and vanishing into nothing are thus eliminated; however, in exchange for them another causal anomaly enters the description; the electron travels part of its path backward in time.[27]

To this we must add the following footnote. First, the alleged anomaly, which is supposed to be removed, is much less "anomalous" than Reichenbach believed. It is not true that the process described above involves "creation from nothing" and "vanishing into nothing"; the pair of particles arises from electromagnetic radiation, into which it can be reconverted, instead of coming miraculously from or vanishing into "pure nothing." Thus the whole process appears causally anomalous only to our Democritian logic of solid and permanent bodies which the whole trend of contemporary physics tends to discredit. Second, the whole process is ruled by Einstein's famous equation about the equivalence of mass and energy; the energy of the vanished quantum was converted into the rest mass of both particles and into their kinetic energy. This is only one example of many impressive confirmations of the relativistic dynamics on the microphysical scale. But then in the light of the very close connection between the dynamics and the kinematics of relativity, should we not expect that the relativistic space-time diagram is applicable to the microphysical scale as well; in other words, that the electrons cannot travel backwards in time any more than the macroscopic signals can?

The fallacious belief in the reversibility of time is another example of the unfortunate influence of false spatial analogies: if we can travel over the same road in an opposite direction, why could not we travel backward over the same "path of time"? But, in truth, as Whitehead and even Russell stressed,[28] we never travel over the same road again. When I travel back to South Station in Boston in the afternoon, it is not the "same road" on which I traveled in an opposite direction in the morning; the road itself is a number of hours older! An attentive analysis would show that the idea of unidirectional time re-emerges in the very formulation which purports to deny it; for "reversal of time" is supposed to take place *after* time flowed in its normal forward direction! Now the word "after" means, if it means anything at all, "continuing in the original forward direction of time." In other words, to say that time at a certain one of its moments changes its direction is equivalent to the self-contradictory assertion that "time flows backward while continuing to flow forward."

5   *The Status of "Now" and the Potentiality of the Future*

Insisting correctly on the relativization of simultaneity, our modern neo-Eleatics jump to the conclusion that "Now" does not have any objective status at all, being nothing but "a temporal mode of experiencing ego."[29] In such a view, becoming still exists on the subjective, psychological level; but the physical world outside of the stream of consciousness simply is, it does not become.[30] In other words, "coming into being is only coming into present awareness."[31] The apparent queerness of this view should not prevent us from analyzing it. Not speaking of its enormous epistemological difficulties (about which later), its main defect is that it not only does not follow from the relativity theory, but is even incompatible with it.

My physical Here-Now, which corresponds roughly to my psychological awareness of the present, precedes all events of my causal future and follows all events contained in the backward cone of my causal past. Since this "before-after" relation is invariant in *all* systems, it follows that *in no frame of reference can my particular Here-Now appear simultaneous with any event of my causal future or with any event in my causal past.*[32] This follows from the fact that the succession of the events constituting the world lines can never degenerate into simultaneity in any system; this obviously applies to the world line of my own body. In this sense my Now still remains absolute. It is not absolute in the classical Newtonian sense since it is confined to Here and does not spread instantaneously over the whole universe. Yet, it remains absolute in the sense that it is *anterior* to its own causal future in *any* frame of reference.

On the psychological level this anteriority of the present with respect to its subsequent future moments is embodied in the characteristic feeling which constitutes the central part of our awareness of time. It is the feeling of the present pointing beyond itself toward a not yet realized future. Future situations can be inferentially preconstructed and even imaginatively anticipated with a great degree of vividness, but by their own nature, they can never be perceived; they cannot become present as long as they are absent; after becoming present, they are no longer future. It is this *absence in time* which is symbolically represented by the characteristic "not-yet" feeling. According to the static interpretation of space-time this feeling is illusory; what we call the future are merely those portions of the world lines which have not yet entered our awareness. We must resolutely get rid of the persistent illusion that they are coming into being; for this is merely an illicit objectification of the temporal order of our perceptions. For the correct construction of the objective view of the world our feeling of "now" is as irrelevant as the feeling of "here"; they both are merely accidental and shifting perspectives of the timeless four-dimensional whole.

Despite its superficial plausibility and its widespread popularity, hardly any other view is more incompatible both with the spirit and the letter of the relativity theory. It ignores completely another feature of the relativistic time-space: no event of my causal future can ever be contained in the causal past of any conceivably real observer. By "conceivably real observer" we mean any frame of reference in any part of my causal past or anywhere in my present region of "elsewhere." In a more ordinary language, *no event which has not yet happened in my present Here-Now system could possibly have happened in any other system.* To believe otherwise would mean either to accept the existence either of the actions moving backward in time or at least of those moving with the velocity greater than light: *quod non.* Since the inclusion into the causal past of the observer is the necessary condition for the perceivability of events, it means that the postulated existence of future events is unobservable in principle. If we continue to postulate it, then we face the following dilemma: either to believe in their observability, which would contradict relativity; or to admit their intrinsic unobservability, but still insist on their existence; this would contradict the most elementary rules of scientific methodology. It is far simpler and sounder to place the unobservable future events into the same category as phlogiston, caloricum, mechanical ether, and other discredited and useless fictions.

We can anticipate the following objection: Your denial of the reality of future events is admittedly based on the exclusion of future observers from the category of "possibly existing observers." Thus the whole argument is nothing but *petitio principii*. The division between the past, present and future frames of reference is relative and arbitrary, since it is continually shifting; what is still in the future for me now will be included (or, as it is fashionable to say, is tenselessly included) in the causal past of our posterity. No particular "now" has a privileged character, being merely an accidental and passing perspective of the changeless spatiotemporal whole.[33]

It should be noted that this counterargument has nothing to do with relativity. For *the transiency of "now" is not a discovery of the relativistic physics; it is as old as human awareness of time.* Furthermore, the following points should be noted:

1. The terms "transient" and "arbitrary" are not synonymous. On every individual world line, the Here-Now moment separates unambiguously the past events from the unrealized potentialities of the future events, and *this separation holds in all other possibly existing frames of reference*. It certainly cannot be called arbitrary. In this precise sense each Here-Now is absolute. Its transiency makes it neither ambiguous nor arbitrary. On the contrary, the specific character of each "now" requires its transiency since without it each "now" would lose its temporal characteristics.

2. The reality of the psychological present and of its transiency is recognized, however reluctantly, even by the protagonists of the static view; otherwise their claim that "now is a temporal mode of experiencing ego"[34] would be devoid of any meaning. Since they claim at the same time that our transient now does not have any objective status in the physical world, they face embarrassedly the following question: How is the transient psychological present intelligibly related to the allegedly becomingless whole? MacTaggart, who was one of the most vigorous opponents of the reality of time, was uneasily aware of this difficulty when he asked: Why do we not live in the reign of George III?[35] No answer can be given as long as we assume the timelessness of the world which puts all "nows" on an equal, "accidental" footing.

3. This difficulty automatically disappears when we frankly accept

becoming in the physical world which runs, so to speak, parallel to the stream of our experience. In other words, our present psychological present has an objective counterpart not, it is true, in the unreal cosmic Now, but in the present moment of the history of our planet. There is an objective succession of moments in the objective world corresponding to the transient character of our mental present. Then MacTaggart's question is answered; in truth, it is otiose even to raise it. It is important to stress that although Here-Now is not equivalent to the Newtonian Everywhere-Now, its significance is far from local. Thus the virtualities of our future history which our earthly Now separates from our causal past *remain potentialities for all contemporary observers in the universe*. Something which did not yet happen for us could not have happened "elsewhere" in the universe. Similarly, transiency is not confined to any particular region; all world lines are irreversible and they are irreversible in the same sense. The idea of two world lines, no matter how remote from each other, running in the opposite direction always implies the absurdities of the backward-running or self-intersecting causal series.[36]

The discussion of epistemological difficulties inherent in the becoming-less view is beyond the scope of this paper. Besides, the root difficulty of this view was indicated in previous paragraphs; it is the same difficulty which was concisely characterized by H. Lotze long before Bergson, James, Lovejoy, Broad, Meyerson, Eddington, Whitehead and Whitrow: "We must either admit Becoming or else explain the becoming of and unreal appearance of Becoming."[37] This untolerable dichotomy of two completely heterogeneous and unrelated realms, of changing "appearance" and static "Reality" (with capital R) is avoided if we accept the genuine reality of becoming. The structure of the relativistic time-space certainly does not discourage us from doing it. The present broadening of the concept of causality in microphysics points in the same direction: the Laplacian static determinism yields its place to the view of the open world in which genuine novelties come into being.

Time in Relativity Theory: Measurement or Coordinate?

HERBERT DINGLE

Before one can begin to understand the effect of relativity theory on our notions of time, it is necessary to realize that the theory is concerned *solely* with the relation between the times assigned to events at different places and with the variation of those times with the state of motion which the observer ascribes to himself and his measuring instruments. This disposes of a number of mistaken ideas which have served, not only to make the theory appear unnecessarily mysterious, but also to give it an entirely false aspect.

In the first place, it entails that the theory is a purely objective physical theory of the ordinary type; it has nothing to do with the subjective experiences of different *observers*. For popular descriptive purposes it has been customary to depict two observers in relative motion and to compare their measurements, but that is justifiable only as an expository device. At least until very recent times, it has been impossible to compare the measurements of such observers because they have never existed—at least, to our knowledge; any difference in the measurements of observers moving with the speeds practicable on the Earth would have been far too small to detect, yet the theory has been devised and subjected to numerous tests by what is effectively a single observer—a terrestrial one. It must therefore stand or fall by its conformability to the experiences of such an observer.

What the theory compares are not the measurements of different observers but the measurements which a single observer must adopt when he makes different assumptions about his state of motion. For example, in most physical investigations the Earth is taken to be at rest. For many astronomical purposes, however, it is taken as moving at a speed of 18½ miles a second round the Sun. Obviously, the result of any experiment made by a terrestrial observer will be exactly the same whichever assumption he makes. The theory of relativity allows him to make either, but (among other things) the time which he assigns to a distant event must differ in the two cases, because he must make corrections for any supposed movement of his instruments according to a definite scheme which the theory formulates. It follows that if there were an observer on the Sun who measured the time of a distant event by the same procedure as the observer on the Earth, he would, according to the theory, obtain by a direct measurement to which no correction would need to be applied the value which the terrestrial observer would obtain after correcting for his assumption that the Earth was moving round a stationary Sun; but that is an incidental consequence. The primary purpose of the theory is to relate physical measurements to the *coordinate system* which a single observer adopts (*i.e.*, in effect, to the state of motion which he ascribes to himself), and all statements concerning the experiences of relatively moving observers which cannot be deduced from the effects of a change of mind of a single observer (and many such statements have been made during the last fifty years) are spurious.

Another consequence of our preliminary statement is that the theory is concerned only with the recording of *instruments* (in our case, clocks) and not at all with the subjective experiences of the person who uses the instruments. It follows immediately that the whole of the considerations in this book which relate to subjective experiences of people are irrelevant to the subject of this essay. Furthermore, relativity does not prescribe any particular form of clock, except that it must be one according to whose indications the velocity of a beam of light keeps a constant value as the beam travels through space. The ordinary clocks to which we have always been accustomed satisfy this condition to a very high degree of approximation, so they are taken in relativity theory as the standard instruments for time measurement. A rigorous definition would be that a standard clock is one according to whose readings a beam of light covers equal distances in equal times. This presupposes that we know how to determine what are "equal distances." Physics deals with this problem, but since we are concerned here only with time, I shall take it to be satisfactorily solved.

These remarks are by way of prologue. They are necessary, not so much for the purpose of explaining the effect of relativity theory on our ideas of time as for the removal of prejudices which would otherwise

456

obstruct an explanation. I turn now to the physical situation at the beginning of this century which presented the dilemma which the relativity theory was created to resolve.

1   *Circumstances in Which Relativity Theory Arose*

The dilemma in question arose from the discovery, not at all unfamiliar in physics, that a theory which had not previously found itself in serious difficulties was suddenly faced with observations directly contrary to its requirements. The theory in question was the electromagnetic theory of Maxwell, amplified by Lorentz, according to which, among other things, light consisted of electromagnetic waves propagated at a constant speed through an all-pervading ether. The standard procedure in such circumstances was, of course, to modify or abandon the theory, but in this case there were special difficulties in taking such a course. No modification was readily conceivable that did not engender other and even greater conflicts with observation, while to abandon the theory would have been to return to a state of almost complete ignorance, for no other theory seemed able to account for the many phenomena with which the Maxwell-Lorentz theory was successful.

A clue to the solution lay in the fact that the successes of the theory were with observations in which all the bodies concerned were either at rest or moving with speeds so small compared with the speed of light that their motion could be neglected, whereas its failures showed themselves whenever this condition was not fulfilled. In particular, the theory broke down when it was necessary to take into account the fact that the Earth was moving round the Sun at a speed of one ten-thousandth of that of light—a small fraction, certainly, but much too large to be ignored. This led Lorentz (and, independently, in part, FitzGerald also) to the idea that bodies moving through the ether were thereby changed; all material bodies were shortened in the direction of motion, and all rhythmical processes, such as those occurring in clocks, were slowed down. The effect increased with speed, and if a body could be made to move as fast as light, its thickness would shrink to nothing and a clock attached to it would "stop," *i.e.*, its hands would move round infinitely slowly.

The result of this would be that, in those cases in which the predictions of the theory disagreed with observations, it was not the theory but the observations that were wrong: the clock and measuring rod with which

457

the observations were made were changed by their motion so that they did not record the true distances and times between events. The amount of the change was such as to make it impossible, with the accuracy of measurement then attainable, to detect the motion through the ether of a single body, such as the Earth, from measurements confined to that body, so that all terrestrial measurements designed to determine the motion of the Earth through the ether would turn out as though the Earth were at rest. It is important to notice, however, that this was, in a sense, accidental. There was a real physical difference between being at rest and moving in the ether, but the changes produced in our measuring instruments by motion were such as to hide the other effects which the theory predicted should be observable—effects which actually occurred but which we were thereby prevented from observing. Furthermore, it was possible that if, in the future, we succeeded in measuring much more accurately, we should then be able to detect the Earth's motion by terrestrial experiments, for the neutralization might be only approximate.

This supposition of Lorentz, however, did give a complete explanation of all the observations that had been made up to that time, but no reason was evident why the postulated changes should occur; they were purely ad hoc devices which enabled the electromagnetic theory to survive. The artificiality of such a procedure made it unacceptable to some physicists—notably Ritz—who felt that a theory forced to such lengths to preserve its existence must be inherently wrong. It was clear that the observations which the original theory could not explain would follow naturally if light kept a constant speed *with respect to its own source* and not with respect to the stationary ether through which it was supposed to travel, and Ritz assumed that this was actually the case. But that meant discarding the electromagnetic theory, for it was fundamental to that theory that the speed of light through the ether depended only on the state of the ether itself: no matter how the source of light might happen to be moving, the light which left it would move at one invariable speed. Ritz, however, pictured light as issuing like a shot from a gun: similar shots fired from similar guns in relative motion travel at different speeds through the air, each traveling at the same speed with respect to its own gun.

Ritz's hypothesis had two advantages over Lorentz's. In the first place, it accounted naturally, without any additional hypotheses, for all those observations which the unmodified Maxwell-Lorentz theory found inexplicable. Secondly, it was preferable, at least aesthetically, even to the Maxwell-Lorentz theory plus Lorentz's ad hoc postulates, for it made the detection of the motion of a body by means of experiments confined to that body impossible *in principle*, and not merely by an accidental coincidence which might break down with improvements in measuring technique. In other words, Ritz's theory conformed to the *principle of relativity*, which required that motion was *essentially* a relation between two (or more) bodies and not a property of either one of them. This was a fundamental requirement of Newtonian mechanics, for Newton's First Law of Motion implies that a single body can be regarded at will as being at rest or moving uniformly[1] with any speed at all. To ask in which of these states it really is would be just as meaningless as to ask whether a body is high or low when there is nothing with which to compare it. One of the most serious problems connected with the electromagnetic theory was, in fact, its violation of the principle of relativity, with its consequent undermining of Newton's First Law. For the existence of an ether implied that there was a difference between a resting and a uniformly moving body of which Newtonian mechanics took no account. Furthermore, if Lorentz's additional postulates were valid, the changes produced by motion should reveal themselves in mechanical as well as electromagnetic phenomena, and therefore when sufficiently high speeds were attained we should expect mechanical laws to break down. All this was very unsatisfactory, and Ritz's theory avoided it in the simplest possible way.

On the other hand, it had one enormous disadvantage: it explained the anomalous observations only at the expense of leaving the normal ones unexplained. The strength of the electromagnetic theory lay in its ability to account for a wealth of observations in situations in which the motions of the bodies concerned were negligible. Of these Ritz could offer only a fragmentary explanation, and his attempt to provide a comprehensive theory in which an ether was not required was cut short by his untimely death. For the relation of the velocity of light to that of its source was in itself a mere detail, the sort of thing that should be deducible from a much more fundamental theory. The Maxwell-Lorentz theory enabled a deduction to be made—namely, that the velocity of light was independent of that of its source. Ritz's supposition was that this was not true, but he constructed no general theory which had this as its consequence.

The practical difficulties of testing the supposition experimentally were at that time insuperable, so the choice was perforce left to each person's taste. Either you followed Lorentz, accepting the universal ether with respect to which motion was meaningful and therefore denying the principle of relativity and with it the basis of Newtonian mechanics; or you followed Ritz, dismissing the ether and preserving the principle of relativity but sacrificing electromagnetic theory.

This is the background against which Einstein's Special Theory of Relativity must be seen in order to be fully understood.[2]

2   *Einstein's Relativity Theory*

The genius of Einstein manifested itself pre-eminently in his perception that the alternatives—the lack of dependence of the velocity of light on that of its source, and the principle of relativity—were not necessarily incompatible. To preserve the principle of relativity it was necessary to deny the ether,[3] but it did not then follow that light could not move in space as though there were an ether. Einstein accordingly made two postulates: 1) *the postulate of relativity*, that there is in nature no "ether," *i.e.*, no absolute standard of rest; 2) *the postulate of constant light velocity*, that beams of light emitted at the same instant from two momentarily coincident bodies in uniform relative motion would travel through space as a single beam—and, furthermore, that the velocity of this beam (the constant represented by c) would be the same with respect to each body. The first postulate accorded with Ritz's theory and the second with Lorentz's, but Lorentz was able to maintain the second only with the aid of an ether. For the velocity of the same beam of light to have the same value when measured from each of two relatively moving bodies, it would seem to be necessary to use different measuring instruments on the two bodies, and this is effectively what Lorentz's hypothesis, with its modification of instruments through motion, stipulated. But if there were no ether to distinguish the moving from the resting instruments, it would be impossible to know which instruments were modified. How, then, could the second postulate hold if the first were accepted?

It is here that the conception of "time" enters the problem. Einstein saw what no one had perceived before, that although physicists had always been accustomed to speaking of the *time* of occurrence of a distant event, they had never defined exactly what that concept meant. Suppose I say that some celestial event—say the attainment of maximum brightness of a variable star—occurred 100 years ago: what do I mean by that? The answer would be that if I made a valid experiment to determine the time, the result I should get would be "100 years ago." But what is a valid experiment? To that one would answer: a determination of the distance of the star and of the time light takes to travel along it. Then, if I find that the star is at such a distance that light takes 100 years to reach me from it, I have determined that the event in question occurred 100 years ago.

But that brings me face to face with the fundamental ambiguity in pre-relativity physics: how do I know how long light has taken to reach me from the star? Suppose, for simplicity, that the star and the Earth are *relatively* at rest. They might still be moving together through the ether at any speed and in any direction. If the common motion is in the direction from star to Earth, then the Earth is, so to speak, running away from the light, and the light must travel further than if Earth and star were stationary, and still further than if they were moving in the opposite direction. And I cannot tell how Earth and star are moving, for all experiments to this end prove abortive. Hence I have no means of discovering how long the light has been traveling, and therefore no means of discovering at what time the star reached maximum brightness.

It follows that the phrase, "the time of a distant event," is indefinite. If A asserts that the event occurred 95 years ago, and B that it occurred 105 years ago, there is no possible means of deciding which, if either, is right. We have therefore a choice of two courses. We can say either that the event occurred at some particular time and no other, but that it is impossible to determine what that time was; or that, so far as nature is concerned, there is no such thing as the *time* of the event, any more than there is such a thing as a natural *name* of an event, for what we cannot by any means discover we can claim no right to suppose to exist. Instinct and tradition had led physicists automatically to choose the first alternative: it seemed to be a necessity of thought that if some instantaneous event occurred, no matter where, it must have occurred at some definite moment of time, even though you could not discover that time. Einstein, however, chose the second alternative. He regarded a distant event as having, in itself, no characteristic at all corresponding to what is called the "time" of its occurrence, and therefore, if we wish to endow it with such a characteristic, we must do so by voluntary definition, just as we confer a name on a star by voluntary definition. Having chosen a definition, we must, of course, stick to it, just as we must always refer to Plato as Plato, and not call him indiscriminately Plato, Aristotle, Beethoven . . . ; but the original definition is, at bottom, entirely arbitrary and is to be determined only by its usefulness in furthering our purpose of learning all we can about nature and expressing our knowledge in the simplest way.

Einstein's definition was this. You are provided with a clock which tells you the times of events at the place where you are, and you want to define the time, according to that clock, of an event at a distance. Place at that distant point a similar clock, stationary with respect to your own, and adjust its readings in the following way. Let a beam of light leave your clock when it reads t_1, travel to the distant clock and be immediately reflected back again, reaching your clock when it reads t_3. Then the distant clock must be adjusted so that it reads $t_2 = \frac{1}{2} (t_1 + t_3)$ at the moment when the beam of light reaches it. The two clocks are then said to be *synchronized*, and thereafter any event which occurs at the distant clock when that clock reads T is defined as occurring at a time T. Clocks may be imagined at all points of space and to be all adjusted in the same manner, so in this way we have a definition of the time of occurrence of any event at all at any point of space.

It is necessary to emphasize that this definition is entirely arbitrary—apart, of course, from considerations of convenience: it is just as arbitrary as calling an unknown quantity x instead of, say, Popocatepetl, though no one would suggest that the latter symbol would be equally felicitous. Prejudices die hard, and there are still physicists who have not entirely rid their minds of the idea that Einstein was *discovering* the right way of timing a distant event, instead of *freely inventing* something to which he gave that name. But he made it perfectly clear in his original paper[4] on the subject that he was choosing a *definition*, and later he found it necessary to correct the false impression to which I have referred. In a course of lectures delivered at Princeton in 1921[5] he said:

The theory of relativity is often criticized for giving, without justification, a central theoretical rôle to the propagation of light, in that it founds the concept of time upon the law of propagation of light. The situation, however, is somewhat as follows. In order to give physical significance to the concept of time, processes of some kind are required which enable relations to be established between different places. It is immaterial what kind of processes one chooses for such a definition of time. It is advantageous, however, for the theory, to choose only those processes concerning which we know something certain. This holds for the propagation of light *in vacuo* in a higher degree than for any other process which could be considered, thanks to the investigations of Maxwell and H. A. Lorentz.

This states quite explicitly two things of fundamental importance for the understanding of Einstein's theory. First, the definition of time, permitting relations to be established between different places, is an entirely voluntary one. "It is immaterial what kind of processes one chooses for such a definition of time." And secondly, the process actually adopted was chosen because Einstein believed that the Maxwell-Lorentz theory had a higher degree of certainty than any other relevant theory: in other words, he had a very strong belief that his second postulate was true.

There is, however, a fundamental difference between the definition and the postulate—namely, that the former, being a freely adopted convention, cannot be tested, while the latter, being a statement about what actually occurs in nature, can be tested, and so is definitely either true or false. The theory, which involves both, is therefore open to proof or disproof; it is not, as it is sometimes represented, a matter of pure logic or mathematics. Beams of light, emitted at the same point at the same instant from two relatively moving sources, will reach a distant point in the direction of the relative motion at the same time, or one after the other, according to the truth or falsity of the second postulate: it matters not at all, for this test, what a clock at the distant point may happen to read when the beams get there. This test does not involve any measurement at all, and so is quite independent of any definition of time measurement; it involves only a comparison of events at the same place.

Unfortunately, this last-named experiment has not yet been made, though it is probably not beyond the resources of modern techniques. In the early days of the theory, however, it was manifestly quite impossible, and recourse was had to indirect arguments. These seemed to show that the velocity of light through space was independent of the velocity of its source, as the postulate required, and accordingly it was held that the postulate had experimental support. This conclusion, however, can now be seen to have been premature.[6] More recently, experiments by Bonch-Bruevich,[7] in which beams of light from opposite limbs of the rotating Sun were compared in the laboratory, showed that the beams covered a given distance in the same time. Unfortunately, however, they were reflected into a horizontal direction before comparison, and in these circumstances Ritz's theory requires that their velocities shall be equal, even though they differed before reflection. Moreover, Fox[8] has called attention to other reasons why light from celestial sources cannot give an indubitable proof of the truth of the second postulate. So far as experiment is concerned, therefore, Einstein's theory must be regarded as still open to confirmation or refutation.

Let us, however, while holding this question in abeyance, proceed with the implications of the theory. Our procedure for defining the times of distant events is a purely practical one, independent of any assumptions at all. The clocks which we distribute throughout space are all at rest relatively to our own clock and to one another, and that is something which has a definite meaning: having adopted a standard space-measuring rod, we know absolutely whether the distance between any two objects is changing with time or remaining constant. But we do *not* know whether the objects are all at rest or all moving together in any imaginable way, and by the first postulate we *cannot* know this, for nature provides no standard of rest which would give meaning to such common motion. If, then, we have in our laboratory two similar clocks, one (A) resting on the bench and the other (B) moving steadily past it, we may choose which we like as our "standard" clock, and distribute clocks in space which are synchronized in the prescribed way. We may thus have two sets of synchronized clocks, each set as a whole moving uniformly with respect to the other, and there is nothing whatever to make us regard one set as more "valid" than the other.

But we can easily see that the time of a distant event will be different in the two cases. For suppose we send out a beam of light from the two laboratory clocks at the instant at which they are both together and both reading t_1. This beam (by the second postulate, it will be the same beam, no matter from which of the clocks it proceeds) will reach a distant object, in the direction towards which B is moving with respect to A, at an instant at which we may suppose some event, E, is occurring, and will be reflected back to the laboratory clocks again. But clearly it will reach B before it reaches A, for B has gone forward to meet it. If the readings of A and B when the beam returns are respectively t_3 and t_3', then the time of the event E according to A will be $t_2 = \frac{1}{2}\ (t_1 + t_3)$, and according to B, $t_2' = \frac{1}{2}\ (t_1 + t_3')$. Clearly, t_2' will be earlier than t_2: which is right?

The answer is: both are right. The definition, and the procedure for putting it into effect, have been faithfully followed in both cases, and we have no grounds at all for preferring one result to the other. A is at rest with respect to the laboratory, it is true, but we cannot know that the laboratory itself is not moving in such a way that B is actually at rest. The state of motion of the object on which the event E occurs does not enter into the matter. Whatever it may be, the event itself is instantaneous, and therefore momentarily coincident with one clock of the A set and one of the B set, and these read differently. Moreover, we are not restricted to two laboratory clocks. We can suppose any number, all moving differently, and then we get an indefinitely large number of times of the distant event, each of which is as good as any other. Our definition therefore commits us to the conclusion that the time of a distant event is a multivalued quantity.

This brings us to an aspect of relativity theory about which perhaps more confusion prevails than about any other; namely, which of its requirements are purely conventional and which objective physical fact? Take the Earth as an example. In our physical and astronomical work we sometimes take the Earth to be at rest, sometimes the Sun, sometimes the local system of stars, sometimes the center of the galaxy, and so on. In all these cases we get different values for the motion of the Earth, but, by the first postulate, they are all equally valid, for the motion ascribed to any one body is entirely conventional. The "time" of any distant event will, as we have seen, vary with our choice, for it is relative to the motion of the standard clock. But if we make an experiment to determine it according to Einstein's prescription, we shall get one single value, for the behavior of the clock and the light will not depend on our mental preference. Hence we must distinguish clearly between the "time" of an event and the result of our experiment: the latter is the "time" only if we have previously decided to regard the Earth as at rest. The experimentally determined value is sometimes called the "proper time," though the term is not used as consistently as one would wish, and the fluctuating value the "coordinate time." One could fill many pages with an account of the false deductions that have been made concerning the readings of clocks which actually are applicable only to coordinate times. The clock-reading at an event, and the time at which the clock has that reading, are two quite distinct things. The former is an unalterable fact of nature; the latter is a convention which we may change as freely as we wish, by changing our coordinate system. When we change it we do not change any clock-reading, but we make the same reading "fast" or "slow" in relation to the coordinate system adopted.

This ambiguity in the notion of the time of a distant event was a great stumbling block in the early days. It would probably have been fatal to the acceptance of the theory but for one tremendous consequence which the theory was able to place on the credit side. *If* Einstein's two postulates were true, and *if* each of two pairs of relatively moving clocks were synchronized according to his prescription, then *either* pair, according to the readings of the other, would appear to be running slow by precisely the amount that Lorentz had postulated for moving clocks, and therefore the necessary corrective to the original Maxwell-Lorentz electromagnetic equations was seen to be required, not as an ad hoc postulate, but as a necessary consequence of the quite natural assumption that if *any* pair of relatively stationary clocks were once synchronized, they would remain synchronized. This seemed to solve the problem that had previously seemed insoluble—namely, that of producing a slowing down of moving clocks relatively to stationary ones without providing an ether to indicate which clocks *were* moving and so without abandoning the principle of relativity. You could take any clock you liked as a standard, and all clocks moving relatively to that would give such times for distant events that such clocks would appear to be running slow.

3   *Breakdown of the Theory*

If this conclusion were indeed valid, and if the second postulate were confirmed experimentally (or, alternatively, if, when synchronized by some agent behaving in the way which the postulate asserts for light, all pairs of relatively stationary clocks remained synchronized), then the theory would be beyond reproach. It can be shown, however, that the foregoing conclusion is not valid: on the principles of the theory it can be deduced, with equal justification, that a moving clock is running both slow and fast compared with the standard stationary one, and therefore that there must be some hidden inconsistency in the theory.[9]

To see this, let us return to the laboratory experiment previously considered, in which the time of a distant event by clock A was defined as $t_2 = \frac{1}{2}(t_1 + t_3)$. This equation may be written, $t_2 - t_1 = t_3 - t_2$, and this means that the time taken by the light to travel *to* the distant event is equal to the time it takes to return *from* it. This is true whatever common state of motion is assigned to the clocks in question. Applying it, then, to clock B, we have $t_2' - t_1 = t_3' - t_2'$. Now the left-hand sides of these equations represent the times, by the two clocks, between *the same* two events, namely, the departure of the light from their common position and its arrival at the distant event E, and we have seen that $t_2 - t_1$ is greater than $t_2' - t_1$. That means that the A clocks, which are synchronized with one another, show a longer time to elapse between the same two events than do the B clocks. Both sets of clocks are, by hypothesis, running regularly, so it must follow that clock B is running slow compared with clock A.

This is not a mere convention. If the postulates of the theory are true, and the clocks are adjusted according to the definition, then their actual readings will be such as to compel this conclusion. This was early recognized by Einstein, and pointed out in his original paper on the theory; the argument which I have given here was, in fact, that which he himself used, though in a more general mathematical form.

But now suppose we consider an event, E', on an object in the opposite direction from A, *i.e.*, in the direction away from that towards which B is moving; and, for convenience, let us suppose that E' is at the same distance from clock A as is E. Let a beam of light proceed from the two coincident clocks at t_1 by each of them, meet the object in question at the instant at which E' occurs, and return to clock A when that clock reads t_3. This beam will reach clock B when B reads t_3'', *after* it has reached clock A. By exactly the same reasoning as before, we find that the time of the event E' by the A clock is t_2, and by the B clock, t_2'', where t_2'' is greater than t_2. Hence the interval between the setting out of the light and the event E' is greater by the B clock than by the A clock, so that the B clock is running *faster* than the A clock.[10]

This is a plain contradiction. Both clocks are running regularly, yet between one pair of events A shows a longer time to elapse, and between another pair a shorter time to elapse, than does B. The reasoning leading to this conclusion is identical in the two cases. It has been applied in one case, and the conclusion has been universally accepted. It has not been applied in the other, presumably because the possibility of contradiction has not been conceived, and, the relative rates of the clocks having been determined by one imaginary experiment, it has not been thought necessary to determine it by another. But it is obvious that there is no reason at all for preferring one pair of events to another, so there is no escape from the conclusion that if we accept the requirement of the theory that "a moving clock runs slow," as it is usually expressed, then we must also accept the opposite conclusion that "a moving clock runs fast," and that invalidates the claim of the theory to reconcile the Maxwell-Lorentz theory with the postulate of relativity.

4 Reasons for the Misconception of Einstein's Theory

It is at first astonishing that so simple a flaw in the theory should not have been noticed very soon, but I think there are a number of reasons for this. In the first place, as I have already said, the objection to the theory that immediately sprang to notice was a different one—namely, that it took away all meaning from the simultaneity of separated events, and so implied that of two such events it was possible, with equal right, to regard one as occurring either before or after the other. This was such an affront to what had been regarded as plain common sense that the attacks on the theory concentrated attention on it, and so no trouble was taken to examine closely the less obvious objections to which the theory might be open. At bottom, of course, this criticism was directed against

Einstein's claim that a distant event had no intrinsic characteristic that could be called the "time" of its occurrence, and that if one wished to use such a term he was free to define its meaning in any suitable way. If that claim were granted, the times of separated events could be defined—and were by Einstein so defined—so that they possessed a number of equally "true" values, and this was regarded as an outrage to reason. As time passed, however, the legitimacy of the claim gradually won acknowledgment, with the result that physicists, and especially philosophers, were much less ready than before to trust their instinctive reaction to apparent anomalies, and ceased to look for them. A theory that had triumphed over so deadly an objection as this could be trusted to defend itself against less direct attacks.

A second reason lay in the habit, to which attention has already been directed, of expressing the requirements of the theory in terms of *observers* rather than *coordinate systems*. An experiment such as that just described became almost automatically conceived as a comparison between two moving observers, A and B, each being provided with a clock. When I have invited consideration of such problems, the invariable response has been, "A will observe . . . but B will observe. . . ." A contradiction in the relative rates of two clocks was thus transformed into a difference between the experiences of two observers: if A made the observations he would find that B's clock, which he would regard as moving, would be running slower than his own, while if B made the observations he would obtain the opposite result. Now this is, of course, not contradictory, any more than it is contradictory to say that B's clock will appear to A to shrink in size as it recedes, compared with his own, while A's clock will appear to B to shrink in size in the same circumstances.

It is clear, however, that the contradiction in this experiment is quite independent of the choice of observer. The necessary readings can be recorded mechanically, and the records examined afterwards by any observer, who may move as he likes: if the records show what the theory requires, he will find in all cases that clock A must be held to run both faster and slower than clock B. Both of the contradictory results are entirely objective, and there is no conceivable reason for preferring one to the other. But the contradiction is lost to sight when we make the invalid transformation of clocks into observers, and so it escapes detection.

A third reason for the oversight is that the theory has had such remarkable success in so many instances that it has become almost impossible to conceive that it might yet be wrong. It is felt—often unconsciously—that even if there is here something that appears contradictory, there must nevertheless be an answer to it; a theory that has proved itself in so many ways can scarcely any longer be doubted. But what is overlooked here is that *all* the confirmations of the theory depend on the accuracy of the electromagnetic equations when the space and time coordinates (x, y, z, t) occurring in them are interpreted in accordance with the Einstein definition of distant times and the corresponding positions of distant points. There has never been a single test of the purely *mechanical* requirements of the theory—the slowing down of moving clocks, contraction of moving rods, increase of mass of moving bodies, result of combining velocities, and so on—and it is in just these requirements that the essential characteristics of the theory lie. The only "clocks" that conform to its demands are atoms, and these are interpreted as clocks only by virtue of electromagnetic theory. The only masses that are shown to increase with velocity are charged particles, which have to be assumed to be moving with the velocities assigned to them by electromagnetic theory. And so on.

All the "confirmations" of the theory are of this character, though some of them depend on electromagnetic theory in a less direct way. And this has so far been inevitable, because all the phenomena by which the theory can be tested depend on the production of very high velocities, which we have no means of producing except by the action of electromagnetic forces on charged particles, and no means of measuring except by the application of electromagnetic theory. What it all amounts to, therefore, is that the theory has experimental support only if the Maxwell-Lorentz equations are true, and if they are not, it provides a satisfactory correcting factor. But only an entirely mechanical experiment, in which no appeal at all need be made to electromagnetic theory, can provide an experimental confirmation of Einstein's theory, and this has never been made.

From all these considerations it is clear that while the equations of the special theory of relativity are a necessary—and, so far as we can see, complete—corrective to the Maxwell-Lorentz equations, their interpretation as a relation between the readings of relatively moving measuring rods and clocks is impossible, since it leads to contradictory results.

5   *Future Possibilities*

We are back, then, at the dilemma that faced us at the beginning of the century. Einstein's brilliant attempt to reconcile his two postulates has been shown to have failed, and we must admit that one or the other must be false. If the first postulate—the postulate of relativity—is false, we are left with Lorentz's theory—that there is an ether with respect to which motion is in principle detectable, but that such motion modifies material rods and clocks in such a way that it can be detected, if at all, only by experiments at such high velocities that the 4th power of their measure relative to the velocity of light is not too small to produce an observable effect. If the second postulate—the postulate of constant light velocity—is false, then we may preserve the relativity principle by adopting Ritz's hypothesis. To decide between these alternatives a really definitive experiment on the relation of the velocity of light to that of its source—an experiment involving no appeal, direct or indirect, to electromagnetic theory—is necessary, and until it is made, speculations are of doubtful value.

This much, however, may be said. Lorentz's supplementary hypotheses are purely ad hoc, and if it should be found that the velocity of light is independent of that of its source, electromagnetic theory would still remain imperfect until some reason was found why motion through the ether should modify even uncharged bodies in the particular way supposed. Ritz's theory would demand a new electromagnetic theory, but this might not be so formidable a task as it seems. Maxwell's original equations for a static system have not broken down, and the problem is not so much the construction of a new electromagnetic theory as a new generalization, applicable also to moving systems, of the original one. Suggestions along these lines have been made elsewhere,[11] and need not be pursued here. A strong point in favor of the first postulate is the success of Einstein's theory of gravitation, which is based on a generalization of that postulate and an ignoring of the second. But it is better to await the verdict of experiment than to proceed with conjectures.

We must, however, consider the alternatives insofar as they affect the problem of time. Although Einstein's theory appears to have failed, his perception that the notion of time at a distance needed definition is still valid and is a mark of the deepest insight. But, that having been said, it does not follow that the notion implicitly held before that fact was realized was necessarily wrong. It was believed that the time of a distant event was a unique quantity, independent of the place of the event and of the state of motion of the standard clock, though no clear idea existed of how to determine its unique value. Einstein showed the necessity of a definition, and exercised his freedom by choosing one which made it multivalued. That choice has failed (*i.e.*, the quantity so defined cannot be related to actual clock readings in the way supposed), and we are now free to try a definition which gives a unique value. It has been too hastily assumed that what was undefined was wrong. It was wrong that it should be undefined, but when that is corrected it might turn out to be right—*i.e.*, to be a concept which can be used in the description of nature with the maximum profit and with consistency.

Both Lorentz and Ritz acknowledge a unique time for a distant event, but, according to the former, moving clocks will not show it. Lorentz distinguished between the true time and a "local time." His theory showed how to determine the former from the latter, provided that one knew how the recording clock was moving, but it threw no light on the practical problem of discovering what its motion was. On Ritz's theory, clocks were not changed by motion through the ether, for he admitted no ether. Clocks could therefore be moved from place to place as one pleased, and no problem of synchronization arose.

6 *Space-Time*

We can now appreciate the significance of one of the most widely discussed implications of the relativity theory—the so-called discovery that, in the words of Minkowski, "henceforth space by itself, and time by itself, are doomed to fade away into mere shadows, and only a kind of union of the two will preserve an independent reality." This unfortunate remark is probably responsible for more misunderstanding about the wider implications of relativity theory than anything else. There are two distinct, and quite final, reasons why the theory can say nothing whatever about "time by itself." First, it is wholly concerned with "time at a distance." Time in itself—*i.e.*, whatever time may be in one's own experience—is quite outside the scope of the theory; that is simply accepted, and, the process of measuring time *here* being given, the theory then proceeds to consider how to measure time *there* in a conformable way. Consequently, as stated in the introductory remarks to this essay, all the metaphysical and psychological aspects of the problem of time are quite irrelevant to the theory of relativity; they belong to time *here*.

Secondly, it is the first principle of the theory that "time at a distance" is not something objective—a part of nature, so to speak—but a phrase which is meaningless until we give it a meaning. It follows that whatever meaning we give it can imply nothing about "time by itself" because it is our own choice.

The difference between the relativity definition of time at a distance, and that which was unthinkingly implied before, is this. According to the latter, an event at any point in space could be said to occur uniquely at such and such a moment of time, so that for each such moment there was one specific state of the whole physical universe. According to the relativity theory, however, the particular event at some distant point that is simultaneous with the present moment depends on the motion of the standard clock, and that motion can be assigned ad lib. It turns out that the "time" of a distant event becomes more and more indefinite the further away that event occurs, and there is thus an inseparable association between the place and the time of an event. But the theory is such that not only the *time*, but also the *place*, of the event (*e.g.*, its distance from oneself) changes with the arbitrary choice of motion, and the changes are related in such a way that a certain combination of them has the same value for all choices. This combination is described as the "position" of the event in "space-time," and it is this characteristic of the theory that has given rise to the illusion that there is an objective "space-time" in nature, of which what we call "space" and "time" are partial aspects.

7   *Conclusions*

Our conclusion, then, is that the physical phenomena which have led to the theory of relativity have no contribution to make to the solution of problems concerned with time. On the contrary, they require us to know what we mean by time before we can consider them. For events near at hand we accept the readings of clocks constructed in accordance with the principles described by G. M. Clemence and H. A. Lloyd, and for distant events we must await a better knowledge of how those clocks behave when set in motion. All the real problems associated with the notion of time are independent of physics.

Comments—
Of Time and Proper Time

The idea of "proper time," or *Eigenzeit* in its original German, a concept that derives from the development of Special Relativity Theory in four dimensions, has enriched our knowledge of the properties of time in a manner unsuspected before the days of Einstein and Minkowski.

In his famous paper "On the Electrodynamics of Moving Bodies," Einstein observed that "the phenomena of electrodynamics as well as mechanics possess no properties corresponding to the idea of absolute rest."[1] By absolute rest he might have meant some conditions in nature, which, through experiments by any number of relatively moving observers, could have been uniquely and clearly associated with the idea of zero velocity or permanent rest. The cogency and power of Special Relativity Theory derive from Einstein's assumption, or, perhaps discovery that the phenomena of electrodynamics and mechanics do, however, possess a property corresponding to the idea of "absolute speed," to borrow Eddington's phrase, or to the idea of "an absolute motion," in contradistinction to the idea of an absolute rest. By absolute motion in this context we mean the existence of a physical process, which, through experiments by any number of observers in relative uniform translation, can be uniquely and clearly associated with the idea of maximum velocity and permanent motion. The physical process involved is the propagation of light. The idea of permanent motion relates to the impossibility of observing a non-propagating beam of light. The absoluteness relates to the invariance of the speed of light for all observers, while the idea of maximum velocity resides in the impossibility of particle and signal velocities greater than that of light.

The invariance of light velocity, its limiting nature, or the invariance of the quantity τ to be discussed below, may not be derived from prior ideas in physics. Rather, these and other interesting properties of nature disclosed by Special Relativity Theory should be thought of as portions of a complete and self-consistent proposition which describe the physical world accurately, provided we follow the experimental instructions of the theory.

We have seen in the essay on time perception in children that the intuition of speed and distance precedes the ability of forming a notion of time. Motion as described in pre-relativity physics does not emphasize this fundamental nature of the perception of speed, for velocity is formed as the ratio of two other, conceptually different and presumably more fundamental quantities: distances and times. But understanding velocity as a ratio of distances and times, far from being something intuitively obvious, requires a level of training or a mental process of at least three steps. First we must separate the sense datum of speed and distance into its spatial and temporal components; then we have to give meaning to the metrization of these components, and only then can we mathematically recombine these metrized quantities and use their ratio as a measure of uniform translation. In the four-dimensional world of relativistic physics the concept of uniform translation itself assumes a fundamental role comparable to that of straight lines in Euclidian geometry, perhaps as an affirmation of the primitive, intuitive primacy of speed over time. Since, however, we have attained sufficient sophistication to be able to perceive the world in terms of space and time, we require that the measures of these dimensions be incorporated in physical descriptions of motion. In Relativity Theory the recombination of lengths and times into a parameter that we can associate with relative motion of bodies is one step more complex than the method of forming the velocity of classical kinematics.

Minkowski showed[2] that there is a mathematical combination of time and distance measures related to an object moving with respect to an observer such that the combination will have the same numerical value for all coordinate frames in which the Newtonian equations of motion hold at least to first approximation. This quantity, often called proper time, is given by

$$\tau = \sqrt{t^2 - R^2/c^2}.$$

Here the speed of light c is a distance-to-time ratio measured by conventional meter rods, by "local" clocks, that is clocks at rest with respect to the observer, and assuming the isotropy of space with respect to the direction of light propagation.[3] The distance R is the spatial separation of events A and B, also measured locally by rigid rods. Time t is that between the two events A and B, as measured by a local clock, but only after prior knowledge of c is admitted. Insofar as proper time is a simple combination of distances and times that describes motion with respect to the permanent or "absolute motion" of a light beam, it in some sense parallels the hypothetical absolute velocity of pre-relativity kinematics. But whereas velocity is unlike distance or time, the quantity τ for all motions at speeds less than maximum velocity is a time-like quantity. That is, the physical dimension of proper time is the same as that of local time.

The fundamental ontological claim of Special Relativity Theory can be regarded to consist of the following: the locally determined quantity τ which may be associated with the journey of a distant body from event A to event B, and which is computed from units of local time is, in fact, the time which would be measured by a clock on the moving body making the $A–B$ journey along R. The rule of correspondence which we have earlier called for is this: my time as I experience it, and my proper time as others infer it, are physically indiscernible. The difference between the two is that while we seem to possess abilities to sense local clock time, we seem to possess no abilities to sense time of rapidly moving objects;[4] the latter must be constructed from spatial and temporal measurements by logically satisfactory rules such as represented by the $\tau = \sqrt{t^2 - R^2/c^2}$ expression.

Inspection of the simple equation for τ shows that the magnitude of that quantity approximates the amount of time by which a mass point traveling from event A to event B is observed to lag behind the "absolute motion" of a photon that originates at point-event A and is also traveling along the same short distance R. The curious condition arises that what we locally perceive as *time* is identified by all others translating relative to us as a *temporal lag*, and that the magnitude of this lag approaches zero as our own relative speed approaches that of light. Insofar as all experimental evidence supports the relativistic prediction that particle velocities may approach but cannot exceed c, one is impelled to suspect that what we perceive as time is something intimately connected with the existence of ponderable mass. The reason is that for non-ponderable mass proper time, and with it time itself, vanishes. This suspicion is strengthened by the interesting similarity on the one hand, between the philosophical arguments of absolute and relational time, and on the other hand the question of inertia as an intrinsic or absolute property of matter versus inertia as an interaction or relational effect. The concepts of time and inertia seem to depend one on the other in some very fundamental way. It also follows that it would be very difficult to construct a relativistically meaningful definition of time for a universe consisting entirely of radiation.

One may of course observe events for which the quantity τ calculated according to Minkowski's prescription becomes algebraically imaginary. But, we cannot give an acceptable meaning "right here" to time units in multiples of $\sqrt{-1}$, hence by the rule of correspondence neither can we assume such distant events as connectable by the ordinary concepts of "earlier than," "later than," or "simultaneous with." As Quantum Theory is believed by some to delimit the applicability of our idea of time while inquiring "Where is here?" and "When is now?", Special Relativity Theory seems to perform a similar function while seeking the answers to "Where is there?" and "When is then?"

Let the free motion of a body be monitored between point-events A and B by observers at rest in different inertial frames. It follows from the manner in which proper time is constructed that they will all agree on the value of τ to be assigned to the motion of the body, but in general they will obtain different readings on their own clocks. Remembering again the rule of correspondence, the multitude of clock indications must all be taken as equally valid times of the same generic identity. It follows that if the events A and B represent two consecutive encounters of two clocks, in general the clocks will register different elapsed times between meetings. In case of a laboratory or "stay-home," and a traveling or "rocket" clock, under some conditions the stay-home clock is expected to register the longer time, under different conditions the traveling clock does.[5] This interesting result of Relativity Theory is paradoxical because at one hand it is a result of the equations of that theory, on the other hand it seems to contradict the basic assertion of the same theory, that all inertial observers are physically equivalent. The paradoxical aspect of this problem, together with the emotional impact of its popular misinterpretation as a scientific fountain-of-youth, has produced a steady flow of articles since the days of Paul Langevin.[6]

Most discussions of the clock paradox fall into one of three groups: those which propose to show that the differential aging relates to uniform translation and therefore it is a Special Relativistic effect; those which trace its origin to non-inertial motion, hence to the domain of General Relativity Theory; and those which maintain that for any actual test both theories must be called upon, for two coincidences of two clocks must involve non-inertial motion and are likely to involve inertial motions. The majority of arguments seem to end up by implying that time is a feature of the Universe.

I find that the philosophically most satisfying view among those who prefer to regard the effect entirely one of inertial motions is that taken by Prof. Whitrow and presented in connection with a thought experiment. He imagines one clock remaining stationary, while a similar device is circumnavigating a static and finite but unbounded universe of positive curvature. He then shows that at the second encounter a time shift should exist, even though the traveling clock has been in uniform translation during the complete test. Whitrow's conclusion is that "The essential difference between the two clocks concerns their relations to the universe as a whole."[7] Representative of those who believe that different clock indications relate to accelerations are the views of Prof. Čapek,[8] who concludes that the problem must therefore be referred to the structure of the universe-at-large. Also cosmological in their conclusions are those advocating a "detailed model solution,"[9] for they must appeal to General Relativity. A separate view is that held by Prof. Dingle. His interpretation is rejected by most physicists because it requires a massless universe[10] containing only the two clocks. Finally, according to the interaction view of Relativity Theory proposed by Prof. Schlegel in his essay later in this volume, the elapsed times measured by the two clocks would be equal. Insofar as his approach is one of relational time, he also refers the problem to the contents and structure of the universe. For a summary and a critique of the clock paradox with emphasis on the experimental problems, the reader is referred to W. G. Rosser's text on Relativity Theory.[11]

What we have said so far suggests that time from the point of view of Relativity Theory is some property of the world at large and the "sympathy" of the clock to the universe that we have noted earlier still exists. As concerns the identical nature of time (of a local clock, such as an oscillating atom) and proper time (time inferred) when the distances are the vast expanses of the universe, a touch of doubt might have been present in Einstein's mind when, discussing the cosmological red shift, he wrote: "Everything finally depends upon the question: Can a spectral line be considered as a measure of a 'proper time' (if one takes into consideration regions of cosmic dimension)?"[12]

Returning now to the question of proper time, we repeat that all time measurements necessary to infer distant time, that is to construct τ, may be made with conventional mechanical, atomic or biological clocks "right here." Therefore, if the equation defining proper time is admitted as universally valid, the properties of time inferred cannot be different from time locally experienced. There is no reason, therefore, why the time of organic clocks at a distance should be thought of as differing from the time of mechanical clocks at a distance, as long as mechanical and organic clocks are found to remain locally compatible. By a similar argument we must also conclude that the existence of temporal order, unless it can be explained "right here," cannot possibly be explained by the type of insight provided by Special Relativity Theory. Finally, while I find the reasons introduced in support of the idea of a static universe much less convincing than those favoring the reality of time, it is clear that both of these contradictory philosophical views can be argued from the four-dimensional description of the world. This suggests that the contribution of Relativity Theory to the determinism–free-will controversy is peripheral.

J. T. F.

Time and Quantum Theory

E. J. ZIMMERMAN

The non-relativistic quantum theory appears to be our most fundamental theory of the behavior of matter. Its successes have been many and spectacular. Its failures are of omission, not commission: it is silent on some interesting questions, but it does not make false predictions. It describes the behavior of microphysical systems, those systems which compose all matter of whatever size.

We attempt here the task of describing this theory in a precise language, not committing the theory to any more than a minimum of metaphysical interpretation. Certain difficulties of interpretation of the theory do produce some philosophical dissatisfactions, which we shall try to analyze.

The concept of time enters into the theory in such a way that for all verifiable statements "time" refers to something defined by a laboratory clock. It appears that, although we have no more fundamental theory than quantum theory, it leaves without unambiguous answers some very interesting questions concerning the nature of microphysical time and space. Philosophical analysis might possibly resolve these questions, but we do not attempt that here.

1   Summary of the Quantum Theory

Detailed presentations of the quantum theory are available from many sources.[1] But, for completeness, it seems desirable to present a sketch of the theory, describing its general methods and results, and considering its status as a logical structure with implications concerning the nature of the physical world.

Within this century physical theory has been in general preoccupied with what is *observable* and *measurable*. There is no longer any demand (if there ever was) that all theoretical constructs represent directly observable quantities, but one still feels that unobservable constructs should be kept to a minimum.

Reasonable as this attitude is, difficulties arise in the investigation of the behavior of microphysical systems. Individual events involving such systems are certainly observable and, in a certain sense at least, measurable; but these events may be neither *reproducible* (in the laboratory) nor *predictable* (by the quantum or any other theory). What can be reproduced (and predicted) are certain sets of microphysical events, or *ensembles* of events which involve a very large (in principle, unlimited) number of individual microscopic events. The irreproducibility of the single event, coupled with the reproducibility of an ensemble, implies that any successful predictive theory will have an irreducibly statistical element.

However, since ensembles are composed of individuals, a connection between single, observable, irreproducible events and multiple, predictable sets of events must be made. The concept of probability of a single event provides that connection in quantum theory. A discussion of the meaning of "probability" would take us too far afield, although it creates some very difficult and even pertinent problems.[2] The term here will be used in the most common sense of frequency of occurrence within an ensemble. As such, of course, it is not directly observable for a single, individual system.

Therefore, the rather uncomfortable situation in the investigation of the microphysical world is this: the theory cannot predict all observable events and further, the theory contains concepts not representing immediately observed quantities.

Two familiar examples will serve to illustrate these features. The first is the double-slit experiment, and the second is radioactive decay.[3]

Let us imagine a beam of particles all traveling with the same velocity from left to right. If this parallel beam strikes a screen containing a narrow slit, the particles leaving the slit no longer travel all in the same direction; they are *diffracted* so that their new velocities make various angles with respect to their initial directions. If these particles are observed individually—and this can easily be done—as they strike some detecting screen placed some distance to the right of the first, they will not all strike within the area of the slit (parallel projected along the incident direction to the detecting screen), but will cover a much wider area. Where an individual particle will be detected is neither reproducible nor predictable, but there will be a pattern of distribution of hits. This pattern is observable *provided* a large number of particles is used, and it is predictable from the theory.

If the first screen contains not one, but two, narrow slits, separated by a short distance, the distribution of hits on the second screen changes remarkably. It is very different from that which would be obtained by adding together the distributions from two slits separately. In particular, a number of positions appear where there are no hits at all on the detecting screen; the particles have *interfered* to produce a distribution different from a simply additive one. Or, in more customary words, the probability that a given particle leaving the first screen shall have a velocity at a given angle with the initial direction has been altered, and this probability is zero along some directions for which it is not zero if either slit is by itself present.

The second illustration involves the general group of microphysical systems which possess energy which can be radiated. Some such systems are the radioactive isotopes, which radiate energy by the emission of a particle (electron or helium nucleus) or of an electromagnetic quantum (gamma ray). A large number of other atomic and molecular processes are of the same quantum mechanical nature. Two things are generally characteristic of these processes: 1) for each system only certain kinds of radiation and only certain energies may be emitted and 2) a given system will radiate, but the time of emission of the radiation, while it can be observed, cannot in any way (theoretically or experimentally) be predicted in advance. If a very large number of similar systems is investigated, the rate of decay of the whole ensemble is reproducible, and quantum theory is successful in predicting a probability of emission per unit time which leads to the observed rate of decay.

Both the arrival of a single particle on the detecting screen and the decay of a single radioactive atom are easily observable. Neither the position of the particle nor the time of decay are predictable by quantum theory, nor can either be reproducibly defined experimentally. The behavior of large ensembles of these systems is reproducible, observable, and predictable by theory.

Measurements on atomic systems are fundamentally different from measurements made on large, macroscopic objects. Large systems have parts, and we study such systems by measuring the relative positions of the parts as dependent upon time. For example, consider measuring out the orbits of planets in the solar system, or the positions and motion of the parts of a clock. To make these measurements we require some measuring standard which can be subdivided by marks into segments *which must be* small compared to the distances between parts of the larger system. The smaller these segments, the better; for the measurement is performed by determining the number of our smallest segments contained in the measured interval. This idea of measurement works splendidly for solar systems and clocks.

But it is clearly otherwise for atomic systems. There exists no measuring stick with subdivisions small enough to indicate relative positions of parts of atoms, for the spaces between the atoms of the stick are of atomic size. We are therefore forced to investigate microphysical systems not with instruments small compared to them, but in fact with instruments which are relatively very large. The instruments are necessarily *external*, not internal, to the system measured, and this severely restricts the kind of information we can get by measurement. So fundamental is this distinction between microscopic and macroscopic measurement that it can be questioned whether it is valid to conceive of an atom as composed of parts in spatial and temporal relation to each other.[4]

There is a further significant difference. Because of the smallness and fragility of an atomic system, no observation (and hence no measurement) can be made on a single system without seriously disturbing it. Even observation by illumination with light, if it is of a kind which could give information about the interior of an atom, will involve sufficient energy to ionize the atom; that is, to blast it apart. We are thus prevented, in principle and in fact, from making a detailed study of the interior of an atom. We can, and do, make atoms interact in systematic ways, and try to deduce from the catastrophic results something about their undisturbed properties. This yields information, but it can scarcely be called measurement in the macroscopic sense, and it is certainly not surprising that the techniques available for investigating microscopic systems have yielded a very different kind of information from that obtained by study of macrosystems. It should therefore not be surprising that the mathematical structure of quantum theory is rather different from the mathematical structure of macroscopic theories.

Although the mathematical formalism can be stated in various equivalent ways, we feel that that of Heisenberg,[5] as developed and extended in a series of beautiful papers by Schwinger,[6] remains at all stages closest to the laboratory. In this formalism, each physically observable quantity (such as position, velocity, angular momentum, energy) is represented by a symbol called an operator. As physical investigations discover the properties of and relationships among the observables, so the parallel task of theory is to discover symbols whose mathematical behavior appropriately describes these physical properties and their relationships. Once the mathematical nature of the operator has been discovered (or postulated, for a new problem), one then tries to "solve the eigenvalue problem for the operator" which means to find other symbols Ψ_i which satisfy the equation (Q is the operator)

$$Q\Psi_i = q_i\Psi_i \qquad i = 1, 2, \ldots \ldots$$

where q_i is an ordinary real number, and the Ψ_i are mathematical objects called "eigenvectors" or "eigenfunctions." In general, there are many solutions Ψ_i of this equation with correspondingly many q_i, called the eigenvalues; there may be, and usually are, infinitely many such eigenvectors. In the customary language: Ψ_i is an eigenvector of Q belonging to the eigenvalue q_i.

Provided the mathematical properties of Q have been correctly chosen, it is found that if a measurement of the physical quantity represented by Q is performed upon the system, the result will always be one of the real numbers, q_i. Sometimes the set of numbers $\{q_i\}$ contains all the real numbers (as it does for the momentum operator for a free particle), and then this is a weak prediction. But many times it happens that the set of eigenvalues $\{q_i\}$ is discrete. So it is for example for the angular momentum operator, or for the energy operator of a bound system. Then the prediction is quite a strong one, and when predicted and observed values agree very precisely it is most impressive. Certain operators, such as "spin," have not only discrete eigenvalues, but also a finite number of them. These seem to represent atomic properties which do not have even analogs in macroscopic physics, but they are essential to a complete description of microphysical systems.

The system is said to be "represented" by one of the Ψ's, say Ψ_i. But in order to describe all of the possible microsystems, it is found that the set $\{\Psi_i\}$ is too restricted, and in general, systems exist which must be described by a *linear* combination of the eigenvectors of a given operator or by the "state vector" Ψ where

$$\Psi = \Sigma C_i\Psi_i = \Sigma C_i'\Psi_i' = \Sigma C_i''\Psi_i'' = \ldots \ldots$$

where the C_i, C_i', etc., are complex numbers which may be taken to be restricted by $\Sigma C_i^*C_i = 1 = \Sigma C_i'^*C_i'$, etc. Note particularly that the state vector Ψ representing the system can be written as a linear combination of the eigenvectors of any one of the operators, and that such linear combinations are mathematically equivalent to each other. This gives a *unique* representative of the system in the theory, regardless of the selection of any specific operator.

The question now arises, when the system is represented by a linear combination of eigenvectors of Q, what will be the result of a measurement of Q for that system? The formalism does not suggest a unique answer. One therefore can ask nature by examining a system with state vector Ψ and measuring the value of the physical quantity represented by Q. As always, one of the eigenvalues of Q, that is, one of the set $\{q_i\}$, will be obtained from the measurement. If, however, the measurement is repeated, generally a different value of q_i will be obtained; for the system described by state vector Ψ which is *not* an eigenvector of Q, the physical quantity Q is one of those microscopic quantities which can indeed be measured on individual microscopic systems, but which cannot be reproduced by experiment.

If one proceeds to repeat measurements of Q on the system described by Ψ, one finds a pattern or distribution of values of the q_i beginning to appear: for instance, q_1 may result from 50 per cent of the measurements, q_2 from 30 per cent, q_3 from 15 per cent, and other values only very rarely. For a very large number of observations, this distribution of frequency of occurrence over values of q_i for the state Ψ *is* reproducible; and moreover, the distribution is calculable from the theory: the probability that q_i results from a measurement of Q is simply given by $|C_i|^2$, which (for those familiar with the notation) is given by the vector product $(\Psi,\Psi) = |C_i|^2$. (This explains also the requirement that $\Sigma_i \, |C_i|^2 = 1$.)

Thus, having stated that Ψ is to "represent" the microsystem, one finds that it does so only in the sense of defining a statistical ensemble to which the microsystem belongs. Probabilities of events can be calculated; but probabilities are not events and cannot be directly observed; Ψ is not directly observable. As Slater[7] concisely put it, quantum mechanics is not a generalization of classical particle mechanics but of statistical mechanics.

From this point of view, the task of the experimental quantum physicist is to determine precisely what ensembles *can* be produced with a given kind of atomic system, hydrogen atoms, for example. It is the task of the theorist to discover the operator observables of the system and the corresponding eigenvectors. Then, each ensemble can be written in terms of the eigenvectors of a selected Q and the fractions $|C_i|^2$ can be compared with those obtained from the corresponding measurements. For example, what are the $|C_i|^2$ for ordinary hydrogen at room temperature? Same, at very low temperature? At very high temperature? What are they after hydrogen has been bombarded by low-energy electrons? By high-energy electrons? By a given X ray? and so on and on: innumerable questions to be asked of nature for each of many operators Q.

Logically, the determination of the frequency of occurrence of the values $\{q_i\}$ in the ensemble is equivalent to determining the average values of all powers of Q. These average values can be calculated from a certain vector product, namely $(\Psi, Q\Psi)$. The exact mathematical nature of these calculations need not concern us but we should note that the state vector Ψ always enters quadratically into the calculation of probabilities or of averages.

There remains finally predictions about the averages of some operator other than Q, say Q'. Suppose the state is a linear combination of two eigenvectors of Q: thus $\Psi = C_1\Psi_1 + C_2\Psi_2$. The average value of Q' for such an ensemble, $(\Psi, Q'\Psi)$, can generally *not* be calculated as if the ensemble Ψ consisted of two subensembles, namely Ψ_1 and Ψ_2. This is a very subtle but very vital point which we have not space to treat adequately. It means that state Ψ has properties which are not possessed by any conceivable mixture of systems described by states Ψ_1 and Ψ_2; this has been cogently discussed recently by Wigner.[8] The reason for this is that state vectors are *linear* combinations of eigenvectors whereas averages are always calculated so as to involve state vectors *quadratically*. The result is the appearance of terms which involve two different eigenvectors and represent the so-called "interference terms" between states. Thus, in the double-slit interference problem, the positions on the detecting screen are not the simple sum of the probabilities of the positions for each slit taken separately. One has Ψ_1 (from slit one) and Ψ_2 (from slit two), and the resultant state vector *is* $C_1\Psi_1 + C_2\Psi_2$; but because of the interference between these two states the probabilities of striking a given point on the detecting screen may be very greatly changed from the sum of probabilities for the two slits separately.

These "interference terms" have given rise to the use of the term "wave mechanics" as synonymous with "quantum mechanics." It is quite true that in many instances—and particularly in the double-slit experiment—one can calculate the correct result by considering a wave motion, somehow associated with the particles, diffracted by the slits and subsequently interfering. Indeed, it is probably possible *always* to work with operators so chosen as to preserve this analogy. However, our point here is that it is *not necessary* to talk about a wave motion associated with a microphysical system. Interference as such is not only exhibited by some kind of physical system oscillating in some temporal order. It is also exhibited between state vectors which obey the postulated laws of the quantum theory *prior to* the introduction of time into the theory. Indeed, what we have so far presented is simply the quantum mechanical way of describing stationary microphysical systems. Schwinger calls this the "geometry of quantum states," which is "restricted to the realm of quantum statics, which, in its lack of explicit reference to time, is concerned with idealized systems such that all properties are unchanged in time or with measurements performed at a common time."[9]

This view of interference between states, which we believe to be of a much more general nature than that exhibited by wave motion, has been vigorously espoused by Lande,[10] who has shown that quantum theory can be very efficiently developed without the analogy of wave motion.

Of course measurements made upon a system may indeed depend upon the time of measurement. At least some of the quantities in the eigenvector equation $Q\Psi = q\Psi$ must have a time dependence. It is the final fundamental postulate of quantum mechanics that there exists for every microphysical system a definite operator H which governs the development in time of the system. There are two alternate ways of specifying the time dependence; they are equivalent in the sense of making always the same predictions. In the Schrödinger representation, any time dependence is ascribed to the state vector, which is then written as a wave function $\Psi(x,t)$.[11] The dynamics of the system—its time development—is then given in terms of the operator H as

$$\frac{\partial \Psi}{\partial t} = -\frac{i}{\hbar} H\Psi$$

The operators Q are then independent of time. Alternately, according to Heisenberg, one may claim that Ψ, specifying the microphysical system, should not change with time. The basic equation $Q\Psi = q\Psi$ can still be time dependent provided the operator Q varies with time. In the Heisenberg representation the dynamical development of a system is given by

$$\frac{dQ}{dt} = \frac{\partial Q}{\partial t} + \frac{i}{\hbar}[QH - HQ]$$

where again $\hbar$ is Planck's constant and H is the Hamiltonian or time-displacement operator.

It is scarcely surprising that discovering the appropriate mathematical form for the Hamiltonian operator is one of the central problems in physics. For relatively simple systems it can be written down in terms of the total energy of the system; in many problems of current interest, for example, in nuclear physics or in solid-state physics, the correct Hamiltonian is either not yet known or is too complicated to work with, so that an approximate H must be used.

This outline of the structure of quantum mechanics will appear strange to those who have studied only popular expositions which rely upon the space-time Schrödinger equation and the corresponding wave mechanical representation of the quantum theory. No doubt the approach through wave theory is more easily pictured than the symbolic approach, which is, however, always much closer to physical experiment than is the wave picture. Further, the wave approach tends to obscure many of the most important concepts of the quantum theory.

This approach starts with some general notions of wave motion, and in one of several possible ways (all of doubtful logical validity) arrives at the Schrödinger equation

$$ i\,\hbar\,\frac{\partial \Psi}{\partial t} = -\,\frac{\hbar^2}{2m}\left(\frac{\partial^2}{\partial x^2} + \frac{\partial^2}{\partial y^2} + \frac{\partial^2}{\partial z^2}\right)\Psi + V\,(r_2 t)\,\Psi. $$

The solutions Ψ of this equation are functions of the three space co-ordinates and the time coordinate. (The same functions represent the components of what we have called the state vector in the space-time representation. In this representation the eigenvectors for the expansion of the state vector are those of the position operator.) $\Psi^* \Psi\, dx$ is then the probability that, upon measurement of position, the particle will be in the region dx. It is then very tempting to think of the particles *being* in dx with this probability, or to think that, of an ensemble of N particles, a fraction $\Psi^* \Psi\, dx$ are in dx. Thinking in this way leads to correct averages *for all physical characteristics whose operators Q have the same eigenvectors as* does the position operator. Unfortunately, for many other physical characteristics, such as momentum, energy, and angular momentum, this method of predicting *gives the wrong values* even for an ensemble. We therefore consider it unwise to emphasize the space-time representation of a system, even though recognizing that it must be of special interest.

Furthermore, if one avoids the above error, it is still easy to fall into considering $\Psi^* \Psi\, dx$ as the *density of* something material: the microsystem is conceived not as a point particle, but as a "fuzzy particle" distributed throughout space (and changing in time) as specified by the function $\Psi\,(x, t)$. Such an interpretation is demonstrably false; Ψ does not describe any such material distribution.

Finally, emphasis of the space-time Schrödinger equation leads to an oversimplified conception of microscopic space and time. The variables (x, y, z) in the Schrödinger equation are, of course, mathematically continuous. It is tempting, since they are given the same names (position), to identify them with physical space. Now, there are many *persons* who believe this identification should be made, and it may be logically possible.

However, as these variables are used within the quantum theory itself, they are *not* physical variables, but dummy variables *which must disappear from all predictions made by the theory*. The (x, y, z) in Schrödinger's equation *as they are in practice used in quantum theory* correspond to numerical indices *over which a sum* must be taken before any prediction of any observation can be made. In this summing, symbols may be introduced which do refer to physical space, but these are *not* the same variables occurring in Schrödinger's equation. Approaching quantum theory by way of the state vector makes this perfectly clear; approaching it by way of the wave function and wave mechanics frequently obscures this distinction.

2   Implications of the Quantum Theory

For several reasons, the quantum theory has proved vulnerable to a great many different philosophical interpretations. In the first place, the empirical basis for the theory is neither so directly established nor so unambiguous as is that for many theories. For example, once the experimental evidence for the constancy of the speed[12] of light is accepted, the Special Theory of Relativity seems to be almost completely determined by a strictly logical argument. There does not seem to be, for the quantum theory, any similar crucial experiment, or even any small group of experiments, the results of which logically determine the formalism of the theory. Quantum theory has developed rather slowly and painfully, with a number of false starts which even today still somewhat becloud the issue. From the existence of the quantum (Einstein and Planck), to the Bohr atom, then the matter-waves of de Broglie, the wave equation of Schrödinger and finally the transformation theory of Dirac, one sees a progression in learning an entirely different mode of thinking. The difficulty did not arise with a few experiments, but rather with a whole class of experiments. As our experience with microphysical systems grew, refinements were forced upon the original ideas of the theory, and there is no evidence that this evolutionary process has ended.

Furthermore, in spite of many attempts at rationalization, the quantum theory remains primarily a postulated formalism, justified chiefly by the fact that it works. Historically, when the formalism failed, it was simply modified (as for example by the introduction of "spin") until it did work. This is in considerable contrast with, for example, the Special Theory of Relativity, which is involved with a fundamental consideration of the nature of time and space, and therefore, if successful, can claim to have modified these concepts. The success of the quantum theory carries no such implication, for it is not predicated upon any such analysis. The important concepts of quantum theory only indirectly involve ideas of space and time.

We have therefore in the above description used language which implies a minimum of metaphysical interpretation. It is best to proceed cautiously in drawing philosophical implications from a theory. There is in particular a danger that conclusions be drawn from some interpretation of quantum theory which contains elements not vital to the theory itself. The above description seems to include all concepts essential to *all* interpretations: the use of operators to represent physical quantities, their eigenvector equations, the representation of a microphysical system by a linear combination of eigenvectors, the existence of "interference" in the state vectors, the calculation of probabilities from averages taken over ensembles, the existence of a Hamiltonian time-displacement operator for the ensemble. A theory without these constructs is not general enough to describe microphysical phenomena, but little, if anything, needs to be added to them.

We will later summarize what we feel is implied about the nature of space and time by this minimum interpretation, but we must immediately concede that it does not lead to many, nor to any very spectacular, conclusions. But there are several excellent reasons why it also does not provide a philosophically satisfying representation of the microphysical world. We wish to emphasize that many interpretations of the theory are in fact embellishments upon the theory not required by physical evidence. In order to understand the attraction of these alternate interpretations, we must examine some of the reasons for the lack of satisfaction with the minimum interpretation.

In the first place, it doesn't even really provide a theory of microscopic nature at all. It starts to do so, with its talk of operators belonging to a microphysical system, and so on. But then when the theory begins to make contact with reality, one finds that the operators lead to representations not of one microsystem, but of an ensemble of similar microsystems—strictly, an infinite number of systems. Furthermore, all predictions of the theory are made on the (well understood) assumption that laboratory-size instruments will be used to make the observations. The existence of these instruments is a logical presupposition, and quantum theory does not attempt to provide a theory of the construction and operation of these instruments. It *assumes* that the physicist will not be employing any objects as measuring devices whose operation he does not understand or which he cannot properly control.

In particular, the parameter "time" appearing in any predictions of quantum theory is to be read by a macroscopic physicist from a macroscopic clock. Position measurements are to be made with macroscopic rods placed between macroscopically separated marks. It is well that this is so, for microscopic particles obey special relativistic equations, not classical ones, at high speeds. Since microphysical systems are always investigated with macrosystems, quantum theory is not inconsistent with the Special Theory of Relativity.[13] However, the union of the two theories is not completely satisfactory; we confine our attention in this essay to the quantum mechanics of systems which do not exhibit relativistic effects.

The first reason then, that the minimum interpretation is philosophically unsatisfactory is that it describes only experiments in which a very large number of atomic particles interact with macroscopic apparatus. (Of course, it is also true that this is the only kind of experiment which can be reliably repeated.)

The second difficulty comes from an almost irresistible tendency to regard Ψ as somehow defining and describing the structure and time dependence of a single microphysical system, possibly not in any very simple way, but at least in some way. Certainly atomic systems do occupy a volume of space, and within this volume, there is presumably something not completely structureless. The actual procedure of the quantum theorist appears to justify this notion, for at the beginning of a problem, he *draws a picture* of a system of particles, assigning them positions and writing down functions of these positions quite as if he were dealing with a classical system. Never, at this stage, has he in mind the ensemble by means of which his predictions may be verified. This procedure is not so much inconsistent as it is curious, for the pictures he draws are not really meant to be anything other than aids to the easy discovery of a particular mathematical representation of the operators in which he is interested, and he has learned from experience how such "pictures" may be used to arrive at suitable expressions. Furthermore, he realizes that there are certain operators, such as spin, which cannot be written down in this way. Nevertheless, one *feels* that, starting with a description-of-sorts of an individual system, the final result should somehow, even if obscurely, describe the structure of an individual system, and not refer exclusively to properties of an ensemble.

The third reason for dissatisfaction with the minimum interpretation is the logical incompleteness of quantum theory. Certainly, the quantum theory is, in any strict sense, incomplete. It does not describe individual atomic events; it does not in fact predict any events whatsoever. What it deals with is probabilities of events, and there must be clear distinction between probabilities of events and the events themselves. Individual atomic events happen: a radioactive atom—*one* atom—can very easily be observed when it decays; and it *does* decay. Concerning the decay, the theory yields only probability statements—and in real life (atomic as well as human!) the very improbable event does indeed sometimes happen. The quantum theory is certainly not a complete theory; the controversy initiated by Einstein, Podolsky, and Rosen,[14] concerning whether a more complete theory is possible, continues unabated today.[15]

There appear to be two general responses to the recognition that the theory does not deal with individual events. The traditional response accepts the theoretical situation as a reflection of something inherent within the real world. Individual events cannot be predicted, because they are unpredictable. *Nothing* determines where the first particle strikes the screen or when the tenth atom will decay. There is no real, physical, objective difference between the first and the third particle, which adheres to them *before* they strike the screen, which causes the first to hit here, the third there. These questions of individual events are, in the language of conservative quantum theory, meaningless; and therefore, the interpretation goes, they are in fact meaningless. Only questions and statements about ensembles are meaningful. This is a logically defensible attitude; it is, however, not exactly a modest one. In view of the well-known fragility of physical theories, it does appear a bit extreme.

The second response has been to reject on some non-physical grounds the irreducibly statistical characteristic of quantum theory. This requires that the fundamental structure of the quantum theory be altered, but in such a way that all reproducible observations are predicted just as they are by the quantum theory. This presents no difficulties in principle and indeed Bohm has shown that it can be done in a virtually unlimited number of ways.[16] Many other proposals to modify quantum mechanics have also been made. Conservative theorists view them all with some suspicion, for they must be so designed as to agree with the predictions of the statistical quantum theory where that is correct, and as yet they have achieved no more.

Revisionists take a great deal of comfort from the fourth reason for dissatisfaction with the minimum interpretation. The quantum theory is by no means able to deal with all of our experience with the atomic world. It must certainly be modified to deal with high-energy processes, particularly the production of new particles of matter and the quantization of this production into a spectrum of masses with discrete and interrelated properties. Precisely how the theory is to be modified, or even if it can be successfully modified, is certainly not yet clear. (We have for this reason confined attention to the low-energy region of quantum theory. Even there, in several important instances, small, but quite accurately measurable effects appear to be connected with the particle spectrum and the production of particles.) Revisionists point out, correctly, that new ideas which restore the individual particle and event to the theory may also prove capable of providing answers to a large number of significant problems where the quantum theory fails. Traditionalists point out, equally correctly, that no revisionist theory has yet done so or even showed considerable promise of doing so. Meanwhile, much work which is quite consistent with the general structure of the statistical interpretation, which in fact rests upon it, has shown progress, and one really does have some understanding of high-energy processes, even if a complete theory is lacking.

It would be presumptuous here for us to try to resolve this disagreement. It would be equally presumptuous to pretend that the disagreement doesn't exist. And because it does exist, the task of arriving at sound metaphysical conclusions from the success of the quantum theory is difficult.

3   *Time in the Quantum Theory*

We finally discuss the significance of the concept of time in the quantum theory and the question whether, from the success of the theory, we can make any statements of general interest about the nature of time.

In the first place, in the practical application of the theory we have described, the "time" which appears in the equations is not a quantum mechanical observable, which would be represented in the theory by an operator, but rather a parameter external to the microscopic system. This "time" does not refer to something internal to the quantum system, but to something measured by a laboratory device—a macroscopic clock. There is, indeed, a very large group of quantum problems which are essentially timeless. In this group are the determinations of the eigenvalue spectra for all physical operators, as well as all problems involving the scattering of one system by another, which are solved as a steady-state time-independent process. Furthermore, wherever theoretical predictions involve time and where these predictions have been verified, the "time" used in the verification has been read from a laboratory apparatus.

It can be, and has been, shown[17] that no accurate clock can be of really microscopic size. A brief and very elementary summary of this work has been given,[18] but one should refer to the original paper for an appreciation of the power of the argument. The result is simply that no accurate clock can have a small mass; hence, perforce, all accurate time measurements involve macroscopic systems. Even the so-called "atomic clock" requires an apparatus of many kilograms mass and produces electrical oscillations of macroscopic size.

Thus neither theoretically nor practically is there any necessity to consider "time" as a variable which is applicable on the scale of microscopic systems.

It does, however, appear to be possible to introduce time as an operator (and thus possibly as a microphysical observable) into the formalism of quantum theory without damage to the theory (but also without appreciable improvement). Morse and Feshbach[19] introduce a time operator, with the apparent purpose of placing the time-energy uncertainty relation on a basis symmetric with that of the position-momentum uncertainty relation. (We discuss these presently.) However, it does not seem to have been used for any further purpose. Engelmann and Fick[20] have pointed out that, since microphysical variables can be represented by operators which depend on the time (but remember that in the Schrödinger representation they are not!), by solving certain equations for the time in terms of operators, an operator expression for the time may be obtained. This procedure appears to lead to non-physical consequences[21] and has not so far proven useful.

It is unusual to find a descriptive essay on quantum theory which does not emphasize the uncertainty relations, mentioned here for the first time. The reason is simply that, in the minimum interpretation of the theory, the content of the uncertainty relations is unremarkable. One can show that if two operators Q and Q' may have the same eigenvectors, then an ensemble of atoms can (at least theoretically) be obtained such that all atoms will have the same eigenvalue q_i for Q *and* all will have a definite value q_i' for Q'. One says that this ensemble is "sharp" in both Q and Q', since all atoms have the same pair of eigenvalues, or that this ensemble is "dispersion free" in both Q and Q'.

If, however, Q and Q' can*not* have the same set of eigenvectors, then it is not possible to form an ensemble which is dispersion free in both physical quantities. Position, x, and momentum in the x-direction, p_x, cannot "share" the same set of eigenvectors, and the theory predicts that any ensemble of particles will have a spread of eigenvalues for x, say Δx, and a spread of eigenvalues for p_x, say Δp_x, of such size that

$$\Delta x \, \Delta p_x \geqslant \frac{\hbar}{2}.$$

This is one of the uncertainty relations. (There are as many uncertainty relations as there are pairs of operators not having the same set of eigenvectors.) It says that, if a large group of particles is assembled within a very small space (Δx), then this group of particles must have a large spread of x-momenta ($\Delta p_x \geqslant \dfrac{\hbar}{2} \dfrac{1}{\Delta x}$); that is, they must have a large spread in their velocities. Or, if a group of particles is assembled all of which have about the same velocity (Δp_x is small) then they will have to cover a rather large region of space ($\Delta x \geqslant \dfrac{\hbar}{2} \dfrac{1}{\Delta p_x}$). Now this is perhaps surprising, but it is hardly very shocking. It does *not* say (as some would have it) that each "particle" is somehow a "fuzzy" distribution of something within Δx and that this something somehow has an ill-defined velocity. It says that the *ensemble* to which the particle belongs is not dispersion free in both position and momentum. To the extent that this—and any other—uncertainty relation is a verified result of quantum theory, it *must* refer to an ensemble and not to an individual particle or system.

It can also be shown that if a microphysical system can radiate energy (*i.e.*, is radioactive), then an ensemble of such systems will not all radiate precisely the same energy, nor will they all radiate at the same time. The spread of energies ΔE, which will be observed, and the spread Δt at the times of emission of radiation will be related by

$$\Delta E \, \Delta t \geqslant \frac{\hbar}{2}.$$

The theoretical basis for this is that the energy E is closely related to the operator which specifies the time dependence of the operators (or of the wave function). But when this relation is carefully stated in terms of ensembles of particles it does not imply anything very extraordinary.

(The time-energy uncertainty relation has frequently been interpreted as saying that, if the energy of a system is to be measured with an accuracy ΔE, then the experiment to measure E must have a duration at least as long as the Δt given by the inequality. The analysis of Bohm and Ahranov[22] seems to show that this is incorrect, and that an accurate measurement of the energy can be carried out in an arbitrarily short time.)

The time of emission of radiation from a single atom can easily be observed. One intuitively feels that *at that definite time* something happened within the region occupied by the atom. That may or may not be so; as far as quantum theory is concerned, nothing special has happened to the state vector of the ensemble to which the particle belonged. From the viewpoint of the theory, the probability of radiation from the ensemble has (very slightly) decreased (because the ensemble contains one less atom which can radiate!), but *this is just what* the time dependence of the state vector shows, namely a continual decrease in the radiation probability from the ensemble.

It seems therefore very difficult to draw any "strange" conclusions about the nature of time from the energy-time uncertainty relation.

The quantum theory really has not resulted from, nor forced, a change in our interpretation of space and time. It assumes that we generally know the meaning of these concepts and proceeds without analyzing them. However, there are three general features of the theory which suggest, if not compel, some new ideas about the nature of microphysical space and time. These are: the existence of quantization, the essential use of probability statements, and the impossibility of constructing space-time models for microphysical systems.

In a strict sense, all bounded microphysical systems have discrete eigenvalues for the physical variables. Analytically it is frequently easier to let the boundaries limiting a system go "to infinity," which may result in a continuous spectrum over at least some range of some physical quantities. But in any event, the existence of only certain "allowed" values for physical quantities is one of the most striking features of the quantum theory; the name itself reflects this.

One might therefore argue that, since quantization does pervade the microphysical world, then very likely space and time themselves are, in the very small, not mathematically continuous, but discrete. Although discrete eigenvalue spectra by no means always, or even usually, have constant differences between adjacent values, a number of persons have proposed that space and time may be "cellular" in structure with a smallest possible interval of time. These suggestions have been thoroughly discussed recently (together with much other significant material) by Whitrow.[23]

This writer does not find the evidence for space-time quantization very impressive. Reasoning from the characteristics of material systems—which are generally quantized—to the characteristics of time and space—which, whatever they be, are certainly not material systems—seems dangerously like reasoning from an analogy which is quite likely to be false. Nuclear theories, although of limited success, do not seem consistent with space-time quantization, since continuity of subnuclear dimensions is implicitly assumed in all of them. Also, both de Broglie (particle) and electromagnetic wave lengths and periods certainly exist which are much shorter than characteristic nuclear distances and times. Thus if there be smallest units, it seems likely that they are much smaller than the nuclear quantities.

The second general feature of quantum theory is the central and essential role played by probability. Of a microphysical system, one can only say that such-and-such a location is *probable*; of a microphysical event, only that it may *probably* occur within a certain time. Since the very purpose of a space-time frame of reference is to locate particles at points in space, and localize events at definite times, the failure (of the best theory we have) to do this might be taken to suggest that we are asking the impossible. It may be that space and time are irrelevant or meaningless concepts in the very small. There are many important concepts of this sort: temperature, pressure, electrical resistance, and a number of other physical properties are not applicable to microscopic systems but emerge from quite different properties of microsystems when they combine to form a macroscopic object. Each of these also has a fundamental statistical character.

Considerations of this sort have led the author to suggest elsewhere[24] the view that both space and time may be features only of the macroscopic world:

Space and time are not concepts which can be meaningfully applied to single microscopic systems. Such systems are to be described by abstract concepts (charge, spin, mass, strangeness, quantum numbers) which make no reference to space and time. These microscopic systems interact in ways that must also be described abstractly, that is without reference to space and time. When a vast number of such microscopic systems so interact, the simplest and most fundamental result is the creation of a space-time framework which gives validity to the classical notions of space and time, but on the macroscopic level only.[25]

Although arguments for this viewpoint can be adduced, it must be emphasized that it is at least as speculative as the existence of space-time quantization. Both speculations have yet to prove merit from the physical point of view, and can at present be advocated only on other grounds, perhaps philosophical.

The third characteristic of quantum theory is the impossibility of describing it in terms of a single consistent, visualizable model. The basic concepts of the theory—state vectors and probability computations—do not involve *things* whose structure can be pictorially represented. They are mathematical entities of a not essentially simple nature, and it is not (or has not yet been) possible to devise a satisfactory physical model for whatever it is that the mathematics describes. It is some fair approximation of the truth to say that an atom in some circumstances "acts like" a particle, and in others it "acts like" a wave. Indeed, it never acts *exactly* like a particle nor *exactly* like a wave, but rather it acts exactly like a member of an ensemble whose dynamics is given by the quantum theoretical formulation. But the two limits of behavior—particle and wave—do exist. Learning under what circumstances the behavior fits each model is an important part of learning the practical quantum theory; equally important is learning under what circumstances neither model is appropriate and reliance must be placed on mathematical analysis, not models.

This essentially abstract nature of quantum theory may be an indication that our notion of time and space continuous at very small distances and intervals needs revision. The failure of quantum theory to provide for either a visualizable structure or a space-time trajectory for microscopic particles may be taken as showing the impossibility of applying classical ideas of space and time in the microscopic world. Most working physicists would probably agree with this, if they were forced to adopt an attitude in the matter, but with some reluctance. The very fact of their reluctance would be, I believe, affirmation that the theory itself does not compel one to adopt any particular reinterpretation of space and time. Quantum theorists appear to work quite happily with macroscopic notions of time and space.

To an increasing minority of physicists, the failure of quantum theory to provide space-time models is taken to imply that something is wrong, or at least incomplete, in the theory itself. Bohm[26] seems to have been the first to show that theories can be developed which yield all the well-tested results of quantum theory but which do permit models of atomic systems in terms of point particles with continuous space-time trajectories determined by (visualizable) fields of force. It is certainly possible that a theory of this type will some day be used to explain the features of the mass spectrum of fundamental particles. There have been a number of recent attempts along these lines[27] but none has achieved unambiguous success. None contradicts the quantum theory; each is an extension of the theory which actually includes the current theory as a special case. None can adduce any clear empirical evidence for its particular viewpoint, and therefore each remains primarily an exercise in creative logic. It is

497

not proper to claim any of these efforts as physical evidence for a philosophical viewpoint; they are themselves philosophical viewpoints, albeit in many instances expressed in mathematical form.

We feel that these three features are the well-established results of current quantum theory: the general occurrence of quantization, the formulation in statistical terms, and the impossibility of space-time models for atomic systems. They strongly suggest, but do not logically impose, revision of classical space-time concepts at the microscopic level. A possible revision might be the quantization of time and of space themselves; an alternate might be a macroscopic view of space-time. We believe it unlikely that these features will disappear from future developments in the theory of microscopic systems, but even that opinion may be, and is being, validly questioned.

It would therefore seem rather dangerous to rely very heavily upon arguments from quantum theory to support any particular philosophical point of view about the nature of time.

4   Conclusions

We have tried to summarize the structure of quantum theory in a neutral language free from terms evoking such emotional responses as bewilderment, astonishment, or awe. The structure is rational, comprehensible, and strange only in that the mathematical notions are somewhat uncommon. Within the context of what we have called the "minimum interpretation" the quantum theory is logically satisfactory, not paradoxical, and certainly as rigorously developed as any physical theory ever has been. Within that interpretation, the formalism, austere and abstract as it may seem, is admirably adapted to its only function: explaining and predicting reproducible experiments involving microphysical systems. One could wish for more than this; one could wish for predictions of unique (non-reproducible) events, but no theory is available for this. In fact, no theory is available which achieves more than the well-established quantum theory.

This being the situation, one probably should be cautious about going beyond the minimum interpretation for the support of metaphysical views about the nature of time. This interpretation does not seem to point conclusively to any notions which are inconsistent with our macroscopically derived concepts of time. (These macroscopic concepts include of course the knowledge obtained from Relativity Theory, but that is the subject of other essays in this volume.)

498

It may even be as likely that a critical examination of the concept "time" will lead to an improved quantum theory as it is that quantum theory will lead to a deeper understanding of the nature of time. A great deal of work has been attempted in this direction, but without clearly established results. To be successful, a theory based on some metaphysical concept of time will have to do more than just reproduce the results of quantum theory; it will have to solve, in detail and with precision and rigor, at least one problem not satisfactorily treated by quantum theory; and for at least this one problem the predictions will have to be experimentally verifiable. No speculative theory now on the horizon appears likely to do this.

Time and Thermodynamics

RICHARD SCHLEGEL

The science of thermodynamics gives us an elemental insight into the nature of time, although its equations contain no reference to the time variable. Thermodynamics tells us about heat and work, and how they are related to various properties of systems of matter, but it does not explicitly state anything about the time variation of these properties. In contrast, the equations of two other basic physical sciences, mechanics and electromagnetism, give information about the change with time of their variables, such as position or electric field. Nonetheless, the phenomena described by the equations of thermodynamics are inescapably part of the temporal flow of natural processes, and thermodynamics does give conditions on how the states of a system must succeed each other. By taking the succession of states to be the same as the temporal sequence of nature, we gain a significant relation between time and natural process.

Our first aim, then, is to elucidate the parallelism between the temporal flow of our general experience and the particular succession of states that is prescribed in the terms of thermodynamics. Specifically, we shall find that changes in the thermodynamic quantity, *entropy*, are related in a definite manner to the natural time sequence.

1 *The Direction of Time*

We shall begin by establishing what we mean by the time sequence of natural events. The primary temporal aspect of events is that they do not occur "all at once"; instead, there is persistence of a natural entity through many different, non-simultaneous occurrences. We are able, then, to say that there is a basic property of *extension* that is characteristic of time.

Conceivably, the world could be so featureless as to have no natural change by which one state of the universe could be distinguished from any other; in such a quiet world, there would be no basis for a concept of time, for there would be no way of establishing time differences. At the other extreme, the world might be in so chaotic a state of continual change that there would be no regularity by which a time scale, or measure of time, could be established. In actual fact, our natural world presents both processes of progressive, continual change, with no apparent exact total repetition of states, and also processes of uniform, cyclic change that are used as clocks and on which a concept of regular "flow of time" may be based. The nature of our time concept rests ultimately on the existence of these two kinds of processes, one of which makes time the associate of change and novelty, while the other gives recurrence and uniformity for the measure of time. Because there is progressive change, we find the future always to differ from the past, and yet, also, standards do exist by which we can measure the duration of processes of change, all the way from astronomical developments requiring billions of years to physical configurations existing for less than a billionth of a second.

The physical processes which we use for time measure have been discussed elsewhere in this volume by Mr. Lloyd and Dr. Clemence. None of these is, of course, absolute in any ultimate sense, and the question arises, how do we establish that the processes which are used for time measure are in fact uniform? We rest our confidence in uniformity on a kind of mutual consistency; on the fact that nature does give us cyclic processes whose periods bear a constant ratio to each other. There is here, too, in a tacit way, an element of choice; we choose as our basic standard cyclic process for timekeeping that one which is uniform in terms of our accepted physical theory. Thus, we expect a rigid rotating force-free body to rotate uniformly. We might, alternatively, select as a standard process the rotation of a body which does not perfectly meet these requirements, the earth, for example, but we would then find our physical theory to become far more complicated than it is at present.[1]

Given that we have clocks and an established time unit, we may write a time series, using an integral number for each unit time interval. On the usual assumption that time extends indefinitely into the past and into the future, we may represent future time by what mathematicians call an ω series,

$$1, 2, 3, 4, \ldots \ldots \ldots \ldots,$$

i.e., by the infinite series of natural numbers, with each number signifying one more unit time interval than its immediate predecessor. Likewise, we may represent past time by what is called the $*\omega$ series, which consists of the negative integers extending indefinitely from -1, *i.e.,*

$$-1, \ -2, \ -3, \ -4, \ldots\ldots\ldots$$

We may say, then, that the present unit of time is designated by 0, and that it comes after the last member of the $*\omega$ series, representing all past time, and before the first member of the ω series, representing all future time.

In writing the series of time intervals we have assumed a distinction between past and future, and indeed, our usual convention about positive and negative signs indicates that as we go through the $*\omega$ series we are ever going to earlier (smaller) times, and that as we go through the ω series we are always going to later (larger) times. We have made, then, an assumption about the direction of the flow of time, and we want explicitly to justify this assumption.

We might suppose that we have clocks, which measure time *intervals,* but that we do not have any sense of the direction of time, or of its flow from past to future. By means of a standard clock we could give numbers, $\theta_1, \ \theta_2, \ \theta_3, \ \ldots \ldots \theta_i, \ \ldots \ldots$ to unit time intervals, each interval being associated with a physical state of the world. We suppose, however, that we do not know how time flows through the θ_i's—whether from smaller to larger "i" values, or larger to smaller. How, then, can we learn of any group of θ_i's what the time order is among them, and, also, whether they are part of an $*\omega$ (past) or ω (future) series?

We may easily and directly answer this question if we have a way of distinguishing between *that which has happened* and that *which has not happened,* at the time of any interval θ_i. The numerical value of the total number of unit intervals between θ_i and the unit interval of some state that has happened will then tell us how far in the past that state is from θ_i; and, the time to a projected future state may be determined in a similar manner. The θ_i's may then be converted into appropriate elements of an $*\omega$ or ω series. Our human awareness and memory directly provide us with one way of making the required distinction: events which have happened are impressed upon us in a way distinctly different from that for events which are expected in the future. And, within human fallibility our memory can order our experienced past events in an "earlier than" serial relation. *The unidirectional nature of time may quite properly be based simply on the firm and pervasive direct experience that we do have of the past-to-future flow of time.*

It is also possible, however, to assign the forward direction—into the future—by reference to nonsubjective physical records. Any device which indicates the existence of a state and can be associated with a clock will serve this purpose. Thus, our "world," for whose states an ordered series of time intervals is to be set up, might be a moving vehicle plus the area over which it moves. Photographs of the vehicle, associated with various θ_i clock numbers, would constitute a physical record of past states (states which have occurred) of the "world." By knowing the interval between the θ_i's and some θ_p, taken as the present, we could infer the direction of time among the θ_i's, and they could then be formed into an $*_\omega$ series. Also, a state of the world (the location of the vehicle) would be associated with a given past time interval. More realistically, we might think of several rock strata in a certain formation, where depth is a measure of time interval since the formation of a given stratum. Here too, we have a record of past occurrences and can order them in an "earlier than" time series.[2]

It is important to note the role that the "present" plays in our setting up of a natural time series for events. It is not, of course, a fixed point in a given series of θ_i's; rather, it is continually changing, and the 0 between an $*_\omega$ and an ω series for past and future is therefore continually shifting. But it is essentially by reference to the present that we make the distinction between past and future, and hence set up the "earlier than" (or "later than") relation that indicates the direction of time. If for a set of events we merely knew the magnitudes of time intervals with respect to some given time point, we would not know the direction of time among the events. For events that *have occurred*, however, the magnitudes of time intervals with respect to the present must be measures of "how far back" in time, and hence give a temporal direction for the events.

In practice we do not generally have to use an explicit time series with respect to the present for the establishing of a direction in time; we do not have to because we have already learned what is the earlier-to-later course of many natural processes. Thus, given two photographs of a person taken some years apart we can generally say which is the later one because of what we know about growth and aging of people with the passage of time.[3] The association of a thermodynamic concept with time is also one in which information gained from experience will tell us of the direction of time, without reference to time measure and a series of recorded or remembered states. We now turn to that concept.

2   *Entropy*

We consider a system to be an aggregate of matter and/or radiation which is substantially isolated from its surroundings. Suppose the system to have a uniform temperature T (in absolute temperature). If a quantity of heat dQ, so small that the temperature remains effectively unchanged, is added to or taken from the system, we say that for the system there is a resulting change of entropy, dS, given by dQ/T. Entropy change is, then, by definition, the quotient of heat gain or loss, divided by the temperature at which the change occurs. We say dQ is $(+)$ if heat is added, and $(-)$ if removed, so the entropy change is correspondingly an increase or decrease, depending on whether heat is added or removed.

The change in entropy, ΔS, which results from a large heat loss or gain, may be calculated by summing over differential heat quantities dQ:

$$\Delta S = \int_i^f dQ/T$$

where "i" and "f" indicate initial and final states of the system respectively. The calculation indicated by the equation must be carried out for a "quasi-static" process, in which the system is approximately at equilibrium during the transfer of each dQ, so that a temperature T may be established at which the transfer occurs. The entropies of many substances, at various temperatures, have been determined from careful thermal observations and are listed in standard tables; it is assumed that the entropy of a pure substance in a stable state is zero at the absolute zero of temperature.

We see that entropy is not a directly observed physical entity or quantity. It is nonetheless a concept of great usefulness in describing natural processes, and the following universal statement may be made about entropy change (this statement is also one way of writing the Second Law of Thermodynamics): *Changes in any isolated system always occur in a way such that the entropy of the system increases or, at the least, remains constant.* A simple illustration is provided by a system made up of two solids, in contact with each other but initially at different temperatures. Heat will flow from the hotter to the colder, and hence one part of the system gains more entropy than the other loses, since for a given dQ the $dS = + dQ/T$ gained by the cooler part will be calculated with a smaller T than will be the $dS = - dQ/T$ that is lost by the other (hotter) part. The over-all effect then is an increase in entropy. Once the two solids are at the same temperature, there will be no further change of entropy.

It has been found that the law of entropy increase may be applied to mechanical, chemical and electrical processes, as well as to purely thermal ones. Thus, a chemical reaction will not occur unless the sum of the entropies of the products is greater than the entropy of the reactants. In natural processes, however, the course of change is commonly such that the system is not always even close to equilibrium, as it must be if the calculation specified in the defining equation for ΔS is to be valid. It may be shown, however, that the entropy increase calculated through intermediate equilibrium states is no greater than the total entropy increase to be expected as a result of changes occurring through nonequilibrium states. Also, it is important that all associated processes be considered in applying the entropy-increase law. It is clear, for example, that when a kettle of boiling water is taken from the stove and allowed to cool it is losing entropy; there is, however, a compensating increase, and more, by the surroundings. Similar considerations apply to the cooling of a liquid in a mechanical refrigerator, except that now the entropy increase involved both in the operation and in the cooling of the mechanical refrigerating element must be taken into account. By specifying the entropy increase as occurring in a total isolated system, we take into account all the sub-systems, some perhaps with entropy decrease, which may be involved.

3   *Entropy and the Direction of Time*

It is now easy to see the association between entropy increase and the direction of time. A total physical system must change in a way such that the entropy of the system, when it is in a given state, is greater than its entropy in the state from which it has changed. (Or, in the limiting case of perfect equilibrium, the entropy remains the same; but we shall consider this situation to be one of no physical change.) But also, the succession of different physical states of a system is the past-to-future direction of time; for, as each new state appears, the preceding ones are separated from it by time intervals which in principle could be measured and recorded. We may say, then, that the progression of states of continually increasing entropy is the same as a past-to-future series in time. Sir Arthur Eddington vividly expressed the situation by saying that entropy increase gives us the direction of "time's arrow."[4] The course of change in the natural world, as time goes on, is characterized by a general increase in the entropy of the various systems making up the world. Only if there were no change would there be no continual entropy increase; but in this case, we would also not have time.[5]

We have pointed out that there are many specific physical changes which we have learned to associate with time's passing, but in the increase of entropy science has discerned a property that belongs in all cases to the continual alteration which natural thermodynamic systems, taken in their entirety, present to us. This insight may be reckoned as one of the great generalizations by which physical science has contributed to our understanding of the universe. Further, the association of entropy increase with time's arrow is in no sense circular; rather, it both tells us something about what will happen to natural systems in time, and about what the time order must be for a series of states of a system. Thus, we may often establish a time order among a set of events by use of the time-entropy association, free from any reference to clocks and magnitudes of time intervals from the present. In actual judgments of before-after we frequently do this on the basis of our experience (even though without any explicit knowledge of the law of entropy increase): we know, for example, that for iron in air the state of pure metal must have been before that of a rusted surface, or that the clothes will be dry after, not before, they have hung in the hot sun.

We expect to find that any physical process which we study in terms of thermodynamics will illustrate the entropy-time relation. And yet, we emphasize that we do not wish to *define* the direction of time in terms of entropy increase. It seems better to refer the order of time to general experience, as we have done; for we do not want to make the mistake of writing as if man had learned of time only after he had developed the science of thermodynamics and the entropy concept.[6] And also, although we confidently speak of entropy increase for systems as a law of thermodynamics, we do find that there are certain possibilities for exceptions when we take a point of view that begins with the atomic particles that make up a system.

4 *Statistical Limitations*

Thermodynamics is a science of relatively large-scale (macroscopic) systems. The basic observables, temperature and pressure, for example, result from average effects of many millions of particles (thus, we may speak of the energy, but not of the temperature, of a single molecule). Indeed, thermodynamic properties literally emerge only with the formation of aggregates of many particles. The sciences of statistical mechanics and kinetic theory, however, take as their starting points individual atomic particles and the laws of their motion, rather than macroscopic observed quantities. These sciences have given an account, somewhat parallel to that of thermodynamics, of systems made up of aggregates of particles. But because of the great number of particles, a treatment cannot be given of each one, and a statistical approach which culminates in describing the probable behavior of the aggregate is employed.

It has been found that the entropy of a system may be defined in a way that yields a result which is completely equivalent to that obtained from the thermal definition, but is in terms of the degree of order of the particles of the system. Specifically, entropy S of a system is given by

$$S = k \log W,$$

where k is a constant and W is the "thermodynamic probability." To illustrate the latter, we think of a gas in a stoppered flask. At thermodynamic equilibrium the temperature and the pressure are uniform for any macroscopic volume of the gas. Actually, however, there are many, many different microstate arrangements, or "complexions," of the individual gas molecules which would give the macrostate of equilibrium; that is, there are many different combinations of position and motion for the molecules that would give over-all equilibrium. The number of these possibilities, for a given macrostate, is the thermodynamic probability W. For another conceivable state of the gas, one in which, say, all of the molecules were in the upper half of the flask, with a vacuum in the lower half, there would again be many different molecular descriptions compatible with the macrostate, but not nearly so many as before, because half of the previous space would not be available to the molecules. The value of W would then be considerably less than for the equilibrium state.

In general, we say that the lower the value of W, the higher the degree of ordering. It is clearly a more ordered state for all molecules to be in one-half of a container than for them to be distributed throughout the container. The highest degree of order is one in which every molecule has a precise location, with no ordinary molecular motion, as in a perfect crystal at absolute zero. In this case $W = 1$, and we see, since $\log 1 = 0$, that $S = 0$.

The entropy-time association for isolated systems becomes, with our new interpretation of S, a statement about increase of disorder. Since entropy increases for successive states of a system, and since $S = k \log W$, we must find states with larger W as time goes on, and we have seen that larger W for a system corresponds to a greater amount of disorder. We can say, then, that the natural tendency of a complete system is to lose order. For a physical system the state of maximum entropy, or, of maximum disorder, is the equilibrium state, and it is towards this state that systems tend in their physical changes, *if* they are isolated and hence undisturbed by outside influences. The smoothing out of heat differences, the transpiration of chemical potentials into corrosion of pure materials, the distintegration and decay of abandoned buildings and cities: these may all be viewed as examples of the tendency towards increasing disorder in an isolated system with the passage of time.

The calculation of the behavior of a system in terms of statistical probability does show, however, that although the equilibrium state is overwhelmingly the most probable, as compared with any state substantially removed from equilibrium, there is nonetheless a nonvanishing probability for any possible nonequilibrium state. This means that, given an initial nonequilibrium state for a system, we can expect changes in the system to be towards the equilibrium state. But the possibility for fluctuations in an opposite direction remains. The chamber of gas, for example, with all the gas initially in one-half of it, will almost certainly move towards the higher entropy state of filling the entire chamber. But there is the rare possibility —far too rare for any man ever to expect to see it—that the gas molecules would move so as to occupy not one-half but only one-fourth of the flask. In this respect we may recall the famous statement of Eddington,[7] that the likelihood of a fluctuation whereby gas would of itself go into one-half of a containing vessel is considerably less than that an army of monkeys striking typewriter keys at random would type out all the several million books in the British Museum, with not an error from first to last.

The fluctuation which would give any pronounced entropy decrease to an isolated macrosystem is, because of its extremely slight probability, only of interest as a theoretical possibility. But as we go to microsystems containing just a few particles, the fluctuations from a most probable state become physically important. It may even be that fluctuations of this kind play an important role in biological phenomena, as sources of mutations and hence of species variation and evolution.[8] Still, one can justifiably argue that no violation of thermodynamic principle is involved in fluctuations in small-scale systems, because these are not properly thermodynamic systems. The Brownian motion of tiny particles immersed in a fluid, for example, is such that one of the particles may move with considerable velocity, and may be regarded as producing work through contact with a system all parts of which are at the same temperature. And yet, the Second Law of Thermodynamics expressly forbids the possibility of a system's doing work by extracting energy from another system at the same tempera-ture. The point to be made, however, is that a particle in Brownian motion is not a thermodynamic system, since it is not an aggregate of constituent entities for which we can determine over-all thermodynamic variables. A similar remark would apply, for example, if one chose just the unusual relatively low-velocity molecules that one finds in a gas at equilibrium, and asserted that they formed a low entropy system of "cold" gas, in violation of the constant entropy associated with an equilibrium state.

The possibility, however, of genuine thermodynamic entropy decrease for an isolated system—no matter how rare—does raise an objection to the definition of time's direction in terms of entropy. If a large, isolated system did by chance go through an entropy decrease as one state evolved from another, we would have to say that time "went backward" if our definition of time's arrow were basically in terms of entropy increase. But with an ultimate definition of the forward direction of time in terms of the actual occurrence of states, and measured time intervals from the pres-ent, we can readily accommodate the entropy decrease; it would become merely a rare anomaly in the physical processes of the natural world.

5   *Time-reversible Processes*

Systems that contain only a relatively small number of particles, with the possibility of strong fluctuations from equilibrium states, may be regarded as intermediate between true thermodynamic systems and systems with a very small number of entities, perhaps even only one, in which we do not use statistical methods at all. In these latter systems we individually de-scribe the behavior of every element of the system. An oscillating pendu-lum, an electromagnetic wave, or the planets of the solar system would be examples of such a system.

Some of the basic equations that describe individual entities are New-ton's Laws of Motion, in either their relativistic or nonrelativistic form, Maxwell's equations for electromagnetic fields, and Schrödinger's equation for the wave-amplitude of individual particles. These, and other of the basic physical equations, have the property of being "time-reversible"; that is, the behavior they describe may occur in one way, but with a reversal of initial conditions the same behavior could occur equally well in reverse. A perfectly elastic ball bouncing on a smooth surface provides an example. The motion through one cycle, from top and again to top of a bounce, is precisely the same, whether described as "forward" or "back-ward" in time; and indeed, one would get the same cyclic motion (except with a "difference of phase") by throwing the ball upward at a given point in the path as by throwing it downward at the same point with reversed velocity. The time-reversible equations describe behavior which is physically possible for either increasingly positive or increasingly negative values of the time variable. In contrast, the thermodynamic processes of a total system (subject to the fluctuation exception which we have noted) always proceed in only one way: towards states of greater entropy, or, with maintenance of a maximum entropy equilibrium state already achieved. The expansion of a gas to fill the flask, and the time-reversed process of contraction of the gas into one-half the container, are not physical processes equally agreeable to the natural way of things.

It is the case, however, that the motions and interactions of the individual entities making up a thermodynamic system will generally be in accordance with time-reversible equations. How can the collective motion be time-irreversible when individual motions are time-reversible? This difficult problem was essentially solved by the H-Theorem of Boltzmann, by the methods of statistical mechanics, and is discussed in the present volume by Prof. Watanabe.

It has been found useful in the study of certain individual particle physical processes to consider time as being reversed.[9] There is no difficulty about doing this, because no backward flow of time is required for the natural progressively changing processes by which our time is ordinarily defined. The time that is reversed is not the time of clocks, measuring the thermodynamic processes of ourselves and the surrounding world, but rather may be regarded as a "reversible-time" parameter in the equations for the processes, a parameter not correlated with ordinary time measure.

The fundamental question, however, as to whether or not the basic elementary particle processes are truly time-reversible is also currently under consideration in physics. There now is (1965) some tentative evidence that time-irreversibility is a property of certain elementary-particle processes; this result, if established, would indeed be of significance as an indication of temporal irreversibility among the most fundamental of natural processes. The theoretical and experimental approaches to the problem are reviewed in a paper by R. G. Sachs.[10]

6   Cosmology and Entropy Increase

A scientific generalization as universal as the assertion of a parallel increase of entropy and time invites speculation about its cosmic significance. If we follow time along the ω series we should eventually come to a state of maximum entropy; for, the ω series has no end, and hence, if it does represent future time, and if the assertion of entropy increase may be applied to the entire universe, the equilibrium state of maximum entropy should eventually be approached. Considerations of this sort have led to a prediction of an ultimately lifeless universe. A world is envisioned in which all energy or temperature differences have been removed: there are no longer hot stars radiating energy to cooler bodies, or highly structured organisms disparate from their environment. Instead, the universe has become a homogeneous mixture, at a dead level of temperature and uniformity.

This model for the future of the universe, aside from questions to be raised about the cosmical application of the entropy-increase law, presents a difficulty when extrapolation is made backwards in time, along the *ω series. Now at each earlier time the universe must have had a successively smaller entropy, and, since the *ω series is also an unending one, one must come to as small an entropy as desired by going far enough back in time. An alternative, that the universe was at some equilibrium state at an earlier time, is hardly tenable, since the universe is clearly not at present in an equilibrium state, and if it had earlier been in one it should have remained so, having then already reached its maximum entropy. It would appear, then, that even if one takes entropy increase as an adequate rule for the foretelling of an eventual future state, one must postulate some other mechanism for bringing the universe to an arbitrarily low entropy state at past times. In some way, as it has been phrased, the universe must have been wound up.

Ludwig Boltzmann, one of the great founders of statistical mechanics and the author of the statistical interpretation of entropy, discussed an interesting solution of the low-entropy initial-state problem.[11] He suggested (following, he wrote, "an idea of my old assistant, Dr. Schuetz") that the universe is generally at equilibrium, but that some considerable part of it may undergo a fluctuation which carries it far from equilibrium. The probability of such a fluctuation's being present would increase with the size of the universe, and in particular, we do find that our observable universe is such a nonequilibrium region—a region, however, that is, in accordance with the general behavior of systems, on its way to maximum entropy and hence to equilibrium.

If we felt confident that we understood all the major forces operating cosmologically, and if it seemed justified to regard the universe as an isolated thermodynamic system, then considerations of rare fluctuations from equilibrium, but always with eventual return to that state, might seem to give convincing insight into the natural history of our world. In fact, however, there is little support for either of these assumptions. With respect to considering the universe as a thermodynamic system, we can say first that the universe may well be infinite, as far as we know, and we have no strict meanings for thermodynamic principles applied to infinite systems. We are justified in studying some one part of the universe, a galaxy, for example, with the use of thermodynamic reasoning, but then only with the reservation that we are neglecting interaction effects with other parts of the universe. Also, it is the case that in general any large sample of our universe which we consider seems clearly to be in a non-equilibrium state. Thermodynamics is, however, essentially a science of equilibrium states; even though we may calculate what must be the least entropy change for a system that is going between two states *via* non-equilibrium intermediate states, we are not able by thermodynamics to say much about specific changes through such states. More detailed physical information than is utilized in thermodynamics is required for explication of the nonequilibrium behavior of a system, if one wishes to say more than that it should eventually come to equilibrium.

The fact that we do not know the factors which determine the over-all structure of the universe (or even what that structure is!) adds further uncertainty to the use of thermodynamics in cosmology. In particular, the observed expansion of our universe of galaxies is a factor which has not been satisfactorily brought into thermodynamics. An expanding universe would seem indeed to be one in which average densities of energy and entropy are varying, but we do not at present have any way of taking this variation into account on a cosmic scale. We do not know whether expansion is at the expense of, say, internal gravitational energy, or whether some otherwise unknown physical factor is involved. The net effect of expansion could well be to decrease the entropy in our observable universe, in opposition to tendencies that make for increase; and there may be other factors too that contribute to entropy change.[12]

In all, considering the indefiniteness of what is meant by the "system of our universe," our lack of complete knowledge about cosmic processes, and the nonequilibrium state of the observable universe, it seems fair to say that one should, at the least, regard any conclusions reached by cosmological extrapolation of the law of entropy increase with a high degree of caution. We must say this, even though in any closed system, in which only physically understood factors appear to be operating, we have not found any reason to doubt the validity of thermodynamics, or the usual increase of entropy of the system with time.

7 Biological Processes

The question of whether or not there is a violation of thermodynamic principles with the growth of a biological organism, or in the development of the various biological species, is one that has often been asked.[13] A parallel presumption, if an affirmative answer is favored, is that in the realm of living things some factors operate which are able to set aside or circumvent the laws of thermodynamics. The basis for raising the question is the high degree of complexity—of organization of matter—that does exist in the realm of living things. It has seemed to many that the formation of living entities must be a result of some teleological factor, or at the least, of some principle that supersedes, for biological systems, the principles of physics. How else, in the face of the physical prescription that the natural tendency of a system is towards increasing disorder, could the system of a living cell grow out of an inorganic world which lacks that kind of complex ordering?

It should be noted that, strictly speaking, there is a certain vagueness about questions of the entropy of living organisms, because entropy values have not been numerically calculated for them. We can, for example, calculate the entropy for a gas in various states, or for chemical compounds of high degrees of complexity, but biophysics has not yet brought us to the stage where we can state the entropy of a human being, or even of an amoeba. This lack of precise information about the entropy value of biological systems is not, however, a decisive point; for it can hardly be denied that we do have an increase in degree of order, and hence a decrease in entropy, as we go through various states of molecular complexity from a quantity of unmixed chemical elements to a plant or animal that is constituted from matter of that same elemental quantity and composition. Likewise, the degree of ordering of matter in, say, a primitive aquatic animal such as a jellyfish is clearly less than in one of the highly intelligent mammals.

We will first consider entropy increase with respect to the growth of an individual living system. The development, for example, from a single pair of cells to a mature animal is indeed an impressive formation of ordered complex molecular systems, with high ordering on the level of the molecules formed, and again on the level of cells formed from the molecules, and, yet again, on the level of organization of the cells. During the lifetime of the animal there is also the wondrous learning and thought, and accumulation of memories, so especially well developed in our human species, which must also require ordering of molecules within the organism. We cannot but see a natural decrease of entropy as we consider the developing biological entity. The mistake which we must not make, however, is to view a living plant or animal alone, divorced from its surroundings. The entropy-increase law, we recall, holds for an isolated system, and a living organism is always in close physical interaction with its surroundings, from which it extracts the chemical substance and energy necessary for growth and life.

The food which animals eat is generally of higher degree of ordering—of lower entropy—than are the waste products they excrete. As a consequence, in the degradation of food there is an entropy gain which compensates for entropy decrease associated with growth and learning. Hence, the net entropy of the system formed by the animal and those parts of its environment with which it interacts may be fairly taken as increasing as a result of the occurrence of the life processes of the animal. We cannot quote exact measurements in support of this conclusion, because, as already noted, we do not have an exact thermodynamics of living systems. It may be pointed out too that the study of these systems, in which processes operate in a complex network making both for stability and for change, requires more than the traditional thermodynamics of equilibrium systems. Many see promise for understanding of biological processes in the field known as irreversible thermodynamics, in which nonequilibrium states are explicitly investigated.

At its base, of course, the hierarchy of terrestrial living things rests on inorganic nutrients, and somewhere organisms must directly convert these nutrients into organized living tissues. This transformation occurs in photosynthesis, but here too the net effect is plausibly regarded as one with associated entropy increase. Suppose, for example, that a single photosynthetic microorganism be introduced into a flask of water containing carbon dioxide and certain dissolved inorganic materials. If the water is then exposed to sunlight the microorganism will soon multiply many times over and a culture containing millions of the organisms may be produced. There has, obviously, been a great increase in the degree of order among the molecular constituents of the flask, and with a corresponding entropy decrease. But, the solar radiation which effected the photosynthesis resulted ultimately from heat released in nuclear reactions in the sun, and these reactions involved an associated decrease in order among elementary particles, with increase in entropy. There is no reason in physics to think that the over-all balance would not give a net increase.[14]

The evolutionary development of the various species may likewise be regarded as a process that brings increasing order in the biological realm, at the expense of increasing disorder in the surrounding inorganic world.

The living species that have developed are evidence that the organized structures which the chemical atoms can form, and the energy transformations which are possible, give possibilities for living systems of almost incredible variation in adaptation and achievement. Also, variation in specific physical factors may have had an important role in the evolutionary process. Thus, it may be that in the early days of our planet the general temperature level was too high for the stable existence of molecules of the complexity required for living organisms of the kind we know. In our era, temperatures over a large part of the earth clearly do allow development of complex living organisms, and there is, we trust, sufficient richness and complexity in nature that present-day man is in not at all the last or best stage in the formation of organisms who understand nature and seek its potentialities for good.

But although natural terrestrial conditions today do seem to favor the development of increasingly highly organized systems, there may have been periods, on a cosmic time scale, when entropy decrease scarcely could occur in our part of the universe. Our galaxy is now sufficiently tenuous, and isolated from others, that our sun loses radiation into the relatively dark skies which surround the solar system. For the earth, there is daytime heating, with the possibility for activation of various reactions, but also night-time cooling; loss of heat means entropy decrease, and associated possibility for stabilization of molecular structures.[15] If heat such as that from the sun came continuously upon all parts of the earth, we might expect that there would be no opportunity, just because of the unfavorable thermal conditions, for natural elaboration of complex molecular systems. The extrapolation of the expansion of the galactic universe, backwards in time, leads us to a situation in which the average stellar density in the universe would be much greater than it is today. Under these conditions the radiation coming to the earth, or any planet, might be so great as to preclude the complexity of atomic systems required for biological development. The skies, literally, would everywhere be hot, and pouring in radiation, so that molecular systems on the earth could not in any manner settle into the organized, lower entropy systems that are within the natural possibilities for atomic configurations. In contrast, the heavens today provide a "heat sink" into which the sun's radiation may ultimately be lost, with consequent over-all entropy loss and associated possibilities for high degrees of order in natural systems. We seem now to have a favorable intermediate situation; at the other extreme, of not even a sun as a source of energy, there would of course be no energy for activation and maintenance of molecular change.

It is apparent from our discussion that acceptance of an accord between the entropy-increase principle and biological phenomena rests to a considerable degree more on a confidence in the laws of physics than on quantitative data in the biological sciences. One can add, however, that there is no established biological violation of the Second Law of Thermodynamics, and that one can give plausible justification for its validity when total biological systems, comprising living organisms together with their environments, are considered. We do not wish, in making this affirmation of the principles of thermodynamics, to state that biological phenomena can be exhaustively explained in terms of purely physical-chemical concepts and laws. It seems to be reasonable to expect that biological studies will always contain principles which would not be obtained from physical studies alone. In chemistry, for example, we learned laws of stereoisomerism which were not apparent in atomic physics, and yet, we also expect to find the structure of complex organic molecules to conform with the basic ideas of atomic physics. In like manner, failing evidence to the contrary, we have reason to assume that a principle which is so fundamental in the physical world as the Second Law of Thermodynamics must also be valid for biological processes.

8   Thermodynamics and Relativity Theory

The application of the Theory of Relativity to thermodynamic systems has received attention from a number of theoretical physicists,[16] and equations have been obtained for transformation properties of thermodynamic variables under relative motion. It cannot be said, however, that relativistic thermodynamics is a satisfactorily completed science, for the results of various workers are not altogether in agreement, and as yet there has been no opportunity for direct test in observation. In any event, we are not concerned in this essay with relativistic thermodynamics as such. One aspect of relativity theory and thermodynamics, however, is of great relevance to the problem of time. The theory of relativity requires that motion between two systems give rise to relative difference in time rates of processes in the two systems, and there is today perhaps no question of basic physical theory that is of wider popular interest than that of whether or not a biological system, moving at speeds close to that of light, actually will age at a slower rate then will a similar system which is at rest to an observer. This problem may be construed as one which asks about relativistic time rate of change in a thermodynamic system, and it is in this form that we shall consider it.

The Lorentz time transformation equation of Special Relativity Theory,

$$\Delta t = \frac{\Delta t' + (v/c^2)\, \Delta x'}{\sqrt{1 - v^2/c^2}} \quad,$$

states the relation between time intervals Δt and $\Delta t'$ in two systems, K and K' respectively, which are in relative motion with respect to each other. In the equation, the quantity $\Delta x'$ is the spatial separation in K' of the process which is observed from K, c is the speed of light, and v is the relative velocity of K', taken to be along the positive x-axis of K. In the case of a process which may be regarded as being at rest at a point in K', $\Delta x' = 0$ and the equation reduces to

$$\Delta t = \frac{\Delta t'}{\sqrt{1 - v^2/c^2}} \quad.$$

The time-rate change described by the Lorentz transformation equations has been confirmed observationally for many different physical processes, and "time-dilatation" is now accepted as an integral part of physics.[17] The increase in mean lifetime of mu-mesons as a result of relative motion, the relativistic Doppler shift for photons, and the existence of matter waves in accordance with the de Broglie equation are three physical phenomena which give direct evidence that the time rate of a process changes with motion relative to an observer, in the manner prescribed by the Special Relativity Theory. It is still the case, however, that there is no empirical confirmation of relativistic time change in large-scale systems, as literal clocks or biological organisms. In view of the unbroken chain of successes of Relativity Theory, the presumption of most physicists is that such time changes will be observed when the necessary relative high speeds have been achieved in experimental situations for clocks or living organisms. We wish, however, to present an alternative possibility: that the relative motion will not in itself have an effect on the time rates of internal processes of systems, and specifically, not on the processes of living things. We shall give some arguments for this alternative, but the reader should be warned that the point of view which is to be suggested is a speculative one.

9   *The Interaction Interpretation*

The relativistic changes in physical quantities which have been observed are generally on the level of atomic or elementary-particle processes. Further, the changes are observed by means of an interaction between a moving physical entity and the observational apparatus. Thus, the mass m

of a moving particle is increased, over that of its rest mass m_0, in accordance with the relation $m = m_0/\sqrt{1 - v^2/c^2}$; the increase may be detected through changes in the trajectory of the particle in an electric field of the observer's coordinate system, or by measurement of momentum delivered by the particle in a collision with another particle approximately at rest to the observer. Or again, photons showing the relativistic Doppler effect interact directly with an observer's photographic film, or with an absorber in a Mössbauer-effect apparatus. Even in the observation of the dilatation of meson lifetimes, the mesons are in physical interaction with the observer through their decay products, since the decay is detected by means of interactions of these decay particles with electronic counters. Further, not only is there interaction with the physical fields or apparatus of a given coordinate system, but there is physical interaction with only this one system. An electron cannot simultaneously have two different masses, with respect to the physical fields of two different coordinate systems, for it can interact only with a single system at a time. Or, a photon cannot simultaneously display two different frequencies, because it interacts only with one absorber or reflector at a given interaction event.

In contrast, a large-scale clock to which relativistic time-rate changes are hypothetically applied is a public instrument which can be read, at a given instant of the clock, by observers in an indefinite number of relatively moving coordinate systems; a burst of photons could go from the clock hands to many different observers. This fact in itself causes no insuperable difficulty. The time-rate changes of relativity are regarded as arising from a set of relatively different time rates, distributed among all relatively moving systems, and the observers in each system would regard the readings of the clock emitting photons in terms of their own clocks; the comparative readings should be in accord with the Lorentz time-transformation equation as applied between each observer's clocks and the photon-emitting clock. But the public nature of the clock also shows us that there is no physical interaction between the internal clock processes themselves and the observing mechanisms in other, relatively moving coordinate systems. The interaction by which the clock is read, for example, through a light beam, is one that is peripheral to the internal mechanisms which form the essential clock (timekeeping) process.

We suggest that it is only for physical processes which do involve an essential interaction between themselves and the observer that the observer should expect to find a relativistic time transformation. Otherwise, it is to be expected that time rates of the processes of a relatively moving system will show no change as a result of relative motion with respect to an observer.

A simple kind of gas clock may be used as an illustration of the distinction which we propose. Suppose such a clock to be made of two chambers connected by a tube through which gas can slowly pass. There is a pressure gauge on one of the chambers, and at time zero one chamber is filled with gas to a pressure P_0 and the other chamber is evacuated. The decrease in pressure in one chamber, or the increase in the other, may be calibrated against a standard clock, so that pressure readings on the gauge will serve as a measure of clock time. The flow of gas from one chamber to the other is, we suggest, an internal process which would be independent of any state of uniform motion with respect to an observer, and if one of the gas clocks moved with respect to a set of similar clocks kept stationary to the observer it would show the same time reading as any of the stationary set. In contrast, light from the pressure gauge of the moving clock would be in interaction with the observer, and would show a relativistic Doppler shift if the clock moved at a sufficiently high speed.

We can gain support for our proposal from one of the generally accepted results of relativistic thermodynamics, to wit, that the entropy S of a thermodynamic system does not change as a result of uniform relative motion.[18] This means that a system should have the same S value to a relatively moving as to a rest observer. The hypothetical gas clock, for example, would have a calculable entropy at each time (pressure) reading of the clock, and time t so determined could be expressed as a function $t(S)$. Now, if S for a system is independent of the system's state of relative motion, then so is t, and we are led to the conclusion that time too is unchanged as a result of motion. Thus, suppose as before that a group of gas clocks, all similar in construction and calibration, are spaced along a straight line. Yet another clock, similar to the others, is moved along the line of stationary clocks and time readings are compared. The principle of entropy invariance tells us that $dS/dv = 0$; *i.e.*, that the entropy of a clock is independent of its relative velocity. Hence, the moving clock should have the same entropy as any contiguous rest clock with which it is compared, and therefore also the same time reading. One might object that the function $t(S)$ could vary with the state of relative motion of a clock. Such a variation, however, would violate the basic relativity postulate that one cannot detect any state of uniform motion of a system by means of measurements made entirely within the system.

In place of the gas clock we might have, of course, any kind of thermodynamic system, including a living plant or animal, and, on the interaction hypothesis which we proposed, the time rate of the internal processes of the system would be the same to a relatively moving observer as to one at rest with respect to the system. The concept of entropy was introduced into thermodynamics by Rudolf Clausius, and I have suggested that complex-system internal processes which would not undergo relativistic time change with motion be called *Clausius processes*. In contrast, the processes involving physical interaction with an observer, for which we would expect relativistic transformations, could be called *Lorentz processes*.[19]

The key point about whether a physical process be a Clausius or Lorentz one, in the sense suggested, is that of interaction with the observer. In Relativity Theory the location or duration of an event or process in space-time may be described by a so-called four-vector. The fourth component of this vector is the time coordinate. The momentum and energy of the process constitute a concomitant four-vector whose fourth component is the energy. We should find, then, that *whenever an interaction involves a relativistic energy change there should also be a time transformation* (or *vice versa*). In the observed instances of relativistic time changes, there is interaction such that the relativistic energy of the observed entity is brought to the observer's apparatus. Thus, mu-meson decay products bring a relativistically increased energy to the counters with which they interact. It is, however, difficult to identify the energy component, parallel to the time, in the hypothetical clock or biological-organism situations for which the conventional relativistic time change is required. The presence, or absence, of energy interaction with a process may serve in general as a specific criterion for whether or not there is physical interaction with the process, and whether it is therefore a Lorentz process for which we can expect to find relativistic time change, or a Clausius process which is time-invariant.

10   *The Velocity of Light*

The postulate that light has the same speed to all observers, regardless of what their uniform motion may be with respect to the light source, is basic to the Special Theory of Relativity, and there is no evidence against the postulate. The Lorentz transformations of clock times and rod lengths may be regarded as the changes which have the consequence that all observers will measure the same light speed. In our proposed interaction interpretation of the Lorentz transformations, the velocity of light will still have the same measured value for all observers, but now as a consequence of relativistic changes in electromagnetic waves or photons themselves. The waves interacting with an observer will undergo wave-length and frequency changes (Doppler shifts) as prescribed by the Lorentz equations, and hence will always move with a constant speed c with respect to an observer. They will, in other words, show Doppler shifts, and a constant speed c, not because all clock processes at rest in one system are Lorentz-transformed with respect to all the processes at rest in another system, but because the waves are themselves transformed in a physical interaction.

There is a significant difference, in the ways a light signal must be regarded, between the conventional clock-rate change and the interaction point of view. With the clock-rate change view, a spreading light signal is a unique phenomenon, and relatively moving observers look at it with different sets of clock readings (and hence, in spite of their relative motion are able to measure the same light speed). On the interaction view, observers have common clock times (time-invariance of Clausius processes) but the light front will move differently with respect to the relatively moving observers. Photons from the source will interact only at a relative speed c, and hence the electromagnetic waves must be regarded as moving differently by different observers. Potentially, a given photon can present any frequency or wave length compatible with the Doppler equation, and the effect of the Lorentz transformations on the photons is such that the observed values are those which give the constant value c, in a given interaction. There is no problem of paradox with different observers, because each photon interacts only with one observer in a given interaction event.

11   *Relational and Constitutive Time*

The Theory of Relativity has taught us that the time rates of processes are not absolute, but vary with relative motion; and, as we have noted, there are many kinds of observations, involving physically interacting systems, which confirm the relativistic modification of the "absolute time" that Isaac Newton used in physics. There is, however, a still more basic question about time which is raised by Relativity Theory, and which can be answered for certain only when we know in what manner the Lorentz time transformation does apply between relatively moving systems. With the conventional clock-change view, relative motion in itself brings about a change in all time rates between two systems. Physical interaction is not necessarily involved; the astronaut, for example, moving at a high relative speed, would age less rapidly than his fellows on the earth, just because of his relative motion through otherwise empty space. On this view, time is *constitutive* of natural processes, in a manner such that it regulates the rate or flow of these processes.[20] And, time is then intimately related to empty space, independent of events, with the result that velocity, which is time rate of change of spatial position, can change time in the manner prescribed by the Lorentz equations of Relativity Theory.

In contrast, if the distinction between Lorentz and Clausius processes should be a valid one, it could be maintained that time is a concept derived from events, from the processes of the world. On this relational view of time,[21] we do not regard time as physically prior to events, and regulative of them. If time is relational in this sense of arising as a consequence of relations between events, then events do not occur as they do because there is an independent regulative time in which they are embedded. Instead, the changes of nature occur in accordance with the constitution and characteristics of the physical world, and time is a concept, or universal property, that we construct through our experience in that world.

There can hardly be a more fundamental question about time than that of whether it is constitutive or relational, and this question therefore adds to the strong physical interest which we have in knowing how universally the relativistic time transformations may be applied in nature. The present-day techniques of experimental physics are such that we can hope within a few years to have some definite observational evidence. An atomic clock,[22] for example, if installed in an artificial satellite would be a time-keeping process which is not in essential physical interaction with ground observers. Such a clock is sufficiently accurate that the behavior of its time rate when in rapid relative motion should tell us whether or not there is any validity in something like the distinction between Clausius and Lorentz processes.

If experimental evidence is found to favor the interaction point of view, the problem of time will be indicated as one which lies within the domain of ordinary scientific theory and observation. Even relativistic time changes, if they are actually confined to situations in which there is interaction, can be studied in association with other physical transformations (such as that of energy). This attitude toward time, and relativistic changes, is one that is congenial to the methods and outlook of science, for it does not call for presuppositions that go beyond the ken of direct scientific investigation. In the words of Niels Bohr, "Certainly, a leading idea in physical theory has been to seek the ultimate cause of all natural phenomena in the relative displacements of material bodies; . . ."[23]

And yet, the interaction view must be regarded as heretical with respect to accepted Relativity Theory. Many times in this century, surprising and even widely disbelieved consequences of Relativity Theory have been found to be true descriptions of nature. It may well be that empirical evidence will tell us that the relativistic time transformation does apply to *all* processes of a relatively moving system, independently of any state of interaction. In this case, the understanding of time will, I believe, be far more difficult than if time were circumscribed within physics, and biology and psychology. The road will be open for new metaphysical speculation, and nature will seem deeper, and more mysterious, than if time were adequately described as a measure but not the maker of her events.

Note Relating to a Paradox of the Temporal Order

In the physical sciences future and past times are often represented by the infinite series of natural numbers. In the preceding article, in an attempt to fill a logical gap which thermodynamicists tend to leave open, Prof. Schlegel has related such representations to that type of time which is usually thought to correlate with entropy change. Unfortunately, the appealing simplicity and promising completeness of the numerical-linear representation of time, separately or in connection with ideas of entropy change, tends to decrease as soon as we attempt to assign some of the experiential properties of time to a series of numbers. The difficulty may be illustrated by an intriguing paradox that has been discussed during the last fifteen years in various journals.[1] The problem can be stated in the following form.

There are ten closed boxes labeled from 1 to 10, and we are told to open them one at a time, in order of their numbering. We are also informed by a reliable source that one of the boxes contains an unexpected egg. By "unexpected," or "surprising" is meant that we cannot deduce by argument in which box the egg is until the box has actually been opened.

Before beginning the experiment we may argue as follows. The egg cannot be in the tenth box for were I to find the first nine empty, an egg in the tenth box would not be an unexpected one. But, it cannot be in the ninth box either, for were I to find the first eight empty, and were I to remember that it cannot be in the tenth, an egg in the ninth would not be unexpected either. And so forth, down to and including the first box.

Having so argued, we may begin our experiment, and as Martin Gardner has it, we find the egg in the fifth box. The paradox is said to reside in that the argument as well as the experiment are straightforward and unassailable, yet they lead to opposite conclusions.

A number of solutions have been suggested, centering mainly on linguistic analysis and on the credibility of the "reliable source." That by A. Lyon includes an interesting discussion on the subject of prediction and free will. What does not seem to have been pointed out is that the paradox is a powerful illustration of the antinomies of time.

The fundamental difference between the theoretical argument and the experiment is the sequence of assumed events. In the logical presentation we posit the last experimental condition first (the opening of the tenth box), followed by the next-to-the-last experimental condition, and so forth. In an experiment actually performed we could not have reached the first logical test without having opened earlier box 1, then boxes $1 + 2$, then $1 + 2 + 3$ and so on. Clearly, the condition of having eight-boxes-open can follow the condition of nine-boxes-open only if we close the lid of the ninth (empty) box and pretend to regress to a state of lesser information. The paradox is produced when arguments involving temporal order are mixed with arguments disregarding time. A world where the logical test holds must be one of "being," for it must accommodate temporal and reversed temporal orders on equal footing. In that world time is unreal, and the unexpected cannot happen. Contrariwise, a world where the unexpected does happen, is one of "becoming," and time by this test at least is not an illusion. Philosophically, therefore, the paradox harks back to the antinomical views of Heraclitus and Parmenides.

We may replace the boxes by days (or by the series of natural numbers) and the egg by a specific event.[2] Or, we may even decrease the time intervals to arbitrarily small lengths and conclude that Zeno's arrow cannot hit the target at an unexpected instant before tomorrow morning, consequently, the arrow cannot move before that date. In the spirit of the logical argument, an event promised to be unexpected could not take place within a delimited period of time. However, if time has no termination, the paradox cannot be applied, for to argue backwards we would have to first presume to have reached the penultimate condition, a contradiction in terms.

The generalized form of this problem can be called the Judgment Day Paradox. According to Judaeo-Christian and Islamic teachings, the Day of Last Judgment will take place at a definite future date, just before the end of the world, but exactly when, we do not know. If time is assumed to be unreal we may apply the logical test to this proposition and conclude that the Day of Last Judgment cannot come. In a world of becoming, however, the idea of an end of time cannot, on the basis of unexpectedness, be ruled out.

What we have asked ourselves about the unexpected egg bears strong similarity to what Banquo in *Macbeth* has asked of the three "imperfect speakers," the three Fates:

> "If you can look into the seeds of time
> And say which grain will grow and which will not,
> Speak then to me. . ."

Both the paradox and the literary quotation draw attention to a fundamental kinship between our ideas of probability (the elements of beliefs and expectations), our techniques of observation (the examination of the boxes or, in *Macbeth*, listening to messengers), the meaning of temporal order (represented by the numbering of the boxes or of consecutive scenes) and the irreversible nature of learning. The following essay analyzes the intricate relations between these entities.

J.T.F.

Time and the
Probabilistic View of the World

SATOSI WATANABE

With certain qualifications and providing proper interpretations, we may state that the salient features of the temporal development of our experience as compared with its spatial extension is its asymmetry or, more precisely, its anisotropy. In the present article, it is intended to shed a few pencils of somewhat novel sidelight on the proposition that this temporal anisotropy is inextricably related to our probabilistic view of the world. It is true that there are indications that microscopic physical phenomena show a certain degree of chirality[1] (violation of space symmetry), but as we shall show later the true nature of the one-way-ness of temporal development of macroscopic physical phenomena is irretrodictability; this is a more stringent notion than irreversibility, which is in a sense a temporal counterpart of chirality.

In a broad sense, the temporal asymmetry of our experience has actually two major aspects. On the one hand, there is the well-known Second Law of Thermodynamics which states that the entropy of an adiabatically isolated system cannot decrease with time. Various versions of the H-Theorem are attempts to explain this law from the atomistic point of view.[2] But, on the other hand, we are familiar with various aspects of life phenomena which in a certain subtle sense seem to go against this entropy-increasing trend of the physical world.[3] A very limited area of such "entropy-decreasing" phenomena was given a mathematical formulation which I called the inverse H-Theorem.[4] Of course, the entropy in the inverse H-Theorem is not a thermodynamical entropy and is defined in terms of behavioral and cognitive variables. Analysis shows interestingly that the asymmetry of the H-Theorem and that of the cognitive version of the inverse H-Theorem originate both from the same source, namely, the inevitable asymmetric interpretation of a certain formally symmetrical equation in the theory of probability. For this reason a parallel exposition of H-Theorem and inverse H-Theorem would be most enlightening, but, limitation of space in the present article forces us to focus our attention only to the H-Theorem, and this mainly in the framework of classical physics.

We may anticipate our conclusions and state that the temporal development of the outer world as it appears to us has a "privileged" direction which can be "fixed" only with reference to our inner world and that the relation is not merely accidental but necessary. The "necessary relation" between the physical time and the psycho-biological time is a very subtle one requiring a careful analysis; the present article is only the first step in this direction.

1   *Example of the H-Theorem*

In this section, we introduce a very simple model of entropy-increasing phenomena. The entropy and other major quantities are not defined exactly as they would be in thermodynamics. Yet, the model contains all the important factors basic to the H-Theorem.

Let us consider a certain number of balls rolling freely on a rectangular tray. A wall in the middle divides the tray into two equal rectangular areas, A and B, with a small opening or gate G. (See Fig. 1.) As a simple example, let us assume that the motion of balls is not free of friction, but the entire tray is constantly shaken randomly so that the balls keep a certain constant distribution of kinetic energy. Each ball has a position and a velocity at each instant, which constitute its *microscopic state*. But we use also a coarser description; we say whether the ball is in the left half A or in the right half B. This is the *macroscopic state* of the ball. The macroscopic state of the system of balls in this case may be characterized by two numbers, M and N, which are respectively the number of balls in A and the number in B. A variable which characterizes a macro-state is a *macroscopic* variable. In the case of each ball in the present example, we have a single variable which takes one of two values, A or B. In the case of the system of balls here, we have two variables M and N (under the constraint, however, that the sum of the two is a constant). It is important to note that there are numerous possible ways of introducing a macro-state. In a real physical system, the *thermodynamical state* is a special kind of macroscopic state, which makes it possible to establish a set of elegant equations among the corresponding macroscopic variables, which are *thermodynamical* variables. One of the thermo-dynamical variables is entropy. In the present case, we may take as a simple facsimile of that quantity

$$S = -\frac{M}{L} \log \frac{M}{L} - \frac{N}{L} \log \frac{N}{L} \tag{1.1}$$

where $L = M + N$. This S, so to speak, measures the evenness of the distribution of balls in A and B, since it is a monotonically decreasing function of $|M - N|$. It takes its maximum value $\log 2$ when $M = N$ and its minimum 0 when $M = L$, or $N = L$.

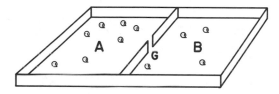

Figure 1. The entire tray is shaken randomly in such a way that the average kinetic energy of the balls remains constant. If the escape-gate G is small, a ball will go through almost all possible positions and all possible directions of flight within one compartment before escaping to the other compartment.

The path of a single ball is unpredictable because it depends on the shaking motion of the tray. But, we may speak of the probability p of a ball in A being found in B after, say, τ seconds. We assume this proba-bility to be independent of the number of balls present in A and B and to be symmetrical with respect to A and B. It is quite possible that the probability of direct flight to the other compartment depends appreciably on the microscopic state of the ball. Some position with some special direction of motion will make it easy to escape directly to the other compartment within a short time. Yet, we claim that if the gate is sufficiently narrow and if τ is sufficiently large, we can use the same p for any ball in A. The justification is that within a time period smaller than τ, any ball in A, before passing to B, runs through all kinds of positions and velocities in A so that the probability in question is the same as the average of such probabilities over all possible microscopic states in A. This is a time average as far as a single ball is concerned, but it is the same as the average taken at one instant of many balls of differ-ent velocities "evenly" distributed in A. Thus, each ball can be considered as a "fair representative" of the balls in general which are in A.

TIME AND MATTER

Now we decide to observe the macroscopic states of the balls every τ seconds. Let us write P_n for the probability of finding a particular ball in A at $n\tau$ seconds, and Q_n for the probability of the ball in B at $n\tau$ seconds. Now we ask what is the probability P_{n+1} of finding the ball in A at $(n+1)\tau$ seconds and what is the probability Q_{n+1} of finding the ball in B at $(n+1)\tau$ seconds. The first event is possible if the ball was in A at $n\tau$ and stayed there or it was in B and moved to A in the last τ seconds. Hence

$$P_{n+1} = P_n(1-p) + Q_n p. \tag{1.2}$$

Similarly

$$Q_{n+1} = P_n p + Q_n(1-p). \tag{1.3}$$

From these two follows

$$P_{n+1} - Q_{n+1} = (1-2p)(P_n - Q_n), \tag{1.4}$$

which means that except when $p=1$ or $p=0$, the difference $|P_n - Q_n|$ gradually decreases with n. Unless $p=1/2$, the difference will never become zero at finite n except in the case where it is zero from the beginning. But, on the other hand, irrespective of the value of p, the P_n and Q_n both will tend to their equilibrium values $1/2$ as $n \to \infty$.

Since we assume that the balls are moving independently of one another, we can use the result (1.4) to represent the average or expected behavior of a number L of similar balls on the tray. If $M_n = LP_n$ and $N_n = LQ_n$ are respectively the number of balls in A and the number of balls in B at instant $n\tau$, then LP_{n+1} and LQ_{n+1} will be the expected numbers at the next instant $(n+1)\tau$. This means that the difference $|M - N|$ gradually decreases, which in terms of entropy means

$$\overline{S}^{(n+1)} \geqq \overline{S}^{(n)} \tag{1.5}$$

where $\overline{S}^{(n)}$ is obtained from (1.1) in which LP_n and LQ_n are substituted for M and N. I put the bar on S to indicate the average behavior we are talking about. The inequality (1.5) is the H-Theorem. The intuitive meaning is clear. If we start with many more balls in A than in B, say, the shaking process will tend to even out the distribution until finally the difference "on the average" disappears.

Here we insert as Fig. 2 an example of the behavior of the *average* entropy as described by (1.1) with the average ball numbers given by (1.4), for the case of $p=0.02$, and $P_0=1$ and $Q_0=0$. It shows a

530

monotonic, one-way increase of entropy. Once having reached the maximum value, the average entropy does not again decrease.

The notion of average used here should be understood primarily in the following sense. We prepare many similar trays (an ensemble of trays) with the fixed initial distribution of balls in A and B. After some time, the trays at the same time will show slightly different numbers of balls in compartments, some more and some less than the expected values given by LP_n and LQ_n. But if we take the average in the ensemble, it will turn out to be given by LP_n and LQ_n. Apart from this primary interpreta-

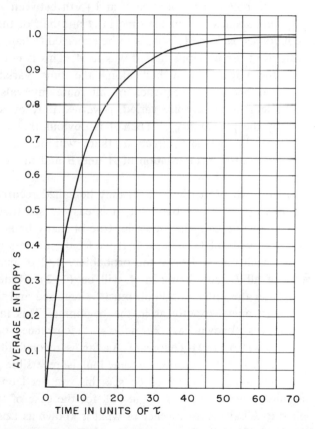

Figure 2. The average entropy for $p = 0.02$ as a function of time. At the initial instant, the probability is unity for one compartment ($P_0 = 1$) and zero for the other compartment ($Q_0 = 0$). An individual case (non-average treatment) shows also the same smooth curve if the number (L) of the balls is extremely large. The logarithm used in computing S for Figures 2 to 5 is to the base 2.

tion, we may also note that if L is very large, the fluctuation of M and N in each individual tray from the average behavior described by LP_n and LQ_n will be percentage-wise smaller.

The unrecoverability of a mixing process is a familiar fact of life, and the one-way increase of entropy (1.5) seems to represent this fact nicely. But we should ask whether this phenomenon is due to the physical nature of the system or to our particular way of performing the experiment and describing it. If we place ourselves at the level of each individual ball there can be no process of aging or any kind of one-way symptom involved. Indeed, each ball is just going back and forth between A and B, and, in the long run, spends half of the time in A and half of the time in B. Every time it comes back to A, what is to be expected to happen thereafter is the same as at the previous time it was in A. This is not all. Suppose we trace the voyage of a ball along the time variable. The time axis will then become a sequence of alternate intervals of stay in A and B. The lengths of stay are varied, governed only by a random law determined by the probability p. Then it is obvious that if we look at this sequence in the opposite direction, there will be no difference whatsoever. It will be also an alternation of A and B and the lengths are governed by the same random law.

I have mentioned two properties, which may be called recurrence and reversibility, with respect to one ball. But, it is easy to see that if these properties are true of one ball, they must be true of many balls. Suppose for instance we start with all L balls in A. If we let the process go on for a very long time, each ball will have the probability $1/2$ of being in A. The probability of all L balls being in A will be $(1/2)^L$. Since this is finite, sooner or later the initial condition must be restored and thereafter everything will start anew, exactly as in the beginning. This fact is the essence of what is known as Zermello's "recurrence objection" (*Wiederkehreinwand*) to the H-Theorem.[5] As far as reversibility is concerned, it is clear that if the motion of one ball is reversible then the motion of two balls is also reversible. To see this, we need only repeat the kind of argument we used for a single ball for the case of two balls. This fact applied to L balls is the essence of what is known as Loschmidt's "reversal objection" (*Umkehreinwand*) to the H-Theorem.[6]

Do these two objections really constitute genuine contradictions to the H-Theorem as expressed by (1.5)? The answer is no. We have to remember that the theorem (1.5) represents the average behavior in an ensemble of infinitely many trays, while the objections concern an individual tray with a finite number of balls. First, the recurrence property.

Let us start with all L balls in A on each tray in the ensemble. After a sufficiently long time, a great majority of trays will have nearly equal numbers of balls in A and B. It is true that occasionally one side gets considerably more balls than the other, and it is even true that all the balls concentrate on one side once in a long while. But, these events happen at different instants from one tray to another. And further, in some trays A may have more balls, and in some other trays B will have more balls. Hence, as far as the average of balls in each compartment at a given instant sufficiently long after the initial instant is concerned, it must be about $1/2$ of L. The H-Theorem claims just this and no more.

As regards the reversal objection, we have to note that the macroscopic, statistical behavior of an ensemble as expressed by (1.5) is of course irreversible. This does not contradict the basic reversibility, because the ensemble we are considering in connection with (1.5) is defined by an initial condition (that the entropy at the initial instant is lower than the maximum) and this condition is already asymmetrical. However, altering the rule of the game it is possible to exhibit the underlying reversibility on the macroscopic level too. The only thing is that the rule of the game thus altered does not correspond well to a laboratory experiment we perform. Suppose we take an extremely large number of trays which have been shaking for a very long time. Specify particular numbers M and N, which also determines a specific entropy value by (1.1). At a particular instant, say $t = 0$, we pick up only those trays which satisfy these conditions, and consider these selected trays as constituting the ensemble. Then, the collective behavior for the positive direction of t will obey the law (1.5). But, at the same time, if we trace back the past history of these trays, the collective behavior in the negative direction of t will be also in the direction of increasing entropy. This is precisely what reversibility entails. M. Yanase has presented a good description of this situation.[7]

The point is that this ensemble is a fair representation of the conditions specified by M and N at $t = 0$ and this is what we need for prediction of the future. But, for the conjecture of the past (retrodiction), a fair representation of the condition M and N at $t = 0$ is not the kind of ensemble we need. The reason is that we often have various additional information about the system beyond the specification M and N at $t = 0$. For instance, the particular instant ($t = 0$) at which M and N take the prescribed values is already known to be in the middle of an experiment. Then, this simple information will be sufficient to make us suspect that the experiment had started with a state that had a still smaller value of entropy. This means that the essence of temporal asymmetry resides in that we cannot apply the same procedure in guessing the past as in guessing the future, and the unique retrodiction is impossible because it depends on different additional conditions. This is what I call irretrodictability—and this is one of the main theses of the present article.

Agreeing that for the purpose of retrodiction we cannot just impose a condition of M and N (or S) at $t = 0$ and trace back what happened on the curve before $t = 0$, we may still have the curiosity to ask ourselves the following question. Were we to trace the entropy curve S of an individual system which is constantly fluctuating up and down as our model consideration seems to suggest, then will the trend (in any one direction of t) of S at any given instant not be just as probably upward as downward? If the answer is in the affirmative, how can we discover an increasing tendency? The best way to answer this question is to refer to the following properties (conjectured first by P. and T. Ehrenfest[8]) of the S-curve along the t-axis. (Figs. 3 and 4 are finite portions of such S-curves in our example.) The S-curve along the infinitely long t-axis remains most of the time very close to its maximum value. When it departs from the near-maximum value it returns very rapidly to its near-maximum value. Suppose you cut the S-curve (of a particular tray, for example) at any specified (but allowed) value of S which is less than the maximum value. Then the intersection must belong to one of the following four cases: 1) it is a local minimum of the curve, 2) it is on an upgrade slope, 3) it is on a downgrade slope, 4) it is a local maximum. The Ehrenfests conjectured that, provided the number of molecules is sufficiently large, Case (1) must be overwhelmingly more frequent than Case (2) or Case (3). (These last two must be equally frequent for an obvious reason.) Case (2) or Case (3) must be overwhelmingly more frequent than Case (4). From these conjectures follow two important consequences. (a) No matter which direction of t one traces, the S-value is increasing with overwhelmingly higher frequency than decreasing. This statement becomes even more reinforced if one allows longer intervals between observations. (b) If the S-value once starts to increase, it will very seldom reverse its trend before it reaches the near-maximum value. The point is, however, that the property (a) can be applied directly to the prediction while it cannot be applied to retrodiction because more information other than the present value of entropy is available in the case of retrodiction.

We shall give a few graphical results which I obtained by a Monte

Carlo simulation (on an IBM 7094 computer) of the shaking tray experiment explained before. This means that whether a ball stays in the same compartment or passes to the other in τ seconds was each time decided by a random number-producing program. Figs. 3 and 4 are the actual values of S from $t = 0$ to $t = 210\tau$ for $p = 0.02$. Fig. 3 is for a small number of balls ($L = 10$, $N_0 = 10$, $M_0 = 0$), as a result it shows violent fluctuation and a short time for recurrence. Fig. 4 is for an intermediate number of balls ($L = 100$, $N_0 = 100$, $M_0 = 0$) and shows a fairly smooth entropy-increase. If we continued this for a very long duration of time, the curve will eventually go down once in a while to its starting minimum value, but we cannot see it here. For a still larger number of L, the curve looks very much like Fig. 2. Fig. 5 shows that the Ehrenfests' conjecture can be confirmed already for $L = 20$. I ran ten series, each of $25,500\tau$ with $p = 0.1$, and the graph is the average of these ten series. The reason why the upgrade curve and downgrade curve do not coincide perfectly is that if a curve stays at the same value of S for two consecutive instants, say, on an upgrade slope, I counted this as two upgrade points. If we increase the number L, the Ehrenfest properties become immediately much more conspicuous, but on the other hand, since the number of less frequent cases becomes so few, the curves become very ugly due to inevitable fluctuations in a finite number of trials. It is interesting to see in Fig. 5 that when S is close to its maximum, the opposite of the

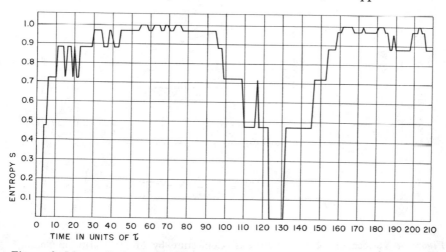

Figure 3. Monte Carlo simulation of H-Theorem with total number of particles $L = 10$ and $p = 0.02$. The experiment started with all ten balls in one compartment ($M_0 = 10$). After 124 τ, all ten balls came back again to the initial compartment.

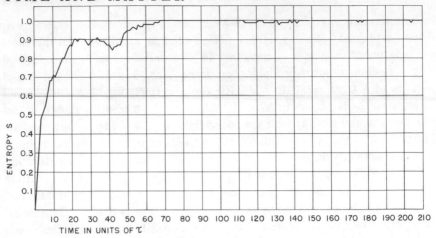

Figure 4. Monte Carlo simulation of H-Theorem with total number of particles $L = 100$ and $p = 0.02$. Initially all the balls were on one side ($M_0 = 100$). The entropy curve is much smoother than in the case with $L = 10$, and the one-way-ness strikes the eye. But this is deceptive, since if we continue this experiment for a very long time the curve will come down to the value zero again.

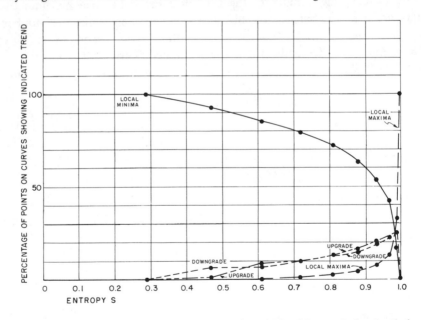

Figure 5. Verification of the Ehrenfests' Conjecture by Monte Carlo simulation, $N_0 = M_0 = 10$, $p = 0.1$. Total time $= 25,500$ τ. Data are the average of ten series of this length. Except for the near-maximum value of S, the points on the curve are overwhelmingly often local minima. This fact becomes more conspicuous for large numbers of balls.

536

Ehrenfest properties becomes the case. This curious phenomenon seems to deserve closer investigation in the future.

2 *Conditional Probability*

In this section we shall introduce the formally symmetric equation of probability theory whose asymmetric interpretation gives rise to the one-way-ness of time.

It is customary to define probabilities for a set of mutually exclusive and exhaustive events. Let us consider the set of events represented by E_i, $i = 1, 2, 3, \ldots, n$. We write $p(E_i)$ for the probability that event E_i occurs. The fundamental axioms of probability theory require that

$p(E_i) \geqq 0$, and that $\sum_{i=1}^{n} p(E_i) = 1$. If we have two sets of possible events, representative members of which are E_i and F_j, we can consider a compound event composed of E_i *and* F_j, where the joint probability of occurrence is written $p(E_i \cap F_j)$. Again we require that $p(E_i \cap F_j) \geqq 0$, and that $\sum_j p(E_i \cap F_j) = p(E_i)$, that $\sum_i p(E_i \cap F_j) = p(F_j)$, and that

$$\sum_i \sum_j p(E_i \cap F_j) = 1.$$

We are often interested in the so-called conditional probability when the two sets of events are not independent. We write $p(F_j \mid E_i)$, which means the probability of F_j given E_i. In general $p(F_j \mid E_i)$ is not equal to $p(F_j)$ unless E_i and F_j are independent.

Clearly, the probability of the compound event $(E_i \cap F_j)$ is given by $p(F_j \mid E_i)p(E_i)$ and also by $p(E_i \mid F_j)p(F_j)$. Hence we can write the two conditional probabilities in the nicely symmetric form

$$p(E_i \mid F_j)p(F_j) = p(E_i \cap F_j) = p(F_j \mid E_i)p(E_i) \qquad (2.1)$$

This relation is upheld for any two arbitrary sets of events $(E_1, E_2, \ldots, E_n)$ and $(F_1, F_2, \ldots, F_m)$, which may or may not be probabilistically dependent. One of the main claims of this article is *that in spite of this formal symmetry between E_i and F_j, applications of this basic formula to many (not all) important problems rely on an asymmetrical interpretation of E_i and F_j, and the temporally unidirectional result mentioned in Section 1 derives its direction from this asymmetry.* We shall now proceed to explain what I mean by asymmetrical interpretation.

537

First, we shall start with some remarks of a mathematical nature. In (2.1) five kinds of probabilities are involved. One joint (unconditional) probability $p(E_i \cap F_j)$, two single (unconditional) probabilities, $p(E_i)$, $p(F_j)$, and two conditional probabilities $p(F_j \mid E_i)$ and $p(E_i \mid F_j)$. The question raised here pertains to which ones of them can be given arbitrary values. According to our derivation it is obvious that four kinds of probabilities $p(E_i \mid F_j)$, $p(F_j \mid E_i)$, $p(E_i)$ and $p(F_j)$ can be derived from $p(E_i \cap F_j)$. Hence insofar as the two basic axioms of probability: $p(E_i \cap F_j) \geq 0$ and $\sum_i \sum_j p(E_i \cap F_j) = 1$, are satisfied, we can arbitrarily choose the values of $p(E_i \cap F_j)$ for all i and all j, and all the rest follows. This is a trivial case. The next case—which incidentally becomes important in our later discussions—is one where one of the conditional probabilities, say, $p(F_j \mid E_i)$ is already determined by the nature of the problem. To what extent can we still choose the values of other probabilities freely and to what extent are they already constrained by the given values of $p(F_j \mid E_i)$? I shall mention some of the facts pertinent to this problem, without giving a rigorous proof.[9]

(a) When only the $p(F_j \mid E_i)$ for all i and all j are given, we can still choose arbitrarily the values of the $p(E_i)$ provided the two axioms are satisfied by the $p(E_i)$. [$p(E_i) \geq 0$, and $\sum_i p(E_i) = 1$]. If the $p(F_j \mid E_i)$ and the $p(E_i)$ for all i and j are given, all the remaining probabilities are determined by them. This is because if the two axioms are satisfied by both $p(F_j \mid E_i)$ and $p(E_i)$, then the two axioms will be automatically satisfied by $p(E_i \cap F_j) = p(F_j \mid E_i)p(E_i)$, and hence also by $p(F_j)$ and $p(E_i \mid F_j)$. When the $p(F_j \mid E_i)$ for all i and all j are given, we cannot freely choose the $p(E_i \mid F_j)$ for all i and all j even complying with the two axioms, because these can contradict $p(F_j \mid E_i)$ as we shall see presently. If there is no contradiction, then the $p(F_j \mid E_i)$ and the $p(E_i \mid F_j)$ determine all the other probabilities, since $p(E_i) = \{\sum_j [p(F_j \mid E_i)/p(E_i \mid F_j)]\}^{-1}$ providing no indeterminate fractions $^0/_0$ are involved in this last expression. When the $p(F_j \mid E_i)$ are given we can very easily choose values for the $p(E_i \mid F_j)$ so that $\sum_i p(E_i) = 1$ is violated.

(b) When the $p(F_j \mid E_i)$ for all i and all j are given, we cannot entirely arbitrarily choose the values of the $p(F_j)$ even though these values satisfy the two axioms. This is because the $p(F_j \mid E_i)$ and the $p(F_j)$ both satisfying the two postulates may contradict each other by giving rise to negative values of $p(E_i)$ according to (2.1). Even if the $p(F_j \mid E_i)$ and the $p(F_j)$ do not involve contradiction of this sort, it can happen that

538

these two sets of probabilities do not lead to definite values of the remaining probabilities. This last happens when we cannot solve $p(F_j) = \sum_i p(F_j \mid E_i) p(E_i)$ for $p(E_i)$.

(c) When the $p(F_j \mid E_i)$ for all i and all j are given, the other conditional probabilities $p(E_i \mid F_j)$ can be derived from the former, first, if, in addition, the $p(E_i)$ are (arbitrarily) given or, second, if, in addition, the $p(F_j)$ are given in such a way as not to contradict $p(F_j \mid E_i)$ and if the simultaneous linear equations involving $p(F_j \mid E_i)$ are solvable. Third, other than those cases where, as in these last two, all the probabilities are determined, there is a very special case where the $p(E_i \mid F_j)$ are determined directly by the $p(F_j \mid E_i)$ while the unconditional probabilities remain undetermined. This happens when the values of $p(F_j \mid E_i)$ are either 0 or 1 and for a given F_j there is only one E_i for which $p(F_j \mid E_i)$ is 1. This means, because $\sum_j p(F_j \mid E_i) = 1$, that if we write the values of $p(F_j \mid E_i)$ as a matrix, there is always one and only one 1 in each column and in each row. We shall call this case a bilaterally deterministic case. It can be seen from the expression, $p(E_i \mid F_j) = [p(E_i)/p(F_j)] \, p(F_j \mid E_i)$, which follows directly from (2.1), that in the bilaterally deterministic case the ratio $[p(E_i)/p(F_j)]$ is determined (equal unity) for a pair for which $p(F_j \mid E_i) = 1$, but $p(E_i)$ and $p(F_j)$ separately are not determined.

These are necessary asymmetric results which arise in the cases where one of the conditional probabilities is determined by the nature of the problem.

Before passing to the next section where we shall discuss this asymmetry more closely, I should like to add an important remark on the notion of conditional probability which will help understand the argument we shall use in a later section. The point I want to make is that the distinction between the unconditional probability [such as $p(F_j)$] and the conditional probability [such as $p(F_j \mid E_i)$] is only relative and depends on the assumptions tacitly or explicitly agreed upon during a discourse or in a discussion. If a certain condition is consistently assumed to be the case during a discussion, we do not count it as a "condition" in the sense of conditional probability. But, if we go out of this framework, and consider probabilities also in the cases where this particular condition is not satisfied, we have to count this condition as a "condition" in the sense of conditional probability. Correspondingly, the unconditional probability reveals itself as a conditional probability in the latter broader context. Most of the unconditional probabilities are actually conditional probabilities in this sense. As a matter of fact, most macroscopic physical phenomena are deterministically decided by the initial state, and if we take all the details of the initial state as the "condition," the probability is bound to become 1 or 0. Thus, the usefulness of the concept of probability stems from the fact that we prefer not to take all the conditions into consideration.

In passing, it may be noted that the (cognitive version of the) inverse H-theorem also stems from the asymmetric interpretation of the symmetric relation (2.1). Namely, when $p(F_j \mid E_i)$ is given $p(E_i \mid F_j)$ has then to be computed by the Bayes formula, and the passage from the prior probability $p(E_i)$ to the ulterior probability $p(E_i \mid F_j)$ results in an entropy decrease.[10]

3   Prediction and Retrodiction—Causality and Freedom

We are told that Newton was of the opinion that all the possible planetary orbits were determined by his laws of mechanics and the law of gravitation, but it was a matter of Providence to have initially placed the planets on some particular orbits. This was Newton's way to express the clear distinction he had in mind between the law-like "essential" factors inherent in nature and the changeable "contingent" factors which are extraneous to nature. One might argue that if we knew the genesis of the solar system it would be, in principle, possible to explain why some of the orbits had to be occupied and some had to be not and that there are no such things as "contingent" factors in nature. But, this argument is missing the point because even if a dynamical explanation for the genesis of the solar system is possible by the use of some temporalistic laws, we shall always have to insert some "initial conditions" into the laws to obtain concrete results. And these initial conditions are precisely the contingent elements. Put the other way, *the need for contingent elements is the price we pay for establishing a universal law.*

This pattern of thought, which is both unavoidably imposed upon us by our inherent nature and invariably successful in the past human experience, is of course intimately bound up with "causality." By causality is meant here our mental habit, probably "reinforced" ontogenetically as well as phylogenetically by its success in life, of considering the invariable properties of nature as describable in terms of implicational propositions of the type: "*If* the object system is in state A, *then* it will be in state B at a later time." In many cases, the implicational proposition has to be replaced by a proposition involving *conditional probabilities*, such as "if A, then probably B" or "if A, then there is such and such probability of B taking place as a consequence." In the non-probabilistic case, we speak of deterministic causality although it may be considered as a special case of probabilistic causality. Now an important point here is that nature does *not* dictate what the initial state A should be. This is a contingent element. This leaves us, humans, the possibility of choosing, or preparing, a particular A. This is extremely significant in connection with the role played by science (or prescientific behavior) in human activities. Man conceives first a goal or "end" which he desires to realize and tries to know the right "means" leading to achievement of this end. The advice science (or prescientific knowledge) gives to him is usually formulatable in the form: if you do A, there is a good probability of your getting B. If B is his "end," then A is an appropriate "means" for him to take. The "means" A is a "cause" leading to the "effect" B which is the "end." Thus, "causality," instead of forcing us humans into a relentless, predestined fate, gives us a guarantee of a certain degree of *freedom*.

The crucial point here is that the basic form of our scientific efforts consists of establishing causal *conditional probabilities* of the type $p(B|A)$ as imposed by nature, where B is usually supposed to take place "later" than A. Thus we recognize that the fact that nature imposes $p(B|A)$ and the fact that B is to the future of A are intimately related. The transition probability p considered in Section 1 is precisely of this type of probability. The H-Theorem is based on this kind of probability.

To cope with the problem in a more general fashion, let us consider two sets of events $E_1, E_2, \ldots, E_n$ and $F_1, F_2, \ldots, F_m$. (For brevity we shall write $\{E_i\}$ and $\{F_j\}$). Instead of "event," we can also speak of "observation" or of "description," since an event is not the entirety of the physical phenomenon which is going on but is the description of the result of an observation of a certain specific aspect of the phenomenon. Now let us assume that the observation of the E's takes place at time t_1 and the observation of the F's takes place at time t_2. Between t_1 and t_2 the system is isolated. To summarize, we may state the following:

(*Postulate 1*) if t_1 is earlier than t_2, $(t_1 < t_2)$, and if $\{E_i\}$ and $\{F_j\}$ are not independent, then the conditional probabilities $p(F_j \mid E_i)$ (for all i, j) are determined solely by nature or by the nature of the physical system.

We see that we thus come back to the asymmetrical case of conditional probabilities discussed in the last section. Suppose that we could state, besides Postulate 1, its partial converse that if $\{E_i\}$ and $\{F_j\}$ are not independent and if the conditional probabilities $p(F_j \mid E_i)$ (for all i, j)

are determined solely by nature, then t_1 is earlier than t_2. This converse of Postulate 1 is equivalent to the statement that if $t_2 < t_1$, and if $\{E_i\}$ and $\{F_j\}$ are not independent, then not all the conditional probabilities $p(F_j \mid E_i)$ are solely determined by nature. Interchanging $\{E_i\}$ and $\{F_j\}$, we can further restate the converse postulate also in the form: if $t_1 < t_2$ and if $\{E_i\}$ and $\{F_j\}$ are not independent, then the $p(E_i \mid F_j)$ are not solely determined by nature. If this last statement were true, then we would be in a very interesting situation, because we could then reduce the direction of time $(t_1 < t_2)$ to the condition that the $p(F_j \mid E_i)$ but not the $p(E_i \mid F_j)$ are given by nature. This is an oversimplification of the true situation. However, with a certain qualification we can in fact assert the second statement.

We have gone far too far ahead of ourselves. Let us place ourselves in the real situation of scientific activities and re-examine the mathematical results we obtained in Section 2. To fix our idea, let us agree (in this paragraph) that t_1 is earlier than t_2 $(t_1 < t_2)$, where t_1 and t_2 refer to $\{E_i\}$ and $\{F_j\}$. When one of the E's, say, E_i, takes place at t_1, and if we try to guess the experimental result of observation $\{F_j\}$ which will be made at t_2, our effort is one of "prediction." The answer to this inquiry is readily contained in the "predictive probabilities," $p(F_j|E_i)$, which are given by the natural law, namely, F_j will happen with probability $p(F_j|E_i)$. On the other hand, it may happen that we know the result, say, F_j, of a later observation $\{F_j\}$ at time t_2 and we want to guess which one of the E's had actually happened at t_1. This effort is what I called "retrodiction."[11] The "retrodictive probability" that E_i happened at t_1 is denoted mathematically by $p(E_i|F_j)$, but this quantity is not necessarily determined by nature (or the nature of the physical system).

Let us now remember item (a) of Section 2. It says in the present case that when only the predictive probabilities are determined by nature, we can freely choose the initial probability. This guarantees the human freedom of action as discussed at the beginning of this section. Once the initial probabilities are determined, all the other probabilities are determined. We shall speak of an *uncontrollable case* when all the probabilities are determined by the nature or definition of the problem, such as the explicitly assumed past history of interaction. Item (a) says further that if the predictive probabilities are determined by nature, the retrodictive probabilities can no longer be entirely arbitrary. However, if the physical system is such that these retrodictive probabilities are determined for some reason without contradicting the predictive probabilities of the natural laws, we have again an uncontrollable case at hand.

Item (b) says that when the predictive probabilities are determined by nature, we cannot entirely freely choose the final probabilities without contradicting the predictive probability. Even if we choose the values of the final probabilities so as not to contradict the predictive probabilities, we have not necessarily determined all the probabilities.

Finally, item (c) states that there are two cases where the retrodictive probabilities are determined: (1) the uncontrollable case, (2) the bilaterally deterministic case. Otherwise, the retrodictive probabilities are not determined solely by nature or the nature of the system. This means that except in the uncontrollable cases and in the bilaterally deterministic case, the converse of Postulate 1 holds. Combining Postulate 1 and its restricted converse, we can now state

(*Theorem 1*) Let two observations $\{E_i\}$ and $\{F_j\}$ be performed respectively at t_1 and t_2, and let it be assumed that the two observational results are not probabilistically independent. Then, except in the uncontrollable case and in the bilaterally deterministic case, the condition $t_1 < t_2$ is equivalent to (*i.e.*, implies and is implied by) the condition that the conditional probabilities $p(F_j|E_i)$ for all i and all j are determined solely by nature or by the nature of the system, and also equivalent to the conditions that at least some conditional probabilities $p(E_i|F_j)$ are not solely determined by nature or by the nature of the system.

From what followed from item (a) and item (b), we can derive

(*Theorem 2*) Let $\{E_i\}$, $\{F_j\}$, t_1 and t_2 be defined as in Theorem 1. In the uncontrollable case, neither $p(E_i)$ for any i nor $p(F_j)$ for any j can be freely chosen. In the bilaterally deterministic case either $p(E_i)$ for all i or $p(F_j)$ for all j can be freely chosen. Except in these two cases, the condition $t_1 < t_2$ is equivalent to the condition that $p(E_i)$ for all i can be entirely freely chosen, and equivalent also to the condition that at least some $p(F_j)$ cannot be freely chosen.

This allows us to state that prediction but not retrodiction can be safely done objectively (*i.e.*, based solely on nature and the nature of the system), but also to invert the argument and to use this distinction for the definition of the direction of time. Theorem 2 relates this one-way-ness to the human freedom of creating the future, but not of creating the past. In Section 1, we saw by a simplified illustration that if $p(F_j|E_i)$ is given, then the H-Theorem concludes that the entropy increases in the direction from E_i to F_j.

4 Reversibility and Retrodictability

It is well-known that all the basic *microscopic* laws of classical physics are symmetrical with respect to the two directions of the time variable.

This does not mean that only small-scale physical phenomena are temporally symmetrical. It is not a question of the scale of the system but it is one of the mode of description. Even large-scale phenomena will show the symmetry if they are described minutely in every detail. This situation raises immediately a pair of twin questions. How is it possible that something essentially symmetric appear asymmetrical, without being distorted in some sense? Which one of the two equivalent directions receives the privileged meaning of "past-to-future"?

What precisely do we mean by saying that the microscopic laws are symmetrical with respect to the two directions of time? We mean by this what is usually called "reversibility" of physical laws.[12] In anticipation of what follows, I may mention that in classical physics reversibility implies retrodictability[13] but retrodictability does not necessarily imply reversibility. That is to say, reversibility is a much more restricted notion.

To explain reversibility, it is probably better to start with the notions of reversed phenomenon and reversed state. Suppose we take a motion picture of a physical phenomenon, and run the film backwards through the projector. The phenomenon one sees on the screen is the reversed phenomenon, and the state of the system at an instant in the reversed phenomenon is the reversal of the state of the original phenomenon at the corresponding instant. Thus, in a reversed state, an object will have the same position but it will be moving in the opposite direction, that is, the velocity changes its sign. In general, all physical quantities can be classified in two classes, those which keep the same sign and those which change the sign in the reversed state.[14] To determine the class of a physical quantity, we have to agree that the spatial positions of all objects keep the same sign, and that the basic attributes of objects, such as mass, electric charge, etc., also keep the same sign.

One may see easily that the reversed phenomenon of the reversed phenomenon is the original phenomenon. Similarly, the reversed state of a reversed state is the original state. Passage from a phenomenon (state) to its reversed phenomenon (state) is called time reversal. Suppose we adjust the origin of time (which is just a matter of convenience[15]) so that the original phenomenon takes place between $t = -\tau$ to $t = +\tau$. Suppose we use t' to designate the time variable of the reversed phenomenon. Then, we can adjust the origin of t' so that the phenomenon takes place between $t' = -\tau$ to $t' = +\tau$. Each pair of corresponding states has the time coordinates t and t' which are related by the simple relation $t = -t'$. Thus, we can include the time variable in the class of those variables which change their signs. In any event, we sometimes refer to the time-reversal as a "change of the direction of time." This simply means the relation $t = -t'$. In the above explanation, we defined two phenomena, which are the reversed phenomena of each other. There is another interpretation of the relation $t = -t'$, and of ensuing change of signs of some physical quantities, namely, there are two observers describing the *same* phenomenon, but one observer uses a variable t and the second observer uses a variable t' which are related to t by $t = -t'$. This does not necessarily mean that the second observer's psychological time from past-to-future is running opposite to that of the first observer. The existence of two interpretations of $t = -t'$ is a special case of the well-known fact that any linear transformation in a vector space allows for two interpretations: either the vectors (points) are considered to have moved or the coordinates are considered to have moved. In the case of time reversal, it is much easier to interpret it as a motion of the points rather than the coordinates. Yanase made some interesting remarks about this matter.[16]

Now we are prepared to define "reversibility." If a phenomenon and its reversed phenomenon are both allowed by a theory under consideration, then the phenomenon is reversible. If a physical theory is such that all phenomena allowed by it are reversible, then the theory is reversible. All the basic theories (mechanical and electromagnetic) of classical physics from which all other laws are supposed to be derivable are reversible. Laws of friction in mechanics, Ohm's Law, Newton's Law of Heat Conduction, which are secondary macroscopic laws supposedly derivable in principle from the more basic laws, are not reversible. All these irreversible laws are in agreement with a general law called the Second Law of Thermodynamics. This law says that if the entropy of an (adiabatically) isolated system changes, it can only increase.

This conflict between the basic reversibility and the macroscopic irreversibility is the classical way of formulating the essential problem of the direction of time. In this connection, I have to make two important remarks. First, we have to note, in agreement with many physicists in the past, that the alleged conflict originates from the applications of statistical consideration in the macroscopic description. Second, we want to point out, in a departure from the accepted view, that the crucial conflict is not between reversibility and irreversibility, but between retrodictability and irretrodictability since the direct consequence of statistical consideration is not irreversibility but irretrodictability. Since, as we shall see, reversibility implies retrodictability in classical physics, irretrodictability implies irreversibility.

Consider two blocks of iron, one hot and the other cold. Bring them into contact, and soon both become of the same temperature. The positions and velocities of all the molecules involved constitute the microscopic description of the phenomenon. Let S_I and S_F designate the microscopically described initial and final states, and let Σ_I and Σ_F be the description of these two states in terms of the temperatures of the two blocks. Using S^* to designate the reversed state, in which all the molecules have the same positions but opposite velocities, we note that S and S^* belong to the same macroscopic state. For instance, temperature is essentially determined by the average of the Gaussian distribution of kinetic energy of the molecules, hence is invariant for time reversal. According to the reversibility of basic laws, if S_I passes to S_F in τ seconds, S_F^* will pass to S_I^* in τ seconds. Macroscopically observed, this will appear, in the first case, as a passage from Σ_I to Σ_F in τ seconds and, in the second case, as a passage from Σ_F to Σ_I in τ seconds. But the passage from Σ_I to Σ_F is accompanied by an entropy increase while the passage from Σ_F to Σ_I is accompanied by an entropy decrease. Hence the second passage is forbidden by the Second Law. (This paradox is Loschmidt's objection to the H-Theorem, as illustrated in Section 1.)

Now, there must be something wrong in this argument. A correct interpretation of the Second Law of Thermodynamics is that an entropy increase is overwhelmingly more probable than an entropy decrease. Hence, Σ_I will pass, *in all probability*, to Σ_F, but Σ_F will, in all probability, not pass to Σ_I. In other words, there are very many microscopic states which belong to Σ_I and the great majority of them are headed for some microstates belonging to Σ_F. Conversely, among many microstates belonging to Σ_F, a very few of them are headed to some microstates belonging to Σ_I. The S_F^* is one of these rarities.

This situation is not unique in thermodynamics. We actually encounter it constantly in life. When we say that "If A, then in all probability, B," we must be extremely careful so that we take a really fair sample of A to start out with. Since neglect of this simple piece of daily wisdom leads sometimes to fantastic misunderstanding, we shall elaborate this point more in detail in the next section.

Coming back to physics, if the probability of a final state (of an isolated system) as a function of the initial state is given by a natural law and not affected by other factors, we say that the phenomenon is predictable. If the probability of an initial state as a function of the final state is given by a natural law and not affected by other factors, we say that the phenomenon is retrodictable. If a phenomenon (say, S_I to S_F) is reversible, it is obviously retrodictable since the passage from S_F^* to S_I^* is determined by the same law, hence if we are given S_F then we can conclude that it must have started from S_I. For retrodictability, however, strict reversibility is not necessary. For instance, if correspondence between S_I and S_F are one-to-one (bilaterally deterministic case), then we can certainly retrodict, even if the passage from S_F^* to S_I^* does not obey the same law. As we have seen, a direct consequence of macroscopic description is that our inference becomes probabilistic. And a direct consequence of probabilistic inference is that it is usually irretrodictable. For instance, if we have Σ_F of the previous example at hand, the initial state might have had a certain temperature difference between the two blocks as in our previous case, or it might have had a uniform temperature. Many different initial states were possible for a given final state, and in order to infer the initial state (even in a probabilistic sense) we have to know the initial probability distribution, which has nothing to do with the Law of Heat Conduction.

As a conclusion of this section, we can repeat that the microscopic reversibility guarantees the basic symmetry in time, but the macroscopic description results in irretrodictability which is, so to speak, a stronger asymmetry than irreversibility, in the sense that irretrodictability implies irreversibility. In Section 7, we shall come back to the two basic questions raised at the beginning of this section, which have only been partially answered so far.

5 *Probabilistic "If-Then" Inference*

It is clear that a deterministic statement: "if B, then A" can be expressed as a special case of conditional probabilistic statement of the type: $p(A|B) = 1$. (We are unconcerned here with the often raised contention that the former statement implies the latter statement, but not vice versa. The two are definitively equivalent when the total number of available samples is finite and $p(B) \neq 0$). Because of this simple relationship, one is often misled to assume a particular axiom which is valid only for the deterministic case to be also valid for general conditional probabilistic statements. The particular axiom in question is syllogism. From two statements: "if C, then B" and "if B, then A," we can derive "if C, then A." On the other hand, from the combination of the two statements: "if C, then B" and "$p(A|B) = a$," it does not follow necessarily that "$p(A|C) = a$" except in the special case where $a = 1$. A rigorous statement and proof of this theorem can be found elsewhere.[17]

This becomes self-evident if we recall with some care the definition of the conditional probability. The conditional probability $p(A|B)$, in the frequency view of probability, is the relative frequency of those objects which are A in the large collection of all those objects which are B. This collection of B is a well-defined subset of the entire set of all those objects which are taken into consideration. We should not take any collection smaller or larger than this particular subset to determine the relative frequency of A, because the relative frequency of A depends of course on the subset in which we are counting the relative frequency of A. If, for instance, C implies B, then the conditional probability $p(A|C)$ is the relative frequency of A in the collection of the C's which is a smaller subset within the collection of the B's. In the cognitive view of probability, $p(A|B)$ is the degree of expectation of an object being A when we take cognizance only of the condition that the object is B. Some philosophers (including Rudolf Carnap) advise that in calculating a probability we should include all the available information about the object in consideration. (Principle of total evidence.) But this is nonsense. Because, if we push this advice to its logical extreme, the situation will often become deterministic and we do not need to use the concept of probability any longer. The usefulness of probability stems from the fact that the description of nature becomes simpler by ignoring intentionally some of the less important factors.

The lesson we draw from the above is that when we encounter a statement like every object which is B is A, $[\forall x B(x) \supset A(x)]$, we must first examine if it really means "every" or it actually means "nearly every." If this latter is meant (which is very often the case), we must be extremely careful to prepare a fair sample of B before applying this statement to an object. In other words, if an adjective is used in an "attributive" way, as B is used here, it has to be understood as meaning "being a fair sample of B," while if an adjective is used in a "predicative" way, as A is used here, it merely means "being A" no matter how odd a sample. For instance, when we say "a man is a right-handed animal," we mean of course that the probability of an object which is a man turning out to be right-handed is high, and we have to prepare the sample correspondingly. For instance, if we examine a random passer-by in a street, this statement may be found to be true. But, if we pick a sample from a group of baseball pitchers, this statement may not be found too appropriate, although "being a pitcher" implies "being a man." If you put all the objects which are B in a bag, and after shaking the bag pick out an object blindly from the bag, this is a fair sample of B. We have to proceed very cautiously to obtain a fair sample when a simple prescription such as bag-shaking does not apply. The reader will note that the explanation of this section can also lucidly resolve the paradoxical situation which Hempel once called "inductive inconsistency."[18]

The remark made in this section becomes extremely important in interpreting a thermodynamical statement of the following type: If a system is in a thermodynamical state B at the initial instant, it will pass to a thermodynamical state A of larger entropy at the final instant, because a thermodynamical or macroscopic statement is a probabilistic one although it is not so stated. According to the lesson we have learned in this section, the initial microscopic state must be a "fair" sample of the macroscopic state: *i.e.*, the microscopic state must be such as might be picked entirely randomly from the collection of all the microscopic states belonging to the macroscopic state B. In other words, the system must be such that no other condition than B can be attributed to it.

6 *Microscopic Deterministic Symmetry and Macroscopic Probabilistic Asymmetry*

In the last sections, we have explained that the essence of the temporal one-way-ness of physical phenomenon resides in its irretrodictability and that the irretrodictability is a probabilistic consequence of "macroscopic" description based on "coarse" observation. But, this still does not explain which one of the two equivalent directions becomes the entropy-increasing direction. As we have seen, if the conditional probabilities $p(F_j|E_i)$ are given by nature alone and the conditional probabilities $p(E_i|F_j)$ are not given by nature alone, then the entropy-increasing direction coincides with the temporal direction from $\{E_i\}$ to $\{F_j\}$, where E_i and F_j are macroscopically described states or events. In order to make the concept clear, let us call the direction from $\{E_i\}$ to $\{F_j\}$ "macroscopic predictive direction," where the distinction between them is defined as above by the asymmetry of conditional probabilities. The crucial question then becomes which one of the two microscopically equivalent directions becomes the macroscopic predictive direction.

The first thing to note is that no answer to this question is possible without making reference to something which can define a direction in time. For instance, you draw a straight line on a large sheet of uniform white paper and want to tell your friend which direction you want to call the positive direction. Very probably, you will have to refer to some object which is not on the paper. Is there anything which can serve as a reference when one wants to give a meaningful definition of the direction? The second important thing to note is that we are not satisfied by just an accidental reference. We want to establish some causal or at least relational link between the referent and the relatum.

Now we have to expect that the answer to the question: which one of the two directions becomes the privileged direction?, will turn out to be disappointing. This is because microscopically the two directions are absolutely equivalent, and the process of macroscopization of observation has no temporalistic connotation. Thus the origin of the asymmetry is bound to be found on the human side, or at least on the interaction between human and its milieu. In principle, it may appear possible to attribute the origin of the preferred direction, not to the physical laws, but to some "contingent" element in nature such as the present expanding state of the universe. But, it is hardly in line with the tradition of physics to conjecture that an isolated laboratory experiment such as the one explained in Section 1 be influenced by the far-off stars. We are not interested in astrological theory. (We shall, however, come back in Section 7 to the possibility of an indirect influence of the accidental temporal direction of development of the universe on us.)

A contention which implies that the origin of the direction of time is to be found on the human side will give an impression that our theoretical enquiry into the direction of time ends in a circular conclusion: we find a privileged direction in nature because we project it on nature. But, we need not fear this kind of circularity. If the situation is circular, it has to be exposed to be so. A circular argument is a bad argument, but knowledge of a circular mechanism is a worthwhile knowledge. Neither can such a conclusion be condemned as sheer subjectivism, because I am not saying that the entropy increases in the direction of my personal psychological time. It is a direction shared by all living beings.

Let us go back to the problem of macroscopic, probabilistic description of nature and see how it is possible, through probabilistic concepts, to give an asymmetric appearance to a basically symmetrical phenomenon.

To make the distinction between the microscopic and macroscopic description clear, let us use $\{E_i\}$ and $\{F_j\}$ to denote the microscopic descriptions of a system at t_1 and t_2 and use $\{\Phi\mu\}$ and $\{\Psi\nu\}$ to denote the macroscopic descriptions of the same system at t_1 and t_2. Thus, more than one, probably very many, E_i's are included in one $\Phi\mu$, and many F_j's are included in one $\Psi\nu$. In classical physics, description is made in terms of continuous variables, but we use discrete variables here. Hence, it is a simplified model, but it will not lose the essentials of the problem. Classical physics is, in its maximal, *i.e.*, microscopic, description bilaterally deterministic. This means that for a given E_i at t_1 there is only one F_j possible at t_2 and for a given F_j at t_2 there is only one E_i possible at t_1.

It is perfectly retrodictable. The probability $p(F_j \mid E_i)$ is 0 or 1, and is equal to $p(E_i \mid F_j)$. Both probabilities are determined by nature or by the nature of the system.

Now, the property of the macroscopic conditional probabilities $p(\Psi_\nu \mid \Phi_\mu)$ and $p(\Phi_\mu \mid \Psi_\nu)$ is the focus of our investigation. Is either or both of these probabilities determined solely by the nature of the system? The contention of this section is that if t_2 is to the future of t_1 according to our biological time, then our mode of setting up the experiment becomes such that $p(\Psi_\nu \mid \Phi_\mu)$ is determined solely by nature, while $p(\Phi_\mu \mid \Psi_\nu)$ is not necessarily determined solely by nature. The reader should not confuse the kind of argument we developed in Section 3 and the proof we are going to give in this section. In Section 3, we showed how intimately this fact is related to the basic premises of human life such as causality and freedom of action. In this section, we raise the question how this fact can become true in the macroscopic description of nature. In other words, we are going to give a proof to Theorem 1 of Section 3.

The macroscopic conditional probability in question is

$$p(\Psi_\nu \mid \Phi_\mu) = \frac{p(\Phi_\mu \cap \Psi_\nu)}{p(\Phi_\mu)} \tag{6.1}$$

which can be written in terms of microscopic probabilities as

$$p(\Psi_\nu \mid \Phi_\mu) = \frac{\sum\limits_{j \in \nu} \sum\limits_{i \in \mu} p(E_i)\, p(F_j \mid E_i)}{\sum\limits_{i \in \mu} p(E_i)} \tag{6.2}$$

where $p(E_i)$ is the probability of the microscopic state E_i at t_1. The summation with respect to i should extend over all those E_i's included in the macro-state Φ_μ, and the summation with respect to j should extend over all those F_j's which are included in the macro-state Ψ_ν. On the right side of (6.2), the probability $p(F_j \mid E_i)$ is determined by the properties of the system and physical laws, while the probability $p(E_i)$ is not.

If we want to conclude that $p(\Psi_\nu \mid \Phi_\mu)$ is determined solely by nature or natural laws, we have to show that $p(E_i)$ actually disappears from the right side of (6.2). This can happen in two ways. First, if all the $p(F_j \mid E_i)$ have the same value within the macro-state Φ_μ, i.e., if $p(F_j \mid E_i) = p(F_j \mid E_k)$ for any pair (E_i, E_k) belonging to the same Φ_μ, then we can take the $p(F_j \mid E_i)$ out of the summation $\sum\limits_{i \in \mu}$, and $p(\Psi_\nu \mid \Phi_\mu)$ will become equal to $\sum\limits_{j \in \nu} p(F_j \mid E_i)$ where E_i is any one of the micro-states in $\Phi\mu$. The second possibility occurs when all the $p(E_i)$'s within each $\Phi\mu$ are equal. Then, we shall have

$$p(\Psi_\nu \mid \Phi_\mu) = \frac{1}{n\mu} \sum_{j\epsilon\nu} \sum_{i\epsilon\mu} p(F_j \mid E_i) \qquad (6.3)$$

where $n\mu$ is the number of terms to be added under $\sum_{i\epsilon\mu}$, *i.e.*, the number of micro-states belonging to the macro-state. The first case can also be written in the form of (6.3), because all $n\mu$ terms to be added under $\sum_{i\epsilon\mu}$ are the same in this case.

The first alternative may be true in some cases, but there is no guarantee that it should be so as a general rule in every case. On the other hand, a situation which can be simulated by the second alternative is the case in general at least in a good approximation in most of the macroscopic laboratory experiments if t_1 is earlier than t_2. The reason is that if $t_1 < t_2$, the first observation $\Phi\mu$ assumes an active character of preparation of a state satisfying the condition $\Phi\mu$, while the second observation is a passive one just to see if Ψ_ν is satisfied. Preparation of a state requires a certain length of time, say τ, which may be very short in the macroscopic scale (compared with, for instance, $t_2 - t_1$), but not very small in the microscopic scale. The microscopic transition probability is such that within a short duration of time, a micro-state passes to another micro-state of the same macro-state with high frequency, particularly when the system is kept under a constraint such that the macro-state cannot change. As a matter of fact, within the order of magnitude of τ, we can expect the system to migrate over a great number of different microscopic states of the same macroscopic state. As a result, the macroscopic transition probability $p(\Psi_\nu \mid \Phi\mu)$ must be considered as a kind of time average over the initial micro-states during the interval of time of the order of magnitude τ. During this time, the micro-state will go over many different E_i's with a kind of natural probability which does not depend on where the migration started. For some of the E_i's, $\sum_{j\epsilon\nu} p(F_j \mid E_i)$ may be large and for some others $\sum_{j\epsilon\nu} p(F_j \mid E_i)$ may be small, but τ is usually large enough so that the time average becomes a constant irrespective of the micro-state at the beginning of the interval τ. The τ is not large enough for the system to go over all the micro-states belonging to the macro-state $\Phi\mu$. (This would be the so-called Poincaré cycle, which is usually much longer than τ.) But τ is large enough for the system to pass through sufficiently many representative micro-states. This may be called pseudo-ergodicity, to make distinction from Boltzmann's quasi-ergodicity. Actually, the time interval for which the pseudo-ergodicity obtains becomes smaller and smaller as the system itself becomes smaller. As a result, in a slowly changing irreversible process in a big system, each small portion of the system can be considered to be in a pseudo-equilibrium.[19] This is the basic assumption of the modern Thermodynamics of Irreversible Processes. All this today is no longer a conjecture, but is in agreement with careful quantitative considerations carried out in modern statistical mechanics. This means that in every case we can approximate the situation by inserting the same weight $p(E_i)$ to each of the unit microscopic states E_i, defining the E_i's in a suitable fashion. (For physicists: each E_i may be supposed to occupy the same infinitesimal volume of the phase space.)

This makes $p(\Psi_\nu \mid \Phi_\mu)$ independent of anything other than the nature of the system. Thus, we have proved Theorem 1 of Section 3. The crucial point is that our laboratory experiment is, thanks to pseudo-ergodicity, such that we can pick up a "fair" sample of Φ_μ at the initial moment, while the observation at the final moment is not. This gives rise to a genuine probabilistic "if-then" statement in the sense of Section 5. Φ_μ is used in an "attributive" role while Ψ_ν is used in a "predicative" role.

To complete our conclusion, we should also show that $p(\Phi_\mu \mid \Psi_\nu)$ is not solely determined by the nature of the system. But this can be proven by an indirect method by showing that we are (macroscopically) neither in bilaterally deterministic case nor in the uncontrollable case, because we know that except in these two extreme cases, if $p(\Psi_\nu \mid \Phi_\mu)$ is determined solely by nature then $p(\Phi_\mu \mid \Psi_\nu)$ is not. We can exclude these two extreme cases since we know first that the macroscopic laws are indeed probabilistic and not deterministic, and second that we can in fact "control" the system in the sense that we can choose its initial condition. A truly macroscopically uncontrollable case (which corresponds to a microcanonical ensemble) occurs when the system becomes so small that the time required by quasi-ergodicity (not pseudo-ergodicity) becomes much smaller than the macroscopic time scale. We can, in fact, imagine a local small portion of a larger system in such an uncontrollable case when the larger system undergoes a slow and steady non-equilibrium process while each small portion is in a quasi-equilibrium. The initial state in a macroscopic experiment is often in a pseudo-equilibrium under the constraints especially designed to realize the initial state, but this state is not necessarily in a pseudo-equilibrium after the constraints have been changed in order to start the transient part of the experiment.

Thus we can conclude that in our experiments, the direction of entropy increase is determined by our special way of preparing the initial state and this is not in nature itself but originates from the side of the experimentor. At the same time, we have to recognize that it is only thanks to the time scale of pseudo-ergodicity which is the property of the object system that this particular way of preparation becomes possible, and as a consequence also that our control of nature based on causality and freedom of action becomes possible.

7   One-way-ness in the Universe and One-way-ness in the Laboratory

I am certain that many readers cannot wait much longer to decry my "anthropocentric," if not "subjective," interpretation of the entropy-increase law. I know the nature of the complaints most of the readers must have nurtured by now. The objection may be formulated somewhat like this. "You may be right about the laboratory experiments in which you prepare the system to be observed, but how about the universe? You perhaps object to applying the concept of entropy to the entire universe. But it is undeniable that there is a consistent temporal direction in astronomical phenomena. The sun, like many other stars, is emitting her energy into the infinite space, which is an entropy-increasing process. Similarly, the earth's rotation is imperceptibly yet steadily slowing down on account of tidal friction, which is another entropy-increasing process. This temporal direction agrees with the direction shown in the conversion of mechanical energy into heat energy taking place as well in the warming-up of an ocean due to the agitation by a storm as in the cooking of food by electricity generated by falling water."

This complaint, which incidentally is perfectly legitimate, has three major ingredients: a) The entropy of the universe at least in and around the solar system is increasing. b) This one-way process has consistently been going on since time immemorial in every smallest fragment of this portion of the universe. c) This entropy increase coincides with the entropy increase in laboratory experiments intentionally contrived by humans. Before commenting on each of these three points, I want to point out an important distinction to make between "the world to be contemplated" and "the world to be acted on." Hereby the mode of description of the world may be physical, psychological or sociological. By "world to be contemplated" is meant a single entity which includes everything and which is developing by itself and is supposedly observed and described without being interacted upon. The observer of this world has to talk about it without voice, write about it without pen and see it without light. This is an imaginary world which exists only as an idealization, and which can only be approximated in reality. When one speaks of an inevitable, objective law of development of the human society, one is imagining this kind of world in which one belongs yet one does not. The "world to be acted on," on the other hand, is the real world, which is the object of our genuine observation and genuine action. This world does not exist independently and outside of us, and is not developing by its own laws only. But, this is the kind of object a true science should handle. The entropy law, too, in its primary sense, must be interpreted, as was done in this paper, in this second context, as a law governing the "world to be acted on."

Let us now go over to the three major objections mentioned above. The first point (a), which asserts an entropy increase without specifying the reference, requires no explanation. The fact, however, that it requires no explanation in itself requires comments. This is somewhat like stating that there is an arrow marked on an infinitely large sheet of blank paper. Being an arrow, it has to be oriented in some direction, but neither identification nor reason can be given to its direction unless something else can be brought in as reference. The statement that the entropy of the universe is increasing (along one of the two possible directions of the time variable) is a tautology and is not an empirical statement.

Point (b) mentions temporal consistency (entropy continues to increase) and spatial consistency (the increase is in unison everywhere). The temporal consistency can be explained basically by the mathematical property of the entropy function. The entropy of an isolated system is most of the time in the very close vicinity of its maximum value. Occasionally it deviates appreciably from the maximum value, but in that case it usually goes back fast and monotonically to the maximum value without starting to decrease again in between. This is true for both directions of time, but we happen to be on one of the two slopes. We should interject here a comment which concerns both point (a) and point (b).

Nobody knows for sure whether the universe is spatially limited (this term "limited" has to be defined first) and hence whether the recurrence property we mentioned in Section 1 applies here. Yet, in view of the complete symmetry of the classical physical laws and of the indefiniteness of time, it seems to me highly implausible that the universe plays only a one-act drama of monotonic change of entropy. This monotonic change must be preceded and succeeded by monotonic changes in the opposite direction. If this is so, then the temporal consistency has only a fleeting validity and the existence of direction (a) loses its point. What appeared to be an arrow in a myopic observation turns out to be a single link in a long chain of many arrows placed alternatively in opposite directions.

Spatial consistency can be explained by the following consideration which may be considered as a corollary ensuing from the mathematical analysis of the preceding section. When a system splits into two at t_1, we have to assume that there occurs a sudden change of constraints at t_1, and the precise behavior of the system after t_1 depends on this change of constraints. However, to show in general that each of the splinter subsystems continues the original entropic trend, we need only to notice that the subsystems are roughly fair representatives of their macro-states in the case where the entropy of the original system has been increasing up to t_1 and would have continued to be increasing after t_1 in the absence of the split. This is explained by the fact that a partial system of a fair sample is a fair sample. This is so because, in agreement with the consideration of the last section, the time necessary for a system to go through a sufficiently large number of representative micro-states within a macro-state (whether it is in equilibrium or in a stationary state) becomes smaller for a smaller system.

Some philosophers[20] have been keen enough to notice the importance of "inheritance" of direction of time at branching, but being unaware of its physical bases, they have mystified the whole issue by attributing an undue importance to this phenomenon. An over-all explanation of this phenomenon can be given by the mathematical nature of the entropy curve, in which the entropy, once on an upgrade, will continue its rise until it reaches the near-maximum. During this process, splitting into similar subsystems can happen, and the entropy being an additive function, the sum of the entropies of similar subsystems and hence the entropy of each splinter system also is likely to rise in unison. But, the mechanism of what happens at the moment of splitting must be somewhat like the explanation given above.

Now coming to point (c), we have to admit that no theory advanced so far explains how biological time "inherits" its direction from the direction of time of the ambient physical world. But, the mechanism of creation and continuation of life must be such that what we feel as the past-to-future direction agrees with the entropy-increasing direction of the portion of the universe we live in. And, in their turn, humans act on the world, as we have seen before, in such a way that the physical processes they create are entropy increasing along their biological time.

To repeat, a problem of entropy increase is worth discussing only when the increase is considered with reference to something. Hence, the entropy increase of a universe-to-be-contemplated is not interesting. The only problem which is worth considering and which can be answered with some clarity is the one of the relation between the world-to-be-acted-upon and the agent which we discussed in the foregoing sections.

8   *The World of Quantum Physics*

If I chided the "objectivistic" view of the world-to-be-contemplated severely in the last section, that was because classical physics, due partly to its peculiar way of concept building and partly to the widespread misconception about science in general, is every so often interpreted in accordance with the strictest version of such an objectivistic view. One of the messages I tried to bring in the present article was that even classical physics leaves ample room for a human agent who executes his freedom of action and pays for it by increasing the entropy of his environment.[21] But, this basically correct "subjectivistic" view of the world-to-be-acted-upon could also be exaggerated to a solipsistic extreme where only person—or his mind—is the agent and all the rest of the universe his puppet show. A real reconciliation of these two ways of thinking would be not to take a happy medium between the two extremities, but to show, so to speak, a pragmatic invariance which is not affected by the angle of view. For instance, in one person's view, the entropy of a system is increasing because he has so prepared its initial state availing himself of his freedom of choice, but a second person who is observing the first person and his experiment might look upon the said entropy-increase perhaps as another case of the "inheritance by division" happening in this portion of the universe. These two descriptions are not necessarily factually contradicting, although different in viewpoint. This kind of invariance or consistency against a rotation of viewing angles should be one of the basic criteria for any valid theory of the world.

In this context, it is extremely instructive to see how the temporal one-way-ness comes into the theoretical structure of quantum physics. It is sometimes contended that Quantum Theory is not a valid theory of the world because of its alleged subjectivism. Subjectivism as such, however, should be no cause for castigation. The main thing is that the principle of invariance for rotation of views as suggested above is upheld in a suitable sense. In particular, it will be of importance for us to see how the same thing is said in different languages of classical and quantum physics.

A summary of the Quantum Theory has already been given in this volume by Professor Zimmerman in his essay, hence the following description of quantum physics is rather sketchy and intends to bring out only one aspect of this theory. Since 1952 I have been emphasizing that quantum physics is probabilistically reversible (or inversible) but essentially irretrodictable,[22] but surprisingly few people have noticed the profound implications of this contention. In Section 4, we noted that reversibility implies retrodictability, but in the case of quantum mechanics, reversibility exists in a certain probabilistic sense, hence the above contention is no contradiction to what has been said in Section 4. On the contrary, probabilistic reversibility actually implies irretrodictability, as we shall presently see.

As in Section 4, let us write S^* to designate the reversed state of the microscopic state S. And let $p(S_2 \mid S_1)$ designate the probability of observing micro-state S_2 at $t = t_2$ when micro-state S_1 was observed at $t = t_1$ ($< t_2$). Then the probabilistic reversibility of quantum mechanics can be expressed as

$$p(S_1^* \mid S_2^*) = p(S_2 \mid S_1) \qquad (8.1)$$

meaning that the probability of observing the reversed state of S_1 at $t = t_2$ starting from the reversed state of S_2 at $t = t_1$ is the same as the probability of observing S_2 at $t = t_2$ starting from S_1 at $t = t_1$. The reversibility discussed in Section 5 can also be expressed as (8.1) but there we limited ourselves tacitly to the case where the probability of the type $p(S_2 \mid S_1)$ is 0 or 1. In quantum physics, the probability becomes occasionally 1 but takes usually a value between 0 and 1 already in the maximal, *i.e.*, microscopic description, and this makes a deep-going difference.

Quantum physics is in perfect agreement with the viewpoint that science should deal with a world-to-be-acted-on rather than with a world-to-be-contemplated. The reason is that quantum physics does not tell how the world *is* but tells only about the result of an observation (which is an interaction) made on an object system by an observer. More precisely, it gives only the probabilities of outcomes in an "ulterior" observation as functions of the outcome of an "anterior" observation, *i.e.*, it gives only $p(S_2 \mid S_1)$, where S_2 and S_1 refer respectively to the ulterior and anterior observations. The wave-function, or state-function, or psi-function which obeys the Schrödinger equation, is a mathematical entity which is ultimately used to calculate $p(S_2 \mid S_1)$.

In spite of the fact that the prescription regarding the usage of quantum mechanics makes it clear that $p(S_2 \mid S_1)$ presupposes a reference of time direction, most of the people carelessly believed that this reference is an arbitrary one introduced only for convenience of verbal interpretation of the quantum theory, and that, thanks to the reversibility (8.1), everything was symmetrical in quantum physics. But, as I pointed out in my 1952 paper, this is a gross mistake. (It was for this purpose that the notion of retrodiction was first introduced.)

The reason is, as we saw in Section 4, that once $p(S_2 \mid S_1)$ with $t_1 < t_2$ (with reference to any time variable) is determined by theory as is the case in Quantum Theory, then $p(S_1 \mid S_2)$ is not determined by theory except in the bilaterally deterministic case and in the uncontrollable case. (The reader should note that in $p(S_1{}^* \mid S_2{}^*)$ of (8.1) $S_1{}^*$ and $S_2{}^*$ are supposed to take place at t_2 and t_1, while in $p(S_1 \mid S_2)$ here S_1 and S_2 are supposed to take place at t_1 and t_2.) Now, we know that in most cases $p(S_2 \mid S_1)$ is not deterministic even in the maximal (microscopic) description, and also that we are, in fact, capable of creating a desired initial state, hence $p(S_2 \mid S_1)$ is given by theory while $p(S_1 \mid S_2)$ is not given by theory, *i.e.*, in Quantum Theory, even in its maximal description, prediction is possible while retrodiction is impossible. Now this shows that the time variable cannot be an arbitrary one in quantum physics. The "anterior" and "ulterior" have to be given a fixed definition, but with reference to what? The way quantum theory is explained and is practiced by physicists indicates that we are supposed to know what is prior and what is posterior, *i.e.*, we are supposed to define the time direction with reference to the psychological or biological time of the observer or of the group of mutually communicating observers.

This will bind us to use probabilistic inference in microscopic and macroscopic physical phenomena as well in laboratory experiments as in astronomical observations along the past-to-future direction as determined by our own biological time. This prescription will turn out to be in agreement in consequence with classical physics as can be seen from the argument of the last section, but this offers a somewhat simpler view of the universe than classical physics. One might argue that this picture will not allow for a "next" period of the universe (in classical sense) where the entropy of the universe, after a long stay at the near-maximum, starts to decrease along the time variable which we are now using. This argument is not well-founded, because all animals will die out long before the period of universal hibernation sets in, and if the entropy really starts thereafter to decrease along the current time variable, then the animals will be certainly living there in the opposite direction of time. That is, if we extend our current time variable, the animals there will start with death and end with birth, but in their psycho-biological time, they will start with birth and end with death. A similar consideration will apply to the deep trough of the entropy—there will be no life in the valley of holocaust either. If we accept this interpretation, there will be no serious pragmatic contradiction in conclusion between the classical, "objective" view and the quantum-theoretical "subjective" view of time, although it may require a considerable speculative tour de force to reconcile the metaphysics behind the two theories.

In passing, we may notice that there are two more aspects of quantum mechanics which have close relations to the aforementioned principle of invariance or consistency. One of them concerns the so-called explanation (such as Heisenberg's) of the uncertainty principle. It is in reality not an "explanation" but it is a verification that what appears to be an inevitable uncertainty in one observer's finding can be compatible with a bystander's description of the combined system of the observer and the observed object. Another point is that by the limiting process $h \to o$ (h: Planck's constant), the flexible world-to-be-acted-upon turns into a rigid world-to-be-contemplated in the microscopic description.[23]

9 *Conclusions*

We are told that space provides room for being and time provides room for becoming. We can accept this adage provided that we interpret the word "becoming" properly. Becoming is not changing, it is fulfilling, it is achieving, it is taking the form which a thing has been intended to assume. Becoming has intent, yet it has no plan. If the entire future development is mapped out already, and if we are shown every day a small fragment of it, then it is not becoming. Becoming is making of the yet-unmade. Becoming is constant death and constant rebirth. It is a simultaneous destruction and creation. Thus, becoming implies creative action, pregnant with purpose, fortified by volition and, above all, nurtured with freedom and guided by knowledge. Our temporal experience of life is precisely a one-way chain of "becoming" in this sense.

How far removed from this real world of life is the physical world of Newton, Maxwell and Einstein! In this frigid world of theory, everything is predetermined in all details from time immemorial to the eternal future. There is no room for creation, no room for freedom, no room for novelty, hence no room for real time. How can a person with heart and blood live in such an impersonal, cruel world? Is the world we live in really such a predestined mechanism? No, we know it is not. We know because we scheme and try, because we succeed and fail and because we hope and despair. Here is, however, a contradiction. Is the rigid and indifferent world of physics a fiction? Or, is the plastic and responsible world of our daily experience an illusion?

We answered to this question by making two major points in this paper. First, the world which we, as an active agent, work upon is not the entire universe but is only its portion which enters into interaction with us, the agent. This portion of world (which by definition cuts out the agent) has no independent, predetermined course of development. It is manipulable and susceptible to our intervention. This idea of plastic world does not contradict the hypothesis—correct or incorrect—that the entire world is predestined. Important, though, is the fact that this latter world is a world to be beheld, but untouched, by a deistic god only, while the former plastic world is *our* world we hate and love because it responds to us.

Second, we pointed out that in such a world-to-be-acted-upon, we can establish, within the framework of orthodox physics, a probabilistic, predictive, causal law, which is of the type: if *A* now, then *B* later with such and such probability. We showed also how inextricably the probabilistic causal law is bound up with the very nature of human action. For, in order to realize a desired goal *B* we prepare the means *A* when we know that the cause *A* leads to the effect *B* with high probability. The utility of a causal knowledge is thus guaranteed by the freedom of choice of the cause (means). We showed how the macroscopic manipulation and observation of the world whose microscopic development is deterministically predestined can make such probabilistic causality and purposive choice of action possible.

The separation of action into goal and means, identical with the separation of development into cause and effect, generates the basic temporalistic polarization that manifests itself as the past-to-future one-way-ness of time. Perhaps lower animals, which do not recognize any causal relations among events and which are blind to the teleological meaning of their own instinct-guided actions, are not living *in time*, as we are. In the same way, the robot and computer are not living *in time*, because they are blindly obeying the predestined programs. Time is awakened when one is faced with freedom of choice.

The most dramatic way to show that the above-mentioned probabilistic causality is the origin of the one-way-ness of time is to consider the interchange of the roles of *A* and *B* in the above definition. We have shown that if probability of *B* as a result of a given *A* is determined by nature, then the probability of *A* preceding a given consequence *B* cannot be determined by nature. This fundamental asymmetry can be summarized in a single statement that nature is predictable but irretrodictable (both probabilistic sense). This means that when we guess the past on the ground of the present, we are relying on something else than a (probabilistic) natural law. My assertion about predictability and irretrodictability of temporal sequences may seem to contradict our daily experience in which the future is uncertain and unpredictable while the past is fixed and recorded. This apparent paradox reflects precisely the point I am making. In the case of guessing the future, we are relying on the probability given by nature, while in the case of deciphering the past record, we are invoking knowledge of an entirely different kind. The former may be represented by a broad probability distribution, the latter by a narrow one. Mathematically speaking, this latter is based on the so-called inverse probability in the sense of Bayesian process of hypothesis evaluation which involves "prior credibilities" expressing *our* extra-evidential evaluation.[24]

A satisfactory feature of our standpoint is that it conforms with the most obvious yet stringent criterion of a valid world view, namely, that there is no pragmatic contradiction between the agent's view and the bystander's view, whereby the bystander is a second observer (onlooker) who includes the first observer-agent as part of his observed world. In a limiting case, we could conceivably push this "transfer of view" to an extreme where the bystander becomes the deistic god I referred to earlier. By doing this, we revive the age-old quandary: Is all that we regard as our free action not an illusion because in the picture which includes both us and our plastic world everything is predetermined? My answer to this is in essence as follows: In the first place, this question involves what may be called a category mistake, since freedom of action belongs to the description by an agent while indetermination of action belongs to the description by a bystander. But, it is also a mistake to claim that these two notions are unrelated. When the environmental circumstances alone determine an action, there is neither freedom nor indetermination. If the action is not uniquely determined by the circumstances alone, but is amply determined by the conjunction of the circumstances, and the history and structure of the agent, then there is freedom but not indetermination. If the action is not determined by this last-named conjunction, then we have both freedom and indetermination. In this last case, we have to resort to a probabilistic viewpoint whether we are agents or bystanders or a deistic god. My conviction is that the true universe belongs to this last case.

In concluding, however, we have also to emphasize the limitations of the probabilistic view of the world. Each individual case becomes a fluctuation in a probabilistic distribution according to this view, and no "meaning" is given to each individual fluctuation. But, for each individual agent, each action chosen has an existential meaning in its relation to the value attached to the future goal. The probabilistic view is blind to this value. Freedom, in the probabilistic view, degenerates into lack of determination. But, freedom acquires a positive aspect in the light of the values system of the agent. We have so implied it in the text, in spite of the restrictions which a superficial interpretation of the essay title may suggest. Thus, we may end the article by introducing an aphorism to replace the adage we mentioned at the beginning of this section: Space is the vehicle of determination and being, time is the vehicle of freedom and value.

Time and the Universe

G. J. WHITROW

In the story of the Fall, as recorded in the third chapter of Genesis, the destroyer of Man's primeval timeless innocence is represented as a serpent. This symbolism is similar to that found in many ancient cultures where the serpent is associated with cycles of endless time, an idea that may have its root in the fact that the serpent periodically sheds and renews its skin.[1] In contemporary thought the concept of time is also often regarded as a kind of serpent that many would like to see banished from our idea of ultimate reality. Instead, I shall argue that time is not only an essential feature of the universe but that these two concepts— time and the universe—are peculiarly and reciprocally related, each being essential to the other.

It must be admitted that both concepts have an elusive quality that we feel sorely tempted to circumvent. In the case of the universe, we would like to explain the whole solely in terms of our knowledge of the part— at best a question-begging procedure. In the case of time, we strive to define it or explain it in terms of other concepts. But, if we rigorously exclude all terms of temporal significance from our assumptions, we cannot account for the continual lapse, or going-on, of time; and, if we admit such terms, then our method is circular, since we introduce, overtly or tacitly, the very thing that we are trying to define or explain. The familiar analogies of time with an indefinitely prolonged line or with an ever-flowing stream are examples of these alternatives. Philosophers like Bergson in the one case and Wittgenstein in the other have stressed the limitations of these analogies. Indeed, Bergson made the shortcomings of the spatial analogy a powerful argument for a philosophy in which the concept of time was central. Many philosophers who have tried to analyze this concept have reacted differently and have been greatly influenced by the apparent inherent illogicality of the temporal process. Logical implication, however, is timeless and one must therefore be on one's guard when ostensibly irrefutable arguments are advanced to establish on purely logical grounds the unreality of time—*eppur si muove*.

1   *The Origin of Time*

One of the most baffling problems is the question of whether the universe could have had an origin in time or not. It is well known that this was one of the main problems that influenced Kant in formulating his philosophy in the *Critique of Pure Reason*. He believed that he could produce apparently indisputable proof for both possibilities, and he came to the conclusion that our idea of time is inapplicable to the universe itself but is merely a part of our mental apparatus for imagining or visualizing the world. It is essential to our experience of things in the world, but we get into trouble if we apply it to anything that transcends all possible experience, in particular to the world as a whole.

First, let us consider his argument that the world must have had a beginning in time. This is established by *reductio ad absurdum* of the assumption that the world had no beginning in time, and that up to every given moment there has passed away in the world an infinite sequence of successive states of things. Kant's discussion is formulated in what strikes us today as imprecise terminology; but, in the language of modern mathematics, he was analyzing the hypothesis that the past sequence of successive states is an open set with no first member. His idea of successive states of things is somewhat vague, but his argument applies to any sequence of discrete events forming a temporal chain, and asserts that in the universe no such chain can be without a first member. This type of physical infinity must be distinguished from the purely mathematical infinity involved in the conceptual analysis of time into a continuum of durationless instants.

Kant argues that the infinity of a series consists in the fact that it can never be completed through "successive synthesis." Hence it is impossible for an infinite world series to have passed away, and consequently a beginning of the world is a necessary condition of the world's existence. This argument has often been seriously misunderstood, because it is thought that it can be automatically rejected merely by appealing to the modern theory of infinite sets and series. This theory, however, is not concerned with temporal concepts, since all reference to time as such has been purged from it. Kant's argument essentially concerns successive events occurring in time. It says nothing about the possibility of an infinite sequence of events in the future, but asserts the impossibility of an infinite sequence of events having occurred already. He maintains that an elapsed infinity of successive events is a self-contradictory concept.

Before considering this question further, let us turn to Kant's counterargument that, notwithstanding this conclusion, the world cannot have had a beginning. For, he argued, if it had then it must have been preceded by an empty time. But no coming-to-be is possible in a completely empty time, since no part of such a time can be distinguished from any other part and "no part of such a time possesses, as compared with any other, a distinguishing condition of existence rather than of nonexistence; and this applies whether the thing is supposed to arise of itself or through some other cause."[2] In other words, the moment immediately before the world began would have contradictory properties: it would be like all other moments of empty time and also unlike them because of its immediate temporal relation to an event in the world.

Now, although I believe, like Kant, that we should reject the idea that the universe was created *in* time, we are not compelled to accept his conclusion that his two arguments together imply that time does not pertain to the universe, for his second argument does not apply if we postulate that the world and time coexist. Kant's analysis was influenced by Newtonian cosmology in which time exists in its own right and is independent of the universe. But, if we reject this idea of absolute time and regard time as the order of succession of phenomena, there can be no time without phenomena. Consequently, the logical dilemma of the world's creation *in* time does not arise.

If we accept this way of circumventing Kant's argument, we must come back and consider further whether or not there was an origin of both time and the universe. In discussing this question we must not allow ourselves to be unduly influenced by the corresponding problem for space. The question of whether the spatial extent of the universe is finite or infinite is distinct from that of its boundedness. Even if finite, world space cannot be regarded as having a boundary, because if we imagine a boundary we can also imagine ourselves going to such a boundary and penetrating it. The situation as regards time is quite different. The difficulty that people often find in imagining an origin of time arises because they tend to think of it as being similar to a boundary of space. But we cannot travel freely in time as we can, at least in principle, in space. If we believe that time and the universe coexist, an origin (or end) of time simply implies a particular temporal restriction on the existence of the universe. There is, however, no *logical* compulsion for us to regard the whole temporal range of phenomena as unlimited. Indeed, if there were any point in so doing, one could regard the universe and all its contents, including simulated memories, as having come into existence at *any* particular moment, past or present. Such a moment would be effectively an origin of time.

If there were no origin of time and the universe, it would mean that, irrespective of whether we can speak of successive "states of the universe," there has been, up to the present moment, at least one temporal chain of events that had no beginning. By a temporal chain of events I mean a sequence of discrete physical occurrences that has the usual properties of linear chronological order. Let E denote any given member of such a temporal chain. From the standpoint of E there is an important difference between an infinite past and an infinite future, each regarded as infinite sets of discrete events: one is potential and the other actual.

An infinite future is only a potential infinity. This means that: 1) with respect to any event occurring in the future of E there will occur future events; 2) any event occurring in the future of E is separated from E by a finite number of intermediate events. Condition (1) states that the future of E is infinite, whereas condition (2) asserts that this infinity is only potential. If, while retaining condition (1), we attempted to reject condition (2), we should have to assert that events will occur that are separated from E by an infinite number of intermediate events. This would imply that, in principle, starting from a particular event all the members of an infinite set (of successive events) could be enumerated by counting. This is a practical impossibility for any infinite set, however long the counting continues.[3] Despite this, it may perhaps still be asserted that events will ultimately occur that are separated from E by an infinite number of intermediate events. This assertion leads, however, to the following difficulty. In a set of discrete events actually happening in sequence following a given event E, at what stage would the number that have occurred cease to be finite and become infinite? Rejection of condition (2) would therefore pose an insoluble problem. By imposing this condition the difficulty is avoided.

Whereas we must therefore regard an infinite future as a potential infinity, an infinite past, if it existed, would necessarily be an actual, or attained, infinity. For, although when we try to think backwards from the present over past events our train of thought merely yields a potential infinity of events thought of, this process does not correspond to the actual successive occurrence of these events. If all the events in a temporal chain culminating in the present are infinite in number, then, because these events actually occurred, the infinity concerned must be an actual, not merely a potential, infinity.

Consequently, if the chain of events forming the past of E is infinite, there must have occurred events that are separated from E by an infinite number of intermediate events. For, if not, then any event in the past of E would be separated from E by only a finite number of intermediate events. This would mean that the set of past events would, like the set of future events, constitute only a potential infinity, whereas it must constitute an actual infinity. It thus follows that, if the past of E contains an infinite number of events in a temporal chain culminating in E, there must have occurred events O in the past of E that are separated from E by an infinity of intermediate events. But this conflicts with our condition that an infinite future with respect to any event, in this case O, is a potential infinity, for E is an event that occurs and O has already occurred. Even if, in this context, we are prepared to forego the Law of Contradiction,[4] we are still confronted with the same insoluble problem that arose earlier in our discussion: when, in the temporal chain from O to E, does the total number of events that have occurred since O become infinite? But this time the difficulty cannot be evaded. If there is an infinite chain of intermediate events extending from O to E, every member of which will have occurred in the future of O and has occurred in the past of E, how was it possible beginning with O ever to reach E?

We conclude that the idea of an elapsed infinity of events presents an insoluble problem to the mind. However strange the idea of an origin of time may seem, the concept of an infinite past is even more puzzling.

2   The End of Time

According to the dying Hotspur, in Shakespeare's *Henry IV, Part I,*

> time, that takes survey of all the world,
> Must have a stop.

This idea came to the forefront of scientific thought during the nineteenth century, as a result of the investigations by Carnot, Clausius and Boltzmann on the Second Law of Thermodynamics. Clausius argued that this law, which he formulated in terms of the abstract concept of entropy, conflicted with the customary view that the general state of the world remains invariable. Boltzmann's definition of entropy in terms of probability made the idea of a "heat-death" of the universe and of an end of time seem almost inevitable. He himself endeavored to escape this consequence of his own principle by speculating on the possibility of there being regions of the universe in which thermal equilibrium has been attained and others in which time runs in the opposite sense to that in our stellar system. He thought that these different regions might be very remote from each other, but that for the universe as a whole the two directions of time would be indistinguishable from one another.

Since Boltzmann's day our knowledge of the distribution of matter in the universe has greatly increased but, although the discovery that to each elementary particle there corresponds an antiparticle has given rise to some bold hypotheses concerning the possible existence of galaxies composed of antimatter, there is little to support his suggestion. Nevertheless, there has been some reaction against the idea of the inevitability of a heat-death of the universe. In particular, in 1931 Milne drew attention to a logical gap in the customary proof that the entropy of the universe as a whole automatically tends to a maximum. He claimed that the proof of the Second Law of Thermodynamics depends on a hidden axiom that should be stated explicitly. According to this postulate, wherever a process occurs in the universe, the universe can be divided into two portions such that one of the portions is entirely unaffected by the process.[5] This axiom would automatically exclude world-wide processes. Milne took care, however, to point out that we cannot conclude that the entropy of the universe is *not* increasing, for every local irreversible process causes such an increase. All we can say is that we have no means of determining change of entropy for the whole universe. Although we can calculate such a change for "closed systems" with something outside them, the universe, by definition, has nothing physical outside it.

Boltzmann's own attempt to identify increasing time with increasing entropy, and hence to produce an explanation of unidirectional time, foundered on the well-known reversibility objection of Loschmidt, but the recent attempt by Reichenbach to rescue this statistical theory of time comes up against the problem of world entropy.[6] The difficulties associated with this concept are therefore of far-reaching significance for the whole problem of time. These difficulties are accentuated by the fact that there is no general agreement concerning the extent of the universe, that is to say, whether it is finite or infinite, and also by the question, not foreseen in Boltzmann's idea, of interpreting the red-shifts in the spectra of the extragalactic nebulae. For, if these are regarded as due to the Doppler effect associated with recessional motion, it would seem that the background conditions of the universe cannot be regarded as invariable.

Thus, although we have no means of disproving the hypothesis of an end of time and the universe, the whole problem has been transformed and there is no longer any case for regarding the question as settled.

3 Time's Arrow and the Expansion of the Universe

It has often been claimed that there is nothing in the external world that corresponds to our sense of the *direction* of time. Indeed, it has been argued that this idea is essentially due to the phenomena of consciousness and memory and that in the physical sciences the concept of "symmetrical" time suffices. None of the fundamental nonstatistical laws of physics (dynamics, electrodynamics, etc.) is affected if we imagine a reversal of time's arrow, and even in the case of entropy we encounter the problem of fluctuations. Is it a miracle or a highly unlikely coincidence that the principal scientific theories somehow manage to cooperate with each other so as to conceal the one-way trend of time from us? According to Mehlberg, "There would be neither a miracle nor an unbelievable coincidence in the concealment of time's arrow from us only if there were nothing to conceal—that is, if time had no arrow. On presently available evidence time's arrow is therefore a gratuitous assumption."[7] I shall argue that this is too sweeping a conclusion to draw.

Apart from human phenomena of consciousness and memory, evidence for the one-way trend of time in the universe comes from the study of processes associated with long intervals of time extending over many millions of years. The irreversibility of organic evolution, often referred to as Dollo's law, is attributed to the comparative improbability of a particular combination of a given set of mutations and a given environment repeating itself, so that the chances of retracing the steps of evolution decrease rapidly with increasing complexity of organisms and environments. According to this view, new mutations lead to new modes of adjustment of organisms to their environments and the subsequent effects of natural selection give rise to that characteristic feature which leads us to think of evolution as a one-way process.

Turning to purely physical phenomena, radiative processes, particularly those associated with the sun and stars, appear to be essentially irreversible. Although we now believe that stars like the sun can continue to shine steadily for thousands of millions of years, the nuclear transformation ultimately responsible for this radiation cannot continue indefinitely. More generally, in the universe at large, this process is repeated on a tremendous scale, localized sources continually dissipating energy into the depths of space. Even if one assumes that all photons are ultimately absorbed so that the radiative process itself becomes time-symmetrical,[8] it is difficult to see how there could be localized sinks of energy corresponding to stellar sources. Thus, the mere fact that stars and galaxies are visible bodies seems to be evidence for unidirectional time, irrespective of whether these sources ultimately become extinct.

Both on the terrestrial and on the celestial scale there is abundant evidence of temporal trend in the universe when sufficiently long intervals of time are considered. But the most intriguing evidence for associating time with the universe as a whole is presented by the spectra of the extragalactic nebulae, or galaxies.

In 1912 V. M. Slipher of the Lowell Observatory at Flagstaff, Arizona, first obtained the spectrum of the great nebula in Andromeda and found that it was displaced to the blue, indicating that this galaxy is moving towards us at the rate of 125 miles per second. We now believe that this relative motion is partly due to the solar system's own motion around the center of the Milky Way. By 1917 Slipher had photographed the spectra of fifteen spirals and found that all but two were receding with average velocities considerably greater than those of the stars. Later work by Humason and others has shown that almost all galaxies exhibit red-shifts in their spectra, corresponding to velocities of recession up to well over a half that of light, and there is no reason to regard even this immense speed as an upper limit. In 1929 Hubble found that out to distances of a few million light-years there was a linear relation between shift and distance so that the greater the distance the larger the shift. Subsequent work has extended the range of Hubble's law, although estimates of distance become increasingly uncertain for the more remote galaxies.

The observed distribution of the galaxies in the sky is extremely irregular. In particular, there is a zone of avoidance in the region of the Milky Way which is thought to be due to the existence of diffuse obscuring matter in the outlying parts. However, when the observed distribution is corrected for this effect, it is found to be much more nearly isotropic. This result is generally regarded as powerful, if not wholly conclusive, evidence that the system of galaxies forms the framework of the whole physical universe. Consequently, the extragalactic red-shifts suggest that the universe as a whole is expanding. Although other interpretations of these spectral displacements have been advocated from time to time, none has received anything like the degree of support that has been

571

accorded to the Doppler effect interpretation in terms of recessional motion, and serious objections have been raised against all other theories that specify a precise physical mechanism. Thus, the only viable alternatives would seem to be that either the universe (or system of galaxies) is expanding or else there is an otherwise unknown law of nature operating on the cosmic scale that has no appreciable effect on the scale of ordinary stellar astronomy. Naturally, astronomers tend to favor an explanation in terms of relative motion rather than appeal to an otherwise unknown effect and will continue to do so until there is good reason to abandon this interpretation of the red-shifts.

One of the first to see that the idea of the expansion of the universe could be used to explore the concept of time was the late E. A. Milne. He argued that, whereas the Newtonian universe *has* a clock, the expanding universe *is* a clock and that time's arrow (to use Eddington's famous expression) is automatically indicated by the recession of the galaxies. In his novel approach to this question, Milne began by drawing an analogy between the expanding universe and a swarm of noncolliding particles moving uniformly in straight lines.[9] If they are contained in a finite volume at some initial instant when they are moving at random, they will eventually form an expanding system. Even if the system is originally contracting, it will eventually expand. On the other hand, an expanding system of this type will never of its own accord become a contracting one. This point can be illustrated even when the system contains only two particles. If initially they are approaching each other, eventually they will be found to be moving apart. But, if initially they are moving apart, they will continue to move apart and will never approach each other. This means that, if we took two film strips and found that in one the particles were approaching and in the other receding, we could immediately tell which picture was taken first.

Nevertheless, just as in the case of entropy the existence of fluctuations makes it impossible to define temporal direction solely in terms of entropic changes, so any attempt to define this direction in terms of world expansion presupposes that there is no reversal of the process. But, although expansion enables us to resolve Olbers' paradox concerning the background brightness of the universe,[10] it is by no means certain that a contracting phase of the universe would be a thermodynamic impossibility. It is, therefore, preferable to regard world expansion as indicating time's arrow rather than as the basic phenomenon that gives rise to it. In any case, the concept of expansion presupposes that there are non-simultaneous states and temporal flux.

4 Cosmic Time and the Expansion of the Universe

The hypothesis of world expansion has played a vital role in the history of the concept of time because it led to the revival of the idea that there are successive states of the universe defining a cosmic time. This idea, which was taken for granted until the present century, fell into abeyance as a consequence of the Theory of Relativity. For, Einstein's concept of the relativity of simultaneity, introduced in 1905, seemed to eliminate from physics the possibility of any objective world-wide lapse of time according to which physical reality could be regarded as a linear succession of temporal states. Instead, each observer was regarded as having his own sequence of temporal states and none of these could claim the prerogative of representing the objective lapse of time. Despite this, theoretical cosmologists studying the expansion of the universe were led, about 1930, to reintroduce the concept of world-wide time, so that the relativity of time became an essentially local phenomenon for observers in motion relative to the cosmic background.

This modern idea of cosmic time has since been challenged by Gödel. In a paper published in 1949, he showed that relativistic world models can be constructed, in which the local times of the so-called "fundamental" observers who follow the mean motion of matter in their neighborhood cannot be fitted together to form a world-wide time.[11] The existence of cosmic time in previous relativistic models of the expanding universe depended on there being no rotational motion in these models, so that at each point the directions of inertial motion are given by the directions of cosmical recession. In Gödel's model, on the other hand, the system of galaxies is seen from each point to be rotating relative to the local compass of inertia. This model was, however, open to the objection that it was theoretically possible for a sufficiently fast-moving observer to travel into his past or future and back again, thus making a complete circuit in time. It is easily seen that such a possibility leads to absurd consequences. More recently, Ozsváth and Schücking have found a locally rotating, expanding world model free from this defect.[12] Consequently, the existence of cosmic time would seem to be the prerogative of homogeneous isotropic world models in which the lines of relative motion of the fundamental "particles," representing idealized galaxies, determine everywhere the local compass of inertia.

5   *The Time Scale of the Universe*

If the universe is expanding, then it can only have existed in its present state for a finite time, although the actual measure of this time will depend on whether the recession of the galaxies is approximately uniform or was significantly different in the past. Hubble's law provides us with empirical information bearing on this question, although its precise interpretation is in dispute. According to this law, and assuming that the extragalactic red-shifts are due to recessional motion, the velocity v and distance r of a galaxy are connected by the relation

$$v = H\,r,$$

where H has the same value for all galaxies investigated. The parameter H, known as Hubble's constant, has the dimensions of an inverse time, but its accurate evaluation has been impeded by the fundamental difficulties of determining a reliable distance scale for extragalactic objects. Hubble's pioneer determination of the distances of the nearest galaxies, notably the Andromeda nebula, depended on the identification of constituent Cepheid variable stars, which it was known could be used as distance indicators. But to obtain a systematic relation between v and r it is necessary to go outside the so-called Local Group of galaxies, and existing telescopes are not powerful enough to reveal stars of this type in more remote galaxies. In some nebulae, however, notably those in the Virgo cluster, Hubble succeeded in identifying particular objects which he believed were the brightest constituent stars. By comparing their apparent magnitudes with those of the brightest stars in galaxies whose distances had already been determined, he was able to estimate how far away they, and the galaxies containing them, were. When he correlated the distances and red-shifts of these galaxies, so as to obtain his law, he found that the reciprocal of H was nearly two thousand million years.

In recent years Hubble's distance scale of the galaxies has twice been drastically revised. In 1952 Baade showed that the distances of the nearest, including the Andromeda nebula, had been underestimated owing to a mistake in the distance scale of Cepheid variables. This discovery affected the estimates of distance of more remote galaxies, since these depended on the distances assigned to the nearest. As a result of Baade's work, which was based on a detailed comparison of the Andromeda nebula and the Milky Way, it came to be generally accepted that all the distances assigned by Hubble to extragalactic nebulae must be multiplied by a factor of rather more than 2. Then, in 1958, Sandage found that the objects in the Virgo cluster, which Hubble believed to be highly luminous stars, were in fact regions of glowing hydrogen which look like bright stars but are believed to be intrinsically brighter. Consequently, these objects are more remote than had been estimated and as a result the distances assigned by Hubble to all galaxies in the Virgo cluster and beyond had to be increased a second time, although the scale factor in this case was rather imprecise. The total effect of the two revisions was to augment the distances of all but the nearest galaxies by a factor of roughly 5. Since these revisions had no effect on the measurement of red-shifts, it followed that the value of H in Hubble's law was diminished by this factor and its reciprocal correspondingly increased, the currently accepted mean estimate being now about ten thousand million years, with a possible range of three thousand million years either way.

The current uncertainty in the value of H, and hence of its reciprocal, is matched by that of its interpretation. If the universe is expanding more or less uniformly, then the reciprocal of H is a measure of the period of expansion from the time when the universe was in an initial state of maximum condensation. If, however, the universe expanded more rapidly in the past than now, then the time of expansion would be correspondingly less; on the other hand, if it expanded more slowly in the past, the time would be increased. Theoretical world models have been devised to cover all three possibilities.[13] Notable examples are Milne's model in which expansion is uniform, the Einstein–de Sitter model in which expansion is slowing down at a rate proportional to the two-thirds power of the time and the Eddington and Lemaître models in which expansion is increasing according to a more complicated law.[14]

As compared with Milne's model which, according to the currently accepted value of H, has an age of about ten thousand million years, the time that has elapsed since expansion began (if the universe is of the Einstein–de Sitter type) is about seven thousand million years. On the other hand, the corresponding figure in the Lemaître universe is not less than twenty thousand million years. All these models are examples of what are sometimes called "exploding universes," but this term is hardly suitable to describe Eddington's model. This is similar in many respects to Lemaître's model, except that it begins as a static universe in a state of unstable equilibrium. A more comprehensive term to describe all these different forms of expanding universe is "evolutionary," because in each of them there is a systematic change in time of the model as a whole due to the effect of expansion.

In 1948 a very different type of world model was introduced by Bondi and Gold.[15] They accepted the Doppler interpretation of the red-shifts and hence the hypothesis that the galaxies are receding from each other, but at the same time argued that the universe as a whole must be in a steady state exhibiting no systematic change in time. To reconcile these otherwise incompatible hypotheses, they assumed that throughout the universe matter is being continuously created *ex nihilo* at a uniform rate. The possibility of any direct observational test of this assumption was ruled out because a simple calculation revealed that the required rate was equivalent to the mass of a hydrogen atom in each liter of volume every five hundred thousand million years. Bondi and Gold made no appeal to the field equations of general relativity, but shortly afterwards a similar world model was constructed by Hoyle, who introduced new terms into these equations to allow for continual creation of matter.[16] These theoretical developments gave rise to considerable controversy, but so far all attempts to discriminate between evolutionary and steady-state models by observational tests have been inconclusive, due to the inadequacies of the observational data at the limits of current instruments.

The most recent and exciting attack on this problem has been made by radio astronomers, whose instruments are thought to be capable of penetrating farther into the depths of space than current optical telescopes. Because all electromagnetic waves travel with a finite velocity and not instantaneously, remote sources of radiation are observed as they were in the distant past when the radiation which we receive now was emitted. Consequently, information thus obtained from the depths of space should give us some clue to the conditions prevailing at an earlier stage in the history of the universe. In particular, if data are collected over a sufficiently wide area of the sky, it should be possible to determine whether the universe was more densely packed in the past than now. Any indication of this would be powerful evidence in favor of evolutionary as against steady-state theories of the universe. Radio astronomers at Cambridge led by Ryle have studied some two thousand radio sources distributed over the sky and conclude that most of them are extragalactic and extremely remote. Moreover, they claim that these objects are more numerous in a given volume of space at the limits of observation than nearer to us. If correct, this would be difficult to explain on any theory of the universe based on the assumption that conditions in the past were similar to those prevailing now. An attempt has, however, been made by Hoyle and Narlikar to reconcile the steady-state hypothesis with Ryle's results.[17] They have constructed a world model that maintains a statistically steady state but with large-scale fluctuations in space and time. To accommodate Ryle's results the scale of the fluctuations must be comparable with the size of the observed universe. Unfortunately, this rules out any possibility of an observational check on the basic steady-state hypothesis and hence of deciding between it and evolutionary cosmology, although in the circumstances the latter might be thought to provide a less artificial interpretation of the phenomena.

Meanwhile an entirely different line of approach to the problem of the cosmic time scale has been developed as the result of investigations on the ages of star clusters and galaxies. According to evolutionary theories of the expanding universe, star clusters and galaxies would be expected to have ages not exceeding that of the universe. On the steady-state hypothesis, although the universe is infinitely old, it can easily be shown that the mean age of matter in a given volume of space is only one-third of the reciprocal of H and that not more than one galaxy in twenty would have an age greater than the reciprocal of H.[18] Consequently, there should be a preponderance of galaxies less than four thousand million years old, whereas in many evolutionary models a considerably higher average age is possible.

In the last few years, ages of star clusters and galaxies have been estimated by studying their components and applying the results of stellar evolution calculations. Many of the ages thus determined have been found to be significantly greater than the reciprocal of H. For example, ages of about twenty-five thousand million years have been assigned to globular clusters surrounding the Milky Way, but some astrophysicists think that these may be overestimates by a factor of about 2.[19] In the case of some galaxies, calculations based on estimates of the mass ratio of diffuse matter to stars and of the stellar formation rate have produced even greater ages.[20] These high values seem difficult to reconcile with almost any theory, evolutionary or steady state, but the whole subject is in such a state of flux that any conclusion that can be drawn at the present time must be regarded as extremely tentative.

Despite this, it is surely significant that many estimates of age of different stellar systems are of the same general order of magnitude as the reciprocal of Hubble's parameter H, suggesting that the time scale of the universe may yet be found to be of the order of ten thousand million years.

6   The Horizon of Time

In his *Essay concerning Human Understanding*, John Locke concluded a chapter on space and time by declaring that "expansion and duration do mutually embrace and comprehend each other; every part of space being in every part of duration, and every part of duration in every part of expansion."[21] Until 1917, no one seems to have had cause to regard this statement as anything but a truism. In that year, however, de Sitter constructed a world model in which cosmic time is subject to a curious and previously unsuspected limitation. In the experience of an observer A located at a given point in the model, there is a finite horizon at which time appears to stand still, as at the Mad Hatter's tea party where it was always six o'clock. This time horizon is, however, only an apparent phenomenon and the time flux experienced by any observer B on this horizon will be the same as that experienced by the observer A. This strange effect occurs because the time required for light, or any other electromagnetic signal, to travel from B to A is infinite.

In de Sitter's original formulation of his world model there was no cosmic time, but it has since been discovered that the most appropriate way to describe his universe is to regard it as a limiting form of the expanding universe of Lemaître and Eddington when the mean density becomes vanishingly small. In this case, since gravitational attraction is effectively zero, the only world force is cosmical repulsion. A different world model with the same space-time geometry is possible, however, if the field equations of General Relativity are no longer imposed. This is the steady-state model with continual creation of matter, in either the Bondi-Gold or Hoyle formulation. This model involves the concept of cosmic time, but it is associated with the idea of an apparent horizon of time in the description of the universe by an observer located in any galaxy.

In some expanding world models there exists another type of horizon at any instant t_0, which divides all galaxies into those already observable from a given galaxy at t_0 and those not yet observable. The Einstein–de Sitter universe has this type of horizon. Some world models, for example, Lemaître's, possess both types of horizon. As an aid to visualizing the two concepts, we can picture the universe as an expanding balloon. Strictly speaking, only the surface of the balloon should be imagined, and although the analogy suggests a closed universe, the argument for open universes is similar. The galaxies can be represented by large dots distributed uniformly over the fabric of the balloon. One particular dot may be associated with a given observer A. Light signals can be represented as small dots moving over the balloon at a constant speed relative to the surface of the balloon. A time horizon of the first type will exist for A, and for all other similar observers, in models where the rate of expansion is, and remains, sufficiently great for some of the small dots moving towards *A* never to reach *A*. In Eddington's graphic phrase, light is then "Like a runner on an expanding track with the winning post receding faster than he can run."[22] On the other hand, a horizon of the second type exists for *A*, if, for example, the balloon expands from an initial state so that at first the rate of expansion exceeds the speed of the small dots, with the result that a finite time must elapse before any given one of them can reach *A*. None will reach *A* unless the rate of expansion decreases from its initial value, and some will never reach *A* if the rate of expansion, after first decreasing, increases again suitably. This happens in a model with both types of horizon, such as Lemaître's universe, which explodes violently and later passes slowly through an unstable equilibrium state (corresponding to the initial state of Eddington's universe) and then expands slowly at an ever-increasing rate.

So far we have assumed that *A* remains anchored to a particular galaxy so that he keeps cosmic time, but if he is allowed to move through the universe (with local speed less than that of light, of course) then the class of events observable by him is increased. If the model possesses a horizon of the first type for *A* before he moves, then he can never move so as to be able to observe every event in the universe. His time horizon will change but can never be wholly abolished.

Most of the world models that have been actively studied as possible forms of the expanding universe possess one or other type of horizon. A notable exception is the uniformly expanding universe studied by Milne. In this sense, the latter possesses a unity in time which is not shared by the other models. In the actual universe we must consider seriously the possibility that events can occur, knowledge of which can never be brought, even in principle, to a given observer, however long he lives, and so can never enter his temporal experience.

7   *Time and Physical Reality*

At the beginning of this essay I said that time and the universe should be regarded as closely related concepts. The fundamental significance of the *passage* of time in relation to the universe has, however, often been disputed by philosophers, and also by many physicists since the general acceptance of the Special Theory of Relativity. For, according to this theory, the time order assigned to certain events is not the same for all observers in relative motion. The conclusion is therefore often drawn that there can be no objective time order for such events.[23] Moreover, in this theory the world is depicted as a four-dimensional manifold in which space and time are linked together so intimately that any separation of the two depends essentially on the observer and is purely subjective. Consequently, many relativists take the view that physical events are already there and do not happen. We merely encounter them in the course of our experience. In the words of Hermann Weyl, "Only the consciousness that passes on in one portion of this world experiences the detached piece which comes to meet it, and passes behind it as history, that is as a process that is going forward in time and takes place in space."[24]

There is no need, however, to accept this basically timeless view of physical reality. Important as the theory of relativity is for our understanding of the concept of time, it does not provide us with a complete account of the subject. If we reject the Newtonian idea that time exists in its own right and agree that it depends on phenomena, we ought not to conclude that it has no objective existence without taking into account the actual pattern of events in the universe. In other words, we must not overlook cosmological considerations, including the role of privileged observers associated with the bulk distribution of matter. These "fundamental" observers form a more restricted class than those contemplated in special relativity. In many world models, as we have seen, they keep a common cosmic time and all events have an objective time order for them. Only when we consider other observers moving through the models with different local speeds do we encounter the discrepancies in temporal order of events that I have mentioned. It is clear that, in this cosmological perspective, these anomalies are due to the motion of the observers in question and not to the events themselves.

There is, however, an important exception to the statement that, for observers keeping a common cosmic time, all events have an objective time order. Strictly speaking, in this context one should speak only of "macroscopic" events. There is evidence that microscopic phenomena have no intrinsic time direction from a macroscopic point of view. Thus a positron (the antiparticle of opposite charge to the electron but of the same mass) can be regarded as an ordinary electron "traveling backwards in time," the physical effects of this time reversal being counteracted by the change in sign of the electric charge. This idea has been invoked to account for the curious phenomena of pair-production and pair-annihilation,[25] but it implies that not all sequences in time can be subsumed under a universal time order. Reichenbach has described this as "the most serious blow that the concept of time has ever received in physics."[26] He concludes that time as we normally understand it must be essentially statistical in character, and that ordered and directed time arises only because positrons (and other antiparticles) are short-lived in the presence of those particles, such as electrons, which conform to the rules of ordered and directed time.

The association of time with statistics has been studied by many physicists and repeated attempts have been made to account for time's arrow on statistical grounds. As we have already seen, any attempt to identify increasing time with increasing entropy comes up against the problem of world entropy. Similarly, the recent work of Penrose and Percival[27] on the direction of time depends on what is essentially a cosmological postulate concerning the existence of real independent events in the universe.

Whether statistics can account for the arrow of time, or whether, as I prefer to believe, time must be regarded as an ultimate feature of the universe, and therefore not dependent on something even more fundamental, it is clear that time and the universe are closely associated. Time, as we normally understand it, is a macroscopic feature of the universe. The essence of time is its *transitional* nature, and no theory of time can be complete that does not account for the fact that everything does not happen at once. But whether everything can be subsumed under a universal time order is an open question. We must be careful to distinguish between properties of the universe and universal properties.

There is, I believe, a profound connection between the temporal and contingent aspects of the universe. The peculiar situation that most of the general laws of physics conceal the one-way trend of time from us may be due to the fact that these laws are essentially abstract general statements which have to be supplemented by so-called initial conditions when we apply them in practice. These conditions refer to individual events whose temporal characteristics are a feature of the universe itself and not of the theories (dynamics, electrodynamics, etc.) in question. These enable us to calculate chains of events but give us no clue as to which of these chains in fact occur. The actual occurrence of any event, even if correctly predicted by a theory, always depends on one or more contingent factors in our calculations, that is to say, factors that cannot be found in the theory or theories to which we appeal. These theories stand at one remove from physical reality and have a timeless quality about them insofar as we assume that we can apply them at any time. The fact that certain fundamental laws of nature appear to be compatible with temporal reversibility should not, therefore, be interpreted as conclusive evidence that there is no objective time order and that any apparent asymmetry of time is simply due to the way the human mind operates. The world and time coexist and each depends upon the other.

The Study of Time

The voices of time are numerous; on the preceding pages we have recorded a few, omitted many others. Here I shall confine my remarks to certain observations, opinions, and reflections which in aggregate will suggest that a successful study of time, because of the nature of the problems involved, must be interdisciplinary in approach.

1   Time and Its Measurement

Although the practical measurement of time is primarily a matter of skills and conventions, questioning the scope of time measurement leads with impressive directness to the fundamental problems of thought, nature, and man.

Consider first two local events and a single local clock: mechanical, atomic, biological or a person making judgments of durations. What properties of the physical world give rise to or correlate with the local time whose passing between the two reference events the clock is said to indicate? At least three such properties have been seriously considered: the tendency of entropy increase of adiabatically isolated systems, the observed recession of distant galaxies, and the statistical lack of antiparticles.[1] In principle, any physical variable which can be suspected to change consistently with time, such as the Newtonian Gravitational Constant,[2] or the presumed accumulation of angular momentum with age[3] might also be suspected of "cluing" the clock. If it is to be taken seriously, a proposal put forth as representing that basic feature of the world which controls the indications of local clocks must be shown to be consistent with our idea of processes in time: for instance, that the universe is always expanding. It must also be demonstrated that a link exists between the selected physical property of things and our clocks, or generalized clocks: for instance, between the thermodynamical "arrow of time" and organic evolution. Finally, the relationship (or its absence) of the favored physical property to other physical parameters of nature that have been proposed for similar reasons must be clarified: for instance, the increase of the constant of gravitation (if that be the case) to the statistical lack of anti-particles.

Assuming now that we have two or more clocks rather than one, and without questioning the reasons of their operations, we may ask: What do we know of the phenomenology of their relative indications? We will find it suitable to think of two visually detectable distant events anywhere in our galaxy and to pretend that we measure the time that has elapsed between them by clocks of various constructions and in various states of relative motion. Subsequently, we will vary the spatial and the temporal separation of the two reference events according to simple rules.

Two identical mechanical clocks will give generally different indications of the elapsed time if the clocks are in relative translational motion or if they are at relative rest but in gravitational fields of different magnitudes. The first of these effects would not be expected to occur if we were to reinterpret the Lorentz transformations of Special Relativity Theory according to Schlegel's prescription. It is not clear how the readings would compare if one clock was at rest, say, on the surface of a star, while the other one was in an accelerating frame, because for a variable velocity the relativistic proper time has no direct physical meaning.[4] We may replace one of the clocks by a timepiece controlled by atomic resonance. We might then find that, even if located next to one another, they would get out of step with their readings related through a logarithmic transformation.

We may employ as clocks the chirping frequency of crickets or the periodic loss of water by lemon cuttings. These and many other biological clocks are remarkably precise, though they best demonstrate their usefulness as time-keepers when acting in aggregate. Assemblages of physiological clocks controlling the great variety of rhythm in the living proved sufficiently reliable to make animal ecology possible and must therefore be considered from the point of view of time measurement on equal rank with inorganic clocks. Insofar as the average of the multitudes of physiological rhythms operative in man may be regarded as equal to the psychological time sense, we may also employ as clocks the judgment of duration of people other than ourselves, or, preferably, construct a clock by using time judgments of groups of people. In all cases, stores of biological and psychological clocks are expected to be subject to our motional changes identical with those of the inorganic clocks discussed above. Superimposed on such changes we will find that, as compared with the inorganic clocks, the accuracy of our living timepieces will depend on such a vast and enormously complex array of physiological and psychological conditions that the very idea of grouping them together in the first place may be seriously challenged.

As we increase the spatial separation of the reference events by imagining more and more "rigid" rods placed between the places where they happen, we must face the problem of cosmic time. If Special Relativity Theory holds only for each "fundamental particle" or galaxy of the universe as is suspected it may, and if time is determined for each galaxy by the average configuration and motion of its stars, then a cosmic time may well permit similar readings of similar clocks at rest in different galaxies, regardless of the basic relative motion of those galaxies. In all cases, however, we must face the problem of time horizons and the possibility that one or both of the events we have thought of could never enter the history of some or all of our clocks. As we decrease the spatial separation between the events, we will reach the limiting condition when they happen at the same place with respect to ourselves. What we then measure between their occurrence is a proper time interval as defined in Special Relativity Theory. As we have seen it in the Comments about time and proper time, the unique properties of proper time intervals make that concept essential for the description of the four-dimensional physical world.

We may now vary the proper time span between the two events and face some fundamental minimal and maximal problems. How short can an interval become and still remain of the nature of time? Quantum Theory suggests that a lower limit may exist in terms of natural limitations to conceivable causes and effects. But, the temporal atomicity underlying this statement is not generally accepted and the model of time used, at least in the physical sciences, is likely to remain that of infinite divisibility. In any case, here we join the classical dialogue of atomicity versus continuity. On the maximal extension of proper time intervals we come upon the questions of the origin and end of time, and must broaden our inquiry to cosmology and philosophy, while not forgetting the age-old concern of man with the presumed supernatural aspects of the beginning and end of the world. Whatever conclusions are reached regarding the finity or infinity of the age of the world, or regarding the microstructure of time, they are bound to have profound influence on man's views about his place and purpose on earth.

An intellectually satisfying theory of time with respect to its measurement will have to accommodate those astronomical and microscopical properties of the physical world which can be measured by conceivable clocks, include the nature of organic or inorganic clocks in its purview and acknowledge the nature of man, the formulator of the problem.

2 The Many Aspects of Permanence and Change

From the classical antinomies of time I have selected that of permanence and change as one example of the interdisciplinary nature of these famous paradoxes.

It is a common observation that a physiological or existential knowledge of "time," a type of expectation expressed, for instance, by seasonal preparations, is operative in all plants and animals. The concept "existential" is used here in a sense similar to that suggested by Polanyi,[5] designating an understanding of something that has a meaning in itself. For organisms above the level of bacteria and viruses, cyclic repetition is the rule of the living, being essential for survival through the maintenance of ecological routine. The recognition of the permanent relations in reality, on the other hand, is more of a cultural prerequisite. Human societies, based as they are on tool-making, record-keeping, and planning, require belief in existents which do, or appear to transcend change. The insight afforded by the comparative study of religions and by the historical study of civilizations lends support to the view that recognition of our passing and the simultaneous desire for permanence have played a central role in creating religions and civilizations.

Although experimental work on the psychology of time sense has thrown light on the dependence of time perception on environmental conditions and mental attitudes, there are very few theories regarding the origin of the sense of time. Perhaps the most complete one is that put forth by theoretical psychoanalysts, who view our knowledge of time as an expression of an instinctual ambivalence which stems from a separation of life and death into a dichotomy, in place of their biological unity. Through this type of dualism the certainty of death that is permanence, combines with our experience of life that suggests change, into an unconscious basis of our knowledge of time.

Scientific laws are regarded as principles of permanence; it is for this reason that they reach their most convincing forms when expressed in the time invariant language of mathematics. If, however, such laws are to predict individual events or chains of events, they require the introduction of contingent factors in the form of initial or boundary conditions. Inasmuch as our experience of change must precede our theories of permanence, it is convenient to regard the former as epistemologically more primitive than the latter. When examined through scientific laws, the fullness of the world appears to us as properly represented only if formal expressions can be found embodying the permanent and accommodating the changing nature of reality.

In the arts also, time experience involves perceptions of change and permanence. Artistic rhythm is not simply a cyclic repetition, for it requires rather that the artist as well as his audience be capable of grasping and responding to the enduring and the varying contents of the work of art, its structure and becoming, as it were. Rhythm in the many arts derives its force not from reiteration, but, in the vocabulary of the aesthetes, from being an image and a reminder of time.

We may speculate that the ability to integrate an existential knowledge of change with a belief in permanent existents is that specific property of the mind which gives rise to our knowledge of time. This hypothesis does not contradict the understanding of man's time sense as an evolutionary adaptation. On the contrary, it permits the successive emergence of three major steps of increasing sophistication: 1) concern with the future, corresponding to expectation of cyclically repetitive events; 2) memory of the past, that is, realization of continuities in the world, and 3) awareness of the present, that is, integration of expectation and memory. In contrast, what we hold to be the time ignorance of animal life consists of the animals' inability to act as though able to distinguish the inevitable from the contingent elements in matters which appear to be of interest to them. Consequently, they do not attempt to influence only the changeable and try to live with the inevitable. The power and skill of man which arises out of his labors to do precisely these is not available to other living creatures. To perceive the world in logically contradictory terms, as embodied, *e.g.*, in the idea of time, seems to be a privilege of man.

A theory of time with respect to the logical properties of time cannot rest content, therefore, with philosophical analysis only but must delineate the history and account for the evolution of man's knowledge of time, and place the idea in perspective regarding its role in the organization of life and in the design of the physical universe.

3   Time, Space, and Personal Identity

The success of the spatiotemporal formalism of Relativity Theory in producing an accurate and satisfactory model of the physical world is sometimes regarded as evidence that time is an illusion, for that theory emphasizes the extensive rather than the temporal aspects of things. Indeed, there is a clear temptation to speculate that the time ignorance of the unconscious, the time ignorance of animals, or even the ill-defined mental present of alleged precognition[6] reflect the same fundamental world structure as does the time invariance of four distances. Consequently, contemporary idealistic thinkers following the lead of Parmenides and Kant, sometimes conclude that the universe is a static, four-dimensional hyperspace already containing all events, of which we become aware as our attention shifts along a coordinate axis of timeless time. From such a universe time has been "eliminated."[7] But if time is unreal and it constitutes not only formally, but ontologically as well, an extensional dimension of a four-space, who will tell us why we can still ask with Villon, "Mais ou sont les neiges d'antan?"

As long as time was thought to be independent of other variables of physics, the properties of time could not have been subjected to experimental test. Empowered, however, by the insight of Relativity Theory, we are now able to think of time as an organic and functional part of the universe. The theory revealed the curious situation that nature permits only certain almost-symmetries between spatial and temporal measurements.

One such almost-symmetry is the requirement for opposite algebraic signs for the squares of the spatial and temporal quantities in the expression of proper time. Thus, proper time, the invariant element of relativistic space-time, carries with it the hallmark of some deep-seated difference between our experiences of space and time. Combining the requirements posed by the necessary geometrical structure of the brain with some results of classical orbit theory, Whitrow[8] has found convincing reasons why the number of dimensions of physical space must be three. If his view or views similar to it are accepted, there is nothing surprising in the fact that the different experiences man has of time and of space are reflected, *e.g.*, in different mathematical signs when spatial and temporal measurements come to be considered coordinate dimensions of a single four-continuum. The reasons why we prefer this specific type of differentiation between the metrics of distances and times might be elucidated by the life sciences and the sciences of the mind.

While both time dilation and the Lorentz contraction of moving bodies must be considered real, time dilation is, so to say, more real. For, motional contraction leaves no traces on a body; a traveling rod does not return bent; but relative temporal gain if there be such, can be public knowledge. However, if two traveling objects were made up of particles intrinsically indiscernible and identically arranged in each object, the difference in age as the objects return to rest next to one another could not be given meaning. Time itself, as far as these objects are concerned, would be difficult to think of. This difference in the spatial and temporal changes, another almost-symmetry of Relativity Theory, not only reconfirms certain further differences between the nature of space and that of time, but it implies that time must be associated with potentially discernible properties of things. The simplest such discernibles of a world of similar particles resides in different spatial configurations. And here we find one area where the fundamental necessity for time's being relational may originate. The other area is the necessity of referring to the distant masses of the universe for the determination of astronomical time.

Although the significance of speeds larger than that of light remains an open question, all finite objects must, by the requirements of the Special Theory of Relativity, experience a temporal component of the four-dimensional world. There is no indication that any one such temporal section of four-space would be of a nature different from any other; in no spaceship can a tree grow before it is planted. Only a photon-cum-clock could register a world of no temporal (and no spatial) character, if "registering" under such conditions could be given a meaning.

The idea that time is a space-like extension of a static world, and hence essentially illusory, receives, as I see it, no substantial endorsement from physical arguments. However, there are strong suggestions from the study of behavior regarding the existence of time ignorance in certain levels of our mental structure. In this context we must carefully distinguish between the idea that time is an illusion and the attribute of time ignorance. The former is a philosophical view, the latter is the absence of time perception. There are reasons to believe that the unconscious is ignorant of the temporal aspect of the world: it knows nothing of before and after, just as it cannot differentiate between subject and object. Equally ignorant of the existence of time is animal life. The suggestion of the broad mental present of the subliminal self mentioned earlier also implies a degree of time ignorance rather than time independence, representing perhaps a regression of sorts to earlier evolutionary states. The sense of time as we understand that concept in man emerges in successive steps in the growing individual, in the evolving human race, and in organic evolution in general. But, if there is evidence of time ignorance among animals and on the primitive levels of the mind, does this not suggest that time is not at all an image we impose on the world, but rather a discovery which we are capable of making as the tools necessary for that discovery are acquired?

Philosophical analysis of time has concentrated a great deal on the qualities of becoming versus being. This dialogue spans the complete tenure of Western thought. It has been following in broad terms the division between, on the one hand, disbelieving in the reality of time and supporting a deterministic and essentially static universe with emphasis on extensional properties of the world, and, on the other hand, believing in time as an intrinsic feature of the universe, with concomitant endorsement of the concept of free will and of a dynamic universe. Yet, the question most intimately affecting the individual is not so much whether truly new things do or do not come into being, but whether truly existing things, mainly people, do or do not go out of being, as it were, through death. Acts of creation are hard to observe, while acts of annihilation are all too apparent. Insofar as death is our most powerful reminder of finiteness, and finiteness of our concern with time, I think that no school or manner of thought including the philosophy of science can hope to come to grips with the problem of time without offering a satisfactory analysis of the relation of time to death.

It is interesting to speculate that knowledge of space, something I believe we share with much of the animal kingdom, relates to the development of self-powered locomotion. Because such locomotion can be considered an early evolutionary adaptation, so may the knowledge of space be considered a primitive stratum of consciousness. With the growth of man's mental powers, a more sophisticated adaptive machinery emerged in the form of the successive discoveries of future, past and present. This new and powerful tool which we recognize as man's knowledge of time, is coeval and necessary for the establishment of his personal identity. For, only through the combination of expectation and memory can he know in what cumulative way he is different from others. In turn, the discovery of one's personal identity is but a way of recognizing his self-awareness.

The agency that we usually hold responsible for the control and modification of personal identity is referred to in ordinary language as the "mind," which thus appears to be something fundamentally temporal in nature. One consequence which follows is that for a better understanding of the relation of mind to those biological functions of man which he shares with other living creatures that do not exhibit personal identities we should look to studies of the temporal organization in man. The other consequence is that a world model which can accommodate not only our biological functions or bodies, but also the unique operational characteristics of the brain should not be sought through constructions of higher spaces but through an improved understanding of the strictly temporal as well as of the quasi-spatial aspects of time.

4   *The Knowledge of Time*

In the Introduction we have made the assumption that when specialists speak of time, they speak of the various aspects of the same entity. This assumption is not equivalent, however, to saying that a precise definition of this entity can necessarily be given. In fact, we have refrained from trying to propose a single answer to the question "What is Time?"; for that question is only a subtle way of asking a multitude of questions, having the same form, concerning thought, man, matter, life, and death. Consequently, the clarification of the concept of time must be preceded by, or be concurrent with, inquiries designed to understand time in its relation to these diverse entities, somewhat as we have done it in this book. We did hypothesize, however, that our perception of time relates to the combination of an organic or existential knowledge of rhythm with a belief in permanent existents, and that the emergence of this combination in man may be thought of as an evolutionary adaptation to reality. On this view, time is that basis of man's perception of the world which lends him uniquely strong support in his struggle for survival and in his attempts to control his environment, and which makes possible his cultural growth. Through time's kinship to the feeling of personal identity, the idea of time enters all expressions of an individual's concern with himself and with his fellow men as registered, for instance, in the arts and letters. Through the technique of employing time in the intellectually satisfying description of the universe, including ourselves as both observers and objects, the idea of time enters the domains of philosophy and science.

In view of the ubiquity of the concept of time, it is not surprising that it can be described many ways. It can be thought to consist of our peculiar interpretation of the probabilistic laws of the macroscopic world, or believed to derive from the existence of contingencies in the universe, *e.g.*, as manifested by certain formal requirements of physics. It can be argued that time is identical with creativity otherwise assigned as the numinous task of a Supreme Being, or that time is an expression of the unresolved conflict between Eros and aggression in the instinctually ambivalent psyche of man. Regardless, however, of the method we employ to understand the idea or represent the feeling of time, our concern with it is always characterized by a specific human skill or knowledge (*sophia*), pertaining to time (*chronos*), which skill and knowledge may therefore be called *chronosophia*. Although all men seem to possess some degree of such knowledge, it is not unreasonable to reserve the name *chronosophy* to designate the interdisciplinary and normative study of time *sui generis*. It is not within the purview of this book to argue and propose a methodology or try to define the precise content of chronosophy, but only to hint at its possible usefulness and suggest its feasibility. We may, however, speculate freely and list some thoughts that come to mind.

The *purpose* of chronosophy could be constructed, for instance, from the elements listed in the Introduction as the purpose of this book. They can be restated in slightly altered form as five intentions:

1. to encourage the search for new knowledge related to time;

2. to set up and apply criteria regarding which fields of knowledge contribute to an understanding of time, and what they may contribute;

3. to assist in epistemological studies, especially in those related to the structure of knowledge;

4. to provoke communication between the humanities and the sciences using time as the common theme; and

5. to help us learn more about the nature of time by providing channels for the direct confrontation of a multitude of views.

Speculations about the possible methodology and *organization* of chronosophy are more dangerous and difficult than marshaling suggestions of reasonable goals. It seems to me that for a successful study of time we will have to encourage communication across the traditional boundaries of systems of knowledge and seek a framework which, conceivably, may permit interaction of experience and theorizing related to time without regard to the sources of experience and theory. We might create, for instance, categories such as these:

1. surveys of historical and current ideas of time in the sciences and in the humanities;

2. studies of the relation of time to ideas of conceptual extremities such as a) to motion and rest, b) to atomicity and continuity, c) to the spatially very large and very small, and d) to the quantities of singular and many;

3. comparative analysis of those properties of time that various fields of learning and intuitive expressions designate unproblematically as "the nature of time";

4. inquiries into the processes and methods whereby man learns to perceive, proceeds to measure, and proposes to reason about time;

5. exploration of the role of time in the communication of thought and emotion;

6. search for an understanding of the relation of time to personal identity and to death;

591

7. research concerning time and organic evolution, time and the psychological development of man, and the role of time in the growth of civilizations; and

8. determination of the status of chronosophy vis-à-vis the traditional systems of knowledge.

The nature of these categories would require that chronosophy be regarded as an independent system of experiential, experimental, and theoretical knowledge about time.

It has been pointed out that those men who created the upheaval which became known as the scientific revolution called it by a different name; the new philosophy. Likewise, what we now call "science" is but a fruitful derivative of the richer, universal concept of "natural philosophy." The emerging need for interdisciplinary studies of time signifies, perhaps, the arrival of an era of a new natural philosophy, to complement science. The demand is, that for an understanding of time, the lure of specialization be abandoned and the humane and universal study of nature be espoused. It is quite possible that if it is to be logically sound, chronosophy will eventually resemble some ways of knowing more than some others but it cannot, by the character of the problems, lose its interdisciplinary perspectives. Such perspectives are forced upon us by the fact that the temporal nature of the world appears on all levels at which man can perceive reality. A complete and satisfactory theory of time must, therefore, satisfy a wide variety of specifications.

We may say without exaggeration that through the many fields of knowledge and methods of self-expression available to us the best minds of humanity have theorized about time, as did and do to some degree all men. The reason for this impressive historical effort might reside in a general and direct interest in time, comparable with that aiming at the discovery of natural laws or at the exploration of other planets. This and similar reasons I find valid, but I do not believe them to be sufficient. There exist other and more fundamental reasons in harmony with the views and perspectives expressed in this volume. Unlike most matters of interest to the inquiring mind, the problems of time cannot be exhausted or be expected to lose their puzzling character upon detailed analysis and in due course be replaced by other themes. Time occupies a privileged position among concerns of speculative and empirical nature because it is essential to man's search for his individual and social destiny. Therefore, the process of clarification and definition of the problems of time must be universal and continuous.

Epilogue and Prologue

The final thoughts of *The Voices of Time* are conceived under conditions which make an analytical conclusion appear meaningless. Instead, they favor an attitude expressible only through a description of that involvement of man in life which gives rise to the problems of time. While the setting is specific, the commitment implied is universal. Fifty miles south of these hills, down in the City, eight million people rush, run, and shake their fists against the passing of time. Largely deprived of the continuity of history and concentrating mostly on the present, they search for, but do not seem to find, meaning and purpose in life. But up here one may still be close to people and to the earth or just be alone and enjoy a fair opportunity of finding ideals to die for, and hence, good reasons to live for. And for men these are the significant aspects of time.

The Hudson River is barely visible to the northwest; behind its silvery lines the Catskill Mountains wait surrounded by lazy clouds. Except for that view, we see only the forest which contains us in its lonely eminence in a continuous now, in an epilogue and prologue to a present: the past and the future rolled into one. Downhill, the Appalachian Trail passes on its way from Maine to Georgia. This is that part of the late afternoon when the warmth of the sun spreads tranquillity on the rich, brown leaves of Autumn, and gaiety on the still green shrubs. Slowly but quite perceptibly the sun is journeying westward; elsewhere a spring day is about to begin. Once in a while the long, deep voice of a riverboat drifts up here, followed sometimes by an echo from the mountains. Otherwise the noise of the place comes only from the rustling of bushes and of high trees, and from the motion of small unseen creatures beneath the leaves. A warbler with a yellow crown and auburn side has been keeping me company, seemingly unmoved by the patent beauty of these instants. But now, in a moment she will be gone and her presence will be in the past.

We could not have concluded the prior paragraph by recording such a logically and linguistically difficult statement describing an event, were it not for the certainty that everyone who was able to read the preceding pages will also understand what was meant, though inevitably some will understand more of it, some less. When one speaks of time this kind of an appeal is always necessary. The limitation or circularity of man's articulate faculties so implied suggests that the fundamental nature of time is not suitable for discursive cognition, notwithstanding his remarkable skill and knowledge pertaining to time and his manifest desire to continuously clarify and define its problems. For the essence of time, like that of man's existence, is only a permission to partake creatively in a world whose contents and properties we may experience, contemplate, and share but never completely describe or precisely formulate.

> A child said *What is grass?* Fetching it to me
>   with full hands,
> How could I answer the child? I do not know what
>   it is any more than he.

J. T. FRASER

September 10, 1964
Manitou Chappel, New York

594

Notes and References

Introduction

[1] The definition is from Dagobert D. Runes, *Dictionary of Philosophy* (Paterson, N.J.: Littlefield, Adams, 1963).

[2] R. H. Knapp and J. T. Garbutt, *J. of Personality*, v. 26 (1958), p. 426.

[3] The terms *vectorial* and *oceanic* are currently preferred to dynamic-hasty and naturalistic-passive. R. H. Knapp, private communication.

[4] Three examples may be given as vaguely indicative of the idea we try to formulate.

According to Herbert Dingle in *The Scientific Adventure* (London: Pitman, 1952), p. 315, in the study of physical phenomena, if one can identify the variation of some measurable entity with time, one may also construct a manifold of this entity and time, in terms of the general laws of the phenomenon. "By a subsequent identification of the time scales of the various phenomena, the independent sciences thus created may be united into a single comprehensive science of physics." This has been followed, completely, for the study of motion (where the manifold is space-time) and Dingle showed how to apply his idea to radiation (in which the manifold is entropy-time).

In the field of biology, we find that "The analytical basis for distinctions between system and environment in biological systems should be applicable throughout the whole range of phenomena embraced by experimental biology. Perhaps the most obvious criterion to use is the time scale on which a particular field of study operates, and this in fact has been the most commonly used determinant for ordering the biological sciences into a linear array." B. C. Goodwin, *Temporal Organization in Cells* (London: Academic Press, 1963), p. 9. In a less formal and more profound way, Goodwin builds his thesis of a thermodynamic theory of time structure "arising from certain dynamic characteristics of cellular control mechanisms" on the observational background offered by "the experimental study of temporal organization in cells." *Op. cit.*, p. 5.

F. L. Arnot, in an enjoyable small book, has suggested what the title of his book describes: *Time and the Universe, a New Basis for Cosmology* (Sydney: Australasian Medical Publishing Company Ltd., 1941).

[5] The most notable in this respect is G. J. Whitrow's *The Natural Philosophy of Time* (London: Nelson, 1961), a scholarly and erudite work which is likely to become a classic contribution to the study of time. Some of the other works of recent vintage reflecting the same trend include S. G. F. Brandon's *History, Time, and Deity* (New York: Barnes & Noble, 1965); S. A. Toulmin and J. Goodfield, *The Discovery of Time* (New York: Harper and Row, 1965); R. W. Meyer, ed., *Das Zeitproblem im 20. Jahrhundert* (Bern: Francke Verlag, 1964); R. Schlegel's *Time and the Physical World* (East Lansing: Michigan State University Press, 1961); H. Meyerhoff's *Time in Literature*

(Berkeley: University of California Press, 1960); J. Campbell, ed., *Man and Time* (New York: Pantheon, 1957) and G. Poulet's *Studies in Human Time* (New York: Harper, 1956).

Notable among articles is a collection of seventeen papers in *Studium Generale*, v. 8 (1955), nos. 8, 9, 10; W. Tomb's "An Essay on Time," Princeton University, Department of Philosophy (1953); W. Voise's "Le Temps, l'Histoire des Science et la Philosophie," *Archives Internationale d'Histoire des Sciences*, v. 16 (1963), p. 155 and J. T. Fraser's *A Short Essay on Time* (unpublished, 1950), on file with the Tau Beta Pi National Honor Society. We should also take note of two lecture series and one symposium. "Time and Eternity," held at the Warburg Institute, London, 1962–63; "Aspects of Time," offered at Cambridge University during the Fall of 1964, and "Interdisciplinary Perspectives of Time," a symposium scheduled for 1966 by the New York Academy of Sciences.

6 See the essay by A. C. Benjamin in this volume.

Ideas of Time in the History of Philosophy

1 For further examples of this approach to the problem of time see O. K. Bouwsma, "The Mystery of Time (Or, the Man Who Did Not Know What Time Is)," *J. of Philosophy*, v. 51, n. 12 (1954), p. 341.

2 Attributed to Poinsot by Claude Bernard. See A. J. Lotka, *Elements of Physical Biology* (Baltimore: Williams and Wilkins, 1925), p. 17. St. Augustine has a similar passage, often quoted in this context, in his *Confessions*, tr. E. B. Pusey (New York: Dutton, Everyman's Library), ch. 11, p. 14.

3 J. Mc. T. E. McTaggert calls these two ordering characteristics the B-series and the A-series respectively. See his *Nature of Existence* (Cambridge: Cambridge University Press, 1921–27), v. 2, ch. 33. The peculiarities of the A-series permit us to make the somewhat startling, yet perfectly correct, statements that tomorrow today will be yesterday and that yesterday today was tomorrow.

4 The two terms are from William James, *Principles of Psychology* (London: Macmillan, 1901), v. 1, p. 609.

5 C. D. Broad, *Scientific Thought* (London: Kegan Paul, 1923), p. 83.

6 Quoted in M. F. Cleugh, *Time* (London: Methuen, 1937), p. 233. Kant has a similar statement, "Only in time can two contradictorily opposed predicates meet in one and the same object, namely, *one after the other.*" *Critique of Pure Reason*, tr. Norman Kemp Smith (2nd ed.; London: Macmillan, 1929), Transcendental Aesthetic II, sec. 5. By this he means simply that only by virtue of time may we say, for example, of a light that it is red and it is green; for while we are uttering the statement the light may change from red to green.

7 J. E. Boodin, *Time and Reality* (New York: Macmillan, 1904), p. 65.

8 Arthur Schopenhauer, *Fourfold Root of the Principle of Sufficient Reason* (London, 1889), p. 32.

9 A. N. Whitehead, *Concept of Nature* (Cambridge: Cambridge University Press, 1920), p. 73.

10 The numbering of the fragments is that given by John Burnet in his *Early Greek Philosophy* (London, 1908).

11 *Theaetetus*, tr. Benjamin Jowett, 152.

[12] *Cratylus*, tr. Benjamin Jowett, 401.

[13] *The Way of Truth*, fr. 8; Burnet, *op. cit.*, p. 199.

[14] I have followed the Burnet (*op. cit.*, p. 367) paraphrasing of Aristotle. Aristotle's wording is not nearly so clear: "If time is continuous, magnitude is continuous also, inasmuch as a thing passes over half a given magnitude in half the time taken to cover the whole: in fact without qualification it passes over a less magnitude in less time; for the divisions of time and of magnitude will be the same. And if either is infinite, so is the other, and the one is so in the same way as the other; *i.e.*, if time is infinite in respect of its extremities, length is also infinite in respect to its extremities: if time is infinite in respect of divisibility, length is also infinite in respect of divisibility: and if time is infinite in both respects, magnitude is also infinite in both respects." *The Student's Oxford Aristotle*, tr. and ed. W. D. Ross (Oxford: Oxford University Press, 1942), v. 2, *Physics*, 233ᵃ. For an attempted solution of Zeno's paradoxes see Bertrand Russell's *Our Knowledge of the External World* (Chicago: Open Court, 1915), Lect. VI. See also G. J. Whitrow, *The Natural Philosophy of Time* (London: Nelson, 1961), p. 135.

[15] Since, according to the best records, Zeno died seventeen years before Diogenes was born, the tale can hardly be true. But its point remains.

[16] "We must either admit Becoming or else explain the unreal appearance of Becoming." H. Lotze, *Metaphysics* (Oxford, 1887), p. 105. Or again, "Calling a thing an illusion or appearance . . . does no longer free us from the responsibility of how an eternal, static system could produce the illusion or appearance." J. E. Boodin, *op. cit.*, p. 65.

[17] *Republic*, tr. Jowett, *op. cit.*, 515.

[18] *Ibid.*, 610.

[19] *Meno*, 82.

[20] *Timaeus*, 17. For a further exposition of this aspect of Plato's view of time see John F. Callahan, *Four Views of Time in Ancient Philosophy* (Cambridge: Harvard University Press, 1948), ch. 1.

[21] *Aristotle*, tr. Ross, *op. cit.*, 219ᵇ.

[22] *Ibid.*, 226ᵃ.

[23] *Ibid.*, 218ᵇ.

[24] *Ibid.*, 221ᵃ.

[25] *Ibid.*, 219ᵇ.

[26] *Ibid.*, 226ᵃ.

[27] *Ibid.*, 225ᵇ–226ᵃ.

[28] *Ibid.*, 226ᵃ. It is not true, of course, that an infinite series necessarily has no first term; only a *regressively* infinite series has this characteristic. But Aristotle is correct in this part of his argument since it is *logically* regressive, *i.e.*, he is seeking an ultimate substance to which change can be applied. However, he is in error in supposing that an infinite regress necessarily arises if we admit that there can be a change of change. Acceleration, for example, is a change of motion, which is itself a change. But motion is a change of position in space, and position is not itself a change.

[29] Some of these topics are discussed in *Four Views of Time in Ancient Philosophy, op. cit.*

[30] All references are to John Locke's *An Essay Concerning Human Understanding*, collated and annotated with Prologemena, Biographical, Critical and Historical, by Alexander Campbell Fraser (Oxford: Clarendon Press, 1894).

31 *Ibid.,* p. 121.

32 *Ibid.,* p. 144.

33 *Ibid.,* p. 213.

34 *Ibid.,* p. 238.

35 *Ibid.,* p. 251.

36 *Ibid.,* p. 253.

37 Roughly speaking, we might say that for both Newton and Locke relational time is real, but in order to explain the world we need also absolute time. For Newton this is "in the world" and is *discovered*; for Locke it is "in the mind" and is *manufactured*.

38 All references are to *Sir Isaac Newton's Mathematical Principles of Natural Philosophy and His System of the World*, Florian Cajori rev. of 1729 tr. Andrew Motte (Berkeley: University of California Press, 1947).

39 *Ibid.,* p. 6.

40 *Ibid.,* p. 547.

41 *Ibid.,* p. 6.

42 *Ibid.,* p. 398.

43 *Ibid.,* p. 8.

44 See E. A. Burtt's *Metaphysical Foundations of Modern Physical Science* (New York: Harcourt, Brace, 1925), p. 256.

45 *Op. cit.,* p. 545.

46 Locke's *Essay* was published in 1690. By 1704 Leibniz had completed his *New Essays on the Understanding*, which was a direct critique of Locke's *Essay*. Because Locke died this same year Leibniz did not publish his commentary at this time, and it did not appear, in fact, until 1765, almost half a century after Leibniz himself had died. All references are to *The Philosophical Works of Leibniz*, tr. George Martin Duncan, from the original Latin and French (2nd ed.; New Haven: Tuttle, Morehouse and Taylor, 1908).

47 *Ibid.,* p. 165.

48 *Ibid.,* p. 214.

49 See Whitrow, *The Natural Philosophy of Time, op. cit.,* p. 42. "According to Newton the universe *has* a clock, whereas according to Leibniz it *is* a clock."

50 Leibniz, *op. cit.,* p. 215.

51 *Ibid., Letters to Clarke,* p. 354.

52 *Ibid., Letters to Clarke,* pp. 334, 352.

53 Bertrand Russell, *A Critical Exposition of the Philosophy of Leibniz* (2nd ed.; London: George Allen and Unwin, 1937), p. 30. See Samuel Alexander's *Space, Time and Deity* (New York: Macmillan, 1920), v. 1, p. 171.

54 *Op. cit., Letters to Clarke,* p. 363.

55 *Ibid.,* p. 336.

56 Russell, *op. cit.,* p. 51.

57 *Immanuel Kant's Critique of Pure Reason*, tr. Norman Kemp Smith (2nd ed.; London: Macmillan, 1929). Because of the limitations of space I shall restrict my discussion of Kant's treatment to that given in the *Transcendental Aesthetic*, and shall not examine the categories.

58 Letter to Moses Mendelssohn, Aug. 16, 1783. *Immanuel Kant's Werke* (Berlin: Cassirer, 1922), v. 9, p. 231.

59 Norman Kemp Smith, *A Commentary to Kant's 'Critique of Pure Reason'* (London: Macmillan, 1918), p. 137.

60 *Transcendental Aesthetic*, par. 1.

[61] *Ibid.*, sec. II, "Time." Kemp Smith asserts that "the fundamental presupposition on which Kant's argument rests—a presupposition never itself investigated but always assumed—is that universality and necessity cannot be reached by any process that is empirical in character." See his *Commentary*, p. xxxiii.

[62] My omission, in this necessarily brief presentation, of Kant's parallel analysis of space makes his position seem, perhaps, more plausible than it really is. Kant believed space to be also a form of intuition. Moreover, he never questioned the prevailing view that space was Euclidean in character. With the development in the nineteenth century of non-Euclidean geometries, the *a priori* character of space became no longer tenable. Consequently Kant's third argument for the *a priori* character of time is considerably weakened. Perhaps our inability to think of a two-dimensional time or of two coexistent time streams may be of the same kind as our inability to think of non-Euclidean geometries prior to their discoveries by Riemann and Lobachevski. In this case our conception of time could not be *a priori,* for it would be at the mercy of our expanding and developing knowledge.

[63] Henri Bergson, *Time and Free Will* (New York: Macmillan, 1910), p. 100.

[64] *Ibid.*, p. 101.

[65] Mary Frances Cleugh, *Time and Its Importance in Modern Thought* (London: Methuen, 1937), p. 110.

[66] *Ibid.*, p. 101.

[67] *Ibid.*, p. 108.

[68] *Ibid.*, p. 140.

[69] *Ibid.*, p. 176.

[70] Alexander, *op. cit.*, v. 1, p. 35. Alexander asserts that he intends to capitalize the initial letters of "space" and "time" when he considers them in general as wholes, and to use lower-case letters when he speaks of "a space" or "a time" as parts of these wholes. But he is not consistent in this usage, and I shall use lower-case letters throughout, except when I am quoting him directly.

[71] *Ibid.*, p. 44.

[72] "Philosophy proceeds by description; it only uses argument in order to help you to see the facts, just as a botanist uses a microscope." Samuel Alexander, *Mind*, Macmillan, London, New Series 120, Oct. 1921, p. 422. He then goes on to say that C. D. Broad, to whose criticism of his *Space, Time and Deity* he is here replying, has a passion for argument since he does it so well. But he himself (Alexander) "dislikes" it. This is a characteristic intuitionistic belief, and makes any criticism of his position impossible; for he can always insist that his critic does not see the "true" facts.

[73] *Space, Time and Deity*, v. 1, p. 17.

[74] Henri Bergson, *Introduction to Metaphysics*, tr. T. E. Hulme (New York: Putnam's, 1912), pp. 22, 53.

[75] H. Minkowski *et al., The Principle of Relativity*, tr. W. Perrett and G. B. Jeffery (London: Constable, 1923). Alexander was evidently familiar with some of the earlier writings of Minkowski, which he does not mention by name.

[76] *Space, Time and Deity*, v. 1, p. 58.

[77] *Ibid.*, p. 44.

[78] *Ibid.*, p. 48. Two remarks may be made about these units of space-time. In the first place, they are not necessarily microscopic or minimal units since a pure event may be an extended volume enduring through a period of time. In the

second place, by the "purity" of the event Alexander meant that it is totally devoid of qualities of any kind and is merely an "enduring volume." It is not, for example, a moving electron, though the space-time may be occupied by a moving electron; nor is it a red sphere, though it may happen to be the container of a red sphere.

79 *Ibid.*, v. 2, p. 50.
80 *Ibid.*, p. 45.
81 *Ibid.*, p. 346.
82 *Ibid.*, p. 46.
83 *Ibid.*, p. 39.

GENERAL REFERENCES

Dunne, John W. *An Experiment with Time.* 5th ed. London: Faber and Faber, 1938.
———. *The Serial Universe.* New York: Macmillan, 1938.
Grünbaum, Adolf. *Philosophical Problems of Space and Time.* New York: Knopf, 1963.
Gunn, John Alexander. *The Problem of Time.* London: Allen & Unwin, 1929.
Guyau, M. *Le Genèse de l'Idée de Temps.* Paris: 1890.
Health, Louise Robinson. *The Concept of Time.* Chicago: University of Chicago Press, 1936.
The Problem of Time. Berkeley: University of California Publications in Philosophy, 1935.
Sivadjian, Joseph. *Le Temps.* Paris: Hermann & Cie, 1938.
Whitehead, A. N. *Process and Reality.* New York: Macmillan, 1929.
Zawirski, Z. *L'évolution de la Notion de Temps.* Cracow: Gebethner et Wolff, 1936.

Time as Succession and the Problem of Duration

1 This problem is expressly formulated in A. C. Benjamin's essay in this volume. Our contribution may be considered as a reply to it.
2 *Cf.* I. Kant, *Critique of Pure Reason* (2nd. ed.; 1787), p. 230.
3 F. W. J. von Schelling, *Die Weltalter*, ed. Schröter (Munich: Biederstein, 1946), p. 77. See also the concept of duration as *duratio intrinseca* in Francisco de Suarez (1548–1617), *Disputationes metaphysicae* (Nr. 50).
4 Schelling, *Werke*, ed. Schröter (Munich: Beck, 1927), v. 2, p. 476. v. 2, p. 476: "We may define space as time stopped and time as space in flux."
5 For the one-sided conception of time as succession it is significant that the *duratio permanens* as well as the *nunc stans* can only be understood as eternity and no longer as time.
6 In connection with this problem see Aristotle's *Physica*, v. 4, 217^b 29; St. Augustine's *Confessions*, v. 11, c. 14.
7 See also the distinction of a double nyn in Aristotle's *Physica*, v. 4, 222^a
8 *Cf.* Schelling, *Die Weltalter, op. cit.*, p. 175.
9 The German *Entsubstanzialisierung* has no adequate English equivalent. It describes the process or period of thought which replaced the metaphysical interpretation of the human essence as a "substantial spirit" with the view which regards man only in respect to his actions and various relations to the world. This thought was continued in pragmatism.

[10] *Cf.* M. Heidegger's question and analysis of the relation of existence to time in his *Being and Time* (New York: Harper & Row, 1962).

[11] For the literature see the bibliography in F. Kümmel, *Über den Begriff der Zeit* (Tübingen: Niemeyer, 1962).

[12] Intentionality is a term of Scholasticism and means thinking as being directed towards objects. The term was taken over by Franz Brentano (1838–1917), who used it in order to describe the fundamental structure of all psychical acts. In this sense intentionality has become the central concept of phenomenology. Husserl for the first time gave a detailed analysis of the temporal structure of mental acts, and Heidegger based his concept of existence on temporality, which he understands as a specific relationship between future, past and present. The differences between Heidegger's conception and the conception offered here could not be worked out in the context of this inquiry. *Cf.* Herbert Spiegelberg, *The Phenomenological Movement. A Historical Introduction* (The Hague: Martinus Nijhoff, 1960), 2 vols.

[13] See especially H. Bergson's *Time and Free Will* (New York: Macmillan, 1910).

[14] The following thoughts can be formulated only in a very compressed and fragmentary manner since our purpose here is only to clarify the new approach itself. For a more detailed analysis of the relation of time to freedom and of the existential attitudes to time, including a consideration of psychopathological temporal phenomena, see Kümmel, *op. cit.*

[15] See for this above all, O. F. Bollnow, *Das Wesen der Stimmungen* (3rd. ed.; Frankfurt: Klostermann, 1956); also, *Neue Geborgenheit. Das Problem einer Überwindung des Existentialismus* (2nd ed.; Stuttgart: Kohlhammer, 1960).

[16] This double movement of time is very well expressed by the image of the "time loop" in an essay by O. F. Bollnow, "Das Nachholen des Versäumten" in *Mass und Vermessenheit des Menschen* (Göttingen: Vandenhoeck, 1962).

[17] See F. Kümmel, "Zeit und Bewusstsein. Entwurf einer Theorie der Erinnerung als Bewusstwerdung des Bewusstseins," *Zeitschrift für philosophische Forschung*, v. 16 (1962), no. 4, p. 532.

[18] See E. Heintel, "Wie es eigentlich gewesen ist. Ein geschichtsphilosophischer Beitrag zum Problem der Methode der Historie," *Erkenntnis und Verantwortung. Festschrift für Th. Litt.* (Düsseldorf: Verlag Schwann, 1960), p. 207.

[19] See an interpretation of both thinkers in Kümmel, *op. cit.*, p. 44ff.

[20] Franz von Baader, *Works* (ed., Hoffmann, 1851–60), v. 2, p. 69.

GENERAL REFERENCES

Becker, O. *Dasein und Dawesen. Gesammelte philosophische Aufsätze.* Pfullingen: Neske, 1963.

Binswanger, L. *Basic Forms and Knowledge of Human Existence.* Munich: Ernst Reinhardt, 1962. Selections translated in *The Worlds of Existentialism*, ed. M. Friedman. New York: Random House, 1964.

Bollnow, O. F. *Mensch und Raum.* Stuttgart: Kohlhammer, 1963.

von Gebsattel, V. E. *Prolegomena zu einer medizinischen Anthropologie. Ausgewählte Aufsätze.* Berlin: J. Springer, 1954.

Heidegger, M. *Being and Time.* New York: Harper and Row, 1962.

Husserl, E. "Göttinger Vorlesungen über das innere Zeitbewusstsein," *Yearbook for Phenomenology*, ed. M. Heidegger, 1928.

Minkowski, E. *Le Temps Vécu*. Paris: Alcan, 1933.

Strauss, E. W. "Das Zeiterleben in der endogene Depression und in der psycho-pathischen Verstimmung," *Festschrift für Bonhöffer, Monatschrift für Psychiatrie*, v. 68, 1928.

———. "Die Störungen des Werdens und des Zeiterlebens," *Gegenwartsprobleme der psychiatrisch-neurologischen Forschung*, ed. Roggenbau, Stuttgart: 1939.

———. "Disorders of Personal Time," *Journal of Southern Medical Association*, 1947.

———. *The Primary World of Senses*. New York: The Free Press of Glencoe, 1963.

von Weizsäcker, W. *The Unity of Perception and Movement*. 4th ed. Stuttgart: Georg Thieme, 1950. Selections translated in *The Worlds of Existentialism, op. cit.*

Notes Concerning Some Properties of Time

[1] M. Bonaparte, *Int. J. of Psychoanalysis*, v. 21 (1940), p. 427.

Time in Christian Thought

[1] *Cf.* M. Eliade, *The Myth of the Eternal Return*, tr. W. R. Trask (New York: Pantheon, 1954).

[2] Until fairly modern times the Book of Genesis was generally regarded as a factually and chronologically precise historical record. It is now clear that it is not history in the modern scientific sense. Christians in general still believe, however, that it expresses, in a form suited to the mentality of those to whom it was addressed, the essential truth that the human race is in need of salvation and that God has worked in and through history to prepare men for the coming of Christ. More specific questions of Biblical interpretation need not concern us here.

[3] There have been many different interpretations of Christ's teaching on the Second Coming. The more important are discussed by N. Perrin, *The Kingdom of God in the Teaching of Jesus* (London: SCM Press, 1963). As already indicated, I am concerned here simply with the way in which the early Christians normally understood it.

[4] *Cf.* F. C. Copleston, *A History of Philosophy* (London: Burns & Oates, 1950), v. 2, ch. 2.

[5] *City of God (De Civitate Dei)*, especially Bk. 5, chs. 11–21 (London: J. M. Dent, 1935), tr. John Healey. Other aspects of St. Augustine's theory are discussed by: J. Guitton, *Le temps et l'éternité chez Plotin et Saint Augustin* (Paris: Boivin, 1933); H. I. Marrou, *L'ambivalence du temps de l'histoire chez Saint Augustin* (Paris: J. Vrin, 1950); L. Boros, *Les catégories de la temporalité chez Saint Augustin* (Paris: Archives de Philosophie, 1958), p. 323; J. Chaix-Ruy, *Saint Augustin, Temps et histoire* (Paris: Etudes Augustiniennes, 1956).

[6] *Cf.* G. Wingren, *Man and the Incarnation, A Study in the Biblical Theology of Irenaeus* (London: Oliver & Boyd, 1959).

[7] *Cf.* J. Daniélou, *Origen* (New York: Sheed and Ward, 1955).

⁸ E. F. Sutcliffe, *The Old Testament and the Future Life* (2nd ed.; London: Burns & Oates, 1947).

⁹ This aspect is particularly emphasized in the "existentialist theology" of R. Bultmann and others.

¹⁰ In many cases it is uncertain whether such sayings were intended originally to refer to the Second Coming or to the individual's response to God's grace. Christian tradition has generally understood them in both senses.

¹¹ St. Augustine, *Confessions*, Bk. 11, chs. 14–30. See also his: *In Joannis Evangelium*, Tract. 38, no. 10 (Migne, P.L. 35, col. 1680).

¹² E.g., Hugh of St. Victor, *De Vanitate Mundi* (Migne, P.L. *176*, cols. 703ff.). There is an English translation of the first two books in *Hugh of Saint-Victor: Selected Spiritual Writings* (London: Faber & Faber, 1962).

¹³ Plato, *Timaeus*, 37C–38C. See F. M. Cornford, *Plato's Cosmology* (London: Routledge & Kegan Paul, 1948), p. 97.

¹⁴ Boethius, *De Consolatione Philosophiae*, Bk. 5, ch. 6.

¹⁵ *Timaeus*, 37D.

¹⁶ Aristotle, *Meteorologia* 1, 14.

¹⁷ *De Generatione et Corruptione* 2, 10. For Greek theories of the Great Year see P. Duhem, *Le Système du Monde* (Paris: Hermann, 1954), v. 1, p. 65 (Plato); p. 164 (Aristotle); p. 275 (Stoics and Neoplatonists). Greek views on time and eternity are discussed by W. von Leyden, "Time, Number and Eternity in Plato and Aristotle," *Philosophical Quart.*, v. 14 (1964), p. 35.

¹⁸ This was not the view of all Greek philosophers, but it was accepted by Plato (most probably), by Aristotle, and by nearly all their successors.

¹⁹ Cf. G. Verbeke, "Les Stoiciens et le progrès de l'histoire," *Revue Philosophique de Louvain*, v. 62 (1964), p. 5.

²⁰ E.g., in Psalm 102 (Vulgate 101) and perhaps in Exodus 3:14.

²¹ E.g., John 8:58 "Before Abraham was, I am."

²² Origen, *De Principiis*, Bk. 2, ch. 3 (Migne, P. G., v. 11, col. 192).

²³ St. Augustine, *De Civitate Dei*, Bk. 12, chs. 13–14.

²⁴ Duhem, *op. cit.*, v. 7, p. 441.

²⁵ *Ibid.*, p. 443.

²⁶ The most notable exception to this generalization was the Cistercian monk Joachim de Floris (*c.* 1132–1202), who propounded a linear theory of history in which there was a clearly defined theological pattern. For a brief account of this, see *Dictionnaire de Théologie Catholique*, v. 8, col. 1432. Roger Bacon, in the thirteenth century, held that human knowledge could and should increase progressively until the end of the world, *Opus Maius* (London: Williams & Norgate, 1900), pt. 2, chs. 13–14, but such optimism does not seem to have been generally shared by his contemporaries.

²⁷ N. Wiseman, *Twelve Lectures on the Connexion between Science and Revealed Religion* (London: 1836), lect. 5.

²⁸ G. Perrone, *Praelectiones Theologicae* (Louvain: 1839), v. 3, p. 81.

²⁹ An excellent historical survey of the whole problem is to be found in F. C. Haber's *The Age of the World: Moses to Darwin* (Baltimore: Johns Hopkins Press, 1959).

³⁰ Cf. Karl Rahner, "The Resurrection of the Body," *Theological Investigations* (Baltimore and London: Helicon Press; Darton, Longman and Todd, 1963), v. 2, p. 210.

³¹ J. Bossuet, *Discours sur l'histoire universelle* (1681).

32 Condorcet, *Esquisse d'un tableau historique des progrès de l'esprit humain* (1795).
33 *Cf.* P. G. Fothergill, *Historical Aspects of Organic Evolution* (London: Hollis & Carter, 1952).
34 For the historical background to the Darwinian controversies, see C. G. Gillispie, *Genesis and Geology* (Cambridge: Harvard University Press, 1951). Their impact on religious thought during the period 1859–72 has been carefully analyzed by A. Ellegård: *Darwin and the General Reader* (Göteborg: Gothenburg Studies in English VIII, 1958).
35 This is based mainly on his best-known work: *Le phénomène humain*, English trans.: *The Phenomenon of Man* (New York: Harper & Brothers, 1959). I have discussed Teilhard's theory in more detail in four articles in: *The Heythrop Journal*, I (1960) 271–84; II (1961) 3–13; III (1962) 347–57; IV (1963) 32–41.
36 Besides the tendency to complexification there exists an opposite one to simplification, since molecules can decompose and organisms sometimes become simpler in the course of evolution. But taking the world as a whole it is complexification which is the more fundamental. Simplification can only occur in so far as there has been previous complexification. And in the long run, the level of complexity does tend to rise.

GENERAL REFERENCES

von Balthasar, Hans Urs. *A Theology of History*. London: Sheed & Ward, 1964.
Beaucamp, E. *The Bible and the Universe; Israel and the Theology of History*. London: Burns & Oates, 1963.
Brabant, F. H. *Time and Eternity in Christian Thought*. London: Longmans, Green, 1937.
Bultmann, R. *History and Eschatology*. Edinburgh: Edinburgh University Press, 1957.
Cullman, O. *Christ and Time*. London: SCM Press, 1962.
Guitton, J. *Justification du temps*. Paris: Presses Universitaires de France, 1941.
Mouroux, J. *Le mystère du temps, approche théologique*. Paris: Aubier, 1962.
Puech, H. C. "Gnosis and Time," *Man and Time*. New York: Pantheon, 1954, p. 38.
Quispel, G. "Time and History in Patristic Christianity," *Man and Time*. p. 85.
Richardson, A. *History, Sacred and Profane*. London: SCM Press, 1963.

Time in Indian and Japanese Thought

1 "*bhava utpattiḥ sattā vā*" (*Ratnaprabhā* ad *Brahmasūtra*), II, 2, 3.
2 It has been the practice since the Ṛg-Veda to use the past participle instead of the finite verb. Even in the Gāthās of the Jain Scripture, this tendency is found. See Herman Jacobi, *Sacred Books of the East*, v. 22, p. 72n.
3 In the Apabhraṁśa language, perfect, imperfect and aorist are seldom found and the past tense is expressed by a past participle.
4 However, there is no such example in the Ṛg-Veda. It can be first found in the Brāhmanas. See B. Delbrück, *Altindische Syntax*, p. 295.
5 J. S. Speyer, *Vedische und Sanskrit-Syntax* (Strassburg: 1896), p. 67.
6 *vṛsala, upalabdhum tarhi vayam āhutāḥ*" (*Mudrārākṣasa*, III, 127, 6).

[7] *Bṛhadāraṇyaka Upaniṣad*, III, 8, 8–11.

[8] *Bṛhad. Up.*, III, 9, 26; IV, 4, 22; IV, 5, 15, *cf.* II, 3, 6.

[9] *Bṛhad. Up.*, IV, 4, 25.

[10] In the seventh chapter of the *Madhyamaka-kārikā*, the theory which claims *utpāda-sthiti-bhaṅga* is refuted. Formerly, in the *Tseng-i-a-han* XII (Taishō, v. II, p. 607 c), it is explained that there are three characteristics by which the conditioned elements are formed, namely appearance, change, and elimination. The *Abhidharma-jñāna-prasthāna*, v. 3 (Taishō 780 c), also explains the same theory.

[11] This is explained in *Mahāvibhāṣā*, v. 39, the *Nyāyānusāra*, v. 13 the *Abhidharma-kośa* volume (the Chinese translation, by Hsuan-tsang), etc.

[12] *Atharva-veda* XIX, 53.

[13] Praesens, imperfectum, futurum, perfectum, aoristum. In the Latin language aoristum is lacking, but plusquamperfectum and futurum exactum are added.

[14] Speyer, *op. cit.*, p. 53.

[15] H. Oldenberg, *Aus dem alten Indien* (Berlin: Gebrüder Paetel, 1910), p. 93.

[16] In India there is no uniform system of marking historical eras. Their method of determining historical periods differs according to time and place. According to a historian, there are more than twenty ways of marking eras in India. This fact presents a great contrast to the uniform adoption in the West of the Christian era. *Cf.* V. A. Smith, *The Early History of India* (Oxford: Clarendon Press, 1904), p. 20.

[17] *E.g.*, *Nyāyabinduṭīkā*, p. 24, line 11. Samkara on the *Brahmasūtra* I, 604, line 10 (Anandrasma Sanskrit Series).

[18] *Ibid.*, 15. We do not minimize the fact that there are some exceptions, *e.g.*, *viśeṣaṇa-viśeṣya-bhāva* (*Vedāntasāra*, 168 b).

[19] Śaṁkara, *op. cit.* I, p. 603, line 3.

[20] *Majjhima-Nikāya* III, 19.

[21] *Vinaya, Mahāvagga*, I, 23.

[22] *Theragāthā*, tr. Mrs. Rhys Davids in *Psalms of the Brethren* (London: Oxford University Press, 1913), 1159.

[23] *Gleanings from Soto-Zen*, ed. Ernest Shinkaku Hunt (Honolulu: Soto Mission, 1960), p. 25.

[24] *Profound Doctrine of the Lotus*, VIII, Pt. 2 (*Taisho Tripiṭaka*, v. XXXIII, 783 b.).

[25] *Shōbōgenzō*, Shōji (section on Life and Death. *Taishō Tripiṭaka*, v. LXXXII, p. 305).

[26] *Shōbōgenzō*, Busshō (section on Buddha-nature. *Taishō Tripiṭaka*, v. LXXXII, p. 93 a).

[27] Sanshō Dōei (Religious Poems of Umbrella-Like Pine Tree).

[28] *Shasekishū*, X, Pt. 1.

[29] This sentence was composed in China based upon such sentences as "In the milk, there is cream; in patient beings there is Buddha nature." and "If you have the desire to seek, you will find." These are from *Mahāprajñā-pāramitā-sūtra*.

[30] *Shōbōgenzō*, Chapter: Busshō.

[31] *Shōbōgenzō*, Chapter: Uji.

[32] *Ibid.*, Chapter: Setsushin Setsushō.

[33] *Ibid.*, Chapter: Hosshō.

[34] *Ibid.*, Chapter: Setsushin Setsushō.

[35] Yasusada Hiyane, *Nihon Shūkyōshi* (*History of Japanese Religions*), p. 828.

[36] *Gomō Jigi*, v. 1, folio 3.

[37] *Dōji-mon*, v. 2, p. 39.
[38] Junsei Iwahashi, *Sorai Kenkyū*, p. 449.
[39] Masaharu Anesaki, *Nichiren the Buddhist Prophet* (Cambridge: Harvard University Press, 1916), p. 119.
[40] In the *Sado-gosho, in Shōwa Shinshū Nichiren Shōnin Imonzenshū* (Kyoto: Heirakuji Shoten, 1934), v. 1, p. 842.

GENERAL REFERENCES

Brandon, S. G. F. *Man and His Destiny in the Great Religions*. Manchester: The University Press, 1962.
Eliade, Mircea. "Time and Eternity in Indian Thought," *Man and Time*. New York: Pantheon, 1958, p. 173.
Miyamoto, Shoson. *J. of Indian and Buddhist Studies*, v. 7, no. 2 (1959), p. 830.
Nakamura, Hajime. *Ways of Thinking of Eastern People*. Honolulu: East-West Center Press, 1964.
Nakayama, Enji. *Bukkyo ni okeru Toki no Kenkyu* (*Studies on Time in Buddhism*) Kyota: Kokyo Shoin.
de Riencourt, Amaury. *The Soul of India*. New York: Harper, 1959, p. 16.
Runes, Dagobert D. "Indian Philosophy," *Dictionary of Philosophy*. Paterson, N.J.: Littlefield, 1963.
Whitrow, G. J. *The Natural Philosophy of Time*. London: Nelson, 1961, p. 154.

Time and Knowledge in China and the West

[1] The system of romanization of Chinese names adopted in this paper is that of Wade-Giles modified only by the addition of the letter "h" in lieu of the aspirate apostrophe.
[2] *Cf.* H. Zimmer, *Philosophies of India* (New York: Pantheon, 1953), p. 450, and the preceding essay by Prof. Nakamura.
[3] No systematic treatment of this specific subject is available, so far as I know, either in Chinese or a Western language; Marcel Granet only touched upon the question in his *La Pensée Chinoise* (Paris: Renaissance, 1934), p. 90.
[4] *Huai Nan Tzu*, ch. 11.
[5] *Cf. Science and Civilisation in China*, J. Needham *et al.*, 7 vols. (Cambridge: Cambridge University Press, 1954—), v. 4, pt. 1, p. 2; hereinafter abbreviated as *SCC*. This is reminiscent of the absolutist definition of time by Strato of Lampsacus (*fl.* 300 B.C.), one of Aristotle's disciples (*cf.* Simplicius, *Phys.*, 789.35), and foreshadows Newton against Leibniz.
[6] *Cf. SCC*, v. 3, p. 93. This recalls the absolutist definition of space by Strato of Lampsacus (Simplicius, *Phys.*, 618.20) followed by Johannes Philoponus (sixth century A.D., *Phys.*, 567.29).
[7] *Cf. SCC*, v. 4, pt. 1, p. 55.
[8] *Ibid.*, p. 56.
[9] A. Forke, *Geschichte d. alten chinesischen Philosophie* (Hamburg: Friederichsen, 1927), p. 413. *Cf.* Aristotle: "movement is the objective seat of before-and-after-ness" (*Physica*, IV, 11, 219a 8, 220a 1). Aristotle was a relationist as regards time for he felt that time existed only by virtue of the motion of moving bodies. The Mohists would not have said this.

[10] *Cf. SCC*, v. 2, p. 190.

[11] *Ibid.*, p. 193. This is somewhat reminiscent of the recognition by Boethus (*fl.* 50 B.C.) that motion and rest have one thing in common, the change of the independent variable of time. "It is not correct to describe rest in a place as 'place' " (*cf.* Simplicius, *Categ.* 433.30). As his predecessor Strato had said, rest is motion along the axis of time.

[12] Not finally completed till *ca.* A.D. 380 but containing much material of the Warring States, Chhin and Han periods (fourth century B.C. onwards). Ch. 5, *cf. SCC*, v. 2, p. 198; trans. here revised to take account of the interpretations of R. Wilhelm and A. C. Graham.

[13] Ch. 23 (tr. Legge, v. 2, p. 85).

[14] *Cf. SCC*, v. 4, pt. 1, p. 3. There are Indian and Semitic parallels, but not historically likely to have influenced the Mohists.

[15] *Cf.* J. Needham and K. Robinson, "Ondes et Particules dans la Pensée Scientifique Chinoise," *Sciences*, v. 1, no. 4 (1960), p. 65; or alternatively *SCC*, v. 4, pt. 1, pp. 3, 9, 202.

[16] See, *e.g.*, S. Sambursky, *The Physical World of the Greeks* (London: Routledge & Kegan Paul, 1956), pp. 181, 238; also *The Physical World of Late Antiquity* (London: Routledge & Kegan Paul, 1962), p. 9.

[17] *Cf. SCC*, v. 2, pp. 165, 182, 201, 203.

[18] *Cf.* Aristotle, *Phys.*, 217ᵇ 33.

[19] *Cf. SCC*, v. 2, p. 176.

[20] *Shih Chi*, tr. Chavannes (ch. 5, v. 2, p. 45), noted by M. Granet, *Danses et légendes de la Chine ancienne* (Paris: Alcan, 1926), v. 1, p. 104. The incident is dated at 621 B.C.

[21] On these unfamiliar views of causality in Chinese natural philosophy, *e.g.*, "reticulate causation" and "synchronistic causation," see *SCC*, v. 2, p. 288 and the following works of C. G. Jung: R. Wilhelm and C. G. Jung, *The Secret of the Golden Flower; a Chinese Book of Life*, including a trans. of the *Thai I Chin Hua Tsung Chih*, Eng. tr. C. F. Baynes (London: Kegan Paul, 1931), esp. p. 142; C. G. Jung and W. Pauli, *Naturerklärung und Psyche* (Zürich: Rascher, 1952), Studien aus dem Jung Institut, no. 4, Eng. tr. R. F. C. Hull, *The Structure and Dynamics of the Psyche, Collected Works*, v. 8. (London: Routledge & Kegan Paul, 1955), containing "Synchronicity, an Acausal Connecting Principle" by Jung, and "The Influence of Archetypal Ideas on the Scientific Theories of Kepler" by Pauli; C. G. Jung, "Über Synchronizität," *Eranos Jahrbuch*, v. 20 (1952), p. 271. For a discussion of similar ideas from a quite different standpoint read N. R. Hanson, "Causal Chains," *Mind*, v. 64 (1955), p. 289. *Cf.* the article by M-L von Franz in this volume.

[22] Chung Yung, II, ii. *Cf.* Fêng Yu-Lan, *History of Chinese Philosophy* (London: Allen & Unwin, 1937), v. 1, pp. 371, 391. This idea is something like the *idios kairos* ('ἴδιος καιρός) of the early Christian writers, the appropriate or decisive moment for action, divine or human.

[23] *Cf. SCC*, v. 2, p. 455.

[24] *Cf.* Forke, *op. cit.*, p. 42.

[25] One of the Ming scholars, Tung Ku, replied to a questioner that time could be said to have a beginning if you were speaking of a single world period (*yuan*), but not if you were speaking of the endless chain of all the world periods. *Cf. SCC*, v. 3, p. 406.

[26] Ch. 40; *cf. SCC*, v. 2, p. 75.

[27] Granet, *La Pensée Chinoise*, p. 90.

[28] I take this formulation from a paper by Fukunaga Mitsuji and the commentary of Than Chieh-Fu on the Mohist proposition quoted above. It has Peripatetic and Neo-Platonic parallels; *cf.* Sambursky, *Physical World of Late Antiquity*, p. 15.

[29] *Cf. SCC*, v. 2, p. 283.

[30] *Cf. SCC*, v. 3, p. 467.

[31] This subject will be fully treated in *SCC*, v. 6; meanwhile see P. Huard and Huang Kuang-Ming (M. Wong), "La Notion de Cercle et la Science Chinoise," *Archives Internat. d'Hist. des Sciences*, v. 9 (1956), p. 111. A remarkable result of this circulation-mindedness has recently been brought to light by Lu Gwei-Djen and J. Needham, "Mediaeval Preparations of Urinary Steroid Hormones," *Nature*, v. 200 (1963), p. 1047. In Renaissance Europe the philosophy and mysticism of the circle and circulation were immensely important in the developing phases of modern science. This influence has been traced in many papers by W. Pagel, among which we can only mention here "Giordano Bruno; the Philosophy of Circles and the Circular Movement of the Blood," *J. of Hist. Med. and Allied Sci.*, v. 6 (1951), p. 116.

[32] *Cf. SCC*, v. 4, pt. 1, pp. 3, 9.

[33] Granet, *La Pensée Chinoise*; the Chinese "preferred to see in time an ensemble of eras, seasons and epochs" (p. 86); "Time and space were never conceived apart from concrete actions" (p. 88); the Chinese "decomposed all time into periods just as they decomposed all space into regions" (p. 96); the Chinese "never bothered about imagining time and space as homogeneous matrices suitable for housing abstract concepts" (p. 113).

[34] *Ibid.*, p. 89.

[35] Granet often used to call this "liturgical time" because of its connection both with the ceremonies of the imperial cosmic religion and with those which marked the incidents of lives within the individual family. Much importance was attached to the correct vesting of the emperor and his attendants in accordance with the season when performing the rites of the Ming Thang or Cosmic Temple. Full details will be found in W. E. Soothill, *The Hall of Light, a Study in Early Chinese Kinship* (London: Lutterworth, 1951), p. 30. Once Summer had officially begun it would have been an unthinkable affront to Heaven and Earth to wear the green robes appropriate to Wood, the element of Spring; all vestments, banners and cult objects had to be changed to red, the color appropriate to Fire, the element of Summer. Such practices are familiar to us in the form of the liturgical colors of Western Catholic worship, though these do not symbolize "blocks" of time as the ancient Chinese ones did, nor are they charged with superstitious anxiety, fear of forfeiting Heaven's favor. Then there were also the ritual lapses of time in family customs, the "rites de passage" which had to be accomplished in each individual's life-history, the temporary prohibitions, the periodical festivals, the duties performed at fixed intervals. See Granet, *La Pensée Chinoise*, p. 97, and also a famous study of his, "Le Dépot de l'Enfant sur le Sol," published originally in *Revue Archéologique*, v. 14, 5th ser. (1922), p. 10, and reprinted in *Etudes Sociologiques sur la Chine* (Paris: Presses Universitaire de France, 1953), p. 159.

[36] See *SCC*, v. 2, p. 232.

[37] *Cf. SCC*, v. 2, pp. 242, 253, 261, 273.

[38] The four seasons were correlated with four of the elements. The sixth month, however, was placed under the aegis of Earth, thus making the seasons up to the right number, five.

[39] *SCC*, v. 2, p. 288.

[40] Granet, *La Pensée Chinoise*, p. 329; *cf. SCC*, v. 2, p. 289.

[41] It is a little unfortunate that Western writers on time have assumed that these ideas can fully represent Chinese thinking. Thus G. J. Whitrow, *Natural Philosophy of Time* (London: Nelson, 1961), p. 58, has concluded that discontinuous "packaged" time was the only time the Chinese knew of. He kindly acknowledges the help of *SCC*, v. 2, but overlooked footnote (f) on p. 288 which would have directed him to another Chinese world, that of the Mohists. Of course he himself was devoting only a single footnote to China. One could hardly compress all the European ideas about time held through the ages into a single footnote, and few Chinese writers, I am sure, would attempt to do so. Western indifference to Chinese thought has been such that Whitrow deserves praise, not blame, for his effort—yet there may still be a lesson in this contrast.

[42] About A.D. 543 the great Taoist swordsmith and metallurgist Chhiwu Huai-Wên, probably the inventor of co-fusion steel, advised the emperor Kao Tsu of the Eastern Wei to change the color of that dynasty's flags from red to yellow in accordance with five-element theory, in order to conquer the Western Wei.

[43] This if fully worked out might add another chapter to the history of the spontaneous development of scientific criticism in China. Of rationalist skepticism there was never any lack (*cf. SCC*, v. 2, p. 365), and there are other examples of the development of critical scientific thought quite independently of European Renaissance influences (see *SCC*, v. 4, pt. 1, p. 189).

[44] Later on in Chinese history, as we shall see, the Indian ideas of long-term recurrence had considerable success in China, so there was cyclical time too.

[45] *Cf.* Granet, *La Pensée Chinoise*, p. 97.

[46] Something on this important subject will be found in *SCC*, v. 3, p. 390, but I regret that our treatment of it was not really adequate. We felt at that time that the interest of the calendar was rather archaeological and social-historical than scientific, not fully appreciating that each "Calendar" had been a full set of astronomical tables and constants intended always to constitute a substantial improvement over all its predecessors.

[47] *Cf. SCC*, v. 3, p. 406.

[48] Among recent interesting discussions of the problem we may mention F. Jäger, "Der heutige Stand der *Schi-ki* [*Shih Chi*] Forschung," *Asia Major*, v. 9 (1933), p. 25; B. Watson, *Ssuma Chhien, Grand Historian of China* (New York: Columbia University Press, 1958), pp. 70, 220; F. A. Kierman, *Ssuma Chhien's Historiographical Attitude as Reflected in Four Late Warring States Biographies* (Wiesbaden: Harrassowitz, 1962), pp. 4, 48.

[49] And to a large extent, in practice. For an account of this system see Yang Lien-Shêng, "The Organisation of Chinese Official Historiography; Principles and Methods of the Standard Histories from the Thang through the Ming Dynasty," *Historians of China and Japan*, ed. W. G. Beasley and E. G. Pulleyblank (London: Oxford University Press, 1961), p. 44. The contribution of A. F. Hulsewé in the same volume (p. 31), "Notes on the Historiography of the Han Period," shows how it developed from the ancient Thai

Shih organization of astronomer-annalists. By the Thang period it had reached its definitive form.

50 On Chinese historiographic traditions in general, see the introduction to *Historians . . .* and especially the small but now long classical treatise of C. S. Gardner, *Chinese Traditional Historiography* (Cambridge: Harvard Univ. Press, 1938, repr. 1961). Han Yu-Shan's *Elements of Chinese Historiography* (Hollywood: Hawley, 1955) is also a useful reference work, though with many inaccuracies. For those who can read Chinese there is an up-to-date monograph by Chin Yü-Fu, *Chung-Kuo Shih-Hsüeh Shih* (*A History of Chinese History-Writing*) (Peking: Chung-hua, 1962). A brief survey of Chinese historiography is given in *SCC*, v. 1, p. 74.

51 By those at least who have taken the trouble to acquaint themselves with China's historiography. A regrettable example of ignorant judgment is cited by E. G. Pulleyblank, *Historians . . .*, p. 135, and to this, I fear, must be added the lecture by H. Butterfield, "History and Man's Attitude to the Past; their Role in the Story of Civilisation" (Foundation Day Lecture, London School of Oriental Studies, 1961).

52 In an article by O. van der Sprenkel, "Chronology, Dynastic Legitimacy, and Chinese Historiography," contributed to the Study Conference at the London School of Oriental Studies in 1956 and circulated at that time in mimeographed form, but unfortunately not printed in *Historians. . . .*

53 *Cf.* note 172 below.

54 Joseph J. Scaliger founded modern historical chronology with his *Opus Novum de Emendatione Temporum* (*Thesaurus Temporum*) (Paris, 1583). See J. W. Thompson and B. J. Holm, *A History of Historical Writing* (New York: Macmillan, 1942), v. 2, p. 5. Although this work is invaluable for the Western world, and made a creditable attempt to say something about Arabic, Persian and Mongol historians, it deliberately if tacitly excluded China from its survey (hence the stricture of Pulleyblank just mentioned), not hesitating however on that account to affirm in its introduction the pre-eminent historical-mindedness of Christian Europe.

55 Sir Isaac Newton, *The Chronology of Ancient Kingdoms Amended, to which is prefixed a Short Chronicle from the First Memory of Things in Europe to the Conquest of Persia by Alexander the Great* (London, 1728). See on this F. Manuel, *Isaac Newton, Historian* (Cambridge: 1964).

56 For the general background of Liu Hsi-Sou see Yabuuchi Kiyoshi, "The Development of the Sciences in China from the 4th to the end of the 12th Century A.D.," *J. of World History*, v. 4 (1958), p. 330. *Chhang shu* methods had first been used by Tu Yü in the third century A.D. in his studies on the *Chhun Chhiu* period. Liu Hsi-Sou's work, which covered the time from the beginning of the Han to the end of the Wu Tai periods, was extended to the Sung, Yuan and Ming by a great successor, Chhien Ta-Hsin (A.D. 1728–1804). We now possess, of course, chronological tables for Chinese history of great detail and precision, *e.g.*, A. C. Moule and W. P. Yetts, *The Rulers of China, 221 B.C. to A.D. 1949* (London: Routledge & Kegan Paul, 1957).

57 On this see the excellent study of E. G. Pulleyblank, "Chinese Historical Criticism; Liu Chih-Chi and Ssuma Kuang," in *Historians . . .*, p. 135.

58 Jean Bodin, *Methodus ad facilem Historiarum Cognitionem* (Paris, 1566), and L. V. de la Popelinière, *Histoire des Histoires; Premier Livre de l'Idée de l'Histoire Accomplie* (Paris, 1599). These were the first Western books to

discuss the laws of historical causation and development and to lay the foundation for a method and critique of history; *cf.* Thompson and Holm, *op. cit.*, v. 1, pp. 561, 563; v. 2, p. 5.

[59] On these three great undertakings see, besides the paper of E. G. Pulleyblank on Liu Chih-Chi, E. Balazs' "L'Histoire comme Guide de la Pratique Bureaucratique; les Monographies, les Encyclopédies, les Recueils de Statuts," in *Historians . . .* , p. 78. Also Han Yu-Shan, *op. cit.*, p. 60.

[60] As is now well known 'Abd al-Raman ibn Khaldūn worked out a general theory of historical development, embodying climate, geography, moral and spiritual forces, and laws of national progress and decay—sociological history in fact—in his *Kitāb al-'Ibar wa-Dīwan al-Mubtada' wa-l-Khabar-fi Ayyām al-'Arab wa-l-'Ajam wa-l-Barbar* (*Book of Instructive Examples and Register of Subjects and Predicates dealing with the History of the Arabs, the Persians and the Berbers*).

[61] E. Pasquier, *Les Recherches de la France* (Paris, 1560, 1611); Pietro Giannone, *Storia Civile del Regno di Napoli* (Naples, 1723); Louis de Secondat, de Montesquieu, *L'Esprit des Lois* (Paris, 1748). On these see Thompson and Holm, *op. cit.*, v. 1, p. 6, v. 2, pp. 61, 90, 561. Butterfield, *op. cit.*, p. 13, wondered whether any non-European civilization had developed the history of laws and institutions.

[62] Han Yu-Shan, *op. cit.*, p. 49.

[63] Han Yu-Shan, *op. cit.*, pp. 49, 61, and Balazs, *op. cit.*, pp. 84, 90.

[64] Balazs, *op. cit.*, Yang Lien-Shêng, *op. cit.* and Pulleyblank, *op. cit.* The latter article gives a vivid account of the methods of Ssuma Kuang and his collaborators Liu Pin, Liu Shu and Fan Tsu-Yü. Some of the techniques of the medieval Chinese historians were remarkably modern. Thus, for example, in the Sung Bureau of Historiography colored inks of various kinds were used to distinguish the different texts, a practice going back as far as about A.D. 500 when the great pharmaceutical naturalist Thao Hung-Ching used it in editing the pharmacopoeia. Then the historian Li Tao (A.D. 1115–84) was renowned for his elaborate filing system, a row of ten cabinets each containing twenty drawers, all notes and documents relating to a particular year being filed in one of the drawers, and eventually sorted into folders in chronological order according to month and day. This system was used in the production of his *Hsü Tzu Chih Thung Chien Chhang Pien* (*Supplementary Continuation of the Comprehensive Mirror of History, for Aid in Government*), bringing the narrative down to A.D. 1180. How far such filing systems originated with Li Tao is not known, but it is more than probable that they had been developing ever since the days of Wu Chün (Thao Hung-Ching's contemporary), and Ssuma Kuang and his collaborators must surely have used something of the kind. On the life and work of Li Tao see Sudō Yoshiyuki in *Komazawa Shigaku*, v. 6 (1957), p. 1.

[65] See Hsü Shih-Lien, *The Political Philosophy of Confucianism* (London: Routledge, 1932), p. 43; Yang Lien-Shêng, *op. cit.*, p. 52.

[66] There has been much discussion of the objectivity and reliability of the Chinese official historians (apparently unknown to critics such as H. Butterfield) and in general sinologists have reached very favorable conclusions. One can mention H. H. Dubs, "The Reliability of Chinese Histories," *Far Eastern Quarterly*, v. 6 (1946), p. 23; E. R. Hughes, "Importance and Reliability of the I Wên Chih," *Mélanges Chinois et Bouddhiques*, v. 6 (1939), p. 173; and

the debate in *Oriens Extremus*, H. H. Frankel, "Objectivität und Parteilichkeit in d. off. Chin. Geschichtsschreibung," v. 5 (1958), p. 133, with the reply by H. H. Dubs, v. 7 (1960), p. 120. For a warm appreciation of the ideals of Chinese historians see E. Haenisch, "Der Ethos d. Chin. Geschichtsschreibung," *Saeculum*, v. 1 (1950), p. 111. The reliability of the voluminous astronomical records in the Chinese official histories is a separate question, and has been discussed in *SCC*, v. 3, p. 417.

67 I am thinking of course of H. Butterfield's stimulating little book *The Whig Interpretation of History* (London: Bell, 1951).

68 *Cf.* Mencius, II (1), vi, 1–7 and VI (1), vi, 4–7.

69 This was quite different from the *karma* of the Buddhists, for retribution did not necessarily overtake the individual either in this life or in some other, but it did bring about the ruin of his house or family or dynasty or social group in the end. This Confucian form of "cosmic reciprocity" was essentially social, and in origin it antedated by many centuries the entry into China of Buddhism and Indian ideas of reincarnation. One can find an almost epigrammatic statement of it in the *Huai Nan Tzu* book, tr. Morgan (ch. 13, p. 160), *ca.* 120 B.C.

70 According to a saying of Confucius current in Han times, "there is more pungency and clarity in showing the Tao in action, in the facts themselves, than in expressing the Tao in empty words" (*Shih Chi*, ch. 130). The Tao inheres in nature and history, it cannot be looked for outside the world.

71 No one was more explicit about this than the great Chang Hsüeh-Chhêng (see the exposition of Demiéville, *op. cit.*, p. 178). The canonical classics, he said, were really history—there was no distinction between *ching* and *shih*—and by the same token all history had canonical value. Chang Hsüeh-Chhêng in fact canonized history, thus becoming the predecessor of his younger contemporary Hegel and in the next generation Karl Marx, although of course completely unknown to them.

72 These were perhaps almost too brilliant, since they show many traces of dialectical, as opposed to formal, logic, and this tendency was powerfully reinforced by the Indian dialectical logic introduced with Buddhism (*cf. SCC*, v. 2, pp. 77, 103, 180, 194, 199, 258, 423, 458). The natural sciences could hardly benefit by dialectical logic without having passed through the stage of formal logic.

73 It is not clear whether it helped more than it hindered. In any case, the mystical-empirical factor may have been much more important than the logical in the European scientific revolution (*cf. SCC*, v. 2, pp. 89, 200), indeed as much as the mathematical.

74 See H. O. H. Stange's stimulating paper, "Chinesische und Abendländische Philosophie; ihr Unterschied und seine geschichtlichen Ursachen," *Saeculum*, v. 1 (1950), p. 380.

75 Notably the renowned Chi-Hsia Academy in the State of Chhi founded about 325 B.C.

76 One should not conclude from this that ancient Chinese society had no democratic elements—the case is quite contrary, and the effects were far-reaching, but it cannot be discussed here. Stange's emphasis on the role of city-state democracy in the development of formal logic in Greece recalls the similar proposal of Vernant to derive deductive geometry from the same social milieu. He regards the demonstration of geometrical propositions in the

agora as a form of logical mathematics particularly congruent with an assembly of equal disputing participants, a democratical reasoning, just as the later Greek writing was a democratized script. Arithmetic and algebra were more specialized secretarial or bureaucratic techniques, less susceptible of public demonstration, and more like the old linear B script. Thus the preference for algebraic methods, so marked in China as well as Babylonia, would have gone naturally with the bureaucratic system of society, so different from the slave-owning city-state democracies. See J. P. Vernant, *Scientific Change; Historical Studies in the Intellectual, Social and Technical Conditions for Scientific Discovery and Technical Invention, from Antiquity to the Present*, ed. A. C. Crombie (London: Heinemann, 1963), p. 102; and *Les Origines de la Pensée Grecque* (Paris, 1964).

77 On the subject of this section see J. Needham, Wang Ling and D. J. de S. Price, *Heavenly Clockwork; the Great Astronomical Clocks of Mediaeval China* (Cambridge: Cambridge University Press, 1960), as also *SCC*, v. 4, pt. 2, p. 435.

78 On the sundial in China see *SCC*, v. 3, p. 302.

79 On the clepsydra in China see *SCC*, v. 3, p. 313, together with Needham, Wang and Price, *op. cit.*, p. 85.

80 J. L. Synge, "A Plea for Chronometry," *New Scientist*, v. 5 (1959), p. 410, has well said that of all measurements made in physics that of time is the most fundamental.

81 In some of the medieval Chinese astronomical clock towers, mercury was used as the fluid instead of water, since it was not subject to freezing in winter, and the containers could be smaller. Experiments with models made by our collaborator Mr. J. H. Combridge have thrown much light on the precise mechanism of the linkwork escapement.

82 Mr. Lloyd's article in this volume shows how the Chinese water-wheel linkwork escapement was a good deal more accurate than the early verge-and-foliot clocks of Europe. Its level was probably not surpassed until after the introduction of the pendulum about the middle of the seventeenth century A.D. and the further improvements that followed therefrom. See his Fig. 3.

83 Reproduced in *Thu Shu Chi Chhêng*, Jen shih tien, ch. 4.

84 Semi-legendary culture-hero and hydraulic engineer.

85 A character in the *Chuang Tzu* book (see p. 13).

86 On the general problem of the origin and development of the idea of Laws of Nature in the different Old World civilizations see *SCC*, v. 2, p. 518. A slightly revised version of this account will be found in J. Needham, "Human Law and the Laws of Nature," *Technology, Science and Art; Common Ground* (Hatfield: Hatfield Coll. of Technol., 1961).

87 No adequate treatment of the history of biology in Chinese culture either in Chinese or a Western language as yet exists, but we hope to present a balanced review of it in *SCC*, v. 6.

88 For details of the instances referred to in this paragraph see J. Needham and D. Leslie, "Ancient and Mediaeval Chinese Thought on Evolution," *Bull. Nat. Institute of Sciences of India* (Symposium on Organic Evolution), v. 7 (1952), p. 1.

89 *Cf. SCC*, v. 2, p. 21.

90 *SCC*, v. 2, p. 488.

91 *I.e.*, to one's predators. *SCC*, v. 2, p. 78.

92 *SCC*, v. 2, p. 421.

93 See H. Zimmer, *Philosophies of India* (New York: Pantheon, 1953), pp. 224, 226, on the Jaina cycles; and also his *Myths and Symbols in Indian Art and Civilisation*, ed. J. Campbell (New York: Pantheon, 1946), pp. 11, 16, 19. Here he discusses the *kali-yuga*, last and worst of the four world ages.

94 *Cf. SCC*, v. 2, p. 485; also v. 4, pt. 1, p. 11.

95 See *SCC*, v. 3, pp. 598, 603.

96 *SCC*, v. 2, p. 486; v. 3, p. 406.

97 See *SCC*, v. 2, p. 420; v. 3, pp. 120, 408. In A.D. 724 the great monk-astronomer I-Hsing computed the number of years which had then elapsed since the "Grand Origin" (Thai Chi Shang Yuan) or general conjunction of the planets, and obtained a result of 96,961,740 years. The fact that a general conjunction is impossible is irrelevant to the spaciousness of the time periods which the Chinese medieval astronomers were prepared to envisage.

98 This was the celebrated reckoning of the learned James Ussher, Archbishop of Armagh. See his *Chronologia Sacra* (Oxford, 1660), p. 45, and *Annals of the World* (Oxford, 1658), p. 1. I am indebted to Prof. H. Trevor-Roper for assisting me with these references.

99 For example the great second-century B.C. medical classic *Huang Ti Nei Ching Su Wên* (ch. 14) periodizes history into ancient (*shang ku*), middle-old (*chung ku*) and recent (*tang chin*) ages, saying that there had been a gradual decline in men's resistance to diseases, so that stronger drugs and treatments were required as time went on.

100 One entire chapter of the *Huai Nan Tzu* book (*ca.* 120 B.C.) is devoted to proving social change and progress since the most ancient times, with many references to material improvement (tr. Morgan, ch. 13, p. 143). The *Huai Nan Tzu* is very Taoist in many ways, but this was a viewpoint of Han Taoism rather than of that of the Warring States.

101 *Cf. SCC*, v. 2, pp. 86, 99, 104, 115.

102 *Cf. SCC*, v. 2, p. 127.

103 Ch. 9 (tr. Legge, v. 1, p. 364), here modified from *SCC*, v. 2, p. 167. Li Yün may be translated "The Mutations of Social Institutions." The wording of parallel passages in the *Mo Tzu* book, chs. 11, 12, 13, 14, 15 (tr. Mei Yi-Pao, pp. 55, 59, 71, 80, 82), fix the date as fourth century B.C., not first century A.D. But these passages are "progressive" rather than "regressive" in tendency, criticizing the ancient ruler-less times as an age when humanity was all "at sixes and sevens," and placing the Ta Thung state in the future, to be brought about by the practice of universal love (*chien ai*). The actual expression Ta Thung is not used in *Mo Tzu*. A similar account to that in the *Li Chi*, but much shorter, occurs in the *Huai Nan Tzu* book (A.D. 120), ch. 2, where the expression Ta Chih (the Ideal Rule) is used instead of Ta Thung: *cf.* tr. Morgan, p. 35.

104 Lit. "for the general use," i.e. not the property of the emperor, feudal lords and patrician families.

105 This phrase, which we might equally well translate the Great Community, was also used in a rather different sense by the late Warring States philosophers, namely to indicate the parallel of the Microcosm (man) with the Macrocosm (the universe). For an example of this see *Lü Shih Chhun Chhiu* (239 B.C.) (tr. R. Wilhelm, ch. 62, p. 160). But the senses are not so far apart because the ancient Chinese felt that social community was "intended by Nature" and

that class differentiation and all strife were a violation of the natural order, a violation, moreover, which would upset Nature and lead to natural calamities, or at least to unfavorable weather conditions, epidemics, etc. Besides, *thung* as "with-ness" allows the translation "Great Similarity."

[106] *Cf.* the valuable paper of Hou Wai-Lu, "Socialnye Utopii Drevnego i Sred-nevekovogo Kitaia" ("Social Utopias of Ancient and Mediaeval China"), *Voprosy Filozofii*, v. 9 (1959), p. 75.

[107] See Shih Yu-Chung, "Some Chinese Rebel Ideologies," *T'oung Pao*, v. 44 (1956), p. 150.

[108] On the history of the Ta Thung concept in China there is a valuable little book by Hou Wai-Lu, Chang Kai-Chih, Yang Chao and Li Hsüeh-Chin, *Chung-Kuo Li-Tai "Ta Thung" Li Hsiang* (Peking: Kho-Hsüeh, 1959).

[109] The word *phing* has both meanings.

[110] It is now under intensive study by sinologists, historians and social philosophers. The best review of the subject in a Western language is probably that of W. Eichhorn, "Thai-Phing und Thai-Phing Religion," *Mitt. d. Inst. f. Orient-forschung*, v. 5 (1957), p. 113, seconded by T. Pokora, "On the Origins of the Notions of Thai-Phing and Ta-Thung in Chinese Philosophy," *Archiv Orientální*, v. 29 (1961), p. 448.

[111] Ch. 22, tr. R. Wilhelm, p. 56.

[112] Ch. 22, W. Eichhorn, *op. cit.*, p. 116.

[113] *Chhien Han Shu*, Ch. 24A.

[114] Ch. 2 (Chi Wu Lun), tr. Legge, v. 1, p. 176.

[115] *Cf.* Isaiah, 40:3,4. There was also a mystique of topographic leveling in Hindu and Buddhist thought, the "alluvial" flatness left behind by world floods or catastrophes on which the Buddhas and Bodhisattvas pace. On this idea see P. Mus, "La Notion de Temps Réversible dans la Mythologie Bouddhique," *Annuaire de l'Ecole Pratique des Hautes Etudes* (*Sect. des Sciences Religieuses*) (1939), pp. 15, 33, 36.

[116] Ch. 53, tr. Wu, p. 75, tr. Chhu, p. 66, tr. Duyvendak, p. 117.

[117] A good introductory account of this literature will be found in P. van der Loon, "The Ancient Chinese Chronicles and the Growth of Historical Ideals," *Historians . . .*, p. 24.

[118] All three were supposed to derive from the oral teaching of Confucius himself.

[119] This dates from 124 B.C., though the governmental title of Po-Shih (doctor or professor) had appeared already in the third century B.C. and the principle of imperial examinations in 165 B.C. When Han Wu Ti endowed "disciples" (*ti-tzu*) as well as professors, the imperial university may be said to have been established. By 10 B.C. it had as many as 3,000 students, not all, of course, "on the foundation." The term Thai Hsüeh, afterwards borne for centuries by the university, occurs first in a memorial by Tung Chung-Shu himself urging its establishment, though the emperor preferred the plans of Kungsun Hung for the same design.

[120] See Fêng Yu-Lan, *op. cit.*, v. 2, p. 81.

[121] *Ibid.*, p. 83. The passage occurs at the end of the first chapter of the *Kungyang Chuan*.

[122] This phrase has an undertone of the Golden Age theory because *shuai* means decay or decadence as well as weakness and feebleness. But it is doubtful whether this was intended because an alternative form of the phrase found in many texts is Chü Luan, *chü* meaning forcible occupation or possession,

the seizing of lands and goods, rebellion, etc., *i.e.*, the correlate of the weakness; in other words, the state of society described in the Gospel of St. Luke, 11:21, or the "law of the fishes" in Buddhism, unending internecine strife, "Nature red in tooth and claw." *Cf.* Hsiao Kung-Chhüan, "Khang Yu-Wei and Confucianism," *Monumenta Serica*, v. 18, no. 96 (1959), p. 142.

123 If Pokora, *op. cit.*, is right, the ideas of Ho Hsiu were directly derived from the popular progressive apocalyptic, possibly through the intermediation of Yü Chi (*ca.* A.D. 120–200). Yü Chi was a naturalist, physician and thaumaturgist, one of the fathers of the Taoist church, and probably the author of one or more of the books which formed the material of the *Thai Phing Ching*.

124 The corpus has been newly edited with the title "*Thai Phing Ching Ho Chiao*" (Peking: 1960) by Wang Ming, who attempts to reconstitute the original text of the main work. Pokora, *op. cit.*, gives a brief description of his book, and Eichhorn, *op. cit.*, discusses the contents of some of the documents which it includes.

125 On him see W. Eichhorn, "Bemerkungen zum Aufstand des Chang Chio und zum Staate des Chang Lu," *Mitt. d. Inst. f. Orientforschung*, v. 3 (1955), p. 291.

126 The standard modern work on this great but ultimately abortive revolution is by Lo Erh-Kang, *Thai-Phing Thien-Kuo Ko-Ming Chan Chêng Shih* (*A History of the Revolutionary War of the Heaven-Ordained Kingdom of Great Peace and Equality*) (Peking, 1949). There is no satisfactory book on the subject as yet in any Western language. For a related discussion, especially interesting in view of what is said in the present article on the Judaeo-Christian attitude to time, see E. P. Boardman's *Christian Influence on the Ideology of the Thai-Phing Rebellion* (Madison: University of Wisconsin Press, 1952).

127 Note that the title was taken from the *Li Chi* but the progressive content from the traditional development of the *Kungyang Chuan*, and from *Mo Tzu*.

128 By L. G. Thompson, *Ta Thung Shu; the One-World Philosophy of Khang Yu-Wei* (London: Allen & Unwin, 1958).

129 The *Huai Nan Tzu* book points the moral by saying that what rendered the culture-heroes worthy of divine honors was the outstanding service which they rendered to the benefit of mankind (tr. Morgan, ch. 13, p. 178).

130 *Cf. SCC*, v. 1, p. 51.

131 See H. Wilhelm, *Die Wandlung; acht Vorträge zum* I-Ging (I Ching) (Peking: Vetch, 1944), Eng. tr. C. F. Baynes, *Change; Eight Lectures on the* I Ching (London: Routledge & Kegan Paul, 1961). Also, *I Ching* (New Hyde Park, N.Y.: University Books, 1964).

132 H. Wilhelm, "The Concept of Time in the Book of Changes," *Man and Time* (New York: Pantheon Books, 1957), p. 212.

133 *Cf. SCC*, v. 2, p. 336.

134 Tr. R. Wilhelm and C. F. Baynes, v. 1, p. 353. Tabulation in *SCC*, v. 2, p. 327.

135 The Grand Canal, connecting Hangchow in the south with Peking in the north directly, embodied the earliest successful fully artificial summit canal in any civilization, across the foothills of the mountains of Shantung. Planned originally by the astronomer and engineer Kuo Shou-Ching, this section was built by the Mongol military engineer Oqruqči (Ao-Lu-Chhi), with a Chinese colleague, Ma Chih-Chên, in A.D. 1287. But it was not capable of full year-round efficiency, however, until Sung Li in A.D. 1411 succeeded in capturing the waters of certain mountain rivers by dams and subsidiary feeder canals,

thus ensuring adequate water levels in the summit sections at all times. For further details see *SCC*, v. 4, pt. 3.

136 Details will be given in *SCC*, v. 5.

137 G. Daniel, *The Three Ages* (Cambridge: Cambridge University Press 1943), p. 9, citing with approval R. A. S. Macalister, *Textbook of European Archaeology* (London: 1921).

138 See his *Ledetrad til Nordiske Oldkindighed* (Copenhagen, 1836), Eng. tr. *A Guide to Northern Antiquities* (London, 1848).

139 The idea that neolithic flaked and polished stone implements were meteorites is old in Europe, whence probably it passed to China by the eighth century A.D., for the pharmaceutical naturalist Chhen Tshang-Chhi (*fl.* A.D. 713–733) is the first to speak of "thunder-axes" (*phi-li fu*), an appellation afterwards common in Chinese scientific literature.

140 *Cf. SCC*, v. 3, p. 173.

141 *De Rer. Nat.* V, 1283, mod. tr. Leonard.
> . . . Man's ancient arms
> Were hands and nails and teeth, stones too and boughs
> Broken from forest trees, and flame and fire
> As soon as known. Thereafter force of iron
> And bronze discovered was; but bronze was known
> And used ere iron, since more amenable
> Its nature is and its abundance more.

142 Daniel, *op. cit.*, p. 13. These words would be much better applicable to Hesiod's account of five ages (gold, silver, bronze, heroic and iron) in the *Works and Days* (eighth century B.C.), 11.110, which is really not in the running as a description of actual technological periods, though some have sought to make it so. See J. G. Griffiths, "Archaeology and Hesiod's Five Ages" in *J. of History of Ideas*, v. 17 (1956), p. 109, with comment by H. C. Baldry, p. 553; F. J. Teggart, "The Argument of Hesiod's *Works and Days*," *J. of History of Ideas*, v. 8 (1947), p. 45.

143 Ch. 13. Attention was first drawn to it in the West by Friedrich Hirth in 1904, "Chinesische Ansichten über Bronzetrommeln," in *Mitt. d. Sem. f. Or. Spr.*, v. 7,200 (pp. 215 ff.). This was a brilliant pioneer effort in the history of metallurgy and proto-archaeology. Hirth returned to the periodization question in his "Ancient History of China, to the End of the Chou Dynasty" (New York: Columbia University Press, 1908; repr. 1923), p. 236. The dating of the *Yüeh Chüeh Shu* and its putative compiler depends on a statement at the end of ch. 3 that in A.D. 52 a period of 567 years had elapsed since a certain event in Yüeh history, fixing that at 515 B.C. But this is simply an appended sentence and cannot in itself date the whole material of the book. From the internal evidence of the style of the passage in question, Hirth was inclined to place it as early as the fifth century B.C., and while this would be almost impossible to prove it is not an entirely unreasonable estimate. It would correspond rather well with what we know of the iron industry at that time. The *Yüeh Chüeh Shu* passage continues to interest archaeologists; *cf.* Chang Kuang-Chih, *The Archaeology of Ancient China* (New Haven: Yale University Press, 1963), p. 2.

144 "The Chinese," wrote Hirth (*Bronzetrommeln*, p. 215) "began to study the developmental periods of prehistory relatively early, and drew their conclusions from tomb finds and other cultural remains."

145 It was one of the great merits of my teacher, Gustav Haloun, that he brought

out this in his paper "Die Rekonstruktion der chinesischen Urgeschichte durch die Chinesen," *Japanisch-Deutsche Zeitschr. f. Wiss. u. Tech.*, v. 3 (1925), p. 243. It was through my *thung chuang*, Laurence Picken, that I came to appreciate this as it deserved.

146 See, *e.g.*, *Chuang Tzu*, ch. 29; *Kuan Tzu*, ch. 84; *Mo Tzu*, ch. 25; *Shang Chün Shu*, ch. 7; *Han Fei Tzu*, ch. 10; *Lü Shih Chhun Chhiu*, ch. 117. Among works of rather later date *cf. Li Chi*, chs. 5, 9; *Lieh Tzu*, ch. 5; *Ho Kuan Tzu*, ch. 13; *Huai Nan Tzu*, ch. 13.

147 Ch. 10, *cf.* tr. W. K. Liao (v. 1, p. 85); Chang Kuang-Chih (*op. cit.*). There is a parallel passage in the *Shih Chi*, tr. Chavannes (v. 2, ch. 5, p. 40). The incident is attributed to 626 B.C. The Prince of Chhin whom Yu Yü addressed was the same Duke Mu who a few years later died and was buried with human sacrifices, "therefore he had never been able to obtain the hegemony, etc."

148 Hirth's view (*Ancient History*, p. 13) that the legendary Chinese "emperors" should be regarded as "symbols of the . . . phases of Chinese civilisation" and "representatives of preparatory periods of culture" is quite acceptable today, and the criticisms of B. Laufer (*Jade*, p. 70) were wide of the mark. Haloun (*loc. cit.*) believed that they had all been originally cosmological gods of world regions and patron-deities of clans, then gradually they became culture-heroes and at last "rulers."

149 It seems quite strange that no one has collected and studied all the passages in Chou, Chhin and Han literature on remote antiquity from this point of view.

150 Probably the oldest reference is in the *Shan Hai Ching* (*Classic of the Mountains and Rivers*), a Chou text, tr. de Rosny (ch. 4, p. 158).

151 Indeed down to the Chhin and Early Han, for Chhin Shih Huang Ti called in all the bronze he could as a disarmament measure after the empire's unification, and had colossal figures made of it (*Shih Chi*, ch. 6).

152 *Cf.* J. Needham, "The Development of Iron and Steel Technology in China," Newcomen Society, London, 1958 (Dickinson Memorial Lecture), repr. (Cambridge: Heffer, 1964).

153 *Cf.* Chêng Tê-K'un, *Archaeology in China*, v. 3, *Chou China* (Cambridge: Heffer, 1963), p. 246; Chang Kuang-Chih, *op. cit.*, p. 195.

A very remarkable feature of Chinese iron technology is the fact, now generally accepted, that cast iron was made almost as soon as iron itself was known, while in the West it took more than seventeen further centuries to achieve this. A number of technical reasons are available in explanation but great credit is due to the ancient Chinese iron-masters.

154 R. H. Lowie, *The History of Ethnological Theory* (London: Harrap, 1937), p. 13.

155 The importance of establishing this point arose in correspondence with Mr. Arthur Clegg.

156 One striking example may be given; the secular variation in the length of the tropical year. Over the ages the Chinese astronomers came to recognize that there had been a very gradual shortening (*hsiao chang*); this was first computed in A.D. 1194 and confirmed with exceptionally accurate observations in 1282. The value obtained for this very minute term was actually much too large, probably because it was desired to "save" three extant observations made in the first millennium B.C. (and probably very inaccurate), but the

whole story is a remarkable example of the cumulative and progressive Chinese endeavor to improve gradually upon all existing previously accepted values. It also shows how essentially correct conclusions, much more spacious and open-minded than those which Europeans of those times were willing to entertain, could often be reached though the numerical data were wrong. Proper motion is another example of this (*SCC*, v. 3, p. 270). On the secular variation of the tropical year length, see Nakayama Shigeru, *Japanese J. of Hist. of Sci.*, p. 68 (1963), p. 128 and *Abstracts of Communications to the Xth. Internat. Cong. of Hist. of Sci.* (Ithaca, N.Y.: 1962), p. 90, as also *Japanese Studies in the Hist. of Sci.*, v. 2 (1963), p. 101.

[157] The data used are taken from Yen Yü, "Shih-liu Shih-chi-ti Wei Ta Kho-Hsüeh Chia Li Shih-Chen" ("The Great Sixteenth-Century Scientist Li Shih-Chen"), *Chung-Kuo Kho-Hsüeh Chi-Shu Fa-Ming ho Kho-Hsüeh Chi-Shu Jen Wu Lun Chi* (*Chinese Scientific and Technological Discoveries and the Men Who Made Them*), ed. Li Kuang-Pi and Chhien Chün-Yeh (Peking, 1955), p. 314; see also Chêng Chih-Fan, "Li Shih-Chen and his Materia Medica," *China Reconstructs*, v. 12, no. 3 (1963), p. 29. In *SCC*, v. 6, we shall present a full account of the Chinese tradition of pharmaceutical natural history and of Li Shih-Chen's place in it.

[158] J. B. Bury, *The Idea of Progress* (London: Macmillan, 1920).

[159] Cf. *SCC*, v. 4, pt. 2, p. 6.

[160] E. Zilsel, "The Genesis of the Concept of Scientific Progress," *J. Hist. of Ideas*, v. 6 (1945), p. 325. Cf. also S. Lilley, "Robert Recorde and the Idea of Progress, a Hypothesis and a Verification," *Renaissance and Modern Studies*, v. 2 (1959), p. 1.

[161] Naturally this is not my description of the great Greek astronomers, but their contemporaries may well have thought of them in this way; one remembers the anecdote about Thales, recorded by Plato (*Theaet.* 174ᵃ), that he fell into a well while watching the stars. There may be something else in this (cf. *SCC*, v. 3, p. 333) but we may take it at its face value here.

[162] Cf. *SCC*, v. 3, pp. 171, 186.

[163] From S. J. Rigaud, *Correspondence of Scientific Men of the Seventeenth Century* (Oxford, 1841), v. 1, p. 158, quoted by A. F. Titley, "Science and History," *History*, v. 23 (1938), p. 108.

[164] Cf. *SCC*, v. 2, p. 73.

[165] Cf. *SCC*, v. 2, p. 122.

[166] The whole story is given in *SCC*, v. 4, pt. 1, pp. 44, 53; and more fully in A. Beer, Ho Ping-Yü, Lu Gwei-Djen, J. Needham, E. G. Pulleyblank & G. I. Thompson, "An Eighth-Century Meridian Line; I-Hsing's Chain of Gnomons and the Prehistory of the Metric System," *Vistas in Astronomy*, v. 4 (1961), p. 3.

[167] Cf. *SCC*, v. 4, pt. 2, pp. 496 ff.; also J. Needham, Wang Ling and Derek J. Price, *Heavenly Clockwork*, p. 116.

[168] Cf. *SCC*, v. 4, pt. 2, p. 507; also Needham, Wang and Price, *op. cit.*, p. 133.

[169] For a fuller treatment of this whole subject see J. Needham, "China's Scientific Influence on the World" in the *Legacy of China*, ed. R. Dawson (Oxford: Oxford University Press, 1964).

[170] A conversation with Dr. O. Temkin at Baltimore in 1950 was my nucleus of crystallization for this *Fragestellung*, though I had often meditated about it while in China during the Second World War. An interesting fact, and

perhaps significant in the light of what the following pages contain, is that although I talked about the inhibitory factors of modern science in China with a great number of friends from Dr. Kuo Pên-Tao in Chhêngtu to Dr. Liang Po-Chiang in Kuangtung, the one thing they never suggested was that the sense of historical time, or the lack of it, could have had anything to do with the matter.

171 See especially O. Cullman, *Christus und die Zeit* (Zollikon-Zürick: Evang. Verlag, 1945), Eng. tr. *Christ and Time*; F. V. Filson, *The Primitive Christian Conception of Time and History* (London: SCM Press, 1951); P. Tillich, *The Interpretation of History*, tr. N. A. Rasetzki and E. L. Talmey (New York: Scribner, 1936); *The Protestant Era*, tr. J. Luther Adams (Chicago: University of Chicago Press, 1948); R. Niebuhr, *Faith and History; a Comparison of Christian and Modern Views of History* (London: Nisbet, 1949); *The Self and the Dramas of History* (London: Faber & Faber, 1956); H. Christopher Dawson, *The Dynamics of World History*, ed. J. J. Mulloy (London: Sheed & Ward, 1957); *Progress and Religion; an Historical Enquiry* (London: Sheed & Ward, 1929); T. F. Driver, *The Sense of History in Greek and Shakespearean Drama* (New York: Columbia University Press, 1960). *Cf.* H. Butterfield, *op. cit.* The reader must also be referred to the essay by Rev. J. L. Russell, S.J., in this volume

172 It is generally known that the custom of denominating years serially in the A.D. sequence started only from the proposal of Dionysius Exiguus in A.D. 525; it is less well known that the minus series of B.C. years, extending backwards from the birth of Christ, was introduced as late as the seventeenth century A.D., and Bossuet in 1681 may have been the first to use it. This system is now almost universal; the Chinese speak of *kung yuan, chhien* or *hou*, before or after, in the "public" or "international" era, a testimony perhaps to the all-pervadingness of modern technological, rather than specifically Christian, Western civilization.

173 *Revelations* 1:4.

174 See Irenaeus, *Contra Haeresios*, IV, 37, 7.

175 Roman thought was rather different, as witness the "linear" epic of Virgil, the metrical chronicles that preceded it, and the theory of the *urbs aeterna*; *cf.* M. Eliade, *The Myth of the Eternal Return*, tr. W. R. Trask (London: Routledge & Kegan Paul, 1955), p. 201; also C. S. Lewis, "Historicism," *Month*, v. 4, new ser. (1950), p. 230.

176 On these see especially M. Eliade, *op. cit.* Eliade opens his account with a fascinating study of the seasonal rites of ancient and primitive peoples which, he suggests, helped them to protect themselves against the psychological fear of the passage of time and the positive horror of anything really new and irreversible (pp. 80, 128, 184, 217). These repetitions, then, joined with the computational long-term cycles of astronomy, led to the Indo-Hellenic universe of cyclical time.

177 Porphyry, *Vita Pyth.* 19.

178 Chrysippus, frs. 623–27; Zeno, frs. 98, 109; Eudemus, fr. 51.

179 *Medit.*, XI. 1.

180 *Physica*, IV, 14, 223b 21. *Problemata*, XVII, 3.

181 *Politics*, 269c, ff.; *Republic*, VIII, 546.

182 *Ecclesiastes*, 1:9.

183 *Cf.* Driver, *op. cit.*, p. 38.

Although historians of science are never tired of hymning the services of Euclidean deductive geometry to the Western world, I vividly remember a conversation with Dr. Paul Lorenzen of Bonn in 1949 in which he expressed the view that Europe had had more geometry than was good for it. Of course geometry was an essential basis for modern science, but it did have the bad effect of inducing too ready a belief in abstract timeless axiomatic propositions of all sorts supposedly self-evident, and too willing an acceptance of rigid logical and theological formulations. As these became invested with the authoritarian assurance which the Latin clergy inherited from the Roman jurisconsults, the explosion of the Reformation was inevitable when the merchant class rose to power; and the West still suffers from the slogans of that time. China, however, was algebraic and "Babylonian," not geometrical and "Greek," so that opposition tended to be practical and approximative rather than theoretical and absolute, and men did not feel obliged to formulate such timeless axiomatic propositions. Hence empirical, historical, "statistical" ethics, with little ideological fanaticism and essentially no persecution for religion's sake at all (*cf.* J. Needham, "The Past in China's Present," *Centennial Review*, v. 4 (1960), pp. 145, 281).

184 *Cf.* Eliade, *op. cit.*, p. 167. Belief in the periodical destruction and re-creation of the universe goes back to the tenth century B.C. *Atharvaveda* X, 8, 39, 40.

185 *Cf.* H. Zimmer, "Myths and Symbols," pp. 11, 16, 19; Eliade, *op. cit.*, p. 169.

186 I think it is important in all this to speak of "recurrence in time" rather than of "the reversibility of time" in the cyclical world theories (though writers on this subject often use the expression), for there was in fact in Buddhist mythology a true doctrine of reversibility, much more far-reaching and much less well known. It has been discussed by P. Mus in his interesting paper already mentioned, "La Notion de Temps Réversible dans la Mythologie Bouddhique." As a condition of deliverance from *samsāra*, the flux of becoming, the Time-Ogre, Mahākāla, whom we see in the Tunhuang frescoes devouring all things; the aspiring Bodhisattva had to reverse time's flow, retracing all his previous existences in an inverted order (*pratilomam*). Then, after his last rebirth, he could proclaim himself the "Firstborn of Time," and stride triumphantly to the summit of the cosmic mountain before disappearing into Nirvāna. This was like running a cinematograph film backwards through the projector, as (with remarkable effect) in certain films of Jean Cocteau.

187 Eliade, *op. cit.*, p. 172. A complete translation of these from the Pāli Canon exists: "The *Jātaka* or Stories of the Buddha's Former Births, translated by various hands," ed. E. B. Cowell (Cambridge: Cambridge University Press, 1895–1913), 6 vols.

188 A good discussion of this will be found in K. Quecke, "Der Indische Geist und die Geschichte," *Saeculum*, v. 1 (1950), p. 362. See also the preceding essay by Prof. Nakamura.

189 Tillich, *Protestant Era*, pp. 23, 30. *Cf.* Niebuhr, *Faith and History*, p. 15.

190 Indeed time is almost absorbed into space, because if every event is infinitely repeated evanescence is only illusory and there is no irreversible change— each moment is like a "still" from a photographic film which can be run through the projector forwards over and over again (Eliade, *op. cit.*, p. 184).

191 *Cf.* Eliade, *op. cit.*, pp. 185, 191.

192 See particularly the interesting recent symposium by F. d'Arcais, A. Buzzati-Traverso, A. C. Jemolo, E. de Martino, Rev. R. Panikkar and U. Spirito,

"Progresso Scientifico e Contesto Culturale," *Civiltà delle Macchine*, v. 11, no. 3 (1963), p. 19. The stimulating but confused book of L. D. del Corral, tr. H. V. Livermore, *The Rape of Europe* (London: Allen & Unwin, 1959), also touches upon this question. So too does the work of K. Jaspers, tr. M. Bullock, *The Origin and Goal of History* (London: Routledge & Kegan Paul, 1953), which recognizes the rise of modern science and technology as the essentially new and unique power-giving component of Western civilization, but accounts for it in ways curiously tentative and faltering for so eminent a thinker. Unlike other presentations of Western philosophy of history mentioned, this is not a Christocentric book, but emphasizes rather the period of the great religious geniuses of all the Old World civilizations *ca.* 500 B.C. I may mention also an article by P. F. Douglass, "Christian Faith and Political Philosophy," *Religion in Life*, v. 10 (1941), p. 267, which surveys many factors possibly connecting the Christian elements in European culture with the rise of modern science in it. To this I was introduced by the late Prof. Roderick Scott in 1944 at Changting in Fukien where he was lecturing at the Fukien Christian University evacuated from Fuchow. Prof. Scott dwelt much on the way in which Chinese inventions, while powerless to change society in China, had revolutionized the history of Europe time after time. Long afterwards I went fully into this question in my contribution to the *Legacy of China* (Oxford: 1964), but could not adopt the view of Scott and Douglass that Christianity was the main reason for the difference.

193 Could this be connected with the fact that the Greeks experimented so little in comparison with their scientific theorizing?

194 Moreover in a fully cyclical-time matrix there was the doctrine of *karma*, of the automatic recompense of good or evil deeds in kind, covering many successive reincarnations of the individual. This was (and is) fundamental to Buddhists, whatever philosophical school they might adhere to in the matter of time's reality. As already pointed out (*SCC*, v. 2, p. 418), however, the law of *karma* had no stimulating effect on the idea of causality in science, or of Laws of Nature, presumably because the Buddhists were really interested only in the moral part of it, while scientific causality has to be ethically neutral.

195 This was the kind of personal knowledge of the chroniclers, too, the *rerum gestarum scriptores*.

196 *Scepsis Scientifica; or, Confest Ignorance the Way to Science, in an Essay on the Vanity of Dogmatising and Confident Opinion* (London, 1661, 1665). Repr. and ed. J. Owen (London: Kegan Paul, 1885).

197 An interesting glimpse of two phases of this process may be had in W. Baron and B. Sticker, "Ansätze z. historischen Denkweise in d. Naturforschung an der Wende vom 18 zum 19 Jahrhundert; I, Die Anschauungen Johann Friedrich Blumenbachs über die Geschichtlichkeit der Natur; II, Die Konzeption der Entwicklung von Sternen und Sternsystemen durch Wilhelm Herschel," *Archiv. f. Geschichte d. Medizin u. d. Naturwissenschaften*, v. 47 (1963), p. 19.

198 *Cf.* also the current debates among radio-astronomers and others, causing no little stir in Cambridge and elsewhere, concerning the "steady-state" theory of the universe as against the "expansion and contraction" theory with its corollary of creation and re-creation. The "Great Year" conception is by no means dead, though it may be hardly recognizable in its scientific overalls. See also Prof. Whitrow's essay in this volume.

[199] *Cf.* also St. Augustine, *Confessions*, XI.

[200] Perhaps Joachim's greatest disciple was William Blake, 600 years afterwards, who received the mystical apocalyptic "antinomian" tradition by way of the Anabaptists, Brethren of the Free Spirit, "Ranters," and other transmitters of revolutionary Christianity. See the interesting study of A. L. Morton, *The Everlasting Gospel; a Study in the Sources of William Blake* (London: Lawrence & Wishart, 1958).

[201] See here L. Thorndike, *A History of Magic and Experimental Science during the First Thirteen Centuries of our Era* (New York: Columbia University Press, 1947 ed.), vol. 2, pp. 203, 370, 418, 589, 710, 745, 895.

[202] An example of the persistence of resignation to the cyclical "wheel of fate," prosperity and decay (*shêng shuai*), may be seen in the conversation between the Tao-Kuang emperor (*r.* 1821 to 1850) and one of his high officials Pi Kuei, translated and reported by Meadows, *The Chinese and Their Revolutions* (London: 1856), pp. 123, 130, 134. "Alas, in all affairs prosperity is followed by decay!" the emperor kept on repeating.

[203] Pointing up the contrast between the two disciplines, Liu Chih-Chi went so far as to advise the exclusion of the astronomical monograph from the dynastic history pattern, presumably because it dealt so much with "timeless," "unhistorical," recurring cycles. He was in favor of natural history, however; *cf.* Pulleyblank, *Historians . . .* , p. 145. A curious parallel will be found in *SCC*, v. 3, p. 634.

[204] Only someone quite unconversant with Chinese traditions could have written, as C. Dawson did, that the "denial of the significance of history is the rule rather than the exception among philosophers and religious teachers throughout the ages from India to Greece and from China to Northern Europe" (*Dynamics of World History*, p. 271).

[205] D. Bodde, "Dominant Ideas [in Chinese Culture]," *China*, ed. H. F. McNair (Berkeley: University of California Press, 1946), pp. 18, 23.

[206] Tillich, *Protestant Era*, p. 19.

Comments on Time, Process, and Achievement

[1] Early views of an eternal universe are often neglected, though they have been voiced *passim*. Giordano Bruno, for instance, held that the universe is eternal. "The individual worlds are also eternal but in a manner different from that of the universe . . . 'It happens that it [the universe] is eternal, although dissoluble. The necessity for such eternity comes from its extrinsic provider and maintainer and not from its intrinsic self sufficiency.' The stars are destructible by nature, but the universe contains all, is all, and therefore requires a nature of lasting duration. This is guaranteed by its necessary and eternal creation," L. Cicuttini, *Giordano Bruno* (Milano, Publicazioni dell 'Universita Cattolica del Sacre Cuore,' new ser. 31 (1950), p. 121, tr. J.T.F.

[2] Joseph Needham, "Science and Society in China and the West," *Science Progress*, v. 54 (1964), p. 64.

[3] On this, see H. Meyerhoff, *Time in Literature* (Berkeley: University of California Press, 1960), p. 106.

[4] Max Lerner, *America as a Civilization* (New York: Simon & Schuster, 1957),

p. 620; see also ch. XII, sec. 7, *passim*. *Cf.* John Cohen, *Behaviour in Uncertainty* (London: Allen & Unwin, 1964), p. 110.

5 John A. Kouwenhoven, *The Beer Can by the Highway, Essays on What Is "American" about America* (New York: Doubleday, 1961), p. 66.

6 J. Needham, "Science and China's Influence on the World," in *The Legacy of China*, R. Dawson, ed. (Oxford: Oxford University Press, 1964), p. 307.

7 Robert H. Knapp, "Time Imagery and the Achievement Motive," *J. of Personality*, v. 26 (1958), p. 426.

8 An interesting reminder of the role of time in a capitalistic economy is the presence of the following phrase in legal documents which involve stipulations of interest payment or the delivery of goods: "Time is the essence of this contract."

9 Robert H. Knapp, "Attitudes toward Time and Aesthetic Choice," *J. of Soc. Psych.*, v. 56 (1962), p. 79.

10 "Science and Society in East and West," *The Science of Science, Society in the Technological Age.*, ed. M. Goldsmith and A. Mackay (London: Souvenir Press, 1964), p. 128.

11 See also Lewis Mumford's chapter on "The Monastery and the Clock," *Technics and Civilization* (New York: Harcourt, Brace & World, 1962), p. 12.

12 H. H. Anderson, ed., *Interdisciplinary Symposia on Creativity* (New York: Harper, 1959), p. 176.

13 Norman O. Brown, *Life against Death* (New York: Random House, 1959), p. 316.

14 Needham, *op. cit.*, p. 61. See also his "Human Law and the Laws of Nature," in *Technology, Science and Art: Common Ground*, Hatfield College of Technology (1961), p. 24.

15 *Life*, September 11, 1964.

Time and the Destiny of Man

The author's books and articles, referred to below, contain full documentation on the points concerned, as well as discussion of the problems involved.

1 *Cf.* S. G. F. Brandon, *Time and Mankind* (London: Hutchinson 1951), p. 15; *Man and His Destiny in the Great Religions* (Manchester: Manchester University Press, 1962), p. 6.

2 For a photograph and a drawing of this figure, which is partly painted and partly engraved, see H. Breuil, *Quatre Cents Siècles d'Art pariétal* (*Les Cavernes ornées del' Age de Renne*) (Montignac: Centre d'Études et de Documentation préhistoriques, 1954), p. 166, *cf.* 176–7.

3 *Cf.* Brandon, *Man and His Destiny*, p. 18; *Time and Mankind*, p. 17; "The Ritual Perpetuation of the Past," *Numen*, v. 6 (1959), p. 112; *History, Time and Deity* (Manchester: Manchester University Press, 1965), ch. II.

4 *Cf.* Brandon, "The Origin of Religion," *The Hibbert Journal*, v. 57 (1958–59), p. 350; *Man and His Destiny*, p. 6.

5 *Cf. Man and His Destiny*, p. 8; *History, Time and Deity*, p. 13.

6 *Cf. Man and His Destiny*, p. 38; see also p. 32.

7 *Cf. ibid.*, p. 57. See also Brandon, *Creation Legends of the Ancient Near East* (London: Hodder and Stoughton, 1963), p. 64; *History, Time and Deity*, pp. 18, 59.

8 *Cf.* Brandon, "Ritual Perpetuation . . .," p. 115; *Man and His Destiny*, p. 35; "The Ritual Technique of Salvation in the Ancient Near East," *The Saviour God* (Manchester: Manchester University Press, 1963), p. 18.

9 *Cf.* "Ritual Perpetuation . . .," p. 123; *Time and Mankind*, p. 121; "Origin of Religion," p. 128.

10 *Man and His Destiny*, ch. IV; *Time and Mankind*, chs. IV and V; *History, Time and Deity*, ch. V.

11 See the references in note 10 and *History Today*, v. 11 (1961), p. 155.

12 *Cf. Man und His Destiny*, ch. VI; *Time and Mankind*, ch. VIII. See also Brandon, *The Fall of Jerusalem and the Christian Church* (2nd ed.; London: 1957), ch. IV; *History, Time and Deity*, ch. VI.

13 See note 12 and "The Historical Element in Primitive Christianity," *Numen*, v. 2 (1955), p. 156.

14 *Cf.* Brandon, "Ritual Perpetuation . . . ," p. 125; M. Werner, *Die Entstehung des christlichen Dogmas* (Bern/Tübingen, 2 Aufl., 1941), p. 420.

15 *Cf.* "Ritual Perpetuation . . . ," p. 126; *Time and Mankind*, pp. 169, 177, 180, 183.

16 *Cf.* "Historical Element . . . ," p. 156.

17 *Cf. Man and His Destiny*, ch. VII; see also L. Massignon, "Time in Islamic Thought," *Man and Time*, ed. J. Campbell (New York: Pantheon, 1958), p. 108.

18 *Cf.* R. C. Zaehner, *Zurvān: a Zoroastrian Dilemma* (Oxford: Clarendon Press, 1955); *The Dawn and Twilight of Zoroastrianism* (London: Weidenfeld and Nicolson, 1961), chs. 8–12. See also Brandon, *Man and His Destiny*, ch. VIII.

19 *Cf. Man and His Destiny*, chs. III and V; *History, Time and Deity*, p. 92.

20 *Cf. ibid.*, p. 175 and ch. XI.

21 Tr. S. Radhakrishnan, *The Principal Upaniṣads* (London: Allen and Unwin, 1953), p. 391. On the Indian evaluation of human destiny see *Man and His Destiny*, ch. IX.

22 Th. Stcherbatsky, *Buddhist Logic* (The Hague: 1932), v. 1, p. 554; *cf.* E. Conze, *Buddhist Thought in India* (London: Allen & Unwin, 1962), p. 34.

23 Conze, *op. cit.*, p. 50.

24 *Cf. Man and His Destiny*, ch. X; *History, Time and Deity*, p. 101.

25 Bhagavadgītā, XI, 23, 30, 32, trans. L. D. Barnett in *Hindu Scriptures*, ed. N. Macnicol (London: J. M. Dent, 1938), pp. 263, 264.

GENERAL REFERENCES

Aron, R. *Introduction à la Philosophie de l'Histoire*. Paris: Gallimard, 1948.

Baillie, J. *The Belief in Progress*. Oxford: Oxford University Press, 1950.

Brandon, S. G. F. "Time as God and Devil," *Bull. John Rylands Library*, v. 47 (1964).

Burgelin, P. *L'Homme et le Temps*. Paris: Aubier, 1945.

Campbell, J., ed. *Man and Time* (New York: Pantheon, 1958).

Case, S. J. *The Christian Philosophy of History*. Chicago: University of Chicago Press, 1943.

Cassirer, E. *An Essay on Man*. New Haven: Yale University Press, 1945.

Childe, V. G. *History*. London: Watts, 1947.

Collingwood, R. G. *The Idea of History*. Oxford: Clarendon Press, 1947.

Croce, B. *La Storia come Pensiero e come Azione*, 4th ed. Bari, 1943.

———. *Teoria e Storia della Storiografia*, 5th ed. Bari, 1943.

Dawson, C. *Progress and Religion*. London: Sheed and Ward, 1931.

Denton, R. C., ed. *The Idea of History in the Ancient Near East*. New Haven: Yale University Press, 1955.

R. Klibansky and H. J. Paton, eds. *Philosophy and History* (essays presented to E. Cassirer). Oxford: Oxford University Press, 1936.

Löwith, K. "Christentum und Geschichte," *Numen*, v. 2 (1955), p. 147.

Oakeley, H. *History and Self*. London: Williams & Norgate, 1934.

Schoeps, H. J. *Vorläufer Spenglers (Studien zum Geschichtspessimismus im 19 Jahrhundert)*. Leiden: E. J. Brill 1953.

Spengler, O. *The Decline of the West*, 2 vols. New York: Alfred A. Knopf, 1945.

Toynbee, A. J. *An Historian's Approach to Religion*. Oxford: Oxford University Press, 1956.

Introductory Note to Part II

[1] John Cohen, *Humanistic Psychology* (New York: Collier-Macmillan, 1962), p. 111.

[2] G. J. Whitrow, *The Natural Philosophy of Time* (London: Nelson, 1962), p. 311.

[3] *Op. cit.*, p. 112.

[4] P. Fraisse, *Les Structures Rhythmiques* (Paris: Erasme, 1956), ch. 9.

Rhythm in Music: A Formal Scaffolding of Time

Notes by the author and by the translator.

[1] On the whole topic cf. Theodore M. Greene, *The Arts and the Art of Criticism* (Princeton: Princeton University Press, 1940), p. 38; A. Briner, *Der Wandel der Musik als Zeitkunst* (Vienna: Universal-Edition, 1955); M. Rothaermel, *Der musikalische Zeitbegriff seit Moritz Hauptmann* (Kölner Beiträge zur Musikforschung 25) (Regensburg: Gustav Bosse, 1963); G. Brelet, *Le temps musical* (Paris: Presses Universitaires de France, 1949).

[2] Gotthold Ephraim Lessing, *Laokoon oder Über die Grenzen der Malerei und Poesie*, 1766; Eng. tr. Ellen Frothingham, *Laocoon; An Essay upon the Limits of Painting and Poetry* (Boston: Roberts, 1877). Lessing opposed a popular eighteenth-century viewpoint which adopted the definition by Simonides of Ceos that painting was silent poetry and poetry talking painting. *Cf.* also Mary Colum, *From These Roots* (New York: Scribners, 1937), and Irving Babbitt, *The New Laokoon* (Boston: Houghton Mifflin, 1910).

[3] *Cf.* e.g., Carl Nordenfalk in *Early Medieval Painting* (New York: Skira-World, 1957), p. 91.

[4] The importance of space regarding the aural experience of music becomes evident in modern techniques of recording and sound transmission, stereophonic equipment, etc.

[5] *Cf.* the essay by Friedrich Kümmel in this volume.

[6] *Cf.* Henri Louis Bergson, *Réflexions sur le temps, l'espace et la vie* (Paris: Payot, 1920) and his *Durée et simultanéité . . .* (Paris: Alcan, 1923). The

concept of time used in the present paper corresponds roughly with Bergson's *durée pure*. On this see also Susanne K. Langer, *Feeling and Form* (New York: Scribner's, 1953), pp. 112–119; Briner, *op. cit.*, p. 8, Rothaermel, *op. cit.*, p. 124, and Leonard B. Meyer, *Emotion and Meaning in Music* (Chicago: University of Chicago Press, 1956), p. 88.

[7] Theodor Lamm, "Zur experimentellen Untersuchung der rhythmischen Veranlagung," *Zeitschrift für Psychologie*, v. 118 (1930), p. 213.

[8] Ludwig Klages, *Vom Wesen des Rhythmus* (Kampen: Kampmann, 1934), p. 14.

[9] *Cf.* René Dumesnil, *Le rythme musical* (2nd ed.; Paris: La Colombe, 1949); M. Lussy, "Die Correlation zwischen Takt und Rhythmus," *Vierteljahresschrift für Musikwissenschaft*, v. 1 (1885), p. 144; L. Drozynski, "Atmungs- und Pulssymptome rhythmischer Gefühle," Ph.D. Dissertation, Leipzig, 1911; Paul Fraisse, *The Psychology of Time* (New York: Harper & Row, 1963), p. 17. The experimental literature is summarized in James I. Mursell, *The Psychology of Music* (New York: Norton, 1937), chs. 4, 5.

[10] *Cf.* Karl Bücher, *Arbeit und Rhythmus* (1896) (5th ed.; Leipzig: Teubner, 1919), pp. 396 and 454. On Bücher's view and the entire older literature, see Robert Lach, *Studien zur Entwickelungsgeschichte der ornamentalen Melopoeie* (Leipzig: Kahnt, 1913), pp. 525 and 612.

[11] *Cf.* I. Dietze in Wundt's *Philosophische Studien*, II, p. 362, as reported by William James, *The Principles of Psychology* (1890) (New York: Dover, 1950), v. 1, p. 613, and P. Fraisse, *op. cit.*, 89.

[12] William James, *op. cit.*, p. 612, Theodor Lipps, *Ästhetik; Psychologie des Schönen und der Kunst* (Hamburg and Leipzig: L. Voss, 1903–06), v. 1, p. 293; M. Ettlinger, "Zur Grundlegung einer Aesthetik des Rhythmus," *Zeitschrift für Psychologie und Physiologie der Sinnesorgane* (1900), p. 161; E. Schmidt, "Über den Aufbau rhythmischer Gestalten," *Neue Physiologische Studien*, v. 14 (1939), p. 15. According to these studies, which consider only the response of the Western listener, such groupings are usually binary upbeat patterns, *i.e.*, a first unaccented unit followed by a second accented one.

[13] *Cf.*, for instance, the widely quoted passage in Guido d'Arezzo's *Micrologus*, c. XV: "Item ut rerum eventus sic cantionis imitetur effectus, ut in tristibus rebus graves sint neumae, in tranquillis iocundae, in prosperis exultantes et reliqua," in Smits van Waesberghe's ed. (Rome: American Institute of Musicology, 1955), p. 174; in Gerbert, *Scriptores Ecclesiatici de Musica* (1784), v. 2, p. 17. Similarly, Nicola Vicentino states in his *L'Antica Musica ridotta alla moderna Prattica* (Rome, 1555), that one has to "movere la Misura, per dimostrare gli effetti delle passioni delle parole." (fol. 94 v.). Musical terms such as *allegro, adagio*, etc., designate typical mood effects in their original meaning.

[14] It seems short during the act of listening, but long in memory, as time passes quickly in completely filled days, while these days will appear long in retrospect.

[15] See H. Osthoff in *Die Musikforschung*, v. 1 (1948), p. 63.

[16] We find three different viewpoints in the literature. 1) Rhythm is identical with or similar to meter because it is regularly grouped movement in time. This interpretation is favored by a number of students of the philosophy and of the psychology of music. Thus, J. Hoffmeister defines rhythm in his article,

"Rhythmus," as "eine gleichmässige, bestimmt gegliederte Bewegung, bei der trotz alles möglichen Wechsels in gleichen Zeitabständen gleiche Vorgänge oder Zustände wiederkehren," *Wörterbuch der philosophischen Begriffe* (2nd ed.; Hamburg: F. Meiner, 1955); *cf.* also P. Fraisse, *Les Structures rythmiques* (Paris: Erasme, 1956), p. 112. 2) Rhythm is subordinate to meter, yet supplies the metric framework with dynamic content. For Moritz Hauptmann meter is continuous, regular measurement, and rhythm the characteristic motion that occurs within this measurement, *Die Natur der Harmonik und der Metrik* (Leipzig: Breitkopf and Härtel, 1853), p. 223. The same interpretation recurs in Hugo Riemann's more sophisticated differentiation between *Rhythmik*, the study of the musical motives themselves, and *Metrik*, the study of the musical framework established by measure and accent which accommodates the motives as its content, *System der musikalischen Rhythmik und Metrik* (Leipzig: Breitkopf and Härtel, 1903). 3) Rhythm is superordinate to meter. It is "sinnvoll geformter musikalischer Kraftverlauf" for Rudolph Steglich, *Die elementare Dynamik des musikalischen Rhythmus* (Leipzig: C. Merseburger, 1930), p. 6; division of time in portions that are "sinnlich fassbar" for Andreas Heusler, "Deutsche Versgeschichte," *Grundriss der germanischen Philologie*, v. 8, (Berlin: W. de Gruyter, 1925–29), v. 1, p. 17. Steglich's definition still insists on configuration and maintains a relation between rhythm and meter, although "configuration" does not necessarily imply any kind of regularity. Heusler's definition eliminates interpretation altogether: The division of time is not "sinnvoll," it is merely "sinnlich fassbar." On this see also S. K. Langer, *op. cit.*, chs. 7 and 8; esp. pp. 125–32. For a more detailed discussion of the literature see Dürr and Gerstenberg, "Rhythmus, Metrum, Takt," *Die Musik in Geschichte und Gegenwart* (Kassel and Basel: Bärenreiter, 1963), v. 11, col. 383, *cf.* also Paul Creston, *Principles of Rhythm.*

[17] *Cf.* Paul Fraisse, *Les Structures rythmiques*, p. 111; Calvin S. Brown, *Music and Literature* (Athens, Ga.: University of Georgia Press, 1948), ch. 3.

[18] *Cf.* F. Blume's article on "Form" in *Die Musik in Geschichte . . .* (1955), v. 4, col. 523.

[19] *Cf.* C. Hoeweler, *Het Rhythme in Vers en Muziek* (The Hague: Mouton, 1952), p. 215; *cf.* also Leonard B. Meyer, *op. cit.*, p. 103.

[20] Riemann, *op. cit.*, p. 130.

[21] See note 19.

[22] The change from quantitative to qualitative rhythms may be related, in fact, to the emergence of dynamic accents in the vernacular languages during the Middle Ages. *Cf.* C. H. Rawski, "Die zweistimmigen Werke des Roman de Fauvel," Ph.D. Dissertation, Vienna, 1936, v. 1, p. 259.

[23] *Cf.* Hoeweler, *op. cit.*, p. 103. Melody stems from rhythm, as it expresses musical time through tones of varying pitch; harmony results from melody as the coincidence of tones in simultaneously sounding melodies.

[24] "Melodic impulse cannot always be identified with impulse generated by rhythm." Ernst Kurth, *Musikpsychologie* (Berlin: Max Hesses, 1931), p. 250. *Cf.* also C. Dahlhaus' article "Melodie" in *Die Musik in Geschichte . . .* (1961), v. 9, col. 28.

[25] Thrasybulos Georgiades, *Greek Music, Verse and Dance* (New York: Merlin, 1956), p. 11 and Curt Sachs, *Rhythm and Tempo* (New York: Norton, 1953), p. 131. The viewpoints of these two authors are not identical through-

out. Georgiades interprets the "static" additive rhythm of antiquity as directly opposed to the "dynamic" multiple rhythm of Western music. Sachs holds that the two classes are "essentially different" and that ancient rhythmology cannot be gauged by the standards of "divisive" rhythm. According to Georgiades all basic rhythmic units are singular and self-contained, while Sachs points out that "quite a number of metrical feet were never allowed to stand alone—the trochee and the anapaest, for instance," *op. cit.*, p. 132.

26 On the rhythmic theories of antiquity, *cf.* Rudolph Westphal, *Aristoxenus von Tarent* (Leipzig: Ambr. Abel, 1893–95), 2 vols.; F. A. Gevaert, *Histoire et théorie de la musique de l'antiquité*, v. II, Gand, Braeckmann, 1881; Georgiades, *op. cit.*; Sachs, *op. cit.*

27 Upon this identification of verse foot and motive rests Westphal's attempt to apply antique meters to the music of the eighteenth and nineteenth century. *Cf.* R. Westphal, *Allgemeine Theorie der musikalischen Rhythmik seit J. S. Bach* (Leipzig: Breitkopf and Härtel, 1880), and H. Riemann's critique in his *Handbuch der Musikgeschichte* (Leipzig: Breitkopf and Härtel, 1919), v. 1, p. 219.

28 Not all scholars agree that the *basis* indicated the strong accent (the down-beat), although there can be little doubt that this was the case in dance and march music. Sachs summarizes the discussion in *Rhythm and Tempo*, p. 128.

29 The most important rhythmic genera are:

the dactyl ↓ ↑ ᴜ ᴜ   time proportion 1:1

the iambic ↓ ↑   "   "   1:2

the paeonic (cretic) ↓ ↑ _   "   "   2:3

30 *Cf.* Georgiades, *op. cit.*, Sachs, *op. cit.*, pp. 122, 125; S. Baud-Bovy, *Etudes sur la chanson cleftique* (Athens: Institut Français d'Athènes, 1958), p. 7.

31 *Cf.* H. C. Wolff, "Die Sprachmelodie im alten Opernrezitativ," *Archiv für Sprach- und Stimmphysiologie*, v. 4 (1940); Werner Steger, "G. H. Stölzels 'Abhandlung vom Rezitativ,'" Ph.D. Dissertation, Heidelberg, 1962; see also the historical remarks in D. J. Grout, *A Short History of Opera* (New York: Columbia University Press, 1947), v. 1, p. 40.

32 *Cf.* Max Seiffert, "Die Verzierungen der Sologesänge in Händels Messias," *Sammelbände der Internationalen Musikgesellschaft*, v. 8 (1907), p. 581 and the Messiah, edited in vocal score by J. M. Coopersmith, New York, Schirmer, 1947. On Handel's lack of familiarity with English pronunciation and resulting awkwardness in the declamation of the texts, see R. M. Myers, *Handel's Messiah* (New York: Macmillan, 1948), p. 75.

33 Tension means here only the tension existing between one musical motive and the other—not that kind of psychological tension which is produced by repetition and by the expectation of interruption. The listener's tension in that way may also be increased by *ostinato* figures. *Cf.* Kurth, *op. cit.*, p. 283.

34 Both harmony and melody are factors independent of the rhythmic chain, although themselves can engender chains of their own through characteristic harmonic progressions or melodic sequences.

35 *Cf.* the discussion in Sachs, *op. cit.*, p. 168, and G. Reese, *Music in the Middle Ages* (New York: Norton, 1940), pp. 226, where the relevant literature is summarized.

36 Only the last chain differs in this respect, the third and the fourth accent of the first and the first and second of the second group overlap, but this seems to be a mere irregularity.

[37] On the basic theory of the *modus*, see G. Reese, *op. cit.*, ch. 10, and H. Besseler's article "Ars antiqua" in *Die Musik in Geschichte* . . . (1951), v. 1, col. 679.

[38] *Cf.* Sachs, *op. cit.*, p. 280, Leo Schrade, "Sulla natura del ritmo barocco," *Rivista Musicale Italiana*, v. 56 (1954), p. 3.

[39] Regarding such complementary rhythm techniques, *cf.* H. Riemann, *op. cit.*, p. 155.

[40] On serial rhythm in twentieth-century music, *cf.* Oliver Messiaen, *Technique de mon langage musical* (Paris: Leduc, 1944); G. Nestler, "Der Rhythmus in der Reihentechnik der Gegenwart," *Kongressbericht* (Köln: 1958); K. H. Stockhausen, ". . . und wie die Zeit vergeht . . ." *Die Reihe*, v. 3 (1957), p. 13.

[41] *Cf.* Hugo Riemann, *op. cit.*, p. 196. The dialectic character of classic and romantic music is discussed in M. Hauptmann's *Die Natur der Harmonik und der Metrik* (Leipzig: Breitkopf and Härtel, 1853), p. 223, see also Peter Rummenhöller, *Moritz Hauptmann als Theoretiker* (Wiesbaden: Breitkopf and Härtel, 1963), p. 29.

[42] On the history of rhythm, *cf.* Curt Sachs, *op. cit.*, and the article by Gerstenberg and Dürr, *op. cit.*

GENERAL REFERENCES

Becking, Gustav. *Der musikalische Rhythmus als Erkenntnisquelle.* Augsburg: Benno Filser, 1928.

Cooper, G. W., and L. B. Meyer. *The Rhythmic Structure of Music.* Chicago: University of Chicago Press, 1960.

Ghyka, Matila C. *Essai sur le rythme.* Paris: Gallimard, 1938.

Ruckmick, A. "A Bibliography of Rhythm," *Amer. J. of Psych.* (1913, 1915).

Schoen, M. "Bibliography of Experimental Studies on the Psychology of Music," *Proceedings*, Music Teachers National Association, 1941–43.

Wolff, H. C. "Das Problem des Rhythmus in der neuesten Literatur," *Archiv für Sprach- und Stimmphysiologie*, v. 5 (1941).

Cf. also the report on the Seventh International Musicological Congress, Cologne, 1958, General Topic I: Categories of Musical Rhythm in European and Extra-European Music. Kassel and Basel: Bärenreiter, 1959.

A Note on Rhythm and Time

[1] For an interesting discussion of this see J. A. M. Meerloo, "Rhythm in Babies and Adults," *Arch. of Gen. Psychiatry*, v. 5 (1961), p. 169.

[2] A visual and artistic rather than analytical review of the role of dance in living organisms may be found in *Dance and Sacred Dance*, also by Meerloo (London: Peter Owen, 1961).

[3] *Nature*, v. 198 (1963), p. 1033.

Time Perception in Children

Notes are by Emily B. Kirby.

[1] *Cf.* Charles E. Osgood, *Method and Theory in Experimental Psychology* (New York: Oxford University Press, 1953), p. 200, for a discussion of the work of J. F. Brown and others working on similar problems of perception.

2 *Cf.* Osgood, *op. cit.*, p. 203, for a definition of autokinetic movement.

3 The question of certainty of conviction should be noted here as in the other dialogues given. In an early work, Piaget states that this is one of the most striking aspects of the mental development of the young child. " 'I know,' such is the only proof that is used for a long time in childish logic." p. 202, J. Piaget, *Judgment and Reasoning in the Child* (New York: Harcourt Brace, 1928).

4 For Fraisse's view of Piaget's work on time, as well as his own, see *The Psychology of Time* (New York: Harper & Row, 1963), ch. 8.

5 See also J. Cohen, C. E. M. Hansel, and J. D. Sylvester, "Interdependence of temporal and auditory judgments," *Nature*, v. 174 (1954), p. 642.

6 For a full explanation of the thorny problem of Piaget's stages and levels see J. Flavell, *The Developmental Psychology of Jean Piaget* (New York: Van Nostrand, 1963), pp. 301, 316, 443.

7 Helpful in dealing with the concept of time in its relationship with space, physical causality, and objects, are: ch. IV, "The Temporal Field," of Piaget's *The Construction of Reality in the Child* (New York: Basic Books, 1954), and, for the child's view of physics, Piaget's *The Child's Conception of Physical Causality* (Paterson, N. J.: Littlefield, Adams, 1960).

8 *Cf.* Piaget, "How Children form Mathematical Concepts," *Scientific American*, v. 189, no. 5 (November, 1953), p. 74.

A Note on Psycho-Physical Isomorphism

1 J. Piaget, *Le Devéloppement de la notion de temps chez l'enfant* (Paris: Presses Universitaire de France, 1946).

2 P. Fraisse, *The Psychology of Time* (New York: Harper & Row, 1963), p. 276.

3 Piaget, *Traité de Psychologie Expérimentale* (Paris: Presses Universitaire de France, 1963), v. 1, p. 148ff.

Time and Synchronicity in Analytic Psychology

1 In *Le Second Principe de la Science du Temps* (Paris: Editions du Seuil, 1963), p. 130, Olivier Costa de Beauregard tries to integrate the study of cybernetics into the field of general epistemological problems in physics and concludes that physics is being forced to enter into an active dialogue with the science of psychology in order to build up a more comprehensive science. Though using a different approach, he actually reiterates a postulate of Wolfgang Pauli, who, being familiar with the concepts of C. G. Jung's depth psychology, expressed the hope that we might now be able to create a new abstract *neutral* language in which certain psychological and physical discoveries could find their common expression, *Aufsätze und Vorträge über Physik und Erkenntnistheorie* (Braunschweig: Vieweg, 1961), p. 121. In his contribution to the symposium *The Unity of Knowledge*, ed. Lewis Leary (New York: Doubleday, 1955), Niels Bohr proposed to introduce his concept of complementarity into psychology, *i.e.*, he prepared to look at the conscious and the unconscious psyche as a complementary pair of opposites. *Cf.* also Max Knoll, "Transformations of Science in our Age," *Man and Time*, papers from the

Eranos Yearbooks (New York: Pantheon, 1957), p. 264, and Ernst Anrich, *Moderne Physik und Tiefenpsychologie* (Munich: Klett, 1963).

2 Jung calls "psychoid" those areas of the unconscious where psychic phenomena fuse with or transgress into matter (see below). This psychoid area corresponds closely with what P. Vignon and O. Costa de Beauregard call *infrapsychisme, Le Second Principe*, p. 80.

3 A "content" in psychology is a thought, wish, impulse, phantasy, etc., that is, any relatively definite self-identical entity in the field of a psychic structure.

4 *Cf.* C. G. Jung, *The Structure and Dynamics of the Psyche, Collected Works* (New York: Pantheon, 1960), v. 8, p. 186.

5 *Cf.* Pauli, *Aufsätze*, p. 113. See also William James, *Varieties of Religious Experience* (New York, 1902), Lect. X.

6 Leary, ed., *op. cit. Cf.* Pascual Jordan, *Verdrängung und Komplementarität* (Hamburg: Strom Verlag, 1947).

7 Jung prefers, in contrast to others, the term unconscious to subconscious, for we do not know enough about it to introduce terms which imply a localization or evaluation.

8 See, for instance, G. van der Leeuw: *Religion in Essence and Manifestation* (London, 1938), where such recurrent motives are discussed in each chapter.

9 Freud calls these motives "archaic remnants" or "vestiges" in contrast to Jung's "archetypal images."

10 The archetype is in itself irrepresentable and can only be abstractly described, in contrast to its products: images, symbolic gestures, thought forms, etc. Pauli presents an excellent survey of the development of Jung's idea of the archetype in his *Aufsätze*, p. 119.

11 *Cf.* Jung, *The Structure*, p. 201. No instinct is amorphous for it bears in itself the complete pattern of a situation, *e.g.*, the instinct of a leaf-cutting ant includes in its pattern the image of the ant, leaf, cutting, transport, fungigarden, etc. Without *all* these elements the instinct cannot function. Such an image is an a priori type. The human instincts function in a similar way and also include patterns, *i.e.*, the archetypal images.

12 As H. Poincaré, *Science and Method* (New York: Dover, 1952), ch. 3 points out, however, such "flashes of insight" are only fruitful if they meet a prepared and trained mind.

13 *Cf.* C. G. Jung, *The Structure*, p. 199. The consciousness of a psychic content is always only relatively conscious, just as its unconscious is also only relative. A dream, for instance, is not a completely unconscious phenomenon insofar as it is recorded. *Cf.* Jung, *op. cit.*, p. 187, and Pauli, *Aufsätze, op. cit.*, p. 116. As Pauli stresses, a dream is also a psycho-physical fact insofar as it is accompanied by processes in the brain.

14 *Cf.* A. Portmann, "Time in the Life of the Organism," *Man and Time*, p. 310. *Cf.* also p. 318: "Just as the adult animal acts 'instinctively,' so the plasm of the species 'instinctively' regulates the temporal sequence of formative processes."

15 It would perhaps be better to say "train of representations."

16 S. Watanabe, "Le Concept de temps en physique moderne et la durée pure de Bergson," *Revue de metaphysique et de morale*, v. 56 (1951), p. 128. See also his essay in this volume.

17 *Cf.* Costa de Beauregard, *Le Second Principe*, p. 114.

18 I would not dare to give decided opinion on this question with the few facts we

have now at hand, but, I tend to agree rather with Prof. M. S. Watanabe that the "time sense" is not necessarily "caused" by the law of entropy.

¹⁹ *Cf.* Costa de Beauregard, *Le Second Principe*, p. 115: "In brief, granting that the universe of the relativists must be considered at every moment as spread out in all its temporal extent, we would propose the following metaphor: In Minkowskian space-time, Bergson's 'attention to life' can be compared to one's attention to reading a book, the successive 'three-dimensional space-like states' playing the same part as the successive pages of the book. And just as an effort to understand the book requires us to read its pages in the order in which they were written, effective participation in the 'text' of the four-dimensional cosmos requires that the 'attention to life' continually peruse the three-dimensional states of the universe in their order of increasing probability." (This note and notes 21, 22, 82, 85 and 87 have been translated from the French by Prof. David Park.)

²⁰ This does not a priori decide the old philosophical question of whether time is only a category of our conscious mind (Kant) or whether it also has an objective existence. It only means that accurate *awareness* of what we today call time seems to be dependent on the functioning of our conscious mind.

²¹ *Cf.* also Costa de Beauregard, *Le Second Principe*, p. 124: "Similarly, even if it is the nature of consciousness always to explore the temporal dimension of the cosmos in the same direction, without any gaps, and trailing after it a long wake of memory gained from the information that it has acquired, the subconscious mind, which is neither logical nor methodical, has no reason to be subject to the same law. Would one not have to say of the subconscious that it coexists with the temporal extent of the cosmos?"

²² Costa de Beauregard says: "It has thus become impossible to conceive the time of the material universe as an advancing front. The time of the material universe, like its space, is an extension, and the parameter that explores this 'fourth dimension' is a coordinate just like those which parametrize ordinary space." *Le Second Principe*, p. 123. The unconscious would be coextensive with the "étendue actuelle," *op. cit.*, p. 124.

²³ *Cf.* Jung, *The Structure*, p. 231.

²⁴ M. Eliade, *The Myth of the Eternal Return* (New York: Pantheon, 1954). Also see G. van der Leeuw, "Primordial Time and Final Time," *Man and Time*, p. 324.

²⁵ *Cf. ibid.*, p. 326.

²⁶ *Cf.* Jung's Introduction to W. Y. Evans-Wentz: *The Tibetan Book of the Great Liberation* (Oxford: Oxford University Press, 1954), p. xiv.

²⁷ Jung says (*ibid.*, p. xlvi): "Because the unconscious is the matrix mind . . . it is the birthplace of thought forms. . . . Insofar as the forms or patterns of the unconscious belong to no time in particular, being seemingly eternal, they convey a peculiar feeling of timelessness when consciously realised. . . . An introverted attitude, therefore . . . which withdraws its emphasis from the external world . . . necessarily calls forth the characteristic manifestations of the unconscious, namely, archaic thought forms imbued with 'ancestral' and 'historic' feeling and beyond them, the sense of indefiniteness, timelessness and oneness. . . . [This] derives from the general contamination of contents, which increases as consciousness dims . . . it is not unlikely that the peculiar experience of oneness derives from the subliminal awareness of all-contamination in the unconscious."

[28] *Cf. ibid.*, p. lix: "I have already explained this 'timelessness' as a quality inherent in the experience of the collective unconscious. The application of the 'yoga of self-liberation' is said to reintegrate all forgotten knowledge of the past with consciousness. The motif of ἀποκατάστασις (restoration, restitution) occurs in many redemption myths and is also an important aspect of the psychology of the unconscious, which reveals an extraordinary amount of archaic material in the dreams. . . . In the systematic analysis of an individual the spontaneous reawakening of ancestral patterns (as a compensation) has the effect of a restoration. . . . The unconscious certainly has its 'own time' inasmuch as past, present, and future are blended together in it."

[29] *Cf.* W. Y. Evans-Wentz' commentary, *op. cit.*, p. 7.

[30] Poincaré, *op. cit.*, ch. III.

[31] In K-F Gauss' *Werke*, v. 10, p. 25, quoted in B. L. van der Waerden, *Einfall und Überlegung. Drei kleine Beiträge zur Psychologie des Mathematischen Denkens.* (Basel-Stuttgart, 1954), p. 1. This same passage is translated into English in Harold Rugg, *Imagination. An Inquiry into the Sources and Conditions that Stimulate Creativity* (New York: Harper, 1963), p. 9.

[32] J. H. Hadamard, *Psychology of Invention in the Mathematical Field* (Princeton: Princeton University Press, 1949).

[33] Waerden, *op. cit.*

[34] *Imagination, op. cit.*

[35] *Cf.* Lancelot L. Whyte, *Accent on Form* (New York: Harper, 1954), p. 155.

[36] About this see Jung, *The Structure*, p. 231: "Archetypes, so far as we can observe and experience them at all, manifest themselves only through their ability to *organise* images and ideas and this is always an unconscious process which cannot be detected until afterwards. By assimilating ideational material whose provenance in the phenomenal world is not to be contested, they become visible and *psychic.*" Because of this cognitive element in the unconscious P. Vignon even thought of calling his "infrapsychism" a "supraconsciousness." *Cf.* Costa de Beauregard, *Le Second Principe*, p. 120.

[37] *Cf.* J. P. Zubeck's experiments in the Psychological Institute of the University of Manitoba in Winnipeg.

[38] Quoted with the permission of the patient.

[39] Wilhelm von Scholz's *Der Zufall, eine Vorform des Schicksals* (Stuttgart: Koch-Verlag, 1924), is a collection of documented coincidences of this type.

[40] The energetic impulse of a particle, for instance, is ruled by probability only. I agree here with the argument of Costa de Beauregard (*Le Second Principe*, pp. 23, 27. *Cf. id.* pp. 45, 47) that chance must be looked on as an objective factor not only as a "chance of ignorance." *Cf.* also W. Pauli, "Wahrscheinlichkeit und Physik," *Dialectica*, v. 8, no. 2 (1954), p. 114, who stresses that a unique realization of a very improbable event is treated from a certain point on as if it were practically·impossible.

[41] If I sneeze when an airplane crashes before my eyes, this is apt to be a meaningless coincidence. If, however, I have entertained hostile fantasies about that plane and then it crashes before my eyes, it will impress me as a meaningful coincidence, though here too we must assume complete causal independence.

[42] We can, for instance, observe while analyzing creative scientists or artists that their dream images often anticipate in symbolic form their later conscious discoveries or inventions. Harold Rugg (*op. cit.*, p. 49) speaks of this as "the creative preconscious activity of the transliminal mind."

43 *Cf.* Jung, *The Structure*, p. 231.

44 P. Kammerer, *Das Gesetz der Serie* (Stuttgart-Berlin: Deutsche-Verlags Anstatt, 1919), quoted in Jung, *The Structure*, p. 426. Kammerer specialized in collecting "run" manifestations of numbers. *Cf.* also L. Kling, "Irrational Occurrences in Psychotherapy and Tendentious Apperception," *J. of Existential Psychiatry*, v. 4 no. 15 (1964).

45 W. von Scholz, *Der Zufall, eine Vorform des Schicksals* (Stuttgart: 1924).

46 W. Stekel, "Die Verpflichtung des Namens," *Zeitschrift für Psychotherapie und medizinische Psychologie*, v. 3 (1911), p. 110.

47 J. W. Dunne, *An Experiment with Time* (2nd ed.; London: Black, 1938).

48 Charles Richet, Xavier Dariex, Frank Podmore, W. H. Myers, Camille Flammarion. *Cf.* the titles and examples in C. G. Jung, *The Structure*, pp. 127, 430.

49 For the literature see S. G. Soal and F. Bateman, *Modern Experiments in Telepathy* (London: Routledge & Kegan Paul, 1954) and J. B. Rhine, *The Reach of the Mind*, 1954. *Cf.* also Jung, *The Structure*, p. 432.

50 This chapter is mainly a shortened version of Jung's third chapter on synchronicity except for the exposé on Ibn Arabi's views.

51 For details see Jung, *The Structure*, p. 491.

52 *Cf.* W. Pauli, *The Influence of Archetypal Ideas on the Scientific Theories of Kepler*. In Jung and Pauli, *The Interpretation of Nature and the Psyche* (London: Routledge & Kegan Paul, 1952).

53 *De mirabilibus mundi*, first printed Cologne, 1485.

54 In my practice I have found that in psychic borderline states such actual events *and* the acute awareness of synchronistic events rapidly increase, a fact which would again link up with the "excessive affect" dominant in such conditions.

55 In whose *De Anima* (ch. 4) the same ideas are expressed. (*Cf. Avicenne perhypatetici philosophiae medicorum facile primi*: Opera etc. Venetiis, 1508. This was the only Latin translation available to Albert the Great.)

56 *Cf.* Henry Corbin, *L'Imagination créatrice dans le Soufism d'Ibn Arabi* (Paris: Flammarion, 1958), p. 140.

57 *Heccëitas* means here God's pure inconceivable *esse* that is, selfsame substance in the Aristotelian sense of the word, in contrast to existence.

58 H. Corbin, *op. cit.*, p. 150: "An identical, eternal *Heccëitas* assumes existential determination; and this actually happens in an instant (al-ân) of an indivisible moment (even though this moment is divisible in thought), that atom of time which one calls the 'present' (amân hadir, but not the 'nunc' which is an abstract limiting condition of past and future and is an idea of pure negativity) without the senses being able to perceive any temporal interval."

59 The Divine Names in Ibn Arabi's system are pure formal potentialities; they correspond very closely to Jung's concept of archetypes, but have (like the Platonic ideas) a more intellectual character.

60 Corbin, *op. cit.*, p. 150.

61 Corbin, *op. cit.*, p. 146. "Ipsëities" are potential but continuously self-identical aspects of God's selfsame inconceivable substance. They work as intermediary agencies between God and the Creation.

62 Quote from Jung, *The Structure*, p. 427 sq.

63 *Cf.* Jung, *The Structure*, p. 428: ". . . Neither philosophical reflection nor experience can provide any evidence for the regular occurrence of these two kinds of connection in which the same thing is both subject and object." *Cf.* also p. 429.

64 *Cf. ibid.*, p. 498. Arnold Geulincx exposed similar ideas.

65 Only Ibn Arabi escapes this objection. Even Costa de Beauregard still postulates a "Grand Esprit" as Source and "first cause" within his neo-finalistic views.

66 Jung, *The Structure*, p. 500, *cf.* also note 70: "I must again stress the possibility that the relation between body and soul may yet be understood as a synchronistic one. Should this conjecture ever be proved, my present view that synchronicity is a relatively rare phenomenon would have to be corrected." *Cf.* C. A. Meier's observations in: *Zeitgemässe Probleme der Traumforschung*, Eidgenössische Technische Hochschule. Kultur und Staatswissenschaftliche Schriften 75 (1950), p. 22, and "Psychosomatic Medicine from the Jungian Point of View," *The Journal of Analytical Psychology*, v. 8, no. 2 (London: 1963), p. 103.

67 The *unus mundus* idea of Jung should not be confused with Erich Neumann's term of Einheitswirklichkeit "unitarian reality," "Die Psyche und die Wandlung der Wirklichkeitsebenen," *Eranos* Yearbook, v. 21 (1952), p. 169. The latter means a fusion of individuals, into and with their surroundings through their patterns of behavior. The *unus mundus* on the contrary has only a *potential* reality as a conglomerate of archetypal structures which have not yet become dualistically actualized into psychic or material (or both) phenomena. *Cf.* Jung, *Mysterium Conjunctionis, Collected Works*, 1963, v. 14, p. 534.

68 This idea of a foreknowledge originally comes from Hans Driesch, *Die Seele als elementarer Naturfaktor* (Leipzig, 1903), p. 80. *Cf.* also Costa de Beauregard, *Le Second Principe*, p. 137, who postulates a "Source" of Information existing before all cosmic events. *Cf.* also p. 131.

69 *Cf.* Jung, *The Structure*, p. 231, and Pauli's comments *ibid.* Many of our representations can well be explained by sense stimuli, even if unconsciously perceived, but the archetypally ordered structures seem to have no outer causal basis. The cognitive element which is contained in these images underlies sudden intuitive creative insights.

70 *Ibid.*, p. 526.

71 Author's Note: For instance, the existence of primary numbers or the number 2 being the only even primary number or 6 being a "circular" number in a decimal system. These are facts which, though seemingly logical, cannot be reduced to any other facts. They are facts of an a priori or given character similar also to the mathematical characteristics of groups, or the fact that it is impossible to solve algebraically equations of degrees greater than 4.

72 Author's Note: Some scientists prefer the word "system" to organization.

73 J. Needham, "Biochemical Aspects of Form and Growth," *Aspects of Form*, ed. Lancelot L. Whyte (Bloomington: Indiana University Press, 1951).

74 *I.e.*, there exist between them and the causal processes no relations that conform to law.

75 Pauli, *Aufsätze*, p. 114.

76 The inability is related to the so-called decline effect, *i.e.*, repetition seems to tire the emotional participation. Stimulated by the well-known experiments of J. B. Rhine, in which he tried to tackle the problem of precognition on an experimental and statistical basis, Jung also tried to attack the same problem with statistical means. His idea was that one must first find out whether two undeniable facts, which are clearly *not* causally interconnected, might tend to coincide in a frequency which lies above mathematical probability. That

two people are married is an undeniable fact and the astrological marriage constellations Sun-Moon conjunctio and Venus-Mars conjunctio are also facts which can be stated independently of whether one "believes" in astrology or not. It is highly improbable that the planets have an actual causally comprehensible influence on the marriage motivations of human beings. Should the two facts therefore coincide more frequently than chance probability would predict, they would represent synchronistic events. Jung's statistical analysis at first showed highly remarkable results but, as Jung's interest decreased the results became increasingly less striking. A similar effect is known from Rhine's experiments. Only persons whose emotional participation is highly aroused tend to produce high scores of extra-sensory perception results. Perhaps it is on account of this that ESP results evade statistical analysis, for the same experiment is not sufficiently repeatable with *one* and the same person. There are two reasons for this: we cannot always voluntarily activate an intense emotional condition, and secondly, even if it is activated it tends to "cool off" by repetition. Jung's results at first were highly improbably "good" (*i.e.*, far above mathematical probability) but this ultimately seems to have been just a synchronistic event in itself. The outcome of the complete test proved to be a highly improbable but definitely a chance result. It was, however, a meaningful chance result, that is, a synchronistic event, but not the statistically tested regularity which he had hoped for. Jung found himself forced to reflect once more critically on the epistemological basis of causality and statistical probability. He says (*op. cit.*, p. 477): "Statistics would not even make sense without the exceptions. . . . Because the statistical method shows only the average aspects, it creates an artificial and predominantly conceptual picture of reality. This is why we need a complementary principle for a complete description and explanation of nature."

[77] *Ibid.*, p. 515.

[78] *Ibid.*: "Although associated with causal processes and "carried" by them, they [the archetypes] continually go beyond their frame of reference, an "infringe-ment" to which I would give the name 'transgressiveness,' because the archetypes are not found exclusively in the psychic area, but can occur just as much in circumstances that are not psychic . . ."

[79] *The Structure*, p. 517. That is, Jung continues, why he called these events synchronistic.

[80] *Ibid.*, p. 518. Jung continues: "We must of course guard against thinking of every event whose cause is unknown as 'causeless.' This is admissible only when a cause is not thinkable. But thinkability is itself an idea that needs the most rigorous criticism. Meaningful coincidences are thinkable as pure chance. But the more they multiply and the greater and more exact the correspondence is, the more their probability sinks and their unthinkability increases until they can no longer be regarded as pure chance, but, for lack of a causal explanation have to be thought of as meaningful arrange-ments. . . . However their inexplicability is not due to the fact that the cause is unknown, but to the fact that a cause is not even thinkable in intellectual terms. This is necessarily the case when space and time lose their meaning or have become relative, for under those circumstances a causality which presupposes space and time for its continuance can no longer be said to exist and becomes altogether unthinkable."

[81] *Ibid.*, p. 519.

82 *Cf.* Costa de Beauregard, p. 137: "We do not believe like Bergson that 'what happens' (in phylogenesis or in the individual activity of living beings) is necessarily unpredictable, either by the collective consciousness of phylogenesis (if there is such a thing) or by the individual consciousness of living beings." We can confirm this for it is just *there* that the psychologist observes the occurrence of synchronistic events. What Costa de Beauregard, following Vignon, calls "infrapsychisme" coincides very closely with what Jung calls the psychoid nature of the archetypes.

83 I must add a few warning remarks here against the habit of regressive magical causal thinking. Again and again I come across a misunderstanding when Jung's idea of synchronicity is considered. The reason usually turns out to be a relapse into mankind's powerful age-old primitive habit of thinking in terms of magical causality. I have often noted talk of "synchronistic causality" or stress that the archetype *causes* the synchronistic events, instead of becoming manifest within or through a cluster of synchronistic events. In my view, words such as telepathy or telekinesis also imply the vague underlying idea of a "transmitting energy" and are thus misleading.

84 *Science and the Modern World* (New York: Macmillan, 1926), p. 304, and M. Čapek, "Bergson et l'esprit de la physique contemporaine," *Bulletin de la Société Française de Philosophie*, 53 année (1959) special number, *Bergson et nous*, Paris (May, 1959), p. 53.

85 Costa de Beauregard, *Le Second Principe*, p. 133: "What is accomplished in spite of the decreasing negentropy of the cosmos, and at its expense, is the accumulation of information by conscious beings incarnated in matter."

86 *Ibid.*, p. 76.

87 Costa de Beauregard, *ibid.*, p. 79: "Action and organization, in our view, are the same." And p. 78: "We conclude that there is no essential difference between the image which represents what we perceive and the project which represents what we wish to do—that is, between the representation which follows a physical situation and that which precedes it." *Cf.* H. Reichenbach, *The Direction of Time* (Berkeley: University of California Press, 1956), who also identifies these two things.

88 What Costa de Beauregard calls (*Le Second Principe*, p. 95) "conscience volitive" which is based on emotional facts in contrast to the "conscience cognitive" which is based on the observation of causality.

89 As Pauli already proposed, it is barely possible that the "meaningful" mutations (that is, mutations appropriate to a situation as judged by man) postulated by a few evolutionists could be explained as synchronistic events. *Aufsätze,* p. 122. He thought of neofinalists and scientists such as B. Rensch and A. Pauly, *Cf.* also Cuénot, *Invention et Finalité en biologie* (Paris, 1941); P. Vignon, *Introduction à la biologie expérimentale* (Paris, 1930), and Costa de Beauregard, *Le Second Principe*, p. 120. Pauli points out (*op. cit.*, p. 124) that the modern idea of evolution (chance events plus selection) would only be valid if one could show that "fitting" chance formation had enough opportunity to come into existence within the known time of life on our planet. But biologists tend to sidetrack this by using vitalistic ideas, or by pointing out that unfitting mutations would certainly die. Whereas the safe results of genetics are statistical laws, rare or even unique events might also be important in biological evolution. Pauli therefore suggests (*op. cit.*, p. 127) that synchronicity might play a role in evolution.

[90] The problem is reduced to the question: what is the nature of the pre-conscious orderedness of our representations and how does this link up with the apparent orderedness or system-quality of outer facts? The concept of causal determinism seems to be connected with a pre-conscious adaptation of our psyche to the law of entropy and the arrow of time implied by this. (*Cf.* Costa de Beauregard, *Le Second Principe*, p. 112.) The concept of synchronicity, on the other hand, seems to be based on a pre-conscious expression of our psyche in the form of systems, structures, or "Gestalts," defined as a simultaneously present "whole" or "order."

[91] Because this has not hitherto been possible, many works on cybernetics tend to ignore the factor of meaning. But, as Costa de Beauregard points out, we cannot simply eliminate the "résonnance psychique," *op. cit.*, p. 75.

[92] *The Structure*, p. 233.

[93] Marcel Granet, *La pensée chinoise* (Paris: Albin Michel, 1950), p. 150.

[94] *Ibid.*, p. 153. *Cf.* also p. 159 and p. 174.

[95] *Ibid.*, p. 158.

[96] *Cf.* also *ibid.*, p. 173. For instance, "5" in Chinese means the center of a quaternary structure.

[97] "des rapports réguliers des êtres," *ibid.*, p. 174.

[98] As Manfred Porkert points out, "Wissenschaftliches Denken im Alten China— Das System der energetischen Beziehungen," in *Antaios*, v. 2, no. 6 (1961). Jung's concept of synchronicity provides us with a clue with which we may understand some aspects of Chinese medicine and natural science which were hitherto despised as superstitions.

[99] Introduction to the *I Ching* (*The Book of Changes*) (New York: Pantheon, 1949), p. iv.

[100] Author's italics.

[101] Costa de Beauregard, *Le Second Principe*, p. 74.

[102] *Cf.* Hellmuth Wilhelm, "The Concept of Time in the Book of Changes," *Man and Time, op. cit.*, p. 219. Leibniz was familiar with this and also tried to make use of his binary calculating system for such purposes in his own way. See his correspondence with Father Bouvet. *Cf.* H. Wilhelm: *Leibniz and the I Ching*, Collectanea Commissionis Synodalis 16 (Peking, 1943), p. 205. *Cf.* also Joseph Needham, *Science and Civilization in China* (Cambridge: Cambridge University Press, 1954–   ), v. 2, p. 340, and his essay in this volume.

[103] Jung, *The Structure*, p. 456.

[104] *Ibid.*, p. 457. L. B. van der Wärden stresses (*op. cit.*, p. 9) that we must assume on account of the occasional unconscious origin of arithmetical theorems, that the unconscious is even capable of reasoning and judging.

[105] As Pauli (*Aufsätze*, p. 122) rightly stresses that the mathematical a priori intuitions belong in the class of what Jung calls archetypal representations, such as the arithmetical idea of the infinite series of integers or the geometrical continuum, for they are regularly recurrent ideas. As the internal logical consistency of the analysis cannot be formally proved from its own premises, Pauli adds that we must find its roots outside mathematics. It seems to be a natural fact which is connected with the functioning of our mind. One finds here again a pre-conscious orderedness of our representations.

[106] D'Arcy Thompson, *Growth and Form* (Cambridge: Cambridge University Press, 1942). Joseph Needham objects to this in his "Biochemical Aspects of Form

and Growth," in Whyte, *Aspects of Form* (p. 79), because these mathematical forms too often harbor secret "devils of vitalism." I would agree with him here if these "forms" are seen as "causes" but not if seen as patterns which manifest through synchronistic phenomena.

107 I would prefer this word to "information."

108 It seems to me that if the quantitative concept of number in mathematics and cybernetics could be revised (*i.e.*, not used only as "counting instruments") and could, in collaboration with depth psychology, be studied as aspects of pre-conscious qualitative elements of order in our mind (without relapsing into ideas of magical causality as the Chinese numerical concepts, for instance, partly presuppose), we might get closer to a method of grasping a causal orderedness in an accurate form. It might also be useful if biochemists would investigate how far number plays a role in this field of study. We then might get closer to the "material" aspect of the instinctual patterns of behavior and at their concomitant emotional factor.

109 It can thus also help us to avoid the danger of stumbling into the pitfall of that old Aristotelian internally contradictory idea of a *causa finalis* which has turned up again in neofinalistic theories.

A Note on Synchronicity

1 Thornton Wilder, *The Bridge of San Luis Rey* (New York: Grosset & Dunlap, 1927).

2 One of Wilder's other great works, the tender play *Our Town* (New York: Coward, 1938), is a much less subtle exposition of time and its relation to life, death and love.

The Time Sense in Psychiatry

1 C. von Monakow and A. Mourgue, *Biologische Einführung in das Studium der Neurologie und Psychopathologie* (Stuttgart: Enke Verlag, 1930).

2 C. U. Ariens Kappers, *The Evolution of the Nervous System in Convertebratae, Vertebratae and Man* (Haarlem: Bohn, 1930).

3 *Prememory* is the hypothetical biochemical change that is supposed later to evolve into the conscious memory.

4 According to these insights we speak in psycho-pathology of phylogenetic regression when a rudimentary remnant of genetic older biological function begins to function again and of ontogenetic regression when an infantile pattern begins to function again.

5 Wilder Penfield, "Observations on the Anatomy of Memory," *Fol. Neurol., Psychiatrica et Neurochirurgica*, v. 53 (1950).

———, "Symposium on the Brain and Mind," *A.M.A. Arch. Neurol. and Psychiat.*, v. 87 (1952).

6 *Time-binding* is defined as 1) reacting in the light of past or future, as well as of present, conditions: in man, a process usually involving imagining or ideation but possible in some degree without it, 2) the social transmission of experience through successive generations so that what happens in one generation is, to some extent, made available to later generations, H. B. and

A. C. English, *A Comprehensive Dictionary of Psychological and Psycho-analytical Terms* (New York: McKay, 1958).

[7] A. Korzybski, *Science and Sanity* (4th ed.; Lancaster, Pa.: The International Non-Aristotelian Library Publishing Company, 1962).

————, *Time-Binding: The General Theory* (2nd ed.; Lancaster, Pa.: The International Non-Aristotelian Library Publishing Company, 1956).

[8] The substance of this symbolization of time in its various aspects is further discussed in Section 8.

[9] Shakespeare shows keen awareness of the difference between the subjective experience of time loaded with feelings of guilt and the "free" time after the unburdening of guilt. "The times have been," Macbeth says after the murder was committed. "The time is free," justice will flower, "planted newly with time," says Malcolm after the crime is punished.

Hamlet's inner fear and horror degenerate into tedium, and the guilty superego makes him remark: "The time is out of joint." The confessional also makes time "free," as guilt-ridden patients often say.

[10] Norman G. Brown, *Life Against Death, The Psychoanalytical Meaning of History* (Middletown, Conn.: Wesleyan University Press, 1959).

[11] For a scholarly study of attitude toward free time, see S. De Grazia, *Of Time, Work and Leisure* (New York: Twentieth Century Fund, 1962). The problem of leisure as seen from the point of view of individual and group psychology, placed in its historical and cultural settings, is the subject of a paper by John Cohen, "The Scientific Revolution and Leisure," *Nature,* v. 198 (1963), p. 1028. The clue to our problems of leisure, he suggests, may lie in our use of the imagination, for man is more of a creature of imagination than of intellect.

[12] J. A. M. Meerloo, "Rhythm in Babies and Adults," *Archives of General Psychiatry*, v. 5 (1961).

[13] J. A. M. Meerloo, *Suicide and Mass-Suicide* (New York: Grune and Stratton, 1962), pp. 41, 42.

[14] This is revealed in the *Letters of Sigmund Freud* (New York: Basic Books, 1960). Fliess supposes that there are various biorhythms with a different cycle active in man, *e.g.*, a physico-chemical curve, an intellectual curve and an emotional curve.

[15] W. S. Inman, "The Moon, The Seasons and Man," *British J. of Medical Psych.,* v. 24 (1951).

[16] W. and A. Menaker, "Lunar Periodicity in Human Reproduction: A Likely Unit of Biological Time," *Amer. J. of Obstetrics and Gynecology*, v. 77 (1959).

[17] The clinical importance of this change in inner time clock relates not only to passengers but can especially become a hazard in airplane pilots. If their sense of time does not adapt well to the new clock, they can remain sleepless and tired and will be more apt to make mistakes. See also H. Strughold, "The Physiological Clock in Aeronautics and Astronautics," *Lectures in Aerospace Medicine*, USAF School of Aerospace Medicine, Brooks Air Force Base, Texas (1963), p. 387.

[18] Meerloo, "Rhythm in Babies and Adults," *op. cit.*

[19] In the case of adults the strategists of brainwashing proved that artificially creating disorder in the awareness of the diurnal temporal cycle and in the estimation of time spans, they could cause confusion and subsequent submission in the victim and sometimes cause psychotic episodes, Meerloo, *The Rape of the Mind* (New York: World, 1956).

[20] S. Freud, *Collected Papers* (London: Hogarth Press, 1925), v. 4, p. 119.

[21] Historical time, for instance, is reflected in our language and verbal habits.

[22] F. S. DuBois, "The Sense of Time and Its Relation to Psychiatric Illness," *Amer. J. of Psychiatry*, v. 3 (1954).

[23] Meerloo, "Symbol Appeal and Mental Contagion," *J. of Communication*, v. 13 (1963), p. 183.

[24] B. L. Whorf, *Language, Thought and Reality* (New York: Wiley, 1956).

[25] U. Wun in "The Burmese Language," *The Atlantic*, February, 1951, p. 153, explains that Burmese is a tenseless language. Past, present and future are not so important, the language talks about actions regardless of the time they take place in. Past and future find their reality only in the present. Verbs stand for immediately known actions. That is why most Burmese are not so happy if they have to make scheduled appointments. For them time is relative and subjective, the time of cooling a pot of rice, or the time the sun is about to set, changing with season and human intention. See also Prof. Nakamura's essay in this volume.

[26] Stuart Chase, "How Language Shapes Our Thoughts," *Harper's Magazine*, April, 1954.

[27] S. Freud, *op. cit.*, p. 119.

[28] Editorial, "Now and Zen," *Psychiatric Quart.* (January, 1958).

GENERAL REFERENCES

Bergler, E., and G. Roheim. "Psychology of Time Perception," *Psychoanalytic Quart.*, v. 15, 1946.

Board, C., Jane King, A. M. Tierney, and P. Lichtenberg, "Time Perspective and Intimacy," *Arch. of Gen. Psychiatry*, v. 1, 1959.

Bonaparte, M. "Time and the Unconscious," *International J. of Psycho-Analysis*, v. 21, 1940.

Bouman, K. H., and H. Grünbaum, "Einschrümpfung der subjektiven Zeit in einem Fall von postencephalitis," *Monatsschrift für Psychiatrie*, v. 73, 1929.

Chari, C. T. K. "The Psychology of the Time Sense," *The Astrological Magazine*, v. 43, 1954 (India).

Clauser, G. *Die Kopfuhr. Das automatische Erwachen.* Stuttgart: Ferdinand Enke Verlag, 1954.

Cohn, F. S. "Time and the Ego," *Psychoanalytic Quart.*, v. 26, 1957.

Cooper, L. F., and M. H. Erickson, *Time Distortian in Hypnosis.* Baltimore: Williams and Williams, 1959.

Dahl, M. "A Singular Distortion of Temporal Orientation," *Amer. J. of Psychiatry*, v. 115, 1958.

Dooley, L. "The Concept of Time in Defense of Ego Integrity," *Psychiatry*, v. 4, 1941.

Douglas, W. J. H. "The Periodicity of the 'Sevens' in Mind, Man and Nature," *British J. of Med. Psych.*, v. 24, 1951.

Dubois, F. S. "Rhythms, Cycles and Periods in Health and Disease," *Amer. J. of Psychiatry*, v. 116, 1959.

Ehrenwald, J. "Gibt es einen Zeitsinn?" *Klinische Wochenschrift*, v. 32, 1931.

Eisenbud, J. "Time and the Oedipus," *Psychoanalytic Quart.*, v. 25, 1956.

Eissler, K. R. "Time Experience and the Mechanism of Isolation," *Psychoanalytic Rev.*, v. 39, 1952.

Fischer, R., F. Griffin and L. Liss, "Biological Aspects of Time in Relation to (Model) Psychoses," *Ann. N.Y. Acad. Sci.*, v. 96, 1962.

Flink, E. B., and R. P. Doc. "Effect of Sudden Time Development by Air Travel on Synchronization of Adrenal Function," *Proc. Sc. Exp. Biol. and Med.*, v. 100, 1959.

Fodor, N. "Time is the Essence—of What?" *Record of Med.*, August, 1956.

Freud, S. "Fausse Reconnaissance in Psychoanalytic Treatment," *Collected Papers*. London: Hogarth Press, 1949, v. 2.

———. "Disturbance of Memory on the Acropolis," *International J. of Psycho-Analysis*, v. 22, 1941.

Gooddey, W. "Time and the Nervous System: The Brain as a Clock," *Lancet*, 1959.

Gross, A. "Sense of Time in Dreams," *Psychoanalytic Quart.*, v. 18, 1949.

Halberg, F. "Diurnal Rhythm Changes in Blood Eosinophil Levels in Health and in Heart and Diseases," *Lancet*, v. 73, 1953.

Hollos, A. "Über das Zeitgefühl," *Internationale Zeitschrift für Psychoanalyse*, v. 8, 1931.

Hugenholtz, P. Th. *Tyd en Creativiteit*. Amsterdam: North Holland Publishers, 1958.

Inman, W. S. "The Moon, The Seasons, and Man," *British J. of Med. Psychology*, v. 24, 1951.

Kafka, J. S. "A Method for Studying the Organization of Time Experience," *Amer. J. of Psychiatry*, v. 114, 1957.

Kahn, E. "Becoming and Being in Time and Space," *Psychoanalytic Quart.*, v. 33, 1959.

Klages, L. *Vom Wesen des Rhythmus*. Kampen auf Sylt, 1936.

Korzybski, A. *Manhood of Humanity*. New York: Dutton, 1921.

Levenson, E. A. "Changing Time Concepts in Psychoanalysis," *Amer. J. of Psychotherapy*, v. 12, 1958.

Meerloo, J. A. M. "Über Entwicklung und Störung des Zeitsinns," *Zeitschrift Neur.*, v. 153, 1935.

———. "Father Time," *Psychiatric Quart.*, v. 22, 1948.

———. "Creativity and the Urge for Eternization," *Suicide and Mass Suicide*. New York: Grune and Stratton, 1962.

Montagu, M. F. Ashley. "Time-Binding and the Concept of Culture," *The Scientific Monthly*, September, 1953.

Moulyn, A. C. "Reflections on the Problem of Time in Relation to Neurophysiology and Psychology," *Philosophy of Science*, v. 19, 1952.

Oberndorf, C. P. "Time, Its Relation to Reality and Purpose," *Psychoanalytic Rev.*, v. 28, 1941.

Reiman, H. A. "Periodic Disease," *J. of the American Medical Association*, v. 141, 1951.

Schilder, P. "Psychopathology of Time," *J. of Nervous and Mental Diseases*, v. 83, 1936.

Scott, W. C. M. "Some Psychodynamic Aspects of Disturbed Perception of Time," *British J. of Med. Psychology*, v. 21, 1948.

Schlesinger, B. "Time, Neurologically Considered," *Monatschr. Psychiatr. und Neurol.*, v. 129, 1955.

Semon, R. *Die Mneme als erhaltendes Prinzip im Wechsel des organischen Geschehens*. Leipzig: Engelman, 1904.

Spielrein, S. "Die Zeit im unterschwelligen Seelenleben," *Imago*, v. 9, 1923.

645

Van der Horst, L. "De psychologie van het Korsakow syndroom," *Psychiatr. Neurol., Bladen*, 1931.

Westermann Holstyn, A. M. "Time and Death in the Life of the Compulsion-Neurotic," *Netherl. Tschr. Geneesk.*, v. 103, 1959.

Win, Klin Maung, "The Psychology of the Burmese Language," *Main Currents in Modern Thought*, v. 16, 1960.

Yates, S. "Some Aspects of Time Difficulties and Their Relation to Music," *International J. of Psycho-Analysis*, v. 16, 1935.

Comments on Time and the Uncanny

[1] The sway that his interest in the future holds over man is demonstrated by the fascinating history of divination, surveyed by John Cohen in *Behaviour in Uncertainty* (London: Allen & Unwin, 1964), p. 162ff. "It needs an effort on our part today, to realize the enormous influence of divinatory rituals in archaic peoples at all levels of civilization" (p. 190).

[2] Critical studies and evaluation of spontaneous precognitive experiences first appeared in the *Proceedings* and the *Journal of the Society for Psychical Research* (London). For a bibliography covering early, pre-experimental interest in precognition see L. E. Rhine, "Frequency of Types of Experience in Spontaneous Precognition," *J. of Parapsychology*, v. 18 (1954), p. 93. Experimental tests for the existence of precognition began under the direction of J. B. Rhine over twenty years ago and are being continued up to this day. For early research at Duke see J. B. Rhine, *J. of Parapsychology*, v. 6 (1942), p. 111; for more recent work, John Freeman, *J. of Parapsychology*, v. 26 (1962), p. 123. These reports deal with experiments that were specifically designed to test precognition. For other work in which findings incidentally pointed to the interpretation that precognition has occurred, see Carington, *Proc. of the Soc. for Psychical Research*, v. 46 (1940), p. 34 and S. G. Soal and K. M. Goldney, in the same journal, v. 47 (1943), p. 21." Private communication from Dr. J. G. Pratt of the Dept. of Neurology and Psychiatry, University of Virginia.

[3] N. O. Brown, *Life against Death* (New York: Random House, 1959), p. 98.

[4] *Ibid.*, p. 100.

[5] *Ibid.*, p. 109. *Cf.* "Time is the revolt of man against death, whose inevitability is revealed to him by the present, against this flowing away not of time but of its contents, against the fact that nothing remains in his hands. . . ." G. Berger, "Approche phénoménologique du problème du temps," *Bull. Société française Psychologique*, v. 44 (1950), p. 102; tr. J. Leath, P. Fraisse, *The Psychology of Time* (New York: Harper & Row, 1963), p. 283.

[6] S. Kierkegaard, *Works of Love*, tr. D. F. and L. M. Swenson (Princeton: Princeton University Press, 1946), p. 253. As quoted by Brown, *op. cit.*, p. 109.

[7] S. Freud, "The Uncanny," *Collected Papers*, v. 4 (New York: Basic Books, 1959), p. 368.

[8] In German, *unheimlich* (uncanny) is what was once *heimlich* (familiar). Freud observes that the meaning of the word *heimlich* is ambivalent in that it developed to include its own opposite (Freud, *op. cit.*, p. 77). To that we may add that *time* is also ambivalent, for it includes, as we have seen, several pairs of logical opposites.

[9] Jung uses the same example as an illustration of his Principle of Synchronicity, C. G. Jung and W. Pauli, *The Interpretation of Nature and the Psyche* (New York: Pantheon, 1955), p. 14.

[10] G. J. Whitrow, *The Natural Philosophy of Time* (London: Nelson, 1961), pp. 77, 312.

[11] W. Stern, *Psychology of Early Childhood*, tr. A. Barwell (New York: Henry Holt, 1924), p. 112.

[12] An impressive record of men and women who faced death apparently without fear may be found in *Butler's Lives of Saints*, eds. M. Thurston and D. Attwater (New York: Kennedy, 1956).

[13] E. Kris, *Psychoanalytic Explorations in Art* (London: Allen & Unwin, 1953).

[14] R. Schafer, *Assessment of Human Motives*, ed. Gardner Lindzay (New York: Holt, Rinehart and Winston, 1963), p. 123.

[15] See J. Cohen, *Humanistic Psychology* (New York: Collier-Macmillan, 1962), p. 111.

[16] Whitrow, *op. cit.*, p. 54.

[17] S. S. Stevens, ed., *Handbook of Experimental Psychology* (New York: Wiley, 1951), p. 780; see also B. Berelson and G. A. Steiner, *Human Behavior* (New York: Harcourt, Brace & World, 1964), p. 44.

[18] Marie Bonaparte, "Time and the Unconscious," in *International J. of Psychoanalysis*, v. 21 (1940), p. 468. *Cf.* P. Fraisse, *op. cit.*, p. 197. "It is perfectly accurate to say that a sense of time can only exist where there is submission to reality."

[19] See Needham essay in this volume.

Subjective Time

[1] J. B. S. Haldane, "Biological Possibilities in the Next Ten Thousand Years," *Man and His Future*, ed. G. Wolstenholme (London: Churchill, 1963), p. 337.

[2] A. Portmann, "Preface to a Science of Man," *Diogenes*, v. 40 (Winter, 1962), p. 3.

[3] H. Piéron, "Les problèmes psychophysiologiques de la perception du temps," *Année Psychol.*, v. 24 (1923), p. 1.

[4] M. François, "Contribution à l'étude du sens du temps. La température interne comme facteur de variation de l'appréciation subjective des durées," *Année Psychol.*, v. 28 (1927), p. 188; "Influence de la température interne sur notre appréciation du temps," *C. R. Soc. Biol.*, v. 108 (1928), p. 201.

[5] H. Hoagland, "The Physiological Control of Judgments of Duration: Evidence for a Chemical Clock," *J. of Gen. Psychol.*, v. 9 (1933), p. 267; "Consciousness and the Chemistry of Time," *Problems of Consciousness*, ed. H. A. Abramson (New York: Josiah Macy, 1951), pp. 164–178. For a detailed discussion see Dr. Hoagland's essay in this volume.

[6] W. H. Thorpe, "Some Concepts of Ethology," *Readings in Psychology*, ed. John Cohen (London: Allen & Unwin, 1964), p. 379; see also T. H. Bullock, "The Origins of Patterned Nervous Discharges," *Behaviour*, v. 17 (1961), p. 48.

[7] See John Cohen, *Humanistic Psychology*, New York: Collier-Macmillan, 1962, ch. 3.

[8] See the essay by Professor Piaget in this volume.

9 See P. Fraisse, *The Psychology of Time*, tr. J. Leith (New York: Harper & Row, 1963).

10 *Ibid.*, p. 116.

11 E. G. Boring, *Sensation and Perception in the History of Experimental Psychology* (New York: Appleton-Century, 1942), p. 582.

12 J. Cohen, C. E. M. Hansel and J. D. Sylvester, "Mind Wandering," *British J. of Psychol.*, v. 47 (1946), p. 61.

13 See H. Woodrow, "Time Perception," *Handbook of Experimental Psychology*, ed. S. S. Stevens (New York: Wiley, 1951), p. 1224.

14 I. J. Hirsh, R. C. Bilger and B. H. Deatherage, "The Effect of Auditory and Visual Background on Apparent Duration," *Amer. J. of Psychol.*, v. 69 (1956), p. 561.

15 J. M. Stroud, "The Fine Structure of Psychological Time," *Information Theory in Psychology*, ed. H. Quastler (New York: The Free Press, 1955), p. 174.

16 G. A. Brecher, "Die Entstehung und Biologische Bedeutung der Subjektiven Zeiteinheit des Momentes," *Zeitschr. für vergleichende Physiologie*, v. 18 (1932–33), p. 204.

17 I. J. Hirsh, "Auditory Perception of Temporal Order," *J. Acoust. Soc. Amer.*, v. 31 (1959), p. 759.

18 V. Benussi, *Psychologie der Zeitauffassung* (Winter, 1913), p. 335, cited by Fraisse, *op. cit.*, p. 130.

19 S. T. Coleridge, *Biographia Literaria* (London: Dent, 1908), p. 62.

20 R. B. MacLeod and M. F. Roff, "An Experiment in Temporal Disorientation," *Acta Psych.*, v. 1 (1936), p. 381.

21 John Cohen, C. E. M. Hansel and J. D. Sylvester, "An Experimental Study of Comparative Judgements of Time," *British J. of Psych.*, v. 55 (1954), p. 108. See also Ph. Malrîeu, "Le problème de la conscience du passé," *J. de Psych.* (Jan.-June, 1954), p. 91; "Le social et le temps de l'enfant," *J. de Psych.* (July-Sept., 1956), p. 315.

22 Quoted by J. Alexander Gunn, *The Problem of Time* (London: Allen & Unwin, 1929), p. 56.

23 J. P. Guilford, "Spatial Symbols in the Apprehension of Time," *Amer. J. of Psych.*, v. 37 (1926), p. 420.

24 A. von Chamisso, *The Marvellous History of the Shadowless Man* (London: Holden & Hardingham, 1913) p. 43.

25 John Cohen, "The Concept of Goal Gradients," *J. of Gen. Psychol.*, v. 49 (1953), p. 303.

26 C. D. Broad, *Scientific Thought* (London: Routledge & Kegan Paul, 1923).

27 See M. Eliade, *The Myth of the Eternal Return*, tr. Willard R. Trask (New York: Pantheon, 1954). See, in particular, I. Meyerson, "Le temps, la mémoire, l'histoire," *J. de Psych.* (July-Sept. 1956), p. 333.

28 See the essay by Dr. Hoagland in this volume.

29 I. J. Hirsh, R. C. Bilger and B. H. Deatherage, "The Effect of Auditory and Visual Background on Apparent Duration," *Amer. J. of Psychol.*, v. 69 (1956), p. 561.

30 H. Helson and S. M. King, "The *tau*-effect. An example of psychological relativity," *J. of Exper. Psych.*, v. 14 (1931), p. 202.

31 A. Geldreich, *Amer. J. of Psych.*, v. 46 (1938), p. 483.

32 S. Abe, "Experimental Study on the Correlation between Time and Space," *Tohoku Psychologia Folia*, v. 3 (1935), p. 53.

[33] M. Abbe, "The Spatial Effect upon the Perception of Time," *Japanese J. of Exper. Psychol.*, v. 3 (1936), p. 1; "The Spatial Effect upon the Perception of Time: Simultaneous Comparison of Phenomenal size of Two Time Intervals Divided by Three Stimuli," *Japanese J. of Exper. Psych.*, v. 4 (1937), p. 1; "The Temporal Effect upon the Perception of Space," *ibid.*, p. 83.

[34] John Cohen, C. E. M. Hansel and J. D. Sylvester, "Interdependence of Temporal and Auditory Judgements," *Nature*, v. 174 (1954), p. 642.

[35] The same consistency in subjective judgments appears in judging the duration, distance and speed of a moving object. See J. F. Brown, "The Visual Perception of Velocity," *Psych. Forschung*, v. 14 (1931), p. 199, and "On time Perception in Visual Movement Fields," *ibid.*, p. 233.

[36] John Cohen, Peter Cooper and Akio Ono, "The Hare and The Tortoise: A Study of the *tau*-effect in Walking and Running," *Acta Psych.*, v. 21 (1963), p. 387.

[37] E. Mach, *Contributions to the Analysis of Sensations*, tr. C. M. Williams (Chicago: Open Court, 1897), first German ed., 1886.

[38] M. Frankenhaeuser, *Estimation of Time* (Stockholm: Almquist and Wiksell, 1959). The reader will find in this valuable monograph a description of many other experiments on temporal judgment.

[39] This is the "resonance" theory of Claparède. E. Claparède, "La genèse de l'hypothèse," *Arch. de Psych.*, v. 34 (1933), p. 1.

[40] W. Russell Brain, *Diseases of the Nervous System* (4th ed.; London: Oxford University Press, 1951), p. 950.

[41] D. Rapaport, *Emotions and Memory* (Baltimore: Williams and Wilkins, 1942), p. 229.

[42] L. F. Cooper, "Time Distortion in Hypnosis, I," *Bull. Georgetown Univ. Med. Center* (1948), v. 1, p. 214; L. F. Cooper and M. H. Erickson, "Time Distortion in Hypnosis, II," *Bull. Georgetown Univ. Med. Center*, v. 4 (1950), p. 50.

[43] *Cf.* R. Fischer, "Selbstbeobachtungen im Mezkalin-Rausch," *Schweizerische Zeitschrift für Psychologie*, v. 5 (1946), p. 308.

[44] Walter de la Mare, *Desert Islands* (London: Faber & Faber, 1932), p. 91.

[45] *Ibid.*

[46] S. Freud, *New Introductory Lectures on Psycho-analysis* (New York: Norton, 1933), p. 105; see also *Beyond the Pleasure Principle*, tr. C. J. M. Hubback (London: Hogarth Press, 1942), p. 32, where he declares that the Kantian notion of space and time are necessary modes of thought and may be re-examined in the light of psychoanalysis. He claims that "unconscious mental processes" are "timeless." They are not arranged chronologically, time alters nothing in them, nor can the idea of time be applied to them.

[47] Marie Bonaparte, "Time and the Unconscious," *International J. of Psycho-Anal.*, v. 21 (1940), p. 427. See also H. Meyerhoff, *Time in Literature* (Berkeley: University of California Press, 1955), p. 153.

[48] J. Ortega y Gasset, *The Dehumanization of Art* (New York: Doubleday, 1956), p. 74. See also R. Bonnot, "Le roman du temps," *J. de Psych.* (July-Sept., 1956), p. 454.

[49] O. Fenichel, *The Psycho-analytical Theory of the Neuroses* (New York: Holt, 1943), p. 296.

[50] E. Jones, *Essays in Applied Psycho-Analysis* (London: Hogarth Press, 1951), p. 257.

[51] See H. S. Liddell, *Emotional Hazards in Animals and Man* (Springfield, Ill.:

C. C. Thomas, 1959), and also *Discussions on Child Development*, ed. J. M. Tanner and B. Inhelder (London: Tavistock, 1956), v. 2, p. 144.

52 M. P. Nilsson, *Primitive Time-Reckoning* (Lund: Gleerup, 1920).

53 H. Werner, *Comparative Psychology of Mental Development* (New York: Follett, 1948), p. 182.

54 D. Lee, "Being and Value in a Primitive Culture," *J. of Philos.*, v. 13 (1949), p. 401.

55 I. Cunnison, "History on the Luapula," *The Rhodes-Livingstone Papers* (London: Oxford University Press, 1951).

56 In a dispatch from the Governor of Ceylon to the British Secretary of State, see R. Pieris, "Character Formation and the Acquisitive Society," *Psychiatry*, v. 15 (1952), p. 53.

57 E. Panofsky, *Studies in Iconology* (New York: Oxford University Press, 1939), p. 92.

58 *Ibid.*

Experimental Evidence for the Biological Clock

1 W. W. Garner and H. A. Allard, "Effect of the Relative Length of Day and Night and Other Factors of the Environment on Growth and Reproduction in Plants," *J. of Agricultural Research*, v. 18 (1920), p. 553.

2 K. C. Hamner, "Endogenous Rhythms in Controlled Environments," *Environmental Control of Plant Growth* (New York: Academic Press, 1963), p. 215.

3 It is said that the American pioneers located bee trees by setting out a small dish of honey and watching carefully as the honey-laden bees left the dish. The direction of departure provided a line to the bee tree. By pursuing this line and occasionally setting down the dish of honey to get a new line, they could follow the bees directly to their hive.

4 He and his wife were in the habit of having breakfast on the unscreened porch of their house at the same time every morning. Numerous bees came each morning to pick up the leftovers of the meal. He noticed that the bees arrived at the correct time each morning even on the days when he and his wife failed to have an outdoor breakfast.

5 G. Kramer, "Orientierte Zugaktivität gekäfigter Singvogel," *Naturwissenschaft*, v. 37 (1950), p. 188.

6 E. Bünning, "Endogenous Rhythms in Plants," *Ann. Rev. of Plant Physiology*, v. 7 (1956), p. 71.

7 K. C. Hamner, "Photoperiodism and Circadian Rhythms," *Cold Spring Harbor Symposia on Quantitative Biology*, v. 25 (1960), p. 269. Work with other plants and insects is in essential agreement with our work on soybean. See also E. Bünning, "Physiological Mechanism and Biological Importance of the Endogenous Diurnal Periodicity in Plants and Animals," *Photoperiodism and Related Phenomena in Plants and Animals*, ed. R. B. Withrow (Washington: A.A.A.S., 1959), p. 507. Bünning, "Circadian Rhythms and the Time Measurements in Photoperiodism," *Cold Spring Harbor Symposia on Quantitative Biology*, v. 25 (1960), p. 249.

8 A recent paper, W. M. Hamner, "Diurnal Rhythm and Photoperiodism in Testicular Recrudescence of the House Finch," *Science*, v. 142 (1962), p. 1294, on the photoperiodic responses of birds provides strong evidence that animal

photoperiodism involves the same mechanism as does that of plants. The house finches studied (*Carpodacus Mexicanus*) mate in response to the increasing day length to which they are exposed in the spring. During the winter the male birds have very small gonads. If such birds are exposed to long days the gonads promptly enlarge several hundred per cent. Comparable birds exposed to six hours of light during each photoperiodic cycle remain in the winter condition (small gonads) as they also do if exposed to twenty-four, forty-eight, or seventy-two-hour cycles; but they promptly attain the mating condition (enlarged gonads) if exposed to twelve-, thirty-six-, or sixty-hour cycles.

9 J. W. Hastings and B. M. Sweeney, "The Gonyaulax Clock," *Photoperiodism and Related Phenomena . . . ,* p. 567; C. S. Pittendrigh and V. G. Bruce, "Daily Rhythm as Coupled Oscillator Systems and Their Relation to Thermoperiodism and Photoperiodism," *Photoperiodism and Related Phenomena . . . ,* p. 475; B. M. Sweeney, "Biological Clocks in Plants," *Ann. Rev. Plant Physiology,* v. 14 (1963), p. 411.

10 K. C. Hamner *et al.*, "The Biological Clock at the South Pole," *Nature,* v. 195 (1962), p. 476.

GENERAL REFERENCES

Cold Spring Harbor Symposia on Quantitative Biology, "Biological Clocks," v. 25, 1960.

Bünning, E. *The Physiological Clock.* New York: Academic Press, 1964.

Cloudsley-Thompson, J. L. *Rhythmic Activity in Animal Physiology and Behavior.* New York: Academic Press, 1961.

"Photoperiodism and Related Phenomena in Plants and Animals," International Symp. held at Gatlinburg, Tenn., and sponsored by the U.S. National Science Foundation. Publication No. 55, 1959, A.A.A.S.

Time Sense of Animals

Detailed references will be found in one or other of the recent monographs or reviews cited below.

1 J. Aschoff, "Tierische Periodik unter dem Einfluss von Zeitgebern," *Z. Tierpsych.* v. 15 (1958), p. 1.

2 F. A. Brown, Jr., "Extrinsic Rhythmicity: a Reference Frame for Biological Rhythms under So-called Constant Conditions," *Ann. New York Acad. Sci.,* v. 98 (1962), p. 775.

3 See K. C. Hamner essay in this volume, note 10.

4 W. H. Thorpe, *Learning and Instinct in Animals* (London: Methuen, 1956).

5 G. E. Folk, Jr., "Modification by Light of 24-hour Activity of White Rats," *Proc. Iowa Acad. Sci.,* v. 66 (1959), p. 399.

6 H Kalmus, "Über das Problem der sogenannten exogenen und endogenen, sowie der erblichen Rhythmik und über organische Periodizität überhaupt," *Rivista di Biologia,* v. 24 (1938), p. 191.

7 M. S. Johnson, "Effect of Continuous Light on Periodic Spontaneous Activity of White-footed Mice (*Peromyscus*)," *J. of Experimental Zool.,* v. 80 (1939), p. 315.

[8] C. S. Pittendrigh, "On Temporal Organisation in Living Systems," *The Harvey Lectures*, v. 56 (1961), p. 93.

[9] E. Bünning, *The Psychological Clock* (New York: Academic Press, 1964), pp. v, 153.

[10] G. P. Wells, *The Sources of Animal Behaviour* (London: H. K. Lewis, 1955), p. 1; and also *Quant. Biol., Cold Spring Harbour Symp.*, v. 25 (1961), pp. xiii, 524.

[11] J. E. Harker, "Diurnal Rhythms in the Animal Kingdom," *Biol. Rev.*, v. 33 (1958), p. 1; and also *The Physiology of Diurnal Rhythms* (Cambridge: Cambridge University Press, 1964), p. 114.

[12] J. W. Hastings, "Unicellular Clocks," *Ann. Rev. Microbiol.*, v. 33 (1959), p. 297.

[13] A. Hardy, *The Open Sea—Its Natural History. The World of Plankton* (London: Collins, 1956), pp. xv, 335.

GENERAL REFERENCES

Calhoun, J. B. "Twenty-four Hour Periodicities in the Animal Kingdom," Pt. 1, The Invertebrates, *J. Tenn. Acad. Sci.*, v. 19, 1944–46, pp. 197, 252; Pt. 2, The Vertebrates, v. 19, pp. 20, 208, 228, 281, 291, 373; v. 20, pp. 228, 291, 373; v. 21, pp. 208, 281.

Caspers, H. "Rhythmische Erscheinungen in der Fortpflanzung von Clunio marinus (Dipt: Chiron.) und das Problem der lunaren Periodizität bei Organismen," *Arch. Hydrobiol., Suppl.*, v. 18, 1951, p. 415.

Cloudsley-Thompson, J. L. *Rhythmic Activity in Animal Physiology and Behaviour.* New York: Academic Press, 1961, pp. vi, 236; and also "Adaptive Functions of Circadian Rhythms," *Cold Spring Harbor Symposia on Quantitative Biology*, v. 25 (1960), p. 345.

Halberg, F., *et al.* "Physiologic 24-hour Periodicity in Human Beings and Mice, the Lighting Regimen and Daily Routine," *Photoperiodism and Related Phenomena in Plants and Animals.* Washington, D.C., Amer. Assoc. Adv. Sci., 1959, p. 803.

Kleitman, N. "Biological Rhythms and Cycles," *Physiol. Rev.*, v. 29, 1949, p. 1.

Park, O., "Community Organisation: Periodicity," W. C. Allee *et al.*, *Principles of Animal Ecology*, Philadelphia: Saunders, 1949, p. 528.

Some Biochemical Considerations of Time

[1] H. Hoagland, "The Physiological Control of Judgments of Duration: Evidence for a Chemical Clock," *J. of Gen. Psych.*, v. 9 (1933), p. 267.

[2] Hoagland, *Pacemakers in Relation to Aspects of Behavior* (New York: Macmillan, 1935); "Chemical Pacemakers and Physiological Rhythms," *Alexander's Colloid Chemistry* (New York: Reinhold, 1944), v. 5, p. 762.

[3] W. J. Crozier, "The Distribution of Temperature Characteristics for Biological Processes; Critical Increments for Heart Rates," *J. of Gen. Physiol.*, v. 9 (1925–26), p. 531.

[4] C. M. Hinshelwood, *Kinetics of Chemical Change* (Oxford: 1940).

[5] B. S. Gould and I. W. Sizer, "Mechanism of Bacterial Dehydrogenase Activity *in vivo*; Anaerobic Dehydrogenase Activity of *Escherichia coli* as a Function of Temperature," *J. of Biol. Chem.*, v. 124 (1938), p. 269.

[6] Z. Hadidian and H. Hoagland, "Chemical Pacemakers," Pts. 1 and 2, *J. of Gen. Physiol.*, v. 23 (1939), p. 31.

[7] E. F. B. Fries, "Temperature and Frequency of Heartbeat in the Cockroach," *J. of Gen. Physiol.*, v. 10 (1925–26), p. 227.

[8] Crozier, "On the Critical Thermal Increment for Locomotion of a Diplopod," *J. of Gen. Physiol.*, v. 7 (1924–25), p. 123.

[9] See Crozier, note 3.

[10] Hoagland, "Pacemakers of Human Brain Waves in Normals and in General Paretics," *Am. J. of Physiol.*, v. 116 (1936), p. 604.

[11] Herbert Jasper, "Cortical Excitatory State and Variability in Human Brain Wave Rhythms," *Science*, v. 83 (1936), p. 259.

[12] W. P. Koella and H. M. Ballin, "The Influence of Environmental and Body Temperature on the Electroencephalogram in the Anesthetized Cat," *Archives Internationales de Physiologie*, v. 62 (1954), p. 3.

[13] Hoagland, see note 1.

[14] Heinz Von Foerster, "Quantum Theory of Memory," *6th Conference on Cybernetics of the Josiah Macy, Jr. Foundation* (New York, 1949), p. 112.

[15] H. Hydén, in *Macromolecular Specificity and Biological Memory*, ed. F. O. Schmitt (Cambridge: Massachusetts Institute of Technology Press, 1962), p. 55.

[16] Hoagland, "Consciousness and the Chemistry of Time," reviewed in *1st Conference on Problems of Consciousness of the Josiah Macy, Jr. Foundation* (New York, 1950), p. 164.

[17] Heinz Von Foerster, "Consciousness and Symbolic Processes," Von Foerster letter to Hoagland published in *4th Conference on Problems of Consciousness of the Josiah Macy, Jr. Foundation* (New York, 1953), p. 107.

[18] Von Foerster's view is that there are certain hypothetical carriers in the brain which have a finite lifetime after impregnation with information that is learned, *e.g.*, nonsense syllables. These hypothetical carriers decay. In his essay cited in my note 14 he speculates a little about the tunnel effect but in general makes no commitment. In the text he refers to impressions stored on quantized molecules (pseudo-isomeric change of protein molecules?). Since the time he did this work, experimental evidence from studies of single neurones indicates specific RNA synthesis accompanying the use of the neurones during learning and specific protein synthesis from this RNA which may delineate the pathway involved in a learning process. Forgetting appears to be the loss of such ability to synthesize protein and retain it in the nerve cells. All of this is consistent with Von Foerster's earlier approach

[19] R. Fischer, F. Griffin, and L. Liss, "Biological Aspects of Time in Relation to Model Psychoses," *Ann. N.Y. Acad. Sci.*, v. 96, art 1 (1962), p. 44. See also Prof. Fischer's essay in this volume.

[20] Seymour Kety, "Human Cerebral Blood Flow and Oxygen Consumption as Related to Aging," *Neurologic and Psychiatric Aspects of Disorders of Aging*, eds. J. E. Moore, H. H. Merritt, and Rollo Masselink (Baltimore: Williams and Wilkins, 1956), ch. 4, p. 31.

[21] B. J. Luyet and P. M. Gehenio, "Life and Death at Low Temperature," *Biodynamica* (1940), p. 341.

[22] B. J. Luyet and E. L. Hodapp, "Revival of Frog's Spermatozoa Vitrified in Liquid Acid," *Proc. Soc. Experimental Biol.*, v. 39 (1938), p. 433.

[23] L. G. Shettles, "The Respiration of Human Spermatozoa and Their Response to

Various Gases and Low Temperatures," *Amer. J. of Physiol.*, v. 128 (1940), p. 4L8.

24 H. Hoaglind and G. Pincus, "Revival of Mammalian Sperm after Immersion in Liquid Nitrogen," *J. of Gen. Physiol.*, v. 25 (1942), p. 337.

25 H. Hoagland, "The Chemistry of Time," *The Scientific Monthly*, v. 56 (1943), p. 56.

26 H. J. Muller, "Should We Weaken or Strengthen Our Genetic Heritage?", *Evolution and Man's Progress*, eds. H. Hoagland and R. W. Burhoe (New York: Columbia University Press, 1962), p. 22.

27 H. J. Muller, *Out of the Night* (New York: Vanguard, 1935).

28 C. Polge, A. U. Smith, and A. S. Parkes, "Revival of Spermatozoa after Vitrification and Dehydration at Low Temperatures," *Nature*, v. 164 (1949), p. 666.

29 A. U. Smith, *Biological Effects of Freezing and Supercooling* (Baltimore: Williams & Wilkins, 1961).

30 R. G. Bunge and J. K. Sherman, "Fertilizing Capacity of Frozen Human Spermatozoa," *Nature*, v. 172 (1953), p. 767. R. G. Bunge, W. C. Keetel, and J. K. Sherman, "Clinical Use of Frozen Semen. Report of four cases," *Fertility and Sterility* v. 5 (1954), p. 520.

Organic Evolution and Time

1 Aristotle, tr. Wicksteed and Cornfold, 1957, *Physics* 219^b, 2–4.

2 *Ibid.*, 221^a, 2–4.

3 C. Lyell, *Principles of Geology* (London: Murray, 1830–33).

4 F. E. Zeuner, *Dating the Past: An Introduction into Geochronology* (3rd ed.; London: Methuen, 1952).

5 H. F. Blum, *Time's Arrow and Evolution* (Princeton: Princeton University Press, 1955).

6 In undisturbed regions the oldest sediments would be at the bottom and the last and youngest at the top (natural law of superposition), Steno, Bishop Nicolaus (Stensen), *The Earliest Geological Treatise (1667)* (New York: Macmillan, 1958). In other regions this order has been upset by diastrophic processes resulting in faults, folds or overthrusts.

7 B. Hocking, "The Ultimate Science," Canadian Broadcasting Corporation, University of the Air (Toronto: C.B.C. publications, 1963).

8 Zeuner, see note 4.

9 R. A. Stirton, *Time, Life and Man* (New York: J. Wiley, 1959).

10 W. Thomson and P. G. Tait, *Treatise on Natural Philosophy* (new ed.; Cambridge: Cambridge University Press, 1888–90).

11 Should any form of terrestrial life, for instance man and a few plants and animals, be established on another planet—solar or otherwise—they would be deprived of the normal changes of the hours, tides and seasons, and possibly subjected to changes of different periods. Some physiological effects of unnatural time regimes have been studied; see Mary C. Lobban, "The Entrainment of Circadian Rhythms in Man." *Cold Spring Harbor Symposia on Quantitative Biology*, v. 25 (1961), p. 325. But we know nothing about their effects on the evolution of the organisms concerned. Quite possibly it might be necessary to simulate days and seasons. See G. T. Hawty, "Periodic Desynchronisation under Outer Space Conditions," *Ann. N.Y. Acad. Sci.*, v. 98

(1962), p. 1116, and H. Strughold, "Day-Night Cycling in Atmospheric Flight, Space Flight, and Other Celestial Bodies," *Ann. N.Y. Acad. Sci.*, v. 98 (1962), p. 1109.

[12] B. C. Goodwin, *Temporal Organization in Cells* (New York: Academic Press, 1963).

[13] G. G. Simpson, *The Major Features of Evolution* (New York: Columbia University Press, 1953).

[14] J. B. S. Haldane, "Suggestions as to Quantitative Measurement of Rates of Evolution," *Evolution*, v. 3 (1949), p. 51.

[15] J. B. S. Haldane, "Selective Elimination of Silver Foxes in Eastern Canada," *J. of Genetics*, v. 44 (1942), p. 296.

[16] Haldane, see note 14.

[17] Traditionally taxonomists think of two different species when they are able to apportion every specimen unequivocally to one or the other. Biometricians may be content to establish significant differences between population means and leave it at that. Geneticists would consider two species to be separate when they show a high degree of sexual isolation and, in particular, when they cannot produce fertile offspring.

[18] Haldane, see note 14.

[19] The late Professor Zeuner explained these by a temporary excess of species splitting over species extinction and the subsequent evolution of the surviving species into genera.

[20] G. G. Simpson, *Tempo and Mode in Evolution* (New York: Columbia University Press, 1944).

[21] H. H. Swinnerton, *Outlines of Palaeontology* (London: Arnold, 1923).

[22] H. Kalmus, "Origin of social organisation," *Symp. Zool. Soc. London*, v. 7 (1965), p. 1.

[23] Haldane, see note 14.

[24] E. H. Colbert, *Dinosaurs* (London: Hutchinson, 1948).

[25] Haldane, see note 14.

[26] Haldane, see note 15.

[27] H. B. D. Kettlewell, "Further Selection Experiments on Industrial Melanism in the Lepidoptera," *Heredity*, v. 10 (1956), p. 287.

[28] Kettlewell, "A Résumé of Investigations on the Evolution of Melanism in Lepidoptera," *Proc. Roy. Soc. Bull.*, v. 145 (1956), p. 297.

[29] Simpson, see note 13.

[30] *Ibid.*

[31] Simpson, "The Nature and Origin of Supraspecific Taxa," *Cold Spring Harbor Symposia on Quantitative Biology*, v. 24 (1959), p. 255.

[32] Haldane, "The Cost of Natural Selection," *J. of Genetics*, v. 55 (1957), p. 511.

[33] Zeuner, *The Pleistocene Period* (London: Royal Society, 1945).

[34] Simpson, see note 13.

[35] Haldane, see note 14.

[36] Haldane, see note 14.

[37] A. Allison, "Protection Afforded by Sickle Cell Trait against Subtertian Malarial Infection," *Brit. Med. J.*, v. 1 (1954), p. 290.

[38] F. Vogel, H. J. Pettenkofer and G. Helmbold, "Über die Populationsgenetic der ABO-Blutgruppen II. Genhäufigkeit und epidemische Erkrankungen," *Acta. Gen. et Statist. med.*, v. 10 (1960), p. 267.

[39] H. Kalmus, "Selection, Migration and Drift," *Genetics of Migrant and Isolate Populations* (Baltimore: Williams and Wilkins, 1963).

40 F. M. Burnet, *The Clonal Selection Theory of Acquired Immunity* (Cambridge: Cambridge University Press, 1959).

41 F. Hawking, "Microfilaria Infestation as an Instance of Periodic Phenomena Seen in Host-Parasite Relationships," *Ann. N. Y. Acad. Sci.*, v. 98 (1962), p. 940.

42 E. Mayr, *Systematics and the Origin of Species* (New York: Columbia University Press, 1942).

43 C. B. Bridges and K. S. Brehme, *The Mutants of Drosophila melanogaster* (Carnegie Institute Publications, 1944), no. 552.

44 Kalmus, see note 39.

45 R. Woltereck, *Variation und Artbildung* (Berne: Müller, 1919).

46 F. J. Ryan, "Natural Selection in Bacterial Populations," Atti *VI. Congr. Intern. Microbiol.* (Rome, 1953).

47 O. Abel, *Grundzüge der Palaeobiologie der Wirbeltiere* (Stuttgart: 1912).

48 B. Kurtèn, "Return of a Lost Structure in the Evolution of the Felid Dentition," *Soc. Sci. Fennica, Comm. Biol.*, v. 24 (1963), p. 3.

49 H. Grüneberg, *The Pathology of Development* (Oxford: Blackwell, 1963).

50 W. K. Gregory, "On the Meaning and Limits of Irreversibility of Evolution," *Amer. Naturalist*, v. 70 (1936), p. 517.

51 A. S. Eddington, *The Nature of the Physical World*, Gifford Lectures 1927 (Cambridge: Cambridge University Press, 1931), p. 70.

52 M. Planck, *The Philosophy of Physics, Complete Works*, v. 3 (New York: Meridian Books, 1959).

53 J. Needham, "Evolution and Thermodynamics," *Time, the Refreshing River* (London: Allen & Unwin, 1943).

54 J. Willard Gibbs, *Collected Works* (New York: Longman, Green, 1928).

55 R. E. D. Clark, *The Universe and God* (1939); see also *Clark Evangelical Quarterly*, v. 9 (1937), p. 128.

56 H. J. Muller has pointed out that though mutation may be reversible, mutant genes cannot be shuffled backwards, nor can selection sequences be reversed in order, "Reversibility in Evolution from the Standpoint of Genetics," *Biol. Rev.*, v. 14 (1939), p. 261.

57 H. G. Keyl, "Verdoppelung des DNS—gehalts kleiner Chromosomenabschnitte als Faktor der Evolution," *Naturwissenschaft*, v. 51 (1964), p. 46.

58 F. Vogel, "Eine vorkäufige Abschätzung der Anzahl menschlichen Gene," *Azt. menchl. Konstitl.*, v. 37 (1964), p. 291.

59 Eddington, see note 51.

60 A. J. Lotka, *Elements of Physical Biology* (Baltimore: Williams and Wilkins, 1925). Quite recently Karl Popper, in "Time's Arrow and Entropy," *Nature*, v. 207 (1965), p. 233, has pointed out some limitations of the Second Law and in particular that entropy need not necessarily increase in an expanding universe.

61 Blum, see note 5.

62 Lotka, see note 60.

63 Muller, see note 56.

64 C. C. Li, *Population Genetics* (Chicago: University of Chicago Press, 1955).

65 O. Shannon and W. Weaver, *The Mathematical Theory of Information* (Bloomington: University of Illinois Press, 1949).

66 M. S. Bartlett, *Essays on Probability and Statistics* (London: Methuen, 1962) and *An Introduction to Stochastic Processes* (Cambridge: Cambridge University Press, 1955).

[67] The genetic composition of populations forming a lineage may be regarded as approximating the links in a Markov chain (see note 66, Bartlett, 1955).

[68] Li, see note 64.

[69] H. Kalmus, "Periodic Phenomena in Genetical Systems," *Ann. N.Y. Acad.*, v. 98 (1962), p. 1083.

[70] The limitation of reversibility and its improbability in more complex situations can be easily demonstrated by means of a Galton board. There one assumes that the probabilities are equal of a ball's deviating to the left or to the right at any particular pin, that is, $P = \frac{1}{2}$. Thus, after one collision the chance of a left deviation being followed by a right deviation is still $\frac{1}{2}$, but when a ball, during its descent, has deviated far to one side, the chances of its reverting to the middle are rather small. This happens in spite of the fact that the probability of each left or right deviation is $\frac{1}{2}$.

[71] H. L. Carson, "Genetic Conditions Which Prevent or Retard the Formation of Species," *Cold Spring Harbor Symp.*, v. 24 (1959).

[72] T. Dobzhansky and O. Pavlowsky, "Indeterminate Outcome of Certain Experiments on Drosophila Populations," *Evolution*, v. 7 (1953), p. 199.

Comments Concerning Evolution

[1] G. G. Simpson, *This View of Life* (New York: Harcourt, Brace & World, 1964), p. 189.

[2] M. Schlick, "Philosophy of Organic Life," in *Die Philosophie in ihren Einzelgebieten*, ed. Dessoir (Berlin: Ullstein, 1925). Tr. H. Geigl and M. Brodbeck eds., in *Readings in the Philosophy of Science* (New York: Appleton-Century-Crofts, 1953), p. 536.

[3] Simpson, *op. cit.*, p. 63.

[4] G. J. Whitrow, *The Natural Philosophy of Time* (London: Nelson, 1961), p. 277.

[5] H. Hoagland, *The Scientific Monthly* (January, 1943), p. 56.

[6] L. B. Slobodkin, "The Strategy of Evolution," *Amer. Scientist*, v. 52 (1964), p. 342.

[7] The quotation is from Michael Polanyi's *Personal Knowledge* (Chicago: University of Chicago Press, 1958), p. 404. Of interest to the question of time and evolutionary principles is Prof Polanyi's concept of the (temporal) "nearness of discovery," as an agent-at-a-distance, as it were, an internal guiding factor in biological achievement (*ibid.*, ch. 13).

[8] *Cf.* H. Hoagland and R. W. Burhoe eds., *Evolution and Man's Progress, Daedelus* (Summer, 1961).

[9] A. Portmann, "Preface to a Science of Man," *Diogenes*, v. 40 (Winter, 1962), p. 13.

Biological Time

[1] J. Konorski, "The Physiological Approach to the Problem of Recent Memory," *Brain Mechanisms and Learning* (Oxford: Blackwell, 1961), p. 115.

[2] R. P. Feynman, R. B. Leighton and M. Sands, *Lectures on Physics* (Palo Alto, Calif.: Addison-Wesley, 1963), v. 1, p. 17–3.

[3] R. Efron, "Temporal Perception, Aphasia and déjà vu," *Brain*, v. 86 (1963), p. 403.

4 Feynman *et al.*, see note 2.

5 D. L. Drabkin, "The Distribution of the Chromoproteins, Hemoglobin, Myoglobin and Cytochrome in Tissues of Different Species and the Relationship of the Total Content of Each Chromoprotein to Body Mass," *J. of Biol. Chem.*, v. 182 (1950), p. 317; M. Kleiber, *The Fire of Life* (New York: Wiley, 1961), p. 215.

6 R. E. Smith, "Quantitative Relations between Liver Mitochondria Metabolism and Total Body Weight in Mammals," *Ann. N.Y. Acad. Sci.*, v. 62 (1956), p. 403; P. Schollmeyer and M. Klingenberg, "Über den Cytochrom-Gehalt tierischer Gewebe," *Biochem. Zeitschrift*, v. 335 (1962), p. 426.

7 D. L. Drabkin, "Kinetic Basis of Life Processes: Pathways and Mechanism of Hepatic Protein Synthesis," *Ann. N.Y. Acad. Sci.*, v. 104 (1963), p. 469.

8 B. Chance, "Enzyme Mechanisms in Living Cells," *The Mechanism of Enzyme Action*, eds. W. D. McElroy and B. Glass (Baltimore: Johns Hopkins Press, 1954); p. 399.

9 L. Jansky, "Total Cytochromoxidase Activity and Its Relation to Basal and Maximal Metabolism," *Nature*, v. 189 (1961), p. 921.

10 As well as by decreasing mitochondrial ribonucleic acid content; See C. Kaiser, "La loi des surfaces," *Extraits de la Rev. Scientifique*, v. 89 (1951), p. 267. Smaller body space (organism size) of an animal requires a higher intensity of body time (shorter chronological time) to live. Similar relations exist on the organ level; compare, for example, the interdependence between heart size and beating frequency; see L. v. Bertalanffy, *Theoretische Biologie*, v. 2, *Stoffwechsel & Wachstum* (Berlin-Zehlendorf: Verlag von Gebr. Borntraeger, 1942).

11 G. A. Sacher, "Entropic Contributions to Mortality and Aging," *Symposium on Information Theory in Biology* (New York: Pergamon Press, 1958), p. 317.

12 G. A. Sacher, "Relation of Lifespan to Brain Weight and Body Weight in Mammals," Colloquia on Aging, Ciba Foundation, v. 5, *The Lifespan of Animals*, eds. G. E. W. Wolstenholme and M. O'Connor (London: Churchill, 1959), p. 115.

13 R. Fischer, F. Griffin and L. Liss, "Biological Aspects of Time," *Ann. N.Y. Acad. Sci.*, v. 96 (1962), p. 44.

14 L. du Noüy, *Le Temps et la vie* (Paris: Gallimard, 1936), pp. 232 and 268.

15 Fischer *et al.*, see note 13.

16 Hofmann's self-observations in what may be called the first systematic description of time contraction after a dose of 0.25 milligrams of LSD related "the impression of being unable to move from the spot" on his way home from the laboratory on a bicycle. ". . . I had the feeling of not getting ahead, whereas my escort stated that we were rolling along at a good speed." See Hofmann on his experience in 1943 as quoted in an anonymous annotation, "The History of LSD-25 (Lysergic Acid Diethylamide), *Triangle* (Sandoz Pharmaceuticals, Basel), v. 2 (1955), p. 117. It appears that the extension of the environmental space, observed by a person under the influence of this drug, also accounts for Hofmann's feeling of not "getting ahead."

17 Fischer *et al.*, see note 13, and G. Grünewald and H. Mücher, "Über den Einfluss zentraler Funktions-aktivierung auf die Schreibmotorik," *Psychopharmacologia*, v. 5 (1964), p. 372.

18 Handwriting is a motor performance, "stretching out in space." One may look at handwriting also as a form of kymograph record of systemic biological

activity. The execution of the performance can be viewed as a projection of centrally coordinated excitation processes in the peripheral end-organ; see G. Grünewald, Rundbrief No. 21, *Iserlohner Schreibkreis* (Darmstadt: 1958), p. 2. The visual angle of the percept is compensated for by the increase in the apparent size of nearby objects (handwriting space) through a decrease in the apparent size of distant objects; see Enoch Callaway III and S. V. Thompson, "Sympathetic Activity and Perception. An Approach to the Relationship between Autonomic Activity and Personality," *Naval Medical Research Institute* (Bethesda, Md., 1953), p. 407. We do not know the role of concomitant pupillary dilation in the occurrence of this phenomenon produced by pyretogenic drugs.

[19] T. Mátéfi, "Mescalin und Lysergsäurediäthlamid-Rausch. Selbstversuch mit besonderer Berücksichtigung eines Zeichentests," *Confinia neurologica*, v. 12 (1952), p. 146.

[20] G. Grünewald, "Über den Einfluss von Drogen auf die Schreibpsycho-motorik," *Archiv für Psychiatrie, Zeitschrift für die gesamte Neurologie*, v. 198 (1959), p. 687; H. J. Haase, "Das therapeutische Achsensyndrom neuroleptischer Medikamente und seine Beziehungen zu extrapyramidaler Symptomatik," *Fortschr. Neurol. Psychiat.*, v. 29 (1961), p. 245.

[21] H. C. B. Denber, P. Rajotte, C. J. Wiart, "Peinture et effets secondaires des neuroleptiques majeurs," *Ann. Médicopsychol.*, v. 120 (1962), p. 26.

[22] H. Hoagland, "The Chemistry of Time," *Science Monthly*, v. 56 (1943), p. 56 and Hoagland, "Problems of Consciousness," 4th Conference of Josiah Macy, Jr. Foundation (New York: 1953), p. 106. See also Dr. Hoagland's essay in this volume.

[23] Fischer *et al.*, see note 13.

[24] We obtained a taped record of fluctuations in tapping rate after half-wave rectification by speeding up the tape fourfold and feeding it to an integrator whose output was recorded in the form of amplitudes on an Esterline Angus recording milliammeter. The frequency response of the milliammeter extended from 0 to 4 cycles per second. Power density spectra of each 2400 second series were then derived, following the procedure of A. Ralston and H. S. Wief, *Mathematical Methods for Digital Computers* (New York: Wiley, 1960), ch. 19, using a Bendix G-20 digital high-speed electronic computer. Subjects were directed to tap at a rate of their own choosing but as regularly as possible. Eight-minute tapping records were taken at the peak of a Psilocybin experience (two and one half hours after the ingestion of a mild dosage of the drug—115 μg/kg) and one day later at the same hour of the day. The control tapping data taken the second day are reproducible for a particular subject within time intervals from one half hour to six months and therefore are characteristic of him.

We do not, however, wish to deal here with the power density spectra which are to be published elsewhere; see R. Fischer, S. M. England, R. Archer and K. Dean, "Psilocybin Reactivity as Measured by Psychomotor Performance" Arzneimittelforschung, in press, 1965).

[25] When tapping rates are used as a measure of intensification of time, the magnitude of the drug reaction should not be assessed in absolute terms but in relation to the control value. See in this connection Wilder's Law of Initial Value, J. Wilder, "Basimetric Approach (Law of Initial Value) to Biological Rhythms," *Rhythmic Functions in the Living System, Ann. N.Y. Acad.*

Sci., v. 98 (1962), p. 1211. This control value is the baseline for the subject and it is self-evident that subjects in various states of excitation and depression cannot be expected to display changes in time experience in the same way as an average healthy subject, *i.e.*, the subject's experience of time contraction may be different in degree or not occur at all. See in this connection the divergent results of J. C. Kenna and G. Sedman, "The Subjective Experience of Time During Lysergic Acid Diethylamide (LSD-25) Intoxication," *Psychopharmacologia*, v. 5 (1964), p. 280.

26 H. Heimann and P. N. Witt, "Die Wirkung einer einmaligen Largactilgabe bei Gesunden," *Mschr. Psychiat. Neurol.*, v. 129 (1955), p. 126, figs. A, B and C.

27 P. N. Witt and R. Baum, "Changes in Orb Webs of Spiders during Growth," *Behaviour*, v. 16 (1960), p. 309, pl. IX.

28 F. W. Fröhlich, *Die Empfindungszeit, Ein Beitrag zur Lehre von der Zeit, Raum- und Bewegungsempfindung* (Jena: Gustav Fischer, 1929), p. 50.

29 R. Magun, "Die Zeitmessung in der Neurologie," ed. G. Schaltenbrand, *Zeit in Nervenärztlicher Sicht* (Stuttgart: Ferdinand Enke Verlag, 1960), p. 122.

30 R. Efron, "An Extension of the Pulfrich-Stereoscopic Effect," *Brain*, v. 86 (1963), p. 295.

31 If the pendulum movements were to be photographed both when the observer sees the straight line movement and when he sees the elliptoid movement, identical pictures would result in both instances. Some authors would, therefore, term the elliptoid movement of the Pulfrich phenomenon an "illusion." Neither or both are illusions. See C. Pulfrich, "Die Stereoskopie im Dienste der isochromen und heterochromen Photometrie," *Naturwissenschaften*, v. 10 (1922), pp. 553, 569, 596, 714, 735, 751.

32 H. Helson and S. M. King, "The Tau Effect: An Example of Psychological Relativity," *J. of Experimental Psych.*, v. 14 (1931), p. 202. See also the essay by J. Cohen in this volume.

33 J. Cohen, C. E. M. Hansel and J. D. Sylvester, "A New Phenomenon in Time Judgment," *Nature*, v. 172 (1953), p. 901.

34 R. M. Boynton, "Spatial Vision," *Ann. Rev. Psych.*, v. 13 (1962), p. 171.

35 For a detailed discussion of the significance of relativistic time transformations the reader should consult Part IV of this volume.

36 A Sollberger, *Biological Rhythm Research* (New York: Elsevier, 1965), p. 67.

37 C. C. L. Gregory, "Model for a Universe," *Cosmos*, v. 1 (1961), p. 133; v. 2 (1962), p. 111.

38 Feynmann *et al.*, see note 2.

39 A. L. Hodgkin, *The Conduction of the Nervous Impulse*, The Sherrington Lectures VII (Springfield, Ill.: C. C. Thomas, 1964), pp. 14 and 15.

40 See J. F. Fulton, ed., *A Textbook of Physiology* (Philadelphia: W. B. Saunders, 1955), fig. 44, showing the approximately linear plot of the relation between nerve conduction and diameter.

41 R. Fischer and E. Steiner-Maccia, "Cogitationes: Limits of Language," *Experientia*, v. 19 (1963), p. 56.

42 See Section 2 of this essay regarding the Weber-Fechner Law and R. Efron, "The Effect of Stimulus Intensity on the Perception of Simultaneity in Right- and Left-handed Subjects," *Brain*, v. 86 (1963), p. 285.

43 Another example: Most biological variables follow log-normal distribution rather than a normal—symmetrical—distribution.
    The Gaussian distribution of the individual lethal doses of a drug, for

example, reveals that equal increment in biological effect of the drug-dose variable corresponds to equal ratios. The Gaussian log-normal distribution tells us that the effect depends on the logarithm of the stimulus—a statement which is but another formulation of the Weber-Fechner Law. Once more, the log-normal distribution is implicit in our way of perception.

[44] R. Efron, "The Effect of Handedness on the Perception of Simultaneity and Temporal Order," *Brain*, v. 86 (1963), p. 261.

[45] Efron's data (see note 44) are consistent with other similar findings reported in the literature reviewed by him.

[46] M. W. Schmidt and A. B. Kristofferson, "Discrimination of Successiveness; a Test of a Model of Attention," *Science*, v. 139 (1963), p. 112.

[47] G. F. Greiner, C. Lonraux and R. Herbein, "La localisation spatiale du son par l'épreuve de Matzker," *Confin. Neurol.*, v. 23 (1963), p. 343.

[48] G. v. Békésy, "Olfactory Analogue to Directional Hearing," *J. of Appl. Physiol.*, v. 19 (1964), p. 369.

[49] Jiří Zeman, "Information and the Brain," *Nerve, Brain and Memory Models*, eds. N. Wiener and J. P. Schadé (New York: Elsevier, 1963), v. 2, p. 70.

[50] J. J. Gibson, "The Survival Value of Sensory Perception," *Biological Prototypes and Synthetic Systems*, eds. E. E. Bernard and M. R. Kare (New York: Plenum Press, 1962), v. 1, p. 230.

[51] H. Werner, *Comparative Psychology of Mental Development* (Chicago: Follett, 1948), pp. 213–98.

[52] L. G. Augenstine, "Protein Structure and Information Content," *Information Theory in Biology*, eds. H. P. Yockey, R. L. Platzman and H. Quastler (New York: Pergamon Press, 1958), p. 103.

[53] H. Branson, "Information Theory and the Structure of Proteins," *Information Theory in Biology*, ed. H. Quastler (Urbana: University of Illinois Press, 1953), p. 84.

[54] A. J. Clark, *The Mode of Action of Drugs on Cells* (Baltimore: Williams & Wilkins, 1933), p. 298.

[55] G. J. Whitrow, *The Natural Philosophy of Time* (London: Nelson, 1961), pp. 4, 324.

[56] C. C. L. Gregory and A. Kohsen, *The O-Structure* (Church Crookham, Hampshire, Eng.: Gally Hill Press, 1959); C. C. L. Gregory, "A Proposal to Replace Belief by Method in the Pre-mensural Sciences," *Nature*, v. 185 (1960), p. 124.

[57] See Gregory, 1960, note 56.

[58] See Gregory and Kohsen, 1959, note 56.

[59] The terms "primitive," "classical" and "baroque" are not meant as value judgments, but are classificatory in nature. They may be applied to other historical processes, scientific as well as artistic.

[60] D. W. Sciama, "Evolutionary Processes in Cosmology," *Theories of the Universe*, ed. M. K. Munitz (New York: The Free Press, 1957), p. 413, and F. Hoyle, "Continuous Creation and the Expanding Universe."

[61] J. J. Grebe, "Time: Its Breadth and Depth in Biological Rhythms," *Ann. N.Y. Acad. Sci.*, v. 98 (1962), p. 1206.

[62] A. Szentgyörgyi, *Introduction to a Submolecular Biology* (New York: Academic Press, 1960).

[63] R. J. P. Williams, "Possible Functions of Chains of Catalysts," pts. I and II, *J. of Theor. Biol.* (1961 and 1962), v. 1, p. 1; v. 3, p. 209.

64 R. A. Spangler and F. M. Snell, "Sustained Oscillations in a Catalytic Chemical System," *Nature*, v. 191 (1961), p. 457.

65 B. Chance, R. W. Estabrook and A. Ghosh, "Damped Sinusoidal Oscillations of Cytoplasmic 'Reduced Pyridine' Nucleotide in Yeast Cells," *Proc. National Acad. Sci.*, v. 51 (1964), p. 1244.

66 P. D. Boyer, "The Nature and Diversity of Catalytic Proteins," *The Nature of Biological Diversity*, ed. J. M. Allen (New York: McGraw-Hill, 1963), p. 69.

67 G. S. Brown, *Probability and Scientific Inference* (London: Longmans, Green, 1958).

68 See Fischer *et al.*, note 13.

69 G. A. Brecher, "Die Entstehung und biologische Bedeutung der subjektiven Zeiteinheit—des Momentes," *Zeitschrift vergleichende Physiologie*, v. 18 (1932), p. 204.

70 J. W. S. Pringle, "On the Parallel between Learning and Evolution," *General Systems*: Yearbook of the Soc. for General Systems Research (Ann Arbor, Mich., 1956), v. 1, p. 90. Pringle pointed out that the description of an evolutionary system involves complexity—not the complexity of a system, but of our description of it. Such a description cannot be given within the framework of classic (2-valued) ontology, successful as it has proven to be for the development of Western science. G. Günther, *Cybernetic Ontology and Transjunctional Operations*, Tech. Rep. 4, Electric, Engineer. Research Lab., University of Illinois, Urbana, 1962, developed a multi-valued logic containing rejection values, *i.e.*, a logic which describes subjective or self-reflective entities. With this logical subjectivity one may further analyze Pringle's statement.

71 W. H. Thorpe, *Learning and Instinct in Animals* (2nd ed.; London: Methuen, 1963). The six types are habituation, conditioning, trial-and-error, insight learning, latent learning and imprinting.

72 J. Peterson, "Aspects of Learning," *Psych. Rev.*, v. 42 (1935), p. 1, quoted from N. Pastore, "An Examination of One Aspect of the Thesis that Perceiving Is Learned," *Psych. Rev.*, v. 63 (1956), p. 309. See his article as well as M. Wertheimer, "Hebb and Senden on the Role of Learning in Perception," *Amer. J. of Psych.*, v. 64 (1951), p. 133, for representative opposing views.

73 William Molyneux, author of the *Dioptrica Nova* (1692), an early treatise on optics, was a man very much aware of the problems of the blind because of his wife's loss of sight. He proposed what he referred to as a "jocose problem" in a letter to Locke on March 2, 1692:

> Suppose a man born blind, and now adult, and taught by his touch to distinguish between a cube and a sphere (suppose) of ivory, nighly of the same bigness, so as to tell when he felt one and t'other, which is the cube, which the sphere. Suppose then, the cube and the sphere placed on a table, and the blind man to be made to see; query whether by his sight, before he touch'd them, he could now distinguish and tell which is the globe, which the cube. I answer not; for tho' he has obtained the experience of how a globe, how a cube affects his touch, yet he has not yet attained the experience, that what affects my touch so or so, must affect sight so or so; or that a protuberant angle in the cube that press'd his hand unequally, shall appear to his eye as it does in the cube.

J. W. Davis, "The Molyneux Problem," *J. of the Hist. of Ideas*, v. 21 (1960), p. 392.

74 M. v. Senden, *Space and Sight* (London: Methuen, 1960).

[75] For current reviews of the status of learning, and specifically perceptual learning, see J. F. Delafresnaye, ed., *Brain Mechanisms and Learning* (Oxford: Blackwell, 1961) and E. J. Gibson, "Perceptual Learning," *Ann. Rev. Psych.*, v. 14 (1963), p. 29.

[76] M. L. Simmel, "Phantom Experiences Following Amputation in Childhood," *J. of Neurol., Neurosurg., Psychiat.*, v. 25 (1962), p. 69.

[77] J. Piaget, *The Construction of Reality in the Child* (New York: Basic Books, 1954), ch. 4, pp. 320 to 349.

[78] J. Money, "Cytogenic and Psychosexual Incongruities with a Note on Space-Form Blindness," *Amer. J. of Psychiat.*, v. 119 (1963), p. 820.

[79] A. M. Ostfeld, "Effects of LSD_{25} and JB_{318} on Tests of Visual and Perceptual Functions in Man," *Fed. Proc.*, v. 20 (1961), p. 876.

[80] M. G. Weil, "Hallucinations visuelles ches les aveugles," *Rev. Oto-neuro-opthal.*, v. 17 (1939), p. 294.

[81] R. L. Gregory, "Human Perception," *British Med. Bull.*, v. 20 (1964), p. 21.

[82] E. J. Furlong, *Imagination* (London: Allen & Unwin, 1961).

[83] L. S. Palmer, *Man's Journey Through Time* (London: Hutchinson, 1957).

[84] J. B. S. Haldane, "Suggestions as to Quantitative Measurement of Rates of Evolution," *Evolution*, v. 3 (1949), p. 51.

[85] Palmer, see note 83.

[86] G. G. Simpson, *The Major Features of Evolution* (New York: Columbia University Press, 1953).

[87] J. B. S. Haldane, "Time in Biology," *Science Progress*, v. 175 (1956), p. 385.

[88] J. C. Stevens and H. B. Savin, "On the Form of Learning Curves," *J. of Experimental Anal. Behavior*, v. 5 (1962), p. 15. H. v. Foerster and O. Poetzl, *Das Gedächtnis: Eine quantenphysikalische Untersuchung* (Vienna: Franz Deuticke, 1948).

[89] S. Roston, "On Biological Growth," *Bull. Math. Biophys.*, v. 24 (1962), p. 369; F. Krüger, "Über die mathematische Darstellung des tierischen Wachstums," *Naturwissenschaften*, v. 49 (1962), p. 454; K.-H. Engel and A. Raeuber, "Das allometrische Wachstum der Kartoffel," *Zeitschrift für Pflanzenzüchtung*, v. 47 (1962), p. 114; R. Hinchcliffe, "Aging and Sensory Thresholds," *J. of Gerontol.*, v. 17 (1962), p. 45; W. R. Stahl, "Similarity and Dimensional Methods in Biology," *Science*, v. 137 (1962), p. 205.

[90] L. F. Richardson, *Statistics of Deadly Quarrels*, eds. Q. Wright and C. C. Lienau (Chicago: Quadrangle Books, 1960).

[91] A. Rápaport and L. von Bertalanffy, eds., *General Systems*, Yearbook of the Soc. for General Systems Research (Ann Arbor, Mich., 1957), v. 2, p. 55.

[92] S. S. Stevens, "To Honor Fechner and Repeal His Law," *Science*, v. 133 (1961), p. 80.

[93] B. Katz, *Electric Excitation of Nerve* (Oxford: Oxford University Press, 1939). E. D. Adrian, *The Basis of Sensation* (London: Christopher, 1928).

[94] F. R. Ferrando, "Information, Entropy, and the Nervous System," *Perspectives in Biology and Medicine*, v. 5 (1962), p. 296.

[95] G. V. Békésy, *Experiments in Hearing* (New York: McGraw-Hill, 1960).

[96] E. E. Green, "Correspondence between Stevens' Terminal Brightness Function and the Discriminability Law," *Science*, v. 138 (1962), p. 1274.

[97] With respect to the difference between power law and exponential function, see J. Anderson, R. W. S. Tomlinson and S. B. Osborn, "An Interpretation of Radioisotope Turnover Data" *Lancet*, (1962), pp. 949 to 950.

[98] M. Treisman, "Laws of Sensory Magnitude," *Nature*, v. 198 (1963), p. 914.

99 S. T. C. Wright, "An Observation Suggesting that the Molecular Weights of Enzymes Can Be Arranged in Three Geometric Series," *Nature*, v. 193 (1962), p. 334.

100 R. B. Kelman, "A Theoretical Note on Exponential Flow in the Proximal Part of the Mammalian Nephron," *Bull. Math. Biophysic.*, v. 24 (1962), p. 303.

101 J. L. Stewart, "A Law for Loudness Discrimination," *Science*, v. 137 (1962), p. 618.

102 E. Janisch, "Das Exponentialgesetz als Grundlage einer vergleichenden biologie," Abhandlung 2, *Theorie d. organischen Entwicklung*, Heft II (Berlin: Springer, 1927).

103 See Stahl, note 89.

104 The amount of energy involved in mental phenomena is so minimal that no measurable changes can be found in cerebral blood flow, oxygen and glucose utilization and blood chemical constituents to differentiate the characteristic psychological and mental effects of D-Lysergic Acid Diethylamide from mathematical performance or from mental states occurring in schizophrenia; see L. Sokoloff, R. Mangold, R. Wechsler, C. Kennedy and S. Kety, "The Effect of Mental Arithmetic on Cerebral Circulation and Metabolism," *J. of Clin. Invest.*, v. 34 (1955), p. 1101; L. Sokoloff, S. Perlin, C. Kornetsky and S. Kety, "The Effects of D-Lysergic Acid Diethylamide on Cerebral Circulation and Over-all Metabolism," *Ann. N.Y. Acad. Sci.*, v. 66 (1957), p. 468. From the values for oxygen and glucose utilization S. S. Kety in *The Nature of Sleep*, Ciba Foundation Sympos., eds. G. E. W. Wolstenholme and M. O'Connor (London: Churchill, 1961), p. 375, calculated the total energy consumption of the brain as very close to 20 watts.

105 In normal perception we have no direct experience of reality. That part of the perceptual world we identify as our body is also a subjective construction of the brain, the body image, or, more correctly, the phantom body. It only needs, for example, an amputation for part of this phantom body to become a hallucination, a perception without an object. The external world is thus doubly subjective because it is a construct of the subject's brain, and it incorporates sensory experience derived from the subject's body; see R. Brain, The 30th Maudsley Lecture, "Perception and Imperception," *J. of Ment. Sci.*, v. 102 (1956), p. 221. The brain represents impulses or images in perceptual space, and the individual sees in his perceptual space objects that are not in his physical space. Hallucinations, like normal perceptions, are constructions of the subject's brain; see R. W. Medlicott, "An Inquiry into the Significance of Hallucinations with Special Reference to Their Occurrence in the Sane," *International Rec. Med.*, v. 171 (1958), p. 664.

106 D. E. Broadbent and M. Gregory, "Division of Attention and the Decision Theory of Signal Detection," *Proc. Roy. Soc.*, B, v. 158 (1963), p. 222.

107 L. F. Cooper and M. H. Erickson, *Time Distortion in Hypnosis* (Baltimore: Williams & Wilkins, 1959).

108 A. M. Weitzenhoffer, "Explorations in Hypnotic Time Distortions. I: Acquisition of Temporal Reference Frames under Conditions of Time Distortion," *J. of Nerv. Ment. Disease*, v. 138 (1964), p. 354.

109 G. H. Deckart, "Pursuit Eye Movements in the Absence of a Moving Visual Stimulus," *Science*, v. 143 (1964), p. 1192.

110 W. Dement and N. Kleitman, "The Relation of Eye Movements During Sleep to Dream Activity: An Objective Method for the Study of Dreaming,"

J. of Experimental Psych., v. 53 (1957), p. 339; N. Kleitman, *Sleep and Wakefulness* (Chicago: University of Chicago Press, 1963), pp. 92 and 93.

[111] M. and D. Jouvet, "A Study of the Neurophysiological Mechanisms of Dreaming," *The Physiological Basis of Mental Activity*, ed. R. H. Peon (New York: Elsevier, 1963), p. 133.

[112] F. Hebbard and R. Fischer, "Effect of Psilocybin, LSD and Alcohol on Small, Involuntary Eye Movements" (Ms. in preparation, 1965).

[113] W. W. Surwillo, "Frequency of the 'Alpha' Rhythm, Reaction Time and Age," *Nature*, v. 191 (1961), p. 823; Enoch Callaway III and C. L. Yeager, "Relationship between Reaction Time and Electroencephalographic Alpha Phase," *Science*, v. 132 (1960), p. 1765.

[114] P. L. Latour, "Reaction Time and E.E.G.," *Acta Physiol. Pharmacol. Neerland.*, v. 10 (1962).

[115] H. K. Beecher, "Increased Stress and Effectiveness of Placebos and 'Active' Drugs," *Science*, v. 132 (1960), p. 91; Beecher, "Objective Evaluation of Subjective Pain Experience," *Psychopathology of Perception*, eds. J. Zubin and P. Hoch (New York: Grune and Stratton, 1965).

[116] L. E. Hollister, "Chemical Psychoses," *Ann. Rev. Med.*, v. 15 (1964), p. 203.

[117] R. C. Elliott, *The Power of Satire: Magic, Ritual, Art* (Princeton: Princeton University Press, 1960), pp. 285 to 292.

[118] W. B. Cannon, "Voodoo Death," *Psychosom. Med.*, v. 19 (1957), p. 182.

[119] A. Jilek, "Geisteskrankheiten und Epilepsie im tropischen Afrika," *Fortschritte für Neurologie und Psychiatrie*, v. 32 (1964), p. 213.

[120] J. L. Mathis, "A Sophisticated Version of Voodoo Death, Report of a Case," *Psychosom. Med.*, v. 26 (1964), p. 104.

[121] L. L. Vasiliev, *Experiments in Mental Suggestion*, tr. A. Kohsen and C. C. L. Gregory (Church Crookham: Gally Hill Press, 1963), pp. 152 and 153.

[122] F. W. Schueler, *Chemobiodynamics and Drug Design* (New York: McGraw-Hill, 1960).

[123] See Efron, note 3.

[124] G. Poulet, *Studies in Human Time*, tr. E. Coleman (New York: Harper, 1959), p. 37.

GENERAL REFERENCES

Eakin, R. E. "An Approach to the Evolution of Metabolism," *Proc. Natl. Acad. Sci.*, v. 49, 1963, p. 360.

Fischer, R. "Selbstbeobachtungen im Mezkalin Rausch," *Schweizerische Zeitschrift für Psychologie*, v. 5, 1946, p. 308.

Green, D. E., and Y. Hatefi. "The Mitochondrion and Biochemical Machines," *Science*, v. 133, 1961, p. 13.

Gruber, H., W. King and S. Link. "Moon Illusions: An Event in Imaginary Space," *Science*, v. 139, 1963, p. 750.

Horowitz, N. H., and S. L. Miller. "Current Theories on the Origin of Life," *Fortschritte der Chemie organischer Naturstoffe*, v. 20, 1962, p. 423.

Krebs, H. A. "The Regulation of Metabolic Processes," *Texas Reports Biol. and Med.*, v. 17, 1959, p. 16.

Moyed, H. S., and H. E. Umbarger. "Regulation of Biosynthetic Pathways," *Physiol. Rev.*, v. 42, 1962, p. 444.

Oparin, A. I. "Origin and Evolution of Metabolism," *Comp. Biochem. Physiol.*, v. 4, 1962, p. 371.

Segall, M., D. Campbell, and M. Herskovitz. "Cultural Differences in the Perception of Geometric Illusions," *Science*, v. 139, 1963, p. 769.

Stevens, S. S. "The Quantification of Sensation," *Daedalus*, v. 88, 1959, p. 606.

Taylor, J. G. "Experimental Design: a Cloak for Intellectual Sterility," *Brit. Jour. Psychol.*, v. 49, 1958, p. 106.

West, L. J., ed. *Hallucinations*. New York: Grune and Stratton, 1962.

Introductory Note to Part IV

[1] J. A. Carroll, *Nature*, v. 184 (1959), p. 260.

[2] Such as the gas clock suggested by R. Schlegel in his essay in this volume.

Timekeepers—An Historical Sketch

[1] F. Thureau Dangin, *Equisse d'une histoire sexagesimal* (Paris: P. Geuthner, 1932).

[2] L. Borchardt, ed., *Altaegyptische Zeitmessung* (Berlin: W. de Gruyter, 1920).

[3] Manuel Rico y Synoba, *Libros del Saber de Astronimica del Rey Don Alfonso X de Castilla* (Madrid, 1866), v. 4.

[4] *Chronica Jocelini de Brakelonda de rebus gestis Samsonis Abbatis Monasterii Sancti Edmundi* (London: Camden Society, 1840).

[5] The period of the verge-and-foliot escapement may be deduced from elementary mechanical considerations, for which see J. Needham, Wang Ling and D. J. de Solla Price, *Heavenly Clockwork* (Cambridge: Cambridge University Press, 1960), p. 113.

[6] *Atti del Real Instituto Venito de Scienza, Lettera Arti,* v. 2. S. viia. Plate 1.

[7] The word *fusee* comes from the Latin *fusata*, a spindle wound with thread. Probably because of its resemblance to winding stairs or spirally wound organs, the *fusee* is also referred to as a cochlea.

[8] The derivation of the word *stackfreed* is unknown.

[9] The cycloid is the locus of a point on the circumference of a circle rolling in a plane along a straight line in the plane.

[10] The precise date of Mudge's lever escapement has not yet been established and it still forms the subject of current research.

GENERAL REFERENCES

Asimov, Isaac. *The Clock We Live On.* New York: Collier-Macmillan, 1963.

Baillie, G. H. *Clocks and Watches, an Historical Bibliography.* London: N.A.G. Press, 1951.

———. *Watchmakers and Clockmakers of the World.* London: N.A.G. Press, 1947.

Barr, Lockwood. *Eli Terry and Scroll Shelf Clocks.* 1952.

Bassermenn-Jordan, Ernst von. *The Book of Old Clocks and Watches,* rewr. H. von Bertelle, tr. H. A. Lloyd. 4th ed.; New York: Crown, 1964.

Chamberlain, Paul M. *It's About Time.* 1941. Repr. London: The Holland Press, 1964.

Drepperd, Carl W. *American Clocks and Clockmakers.* New York: Doubleday, 1947.

Eckhardt, G. H. *Clocks of Pennsylvania and Their Makers.* New York: Devin-Adair, 1955.

Fleet, Simon. *Clocks.* New York: Putnam's, 1961.

James, Arthur E. *Chester County Clocks and Their Makers.* 1942.

Johnson, Chester. *Clocks and Watches.* New York: Odyssey, 1964.

Jones, Leslie Allen. *Eli Terry, Clockmakers of Connecticut.* 1942.

Lloyd, H. Alan. *A Collector's Dictionary of Clocks.* New York: A. F. Barnes, 1965.

———. *Old Clocks.* London: Ernest Benn, 1958; 3rd ed.; 1964.

———. *Some Outstanding Clocks over 700 Years, 1250–1950.* London: Leonard Hill, 1958.

Marshall, Roy K. *Sundials.* New York: Macmillan, 1963.

Milham, Willis I. *Time and Timekeepers.* 2nd ed.; New York: Macmillan, 1945.

Palmer, Brooks. *The Book of American Clocks.* New York: Macmillan, 1950.

———. *The Romance of Time.* 1954.

Singer, C., E. J. Holmyard, A. R. Hall and T. I. Williams. *A History of Technology.* Oxford: Oxford University Press, 1957, v. 3, in which see: "The Calendar," p. 558, "Portable Sundials," p. 594, "Water Clocks," p. 601, and "Mechanical Timekeepers," p. 648.

Ward, F. A. B. *Time Measurement.* Pt. 1. Science Museum, London: 1947; Pt. 2, 1950.

Time Measurement for Scientific Use

[1] One may argue that the Astronomical Unit, for instance, is not less natural as a unit of distance than the year is as a unit of time. Both are, of course, arbitrary in the sense that others might have been selected. Both are natural in the sense that they are based on phenomena occurring in nature. But the Astronomical Unit is in fact related to the year by an invariant formula and is, in my view, only a convenient way of avoiding continued reference to the year, having no physical existence in the sense that the meter bar has. It is not even equal to the mean distance between earth and sun.

[2] For references on calendars see: F. K. Ginzel, *Handbuch der mathematischen und technischen Chronologie* (Leipzig, 1906–14). Repr. (Leipzig: 1958). This is the most comprehensive and detailed general treatise on calendars and chronological systems; it is extensively documented and is a reliable source of information as far as historical knowledge extended at the date of its publication. For tables of the principal calendars of ancient and modern times see R. Schram, *Kalendariographische und chronologische Tafeln* (Leipzig, 1908). See also R. A. Parker, "The Calendars of Ancient Egypt," Oriental Institute of the University of Chicago, Studies in Ancient Oriental Civilization, no. 26 (Chicago: University of Chicago Press, 1950); and S. G. Barton, "The Quaker Calendar," *Proc. Amer. Phil. Soc.*, v. 93 (1949), p. 32.

[3] See F. H. Colson, *The Week* (Cambridge: Cambridge University Press, 1926).

4 See, for instance, G. M. Clemence, "Time and its Measurement," *Amer. Scientist*, v. 40 (1952), p. 260; "Standards of Time and Frequency," *Science*, v. 123 (1956), p. 567; and "Astronomical Time," *Rev. Modern Physics*, v. 29 (1957), p. 2.

5 See G. M. Clemence, "Relativity Effects in Planetary Motion," *Proc. Amer. Phil. Soc.*, v. 93 (1949), p. 532.

6 Certain areas are offset by a difference of one half-hour from the world-wide minute and second routine. They include the whole of India, Iran, British Guiana, Cook Islands, parts of Central Australia and some other areas.

7 *Monthly Notices Roy. Astron. Soc.*, v. 99 (1939), p. 541.

8 See *Proc. International Conference on Weights and Measures* (Paris, 1956).

9 On this, see L. Essen, "The National Physical Laboratory's Caesium Standard," *Proc. Roy. Soc.* (1947).

10 *Cf.* S. Aoki, "Note on Variability of the Time Standard Due to Relativistic Effect," *Astron. J.*, v. 69 (1964), p. 221.

11 See P. A. M. Dirac, *Proc. Roy. Soc.*, A-165 (1938), p. 199, and P. Jordan, *Zeitschrift für Physik*, v. 157 (1959), p. 112.

Comments—Relativistic Dialectics

1 G. M. Clemence, *Rev. Modern Phys.*, v. 29 (1957), p. 7.

2 L. I. Schiff, *Rev. Modern Phys.*, v. 36 (1964), p. 510.

3 A. Einstein, *Sidelines on Relativity* (London: Methuen, 1922), p. 36.

4 See, *e.g.*, R. M. Gale, "Is it now now?", *Mind* (January, 1963).

5 H. Dingle, *Phil. Sci.*, v. 27 (1960), p. 233.

6 A. Einstein, *op. cit.*, p. 35.

Time in Relativity Theory: Arguments for a Philosophy of Being

1 The word "equivalence" is taken in the technical sense it has in physics, to which we shall come back later. That the time concept, in one of its fundamental aspects, depends closely on the analysis of motion and is thus referred to spatial measurements, has long been recognized. One may recall, in this respect, Aristotle's "Time is the measure of movement" and "We measure time by means of motion and motion by means of time." True, Aristotle's movement was what we call change, our movement being for Aristotle local movement; let us then say that our quotations from Aristotle are restricted to the special case of local movement.

2 "Absolute space, by its own nature and without relation to anything external, remains always uniform and unmoved. Relative space is some movable scale or magnitude of absolute space, defined by our sensory perception of the locations of solid bodies, and commonly taken for immovable space. Thus aerial or celestial dimensions are defined with respect to positions on the earth." *Philosophiae Naturalis Principia Mathematica*, 3rd ed. (London, 1726), p. 6, here tr. D. Park.

3 See the essay by A. C. Benjamin in this volume.

[4] Absolute translational accelerations produce ordinary inertial forces, absolute rotations produce Coriolis' inertial forces in addition. The deeper view that, according to general relativity, both inertial and gravitational forces must be thought of as relative is beyond the scope of the present study.

[5] The common orientation of the Galilean frames is such that it entails no over-all acceleration of the so-called fixed stars; besides, the mass center of the solar system is at rest in one of the Galilean frames (this property being deducible from theoretical dynamics).

It is interesting to note that Galileo's opponents were sticking alternately to an absolute space principle (the earth being considered as absolutely at rest) and to a relative motion principle (the earth and sun being taken as kinematically equivalent frames). Galileo was implicitly sustaining the (utterly new) relativity principle of dynamics in that he referred celestial motions to the Galilean frame in which the sun's center is (practically) at rest.

[6] W. Thomson and P. G. Tait, *Treatise on Natural Philosophy* (Cambridge: Cambridge University Press, 1888–90), pt. 1, p. 241.

[7] An interesting application of epistemological views put forward by P. Duhem and by H. Poincaré is found here. According to Duhem, *The Aim and Structure of Physical Theory* (Princeton: Princeton University Press, 1954), chs. VI and VII, new experimental facts never *impose*, but only *allow* and *suggest* the formulation of adequate postulates; and, according to Poincaré, *La science et l'hypothèse* (Paris: Flammarion, 1905), ch. III, "De la Nature des axiomes"; see also ch. VIII, a system of postulates is nothing else than a system of definitions. Here it turns out that the very form of the laws of inertial motion discovered by Galileo is such as to allow a joint definition of spatial reference frames and of a time scale which is highly privileged.

[8] It is true that Kepler's laws proved the coherence of the various astronomical time scales associated with the motions of the planets; but Kepler's kinematical laws are so near to Newton's theory that we may regard them as some of its immediate empirically verifiable consequences.

[9] Interesting remarks on the interpretation of the universal constant implicit in Newton's formula have been made independently by D. W. Sciama, *Monthly Notices Roy. Astr. Soc.*, v. 113 (1953), p. 34, and by D. Park, *Journal de Physique*, v. 18 (1957), p. 11; see especially p. 16.

[10] Galileo points out that a package of bricks must "evidently" exhibit the same law of free fall as would the individual bricks contained in it.

[11] This important point is explicitly stated by Fresnel (letter to Arago, September, 1818, published in *Oeuvres Complètes*, v. 2, p. 627). For the foregoing experiments and theories see E. T. Whittaker, *History of the Theories of Aether and Electricity*, 2nd ed. (London: Thomas Nelson, 1951), v. 1.

[12] *Ann. d. Phys. u. Chemie*, v. 150 (1873), p. 497.

[13] *Journ. de Phys.*, v. 3 (1874), p. 201.

[14] In Potier's presentation of Fresnel's law, the refractive index is eliminated.

[15] *Comptes Rendus*, v. 235 (1952), pp. 1007, 1009; *Vitesse et univers relativiste* (Paris, 1954), SEDES. The other general postulates of Abelé and Malvaux are that the postulated group is continuous, and that the velocities form an ordered set.

[16] This conclusion may seem at first sight surprising, since Veltmann and Potier proved that, as a consequence of the Fresnel ether-drag formula, classical and relativistic kinematics are equivalent in the first order in v/c. The answer

is, of course, that they are equivalent only insofar as the group property is not postulated for the velocity composition law. If the group property is postulated, then relativistic kinematics is deduced, as shown by Abelé and Malvaux.

17 H. Poincaré, *Rendiconti Circolo Mathematico di Palermo*, v. 21 (1906), p. 129. In Einstein's 1905 paper, *Annalen der Physik*, v. 17 (1905), p. 891, the group concept is implied in the postulated invariance of the laws of physics under the Lorentz transformations.

18 *Annales scientifiques de l'Ecole Normale Supérieure*, v. 3 (1874), pp. 375, 420. Mascart was one of the experimentalists who accumulated "negative" first-order experiments during the nineteenth century.

19 *Electricité et Optique*, 2nd ed. (Paris: Gauthier Villars, 1901), p. 613. See also *La Science et l'hypothèse, op. cit.*, ch. X, p. 201, and *La Valeur de la science* (Paris: Flammarion, 1905), ch. VII, p. 175; ch. VIII, p. 185.

20 The treatise by Lie and Engel appeared in 1888.

21 It remains a psychological mystery why Poincaré, who had contributed very profoundly to both elements of the problem, did not follow it through to what now appears as its logical conclusion. Even the remark that the Lorentz formulas may be interpreted as a tetrad rotation in a four dimensional space can be found in his 1905 paper. The reason is perhaps that Poincaré was more unwilling than Einstein to give up the idea of an ether; if one believes in an ether, then the fact of relativity emerges as an extraordinary consequence of the equations of electrodynamics, but one is apt not to perceive its fundamental character.

22 Einstein, see note 17.

23 It is easily verified that if one imposes the condition

$$x^2 + y^2 + z^2 = c^2t^2$$

the corresponding primed variables are such that

$$x'^2 + y'^2 + z'^2 = c^2t'^2;$$

that is, the velocity of light is found isotropic, with the same value c, in both the x, y, z, t and the x', y', z', t' reference frames. Moreover, one verifies easily that the Lorentz transformation has the group property, and that the expression

$$s^2 = x^2 + y^2 + z^2 - c^2t^2$$

is an invariant of the group.

24 This is an abbreviated way of describing what happens when a Lorentzian observer attaches x' values to moving points at a given instant t of his own proper time. The physical problem of the optical appearance of moving objects at relativistic velocities (that is, velocities not negligible compared to c) is far more sophisticated, as explained in the papers of J. Terrell, *Phys. Rev.*, v. 116 (1959), p. 1041, and V. F. Weisskopf, *Phys. Today*, v. 13 (September, 1960), p. 24.

25 Duhem and Poincaré, see note 7.

26 These techniques, inaugurated in 1892 by Michelson and Benoit, use as their scale of distances the wave length of a standing wave emitted by a monochromatic source. In 1960, the Eleventh International Conference on Weight

and Measures decided that the standard meter should henceforth be defined as a multiple of the vacuum wave length of a particularly narrow emission line of Krypton 86.

[27] Hertzian chronometry was inaugurated in 1948 by Lyons; today, "atomic clocks" stabilized by the hyperfine transition of atomic hydrogen yield precisions better than 10^{-12} over long intervals of time.

[28] Before relativity, this wave would have been characterized as a function of the time in a space of $3N$ dimensions (N is the number of particles in the solid). But we require a relativistic description. This is possible in principle in terms of a second-quantized theory, but there is not yet a relativistic formulation of many-body theory, and the problem is still unsolved.

[29] Technically speaking, the Klein Gordon equation

$$\left(\frac{\partial^2}{\partial x^2} + \frac{\partial^2}{\partial y^2} + \frac{\partial^2}{\partial z^2} - \frac{1}{c^2} \frac{\partial^2}{\partial t^2} \right) \psi = -k^2 \psi$$

and d'Alembert's equation

$$\left(\frac{\partial^2}{\partial x^2} + \frac{\partial^2}{\partial y^2} + \frac{\partial^2}{\partial z^2} - \frac{1}{c^2} \frac{\partial^2}{\partial t^2} \right) \psi = 0$$

are both invariant under the Lorentz transformations.

[30] Differential techniques of measurement like interferometry are normally much more exact than direct measurements. Thus a precise recent version of Michelson's experiment, Jaseja, Javan, Murray and Townes, *Phys. Rev.*, v. 133 (1964) p. A1221, would have permitted the detection of an ether wind of 1 km/sec, while a precision of about one part in 10^8 in direct measurements would only allow a wind of 50 km/sec to be detected by kinematical measurements. Thus the situation is perhaps less serious than implied above, but qualitatively it is unchanged. A recent measurement using the Mössbauer effect, Champeney, Isaak and Khan, *Rev. Modern Phys.*, v. 36 (1964), p. 469, would have detected a wind of 5 m/sec.

[31] For brevity, we speak only of inertial motion. A more detailed discussion would involve the Newtonian formula $F = ma$ and its analogue in quantum mechanics.

[32] The fact that the waves used in practice to measure lengths and times are often of very different frequencies is purely incidental. Perhaps the development of Laser techniques will someday permit a single optical atomic transition to be used for both purposes.

[00] Klein-Gordon, see note 29.

[34] *Proc. Roy. Soc. A*, v. 194 (1948), p. 348.

[35] *Phys. Rev.*, v. 80 (1950), p 298.

[36] Plyler, Blaine and Connor, *J. Opt. Soc. Amer.*, v. 45 (1955), pp. 102, 1115.

[37] Rank, Shearer and Wiggins, *Phys. Rev.*, v. 94 (1954), p. 575.

[38] D'Alembert and Klein-Gordon, see note 29.

[39] Euclidean because the invariant s^2 is written in terms of only the squares of coordinate intervals; pseudo-Euclidean because of the minus sign before the term $c^2 t^2$.

[40] The Lorentz transformations as defined by the formulas given above *are* continuous.

[41] That is, in the particular formula used for s^2. In classical kinematics there is no such formula involving four squares, for distance is not defined in the classical geometry of space and time. This geometry is thus said to be not *metric*, but only *affine*. (In this way it resembles, for example, the geometry of a thermodynamic diagram.)

[42] S. Tomonaga, *Prog. Theor. Phys.*, v. 1 (1948), p. 27. J. Schwinger, *Phys. Rev.*, v. 73 (1948), p. 1439. R. P. Feynman, *Phys. Rev.*, v. 76 (1950), pp. 739, 769.

[43] An example of the utility of relativistic considerations in this domain was given long ago by W. Pauli, *Zeitschrift Physik*, v. 18 (1923), p. 272; v. 22 (1924), p. 261.

[44] Tomonaga, see note 42.

[45] An extension in space but not in time is possible relative to a particular set of coordinate axes, but such a property cannot be described in terms of covariant geometrical concepts and so has no objective character.

[46] H. Minkowski in A. Einstein *et al.*, *The Principle of Relativity* (London: Methuen, 1923), tr. W. Perrett and G. B. Jeffery, p. 75: "Henceforth space by itself and time by itself, are doomed to fade away into mere shadows, and only a kind of union of the two will preserve an independent reality."

[47] *Dinglers polytechnisches J.*, v. 345 (1930), p. 122.

[48] H. Weyl, *Philosophy of Mathematics and Natural Science* (Princeton: Princeton University Press, 1949), p. 116: "The objective world simply *is*; it does not *happen*."

[49] L. Fantappiè, *Nuove Vie per la Scienza* (Rome: Sansoni, 1961), paragraph heading, p. 33: "Essistenza del passato e del futuro, insieme al presente."

[50] R. P. Feynman, "The Theory of Positrons," *Phys. Rev.*, v. 76 (1949), p. 749: "It is as though a bombardier flying low over a road suddenly sees three roads and it is only when two of them come together and disappear again that he realizes that he has simply passed over a long switchback in a single road. The over-all space-time point of view leads to considerable simplification in many problems."

[51] Several authors have tried to connect this law to the "generalized Carnot principle" of cybernetics, $\Delta I < \Delta S$, where I is information and S is entropy. See notably, N. Wiener, *Cybernetics* (Cambridge: Massachusetts Institute of Technology Press, 1948), Introduction; O. Costa de Beauregard, *Le second Principe de la science du temps* (Paris: Editions du Seuil, 1963), ch. 4 §4; A. Grünbaum, *Philosophical Problems of Space and Time* (New York: Knopf, 1963), ch. 9. *Cf.* the essay by Prof. Watanabe in this volume.

[52] O. Costa de Beauregard, *Le Second Principe. . . .* See also K. Popper, *Nature*, v. 177 (1956), p. 538, v. 178 (1956), p. 382, v. 179 (1957), p. 1296; J. MacLennan, Jr., *Phys. Rev.*, v. 115 (1959), p. 1405; O. Penrose and I. C. Percival, *Proc. Phys. Soc.*, v. 79 (1962), p. 605.

[53] N. Bohr, *Atomic Theory and the Description of Nature* (Cambridge: Cambridge University Press, 1934).

[54] *Ibid.*, p. 116; see also p. 100.

[55] This problem and others are considered at some length in the author's two books, *La Notion de temps: Equivalence avec l'espace* (Paris: Hermann, 1963), see especially ch. IV, C, D, E and *Le Second Principe . . .* , see especially ch. IV, B.

Time in Relativity Theory: Arguments for a
Philosophy of Becoming

[1] E. Cunningham, *The Principle of Relativity* (Cambridge: Cambridge University Press, 1914), p. 191.

[2] L. Silberstein, *The Theory of Relativity* (London: Macmillan, 1914), p. 134.

[3] *Bulletin de la société française de philosophie* (April, 1922), p. 108.

[4] *Ibid.*, p. 111: "Dans le continuum à quatre dimensions il est certain que toutes les directions ne sont pas équivalentes."

[5] E. Meyerson, *La Déduction relativiste, cf.*, in particular, ch. VII, "Le Temps"; on Einstein's different views *cf.* pp. 100 and 104.

[6] A. Einstein, in the article, "A propos de *La Déduction relativiste* de M. E. Meyerson," *Rev. philosophique*, v. 105 (1928), p. 161.

[7] K. Gödel, "A Remark about the Relationship between Relativity Theory and Idealistic Philosophy," *Albert Einstein, Philosopher-Scientist*, v. 7, The Library of Living Philosophers, ed. Paul Schilp (Evanston: 1949), p. 557.

[8] A. Einstein, "Remarks Concerning the Essays Brought Together in This Cooperative Volume," *op. cit.*, p. 687.

[9] *Op. cit.*, p. 688.

[10] *La Déduction relativiste, op. cit.*, p. 97. I extended Meyerson's survey of the divided opinion on this matter in the article "Relativity and the Status of Space," *Rev. of Metaphysics*, v. 9 (1955), p. 160, where I also pointed out that the quotation from A. S. Eddington was taken by Meyerson out of the context. The hesitancies on this point in H. Reichenbach's thought can be seen most clearly in his posthumous book, *The Direction of Time* (Berkeley: University of California Press, 1956); after rejecting the static interpretation of space-time which he traces correctly to the influence of the Eleatic tradition (p. 11) he virtually eliminated time later when he claims that there is no unique direction of time either on the cosmic scale (p. 128f.) or on the micro-physical level (p. 262).

[11] W. V. O. Quine, "The Myth of Passage," *J. of Philos.*, v. 48 (1951), p. 457; Quine, *Word and Object* (Cambridge: M. I. T. Press, 1960), p. 172.

[12] O. Costa de Beauregard, *Le Second Principe de la science du temps* (Paris: Editions du Seuil, 1963), p. 132; A. Grünbaum, *Philosophical Problems of Space and Time* (New York: Knopf, 1963), p. 329.

[13] K. Gödel, *op. cit.*, p. 557.

[14] See essay by O. Costa de Beauregard in this volume.

[15] *Cf.* the distinction between "determinism" and "causality" made by Louis de Broglie prior to his reconversion to determinism: *Continu et Discontinu en Physique Modern* (Paris: Editions A. Michel, 1959), p. 61. *Cf.* also my articles "The Doctrine of Necessity Re-examined," *Rev. of Metaphysics*, v. 5, no. 1 (September, 1951), p. 11, and "Toward a Widening of the Notion of Causality," *Diogenes*, no. 28 (Winter, 1959), p. 63.

[16] The only way of escaping this conclusion would be to hope, rather unrealistically, that some future discovery would establish the existence of the velocities larger than that of light and gravitation; but this would falsify one of the central ideas of the relativity theory.

[17] Quoted by H. Poincaré, "Science and Method," *The Foundations of Science* (Ephrata, Pa.: The Science Press, 1946), p. 400.

[18] P. Langevin, "Le temps, l'espace et la causalité dans la physique moderne," *Bull. de la Société française de la philosophie* (October, 1911); "L'évolution de l'espace et du temps," *Rev. de métaphysique et de morale*, v. 19 (1911), p. 455; *La Physique depuis vingt ans*, Bibliothèque d'Histoire et de Philosophie des Science (Paris: 1923), p. 265.

[19] Even such a careful thinker as Philip Frank makes the following slip: "We find that the spatial distance *S*, as well as the temporal distance *T*, of two events, depends on the system of reference. *Either of them can even disappear if we choose a certain system of reference*"; cf. *Philosophy of Science. The Link between Science and Philosophy* (Englewood Cliffs, N.J.: Prentice-Hall, 1957), p. 161. Italics are mine. The italicized statement is certainly *not* generally true; *S* cannot be eliminated in the case *a* nor can *T* be eliminated in the case *c*! Professor Frank discussed both these cases *a* and *c* on the previous page (p. 160) without, however, perceiving the consequences drawn by Langevin in 1911. In fairness to him we must stress that he is as much opposed to the static interpretation of space-time as we are; cf. the whole chapter of the same book ("Is the World 'Really Four-Dimensional'?"), p. 158, especially p. 162; also his polemic against Sir James Jeans' static interpretation in *Interpretations and Misinterpretations of Modern Physics* (Paris: Hermann & Cie, 1938), p. 46.

[20] The error in Laplace's assumptions underlying his calculations were pointed out by W. Wien, "Über die Möglichkeit einer elektromagnetischen Begründung der Mechanik," *Wiedemann's Annalen* (1901), p. 501.

[21] A. Einstein, "Autobiographical Notes," *Albert Einstein: Philosopher-Scientist, op. cit.*, p. 61.

[22] *Newtoni Opera*, ed. Horsley, v. 3, p. 8: "unumquodque durationis indivisibile momentum ubique." Similarly Pierre Gassendi: "Et quodlibet Temporis momentum idem est in omnibus locis," *Opera omnia* (Florentia, 1728), p. 198.

[23] A. S. Eddington, *The Nature of the Physical World* (Cambridge: Cambridge University Press, 1933), p. 42.

[24] A. N. Whitehead, *Science and the Modern World* (New York: Macmillan, 1926), p. 172.

[25] A. A. Robb, *The Absolute Relations of Time and Space* (Cambridge: Cambridge University Press, 1921), p. 12.

[26] Robb, *Geometry of Time and Space* (Cambridge: Cambridge University Press, 1936), p. ˙5.

[27] H. Reichenbach, *The Direction of Time* (Berkeley: University of California Press, 1956), p. 265.

[28] Whitehead, *The Concept of Nature* (Cambridge: Cambridge University Press, 1920), p. 116; B. Russell, *The Analysis of Matter* (New York: Dover, 1954), p. 61.

[29] *"Jetzt" ist der Zeitmodus des erlebenden Ichs. Cf.* Hugo Bergmann, *Der Kampf um das Kausalgesetz in der jüngsten Physik* (Braunschweig: Vieweg, 1929), p. 28; quoted approvingly by A. Grünbaum, *Philosophical Problems of Space and Time, op. cit.*, p. 323.

[30] H. Weyl, *Philosophy of Mathematics and Natural Science* (Princeton: Princeton University Press, 1949), p. 116.

[31] Grünbaum, *op. cit.*, p. 329.

[32] This is rarely sufficiently emphasized. *Cf.* my article "Note on Whitehead's Definitions of Co-Presence," *Philos. of Sci.*, v. 24 (1957), p. 79.

[33] This was in substance Hugo Bergmann's argument. *Cf.* Grünbaum, *op. cit.,* p. 323.

[34] See above note 29.

[35] J. E. MacTaggart, *Studies in the Hegelian Dialectic* (2nd ed.; New York: Russell & Russell, 1964), p. 160.

[36] At least as long as we assume their interaction.

[37] Quoted by G. J. Whitrow, *The Natural Philosophy of Time* (London: Nelson, 1961), p. 311.

Time in Relativity Theory: Measurement or Coordinate?

[1] The question of non-uniform motion belongs to the *general* theory of relativity, which is outside the considerations of this essay. Nothing in that theory affects what is said here about the concept of time.

[2] Historically, Einstein's theory appeared before Ritz's, but it made little impression at first, and by the time the nature of the choice was clearly seen, it was recognizable as a reconciliation of what seemed to be incompatible elements of the other theories.

[3] By "ether" is here meant any universal medium with respect to which a body can be said to be at rest or in motion, such as Maxwell postulated. Nothing is implied concerning any other meaning that may be attached to the word.

[4] *Ann. d. Phys.*, v. 17 (1905), p. 891.

[5] Published in *The Meaning of Relativity* (5th ed.; 1955), p. 28.

[6] H. Dingle, *Monthly Notices of the Royal Astronomical Society*, v. 119 (1959), p. 67.

[7] A. M. Bonch-Bruevich, *Optika i Spektroskopia*, v. 9 (1960), p. 134. Other more recent experiments also are inconclusive. See H. Dingle, *Brit. J. Phil. Sci.*, v. 15 (1964), p. 41.

[8] J. G. Fox, *Amer. J. of Physics*, v. 30 (1962), p. 297.

[9] I pointed this out first in an article in *Philos. of Sci.*, v. 27 (1960), p. 233, and again in Appendix II of *A Threefold Cord*, Samuel and Dingle (London: Allen & Unwin, 1961). See also *Nature*, v. 195 (1962), p. 985; v. 197 (1963), pp. 1248, 1287.

[10] The result obtained here is confirmed and made precise by use of the equations of the theory.

[11] *Phil. of Sci.*, v. 27 (1960), p. 233.

Comments—Of Time and Proper Time

[1] *Annalen der Physik*, v. 17 (1905). Repr. in H. A. Lorentz, A. Einstein, H. Minkowski and H. Weyl, *The Principle of Relativity* (New York, Dover), p. 37.

[2] H. Minkowski, "Space and Time—Address delivered at the 80th Assembly of German Natural Scientists and Physicians, at Cologne, 21 September 1908," repr. in H. A. Lorentz, *op. cit.*, p. 75.

[3] On this see the two chapters entitled "The Determination of Time at a Distance," in G. J. Whitrow's *The Natural Philosophy of Time* (London: Nelson, 1961), p. 183; also W. F. Edwards, *Am. J. Physics*, v. 31 (1963), p. 482.

4 An analogous argument holds for the visual appearance of rapidly moving objects, in that we have no "common sense" notion of how such objects may look. *Cf.* note 24 of Costa de Beauregard, and W. F. V. Rosser, *An Introduction to the Theory of Relativity* (London: Butterworths, 1964), p. 163.

5 For an analysis of motional conditions when the traveling clock is expected to register the longer period of time, see E. T. Benedikt, "The Clock Paradox in Vertical Free Fall," 7th Annual Meeting of the American Astronautical Society, 1961, Preprint 61–43. According to J. E. Romain "the definition of 'natural time' in an accelerated frame is by no means a simple and obvious matter," "Time Measurement in Accelerated Frames of Reference," *Rev. of Mod. Physics*, v. 35 (1963), p. 376.

6 The annotated bibliography *The Clock Problem (Clock Paradox) in Relativity*, comp. M. Benton (Bethesda, Md.: U.S. Naval Research Laboratory, 1959), contains references to almost 250 serious articles.

7 G. J. Whitrow, *op. cit.*, p. 222.

8 M. Čapek, *The Philosophical Impact of Contemporary Physics* (Princeton: Van Nostrand, 1961), p. 201. See also Section 2 of Costa de Beauregard's essay in this volume.

9 The apt expression is from E. P. Ney, *Electromagnetism and Relativity* (New York: Harper & Row, 1962), p. 87.

10 *Ibid.*, p. 88.

11 W. F. V. Rosser, *op. cit.*, p. 397.

12 *Albert Einstein: Philosopher-Scientist*, ed. P. Schilp (New York: Tudor, 1951), p. 685.

Time and Quantum Theory

1 Among many others: David Bohm, *Quantum Theory* (Englewood Cliffs, N.J.: Prentice-Hall, 1951); P. A. M. Dirac, *Principles of Quantum Mechanics* (3rd ed.; Oxford: Oxford University Press, 1947); Eugen Merzbacher, *Quantum Mechanics* (New York: John Wiley, 1961); L. I. Schiff, *Quantum Mechanics* (2nd ed.; New York: McGraw-Hill, 1955); G. Temple, *General Principles of Quantum Theory* (London: Methuen, 1934), a highly recommended little book.

2 The literature is vast. For a general survey, see Ernest Nagel, "Principles of the Theory of Probability," *International Encyclopedia of Unified Science* (Chicago: University of Chicago Press, 1939), v. 1, no. 6; H. Jeffries, *Theory of Probability* (Oxford: Clarendon Press, 1939) and *Scientific Inference* (2nd ed.; Cambridge: Cambridge University Press, 1957) are rewarding. David Bohm's *Causality and Chance in Modern Physics* (Princeton: Van Nostrand, 1957) and ch. VIII Karl R. Popper, *The Logic of Scientific Discovery* (New York: Science Editions, 1961), are also of interest, among many others.

3 Nearly every introductory text in atomic or modern physics discusses the significance of these experiments. The Bohr-Einstein papers in *Albert Einstein: Philosopher-Scientist*, ed. P. A. Schilp (New York: Harper, 1959), v. I, are perhaps the best-known discussions of this material.

4 Julian Schwinger, "Algebra of Microscopic Measurement," *Proc. of National Acad. of Sci.*, v. 45 (1959), p. 1542, and "Geometry of Quantum States,"

op. cit., v. 46 (1960), p. 257. Subsequent papers, op. cit., v. 46 (1960), pp. 570, 883, 1401; v. 47 (1961), pp. 122, 1075; v. 48 (1962), p. 603.

[5] Werner Heisenberg, *Physical Principles of the Quantum Theory* (New York: Dover, 1930). See also references cited in note 1.

[6] Schwinger, see note 4.

[7] J. C. Slater, *J. Franklin Inst.*, v. 207 (1929), p. 449.

[8] E. P. Wigner, *Amer. J. of Physics*, v. 31 (1963), p. 6.

[9] Schwinger, op. cit., v. 46 (1960), p. 257.

[10] A. Lande, *Foundations of Quantum Theory* (New Haven: Yale University Press, 1955) and *From Dualism to Unity* (Cambridge: Cambridge University Press, 1960).

[11] Our language is a little inexact. For the precise relation between the state vector and the Schrödinger wave function see ch. 5 of Dirac, op. cit., or B. Kursunoglu, *Modern Quantum Theory* (San Francisco: Freeman, 1962), ch. 5.

[12] Einstein's classic paper "On the Electrodynamics of Moving Bodies," *Annalen der Physik*, v. 17 (1905), p. 891, is still unexcelled as a masterpiece of logical deduction and exposition.

[13] E. P. Wigner, *Rev. of Mod. Physics*, v. 29 (1957), p. 255.

[14] The controversy concerning whether a more complete quantum theory is possible began with the celebrated paper by Einstein, Podolsky, and Rosen, "Can Quantum-Mechanical Description of Physical Reality Be Considered Complete?", *Physical Rev.*, v. 47 (1935), p. 777, and has continued almost unabated for nearly thirty years.

[15] A recent special conference at Xavier University, October, 1962, dealt specifically with this subject. See the report by F. G. Werner, *Physics Today*, v. 17, no. 1 (January, 1964), p. 53.

[16] D. Bohm, *Physical Rev.*, v. 85 (1952), pp. 166, 180. Bohm, whose recent critical analyses and contributions to the fundamental understanding of quantum theory are second to none, no longer adheres to the point of view of these older papers. See, for example, his article in *The Scientist Speculates*, ed. I. J. Good (New York: Basic Books, 1962).

[17] H. Salecker and E. P. Wigner, *Physical Rev.*, v. 109 (1958), p. 571.

[18] E. J. Zimmerman, *Amer. J. of Physics*, v. 30 (1962), p. 97.

[19] P. M. Morse and H. Feshbach, *Methods of Theoretical Physics* (New York: McGraw-Hill, 1953), p. 247.

[20] F. Engelmann and E. Fick, *Supp. del Nuovo Cimento*, v. 12 (1959), p. 63.

[21] H. Paul, *Annalen der Physik*, v. 9 (1962), p. 252.

[22] D. Bohm and Y. Ahranov, *Physical Rev.*, v. 122 (1961), p. 1649.

[23] G. J. Whitrow, *Natural Philosophy of Time* (New York: Harper & Row, 1963). Another thoughtful study has been done by R. Schlegel, *Time and the Physical World* (East Lansing: Michigan State University Press, 1961).

[24] Zimmerman, see note 18.

[25] *Ibid.*, p. 101.

[26] Bohm, see note 16.

[27] An extensive summary, with bibliography, of work prior to 1957 was given by H. Freistadt, *Supp. del Nuovo Cimento*, v. 5 (1957), p. 1. Later formulations have been given by J. P. Wesley, *Physical Rev.*, v. 122 (1961), p. 1932, and others.

Time and Thermodynamics

[1] I have discussed clocks, cyclic and noncyclic processes, and natural law in my book, *Time and the Physical World* (East Lansing: Michigan State University Press, 1961), ch. 1. Also, see the paper by G. M. Clemence, *Amer. Sci.*, v. 40 (1952), p. 260.

[2] The relation between time sense and observation of physical systems has been discussed by L. Rosenfeld in his paper, "Questions of Irreversibility and Ergodicity," *Proc. International School of Physics "Enrico Fermi,"* XIV Course (New York: Academic Press, 1961), p. 3.

[3] But there are counterexamples. A man may tell you that his two grown sons are two years apart, and yet you may not know the direction of time flow between their birth dates until you do establish the interval between each date and the present (or, equivalently, some time which you can relate to the present).

[4] A. S. Eddington, *The Nature of the Physical World* (New York: Macmillan, 1929), ch. IV.

[5] For a more detailed discussion of the relations between entropy change and our time concept see R. Schlegel, *op. cit.*, p. 25.

[6] The use of entropy increase as the *primary* criterion of time's direction was effectively criticized by L. Susan Stebbing in her book, *Philosophy and the Physicists* (London: Methuen, 1937), ch. XI.

[7] A. S. Eddington, *op. cit.*, p. 72.

[8] The view that gene mutation is commonly a result of a fluctuation in the energy of a molecule was expressed by Erwin Schrödinger, following a suggestion of Max Delbrück, in his *What Is Life?* (Cambridge: Cambridge University Press, 1945), p. 63.

[9] See H. Margenau, "Can Time Run Backwards," *Philos. of Sci.*, v. 21 (1954), p. 79. Also, discussion and references in R. Schlegel, *op. cit.*, p. 62.

[10] Robert G. Sachs, "Can the Direction of Flow of Time Be Determined?", *Science*, v. 140 (1963), p. 1284.

[11] L. Boltzmann, *Nature*, v. 51 (1895), p. 414.

[12] Aside from considerations of entropy increase, the suggestion has been made that the expansion of the universe does itself give a physical phenomenon that defines the direction of time. This definition would indeed associate time with what appears to be a universal feature of our universe. It is also true, however, that expansion is a phenomenon that as yet has not been so closely related to the body of physical theory as has entropy increase, and it may not be a comparably basic principle. The difference is illustrated by the fact that cosmological models have been proposed in which expansion of the universe would be only one phase of a cyclic expansion-contraction behavior (although with, of course, a time extent of billions of years for an expansion period of the cycle). Cyclic models of this kind are quite as tenable, on the basis of what we know today, as are cosmological models which show only progressively expansive behavior.

[13] See, *e.g.*, Joseph Needham's essay, "Evolution and Thermodynamics," *Time: the Refreshing River* (London: Allen and Unwin, 1943). A comprehensive recent statement of possible inadequacies of purely physical principles is given by Peter T. Mora, "Urge and Molecular Biology," *Nature*, v. 199 (1963), p. 212. In quite a different way, Karl R. Popper has argued on the basis of our lack of definite knowledge of cosmological structure that we

should not give any cosmic significance to entropy decrease, or to connect that decrease with the arrow of time. See his "Time's Arrow and Entropy," *Nature*, v. 207 (1965), p. 233.

[14] I have taken this illustrative situation from Harold F. Blum's *Time's Arrow and Evolution* (New York: Harper, 1962), p. 205. In this book Blum repeatedly emphasizes the importance of taking into account the total system of process and surroundings when applying the Second Law of Thermodynamics to biological processes. A numerical calculation of the entropy increase resulting from energy-density dilution in radiation that comes from the sun and affects photosynthesis, as compared with entropy decrease in the photosynthetic process, has been given by Wesley Brittin and George Gamow in their paper, "Negative Entropy and Photosynthesis," *Proc. National Acad. of Sci.*, v. 47 (1961), p. 724.

[15] The possible role of night-day temperature alteration in biological evolution has been discussed by J. Lee Kavanau, *Amer. Naturalist*, v. 81 (1947), p. 161.

[16] See, *e.g.*, M. Planck, *Ann. der Physik*, v. 26 (1908), p. 1; R. C. Tolman, *Relativity, Thermodynamics, and Cosmology* (Oxford: Oxford University Press, 1934); C. Eckhart, *Physical Rev.*, v. 58 (1940), p. 919; M. von Laue, *Die Relativitätatheorie*, v. I, 5th ed., (Braunschweig: Vieweg, 1952), p. 145; G. A. Kluitenberg and S. R. de Groot, *Physica*, v. 20 (1954), p. 199, and v. 21 (1955), p. 148.

[17] This is a particular aspect of the relatedness of space and time that is discussed by O. Costa de Beauregard in his essay in this volume.

[18] See, *e.g.*, W. Pauli, *Theory of Relativity*, tr. G. Field (New York: Pergamon Press, 1958), p. 135.

[19] A detailed discussion of the proposed distinction for relativistic transformations is given by R. Schlegel, *op. cit.*, ch. VIII.

[20] A. C. Benjamin in his essay in this volume has discussed constitutive ideas of time in philosophical thought, particularly as elucidated by Samuel Alexander.

[21] Also discussed in A. C. Benjamin's essay, with respect to Locke, Berkeley, and Hume.

[22] See G. M. Clemence's essay in this volume.

[23] Niels Bohr, Faraday Lecture, *J. of the Chem. Soc.* (1932), p. 349.

Note Relating to a Paradox of the Temporal Order

[1] M. Scriven, *Mind*, v. 60 (1951), p. 403; A. Lyon, *Mind*, v. 68 (1959), p. 272; T. H. O'Beirne, *The New Scientist* (May 25, 1961); M. Gardner, *Scientific American* (March, 1963), p. 144.

[2] This has been done in that form of the paradox which relates to a condemned man. "You will be hanged," said the judge, "within one week, one noon. But which day, you will not know until you are so informed on the morning of the day of the hanging."

Time and the Probabilistic View of the World

[1] S. Watanabe, *Physical Rev.*, v. 6 (1957), p. 1306.

[2] L. Boltzmann, *Vorlesungen über Gastheorie*, 2 vols. (Leipzig, 1896–98).

[3] S. Fujiwara, *Japanese J. of Astro. Geophysics*, v. 5 (1923), p. 143; *J. of Roy. Meteorological Soc.*, v. 49 (1923), p. 89; H. Bergson, *Evolution Créatrice*

(Paris, 1907); N. Wiener, *Cybernetics* (2nd ed.; Cambridge: Massachusetts Institute of Technology Press, 1963); S. Watanabe, *Time* (Tokyo: Hakujitsu, 1947).

4 S. Watanabe, *IBM J. of Research and Development*, v. 4 (1960), p. 66; *IRE Transactions Information Theory*, v. IT-8 (1962), p. 248; S. Watanabe, an article in the volume dedicated to the memory of Professor Norbert Wiener (in preparation); S. Watanabe, *Proc. of International Symp. on Foundation of Mathematics* (Tihany, 1962).

5 E. Zermello, *Ann. d. Phys. u. Chem.*, v. 57 (1896); P. and T. Ehrenfest, *Encyklopädie der mathematischen Wissenschaften*, v. IV-4 (1907–11).

6 P. and T. Ehrenfest, see note 5; J. Loschmidt, *Sitzungsber. Akademie Wissenschaften*, v. 73 (1876), p. 134, v. 75 (1877), p. 67.

7 M. M. Yanase, *Annals of the Japan Assoc. for Philos. of Science*, v. 1 (1957).

8 P. and T. Ehrenfest, see note 5.

9 S. Watanabe, *Knowing and Guessing* (New York: John Wiley, in preparation).

10 S. Watanabe, see note 4; also see an article in the volume dedicated to the memory of Professor Norbert Wiener, *op. cit.*

11 S. Watanabe, article in *Louis de Broglie, physicien et penseur* (Paris: Albin Michel, 1952); see also Watanabe, *Rev. Mod. Phys.*, v. 27 (1955), p. 179.

12 There is in classical physics, besides reversibility—and probably more basic than reversibility—what is sometimes called "inversibility," which is invariance for the combination of time-reversal and space-reflection. This inversibility is formally inherited in quantum physics, but it receives an interpretation as an invariance for the combination of time-reversal (T), space-reflection (P) and interchange of positive and negative change (C). S. Watanabe, *Physical Rev.*, v. 5 (1951), p. 1008; S. Watanabe, *Rev. of Mod. Phys.*, v. 27 (1955), pp. 26, 40. This theorem received the name of C-T-P Theorem later. Some unconfirmed indications of breakdown of C-P invariance have been reported recently, which would in the presence of C-T-P Theorem imply a breakdown of the pure reversibility (T invariance). There are serious thinkers who conjecture that the alleged breakdown of microscopic reversibility has something to do with the macroscopic one-way-ness. This, however, involves a confused argument since breakdown of any invariance of this kind does not necessarily imply irretrodictability. *Cf.* also S. Watanabe, *Le deuxième théorème de la thermodynamique et la mécanique ondulatoire* (Paris: Hermann, 1935), also D. Sc. thesis, University of Paris, 1935; S. Watanabe, *Scientific Papers, Institute for Physico-Chemical Research*, v. 31 (1937), p. 109; S. Watanabe, Supplementary Issue in Commemoration of Thirtieth Anniversary of the Meson Theory, *Progress of Theoretical Phys.* (1965), in preparation.

13 Watanabe, see note 11.

14 Watanabe, *Rev. Mod. Phys.*, v. 27 (1955), pp. 26, 40.

15 This is because all physical laws are invariant for displacement in time.

16 Yanase, see note 7.

17 Watanabe, see note 9.

18 C. G. Hempel, *Synthese*, v. 12 (1960), p. 439.

19 Landau-Lifshitz, *Statistical Physics* (London: Addison-Wesley, 1958).

20 H. Reichenbach, *Direction of Time* (Berkeley: University of California Press, 1956).

21 S. Watanabe, *Rev. Métaphysique et Morale*, v. 56 (1951), p. 128.

[22] S. Watanabe, see note 11.

[23] A separate paper, along the same line as the present paper, but with more emphasis on physics and less emphasis on philosophical implications, will appear in the Supplementary Issue in Commemoration of Thirtieth Anniversary of the Meson Theory, *Progress of Theoretical Physics* (1965), in preparation. *Cf.* note 12.

[24] S. Watanabe, *Proceedings of Intern. Symp* . . . , see note 14.

Time and the Universe

[1] J. G. Fraser, *Folklore in the Old Testament* (New York: Macmillan, 1923), p. 26. The serpent symbolism is found not only in ancient Babylonia, Persia, India, etc., but also in the Mayan and Aztec cultures it was associated with cycles of endless time.

[2] I. Kant, *Critique of Pure Reason*, tr. N. Kemp Smith (London: Macmillan, 1934), p. 127.

[3] B. Russell, *The Principles of Mathematics* (2nd ed.; London: Macmillan, 1942), p. 459.

[4] E. Frank, *Das Prinzip der dialektischen Synthesis, Kantstudien*, Ergänzungsheft, no. 21 (1911), p. 9, argues that Kant failed to recognize that the Law of Contradiction is restricted to abstract reasoning alone, and that he erred in defending its validity for the phenomena of time, movement and change.

[5] E. A. Milne, *Modern Cosmology and the Christian Idea of God* (Oxford: Oxford University Press, 1952), p. 149.

[6] G. J. Whitrow, *The Natural Philosophy of Time* (New York: Harper, 1961), p. 288.

[7] H. Mehlberg, *Philos. Rev.*, v. 71 (1962), p. 104. Experiments made in 1964 (and since) seem to show, however, that the 'weak' interactions in particle physics are not strictly time-symmetric. See J. H. Christiansen *et al.*, *Physical Rev. Letters*, v. 13 (1964), p. 138.

[8] J. A. Wheeler and R. P. Feynman, *Rev. Mod. Phys.*, v. 17 (1945), p. 157; *ibid.*, v. 21 (1949), p. 425. Mention should also be made of recent work based on this treatment of electromagnetism in terms of direct particle interaction. For, although Wheeler and Feynman have since abandoned their theory (because of the limitations of the particle concept), attempts have been made to apply it to expanding models of the universe and thereby correlate the hypothetical cosmological arrow of time with the electromagnetic arrow associated with the choice of retarded solutions of Maxwell's equations, this choice being dictated by the expansion of the universe. See J. E. Hogarth, *Proc. Roy. Soc.*, A, v. 267 (1962), p. 365; F. Hoyle and J. V. Narlikar, *ibid.*, v. 277 (1963), p. 1; J. V. Narlikar, *Brit. J. Phil. Sci.*, v. 15 (1965), p. 281.

[9] E. A. Milne, *Nature*, v. 130 (1932) p. 9.

[10] G. J. Whitrow, *The Structure and Evolution of the Universe* (London: 1959), p. 99.

[11] K. Gödel, *Rev. Mod. Phys.*, v. 21 (1949), p. 447.

[12] L. Ozsváth and E. Schücking, *Nature*, v. 193 (1962), p. 1168.

[13] Oscillating models that alternately expand and contract have also been studied, but the physics of the contracting phase is still obscure.

[14] A. S. Eddington, *The Expanding Universe* (Cambridge: 1933), p. 72.

15 H. Bondi and T. Gold, *Monthly Notices of the Roy. Astro. Soc.*, v. 108 (1948), p. 252.
16 F. Hoyle, *Monthly Notices of the Roy. Astro. Soc.*, v. 108 (1948), p. 372.
17 F. Hoyle and J. V. Narlikar, *Monthly Notices of the Roy. Astro. Soc.*, v. 123 (1961), p. 133; v. 125 (1962), p. 13.
18 I. King, *The Observatory*, v. 81 (1961), p. 128.
19 A. Sandage, *Astrophys. J.*, v. 135 (1962), p. 349.
20 V. C. Reddish, *Sci. Prog.*, v. 50 (1962), p. 600.
21 J. Locke, *An Essay concerning Human Understanding* (1690), bk. II, ch. XV.
22 A. S. Eddington, *The Expanding Universe* (Cambridge: Cambridge University Press, 1933), p. 73.
23 The events in question are those that lie outside the observers' light cones.
24 H. Weyl, *Space Time Matter*, tr. H. L. Brose (London: Methuen, 1922), p. 217.
25 For further discussion of this question see G. J. Whitrow, *The Natural Philosophy of Time*, p. 280.
26 H. Reichenbach, *The Direction of Time*, ed. M. Reichenbach (Berkeley: University of California Press, 1956), p. 268.
27 O. Penrose and I. C. Percival, *Proc. Phys. Soc.*, v. 79 (1962), p. 605.

The Study of Time

1 See G. J. Whitrow's discussion of the statistical theory of time in *The Natural Philosophy of Time* (London: Nelson, 1961), p. 276.
2 That the Newtonian G is time-dependent has been considered by Dirac, Milne, Dicke and others, mostly in journal articles. That G has a secular variation because of the expansion of the Universe was suggested by P. A. M. Dirac, *Proc. Roy. Soc.*, v. A165 (1938), p. 199. Martin Johnson, *Time, Knowledge and the Nebulae* (New York: Dover, 1947), p. 104; also Pascual Jordan, *Schwerkraft und Weltall* (Braunschweig: Vieweg, 1955), par. 34.
3 Johnson, *op. cit.*, pp. 108, 111.
4 V. Fock, *The Theory of Space, Time and Gravitation*, tr. N. Kemmer (New York: Pergamon Press, 1959), p. 34.
5 M. Polanyi, *Personal Knowledge* (Chicago: University of Chicago Press, 1958), pp. 58, 90.
6 Here I am alluding to the formulation of G. N. M. Tyrrell put forth in *The Personality of Man* (London: Pelican Books, 1946), ch. 9, "What does foreknowledge imply?" Following a suggestion of H. F. Saltmarsh, Tyrrell speculates that for the subliminal self events can be co-present even though for the conscious self they are in the present and in the future.
7 This very apt term describes that trend of idealist thought which regards time as unreal. The origin of the term, the "elimination of time" is credited to Emile Meyerson, see Whitrow, *op. cit.*, p. 1.
8 G. J. Whitrow, *The Structure and Evolution of the Universe* (New York: Harper, 1959), p. 199.

Biographical Notes on the
Authors and Translators

BENJAMIN, A. CORNELIUS (b. 1897, Grand Rapids, Michigan). M.A., Ph.D., University of Michigan. Since 1945, John Hiram Lathrop Professor of Philosophy, University of Missouri (Chairman of Department, 1945–56); Visiting Professor of Philosophy, Baylor University, Spring Semester, 1961. Guggenheim Fellow, 1930–31; President, Western Division American Philosophical Association, 1949; Member, Philosophy of Science Association; Member, Sigma Xi; Fellow, American Association for the Advancement of Science, and Official Delegate of American Philosophical Association to this group (1961–64). Author of *Logical Structure of Science* (1936); *Introduction to the Philosophy of Science* (1937); *Operationism* (1955); *Science, Technology and Human Values* (1965). Contributor to scientific, philosophical and educational journals.

BRANDON, SAMUEL GEORGE FREDERICK (b. 1907, Portsmouth, England). M.A., D.D. Since 1951, Professor of Comparative Religion, University of Manchester; College of Resurrection, Mirfield; University of Leeds. Wilde Lecturer in Natural and Comparative Religion, University of Oxford. War service: Royal Army Chaplains Department, British Expeditionary Force, Flanders Campaign, and Dunkirk Evacuation; North Africa; Italy, mentioned in dispatches; Austria, occupation forces. Author of *Time and Mankind* (1951); *The Fall of Jerusalem and the Christian Church*, 2nd ed. (1957); *Man and His Destiny in the Great Religions* (1962). Contributor to *Myth, Ritual and Kingship* (1958); *Creation Legends of the Ancient Near East* (1963); *The Saviour God*, ed. and contrib. (1963); *History, Time and Deity* (1965).

ČAPEK, MILIČ (b. 1909, Bohemia). Ph.D. in Philosophy, M.S. in Physics, King Charles University at Prague. Post-doctoral studies: Sorbonne; University of Chicago. Fellowships: French Foreign Office; The Fund for the Advancement of Education at Yale University. Teaching of physics at University of Nebraska (1944–46); University of Olmutz, Czechoslovakia (1946–47). Teaching of philosophy at Carleton College (1948–62), Professor of Physics from 1957; Professor of Philosophy of Boston University since 1962. Actively participated at the following international congresses: The Tenth Congress of the French Philosophical Societies of Paris, 1959; The Tenth International Congress of History of Science at Ithaca, 1962; The Thirteenth International Congress for Philosophy at Mexico City, 1963. Author of *Bergson and the Trends of Contemporary Physics* (1938); *The Philosophical Impact of Contemporary Physics* (1961); and of numerous articles in American and French philosophical periodicals.

CLEMENCE, GERALD M. (b. 1908 in Greenville, Rhode Island). Ph.B., Brown University; Sc.D. (hon.), Case Institute of Technology; Dr., Universidad de Cuye (Argentina). Astronomer, U.S. Naval Observatory since 1930; Director, *Nautical Almanac* (1945–58); Scientific Director Naval Observatory (1958–63); Senior Research Associate and Lecturer in Astronomy, Yale University, since 1963. Member, National Academy of Sciences; Fellow, American Academy of Arts and Sciences; Fellow, Royal Astronomical Society; Correspondent and Member of several astronomical and scientific associations. Author of *Methods of Celestial Mechanics* (1961), with Dirk Brouwer; *The Motion of Mercury 1765–1937* (1943); *First Order Theory of Mars* (1949) and *Theory of Mars* (1961); and other books and monographs. Author of chapters of books and articles in encyclopedias (*Britannica, Encyclopedia of Science and Technology*) on astronomical subjects, and of more than sixty research papers, letters, articles and reviews in the general area of relativity, astronomy, time measurement and navigation.

CLOUDSLEY-THOMPSON, JOHN LEONARD (b. 1921, Murree, India). M.A., Ph.D. (Cambridge), D.Sc. (London), F.L.S., F.R.E.S., F.I. Biol., F.W.A. Professor of Zoology, University of Khartoum and Keeper, Sudan Natural History Museum, since 1960. Formerly Lecturer in Zoology, King's College, University of London. Educated Marlborough College and Pembroke College, Cambridge. War Service: Commissioned into 4th Queen's Own Hussars, Transferred to 4th County of London Yeomanry (Sharpshooters). Operation "Crusader" and the relief of Tobruk 1941. Wounded at the "Knightsbridge" tank battle, May 1942. Instructor (Captain) at Sandhurst, 1943. Rejoined regiment for D Day in Normandy, June 1944. Escaped from Villers Bocage; Caen Offensive, etc. Cambridge Iceland Expedition, 1947. Expedition to Southern Tunisia, 1954. Expeditions to various parts of central Africa since 1960. Author of: *Biology of Deserts* (1954); *Spiders, Scorpions, Centipedes and Mites* (1958); *Animal Behaviour* (1960); *Land Invertebrates*, with John Sankey (1961); *Rhythmic Activity in Animal Physiology and Behaviour* (1961); *Life in Deserts*, with M. J. Chadwick (1964); *Desert Life* (1965); *Animal Conflict and Adaptation* (1965); and many scientific articles.

COHEN, JOHN (b. 1911). M.A., Ph.D. (Lond.), F.B. Ps.S. Professor of Psychology in the University of Manchester. Formerly, Adviser, Offices of the War Cabinet; Professor, University of Jerusalem; Lecturer, Universities of Leeds and London. Foreign Member, Centre de Recherches de Psychologie Comparative; Fellow Member, World Academy of Art and Science; Former Member of Council, British Psychological Society. Brit. Editor, *Acta Psychologica*. Author of *Human Nature, War and Society* (1946); *Humanistic Psychology* (1958); *Chance, Skill and Luck* (1960), *Behaviour in Uncertainty* (1946). Co-author of *Risk and Gambling* (1956). Editor: *Readings in Psychology* (1946).

COSTA DE BEAUREGARD, OLIVIER (b. 1911, Paris). Ph.D. in Physics, 1943, with work on the physical interpretation of various aspects of the Dirac electron theory and on the theory of spinning media described by an asymmetrical energy momentum tensor; Ph.D. in Philosophy, 1963. Obtained Licence des

Sciences and completed military service. Research Engineer at the applied aerodynamics section of Societé Nationale de Constructions Aéronautiques du Sud Est. Education: Entered CNRS, Section de Physique Théoretique, 1940; Named Maitre de Recherches at CNRS, 1949. Author of *La Theorie de la Relativite Restreinte* (1949); *Theorie Synthetique de la Relativite Restreinte et des Quanta* (1957); *La Notion de Temps-Equivalence avec l'Espace*, (1963) and *Le Second Principle de la Science du Temps—Entropie, Information, Irreversibilite* (1963).

DINGLE, HERBERT, D.Sc., A.R.C.S. Professor of Natural Philosophy at the Imperial College of Science and Technology, London, 1937–46; Professor of History and Philosophy of Science at University College, London 1946–55; now Emeritus Professor of the latter subject in the University of London; President of the Royal Astronomical Society, 1951–53; and Vice-President of the International Union for the History of Science, 1953–56; member of British Government Eclipse Expeditions; Corresponding Member of the Institute of Coimbra; held various lectureships, including the Lowell Lectureship, 1936; written a number of papers on various aspects of Astronomy, Physics, and the Philosophy of Science. Author of *Relativity for All* (1922); *Modern Astrophysics* (1924); *Science and Human Experience* (1931); *Through Science to Philosophy* (1937); *The Special Theory of Relativity* (1940); *Mechanical Physics* (1941); *Sub-Atomic Physics* (1942); *Science and Literary Criticism* (1949); *Practical Applications of Spectrum Analysis* (1950); *The Scientific Adventure* (1952); *The Sources of Eddington's Philosophy* (1954); *A Threefold Cord: Philosophy, Science, Religion*, with Viscount Samuel (1961). Editor of: *Life and Work of Sir Norman Lockyer* (1929) and *A Century of Science* (1951).

DÜRR, WALTHER (b. 1932). Ph.D., Tübingen. Assistant to the Director of the Musicological Institute, University of Tübingen. Formerly Lecturer in Music History and German Literature at the University of Bologna. Author of: *Zum Verhältnis von Wort und Ton im Rhythmus des Cinquecento—Madrigals* (1958); *Zwei neue Belege für die sogenannte spielmännische Reduktion* (1958); *Die italienische Canzonette und das deutche Lied in Ausgang des XVI. Jahrhunderts* (1960); *La musica del Rococo* (1960). Coauthor, with Professor Dr. Gerstenberg, "Rhythm, Meter, Time and Tempo" in the *Encyclopedia of Music* (1962).

FISCHER, ROLAND (b. 1915). M.A., Ph.D., Senior Investigator, Research Division, Department of Psychiatry, Associate Professor of Psychiatry, Asst. Professor of Physiological Chemistry, the College of Medicine, Ohio State University. Over one hundred and twenty five research publications in the following areas: biological aspects of normal and pathological behavior, "model psychoses," psychopharmacological aspects of time and space; models and mechanism of Gram stain, drug action, central nervous system excitation and depression, gustatory chemoreception and pharmacogenetics. Member of Biochemical Society (Brit.), The Society for Biological Psychiatry, The Philosophy of Science Association, etc., and past chairman of the Chemical Institute of Canada. Chairman of the New York Academy of Sciences sponsoring Intl. Conference on "Interdisciplinary Perspectives in Time" (1966).

Biographical Notes on Authors and Translators

VON FRANZ, MARIE-LOUISE—Ph.D., Lecturer and Training Analyst at the C. G. Jung Institute in Zürich. Worked with Dr. Jung analytically and as a collaborator, for twenty-seven years. Author of *Visionen des Niklaus von Flüe*; Co-author of *Die Graalalegende*; wrote the third volume of C. G. Jung's sequence of books on alchemy, Mysterium Conjunctionis. Theoretical interest in mythology, author of a treatise on the *Dreams of Descartes* in the publication series of the Institute for Analytical Psychology.

FRASER, JULIUS THOMAS (b. 1923). B.E.E., 1950. Senior Scientist, General Precision, Inc., since 1955. Research Associate, Department of Physics and Astronomy, Michigan State University, since 1962. Held research engineering positions with Westinghouse Electric, and Mackay Radio and Telegraph Corporations. Specialist in engineering physics; published a study of inertial fields. Holds several patents related to solid-state microwave components, industrial instrumentation and control, and nuclear gyroscopes. Conceived, developed and edited *The Voices of Time* (1965).

HAMNER, KARL C. (b. 1908, Salina, Kansas). M.S., University of Chicago. Associate Plant Physiologist in the U.S.D.A. (1936–37); Bureau of Plant Industry, Beltsville, Maryland. 1948 to present, Professor of Botany, University of California in Los Angeles. Chairman of the Department of Botany, 1948–57. President, Western Section, American Society of Plant Physiologists, 1950. President, U.C.L.A. Section Sigma Xi, 1961. Gave invited papers before the Society of Experimental Biology, in Oxford, England, (1946); Commonwealth Specialist Conference in Agriculture, Australia, (1949); Australian Academy of Science Symposium on Phytotoroms, (1962). Contributor to scientific journals.

HOAGLAND, HUDSON, Ph.D. Professor of General Physiology and Chairman of the Biology Department of Clark University, 1931–44. Held a Guggenheim Fellowship and in 1944 founded, in collaboration with Gregory Pincus, the Worcester Foundation for Experimental Biology. Since then he has been Executive Director of this institution. Research Professor in Biological Psychiatry at Boston University School of Medicine; recipient of four honorary doctorates and received the 1965 Modern Medicine Award for distinguished achievement; past-President of the American Academy of Arts and Sciences; taught at Harvard and at Cambridge University; member of a number of professional and honorary societies; has served on various committees and boards of private and government agencies. Trustee of the Worcester Memorial Hospital, Woods Hole Oceanographic Institution, and was a member of the Committee of the Harvard Board of Overseers to visit the Harvard Medical School and School of Dental Medicine. Editor or author of several books and has contributed a number of research papers in the fields of neurophysiology, neuropharmacology, adrenal and stress physiology, biochemical aspects of schizophrenia and physiological time.

KALMUS, HANS, D.Sc., M.D. Reader of Biology in the University of London, working at the Galton Laboratory of University College, London. Was Associate Professor of Genetics at McGill University, Montreal, and held visiting Professorships at Indiana and Jerusalem Universities. Co-founder of the Inter-

686

national Society for the Study of Biological Rhythm and the recipient of the Dreyfus prize for Genetics from the University of Sao Paolo. President of the Society for the Study of Animal Behaviour. Author of *Paramecium* (1931); *Genetics* (1948); *Simple Experiments with Insects* (1948); *Variation and Heredity in Man* (1958); *The Chemical Senses* (1960); and also numerous articles on biological subjects including many related to periodicity.

KÜMMEL, FRIEDRICH (b. 1933, Essingen, Germany). Ph.D. Instructor in philosophy at the Institute of Philosophy, University of Tübingen. Teacher in public schools until 1956. After resumption of his studies in philosophy, education, and theology at the Universities of Tübingen and Göttingen, he obtained his Doctor's degree in Philosophy from the University of Tübingen. Author of *On the Concept of Time* (1962).

LLOYD, HERBERT ALAN, M.B.E., F.S.A., F.B.H.I. King's College School and Sheffield University and on the Continent. Served in First World War, 1914–19. Founder of the science of the interpretation of aeroplane photographs through his publication in 1916 of *Notes on the Interpretation of Aeroplane Photographs*. 1920–48, in international business; since retirement, has carried out research on antiquarian Horological Society, London. Corresponding Member of the Academie des Sciences, Belles Lettres & Arts, Besancon, France. Author of: *The English Domestic Clock* (1938); *Chats on Old Clocks* (1951); *Old Clocks* (1958); *Some Outstanding Clocks over 700 Years, 1250–1950* (1958); *A Collector's Dictionary of Clocks* (1964); translated from the German *The Book of Old Clocks and Watches* (1964). Contributed the chapters on "Mechanical Clocks" in *A History of Technology* (1954–58), and in "Clocks and Watches" in the *Connoisseur's Concise Encyclopedia of Antiques* (1954).

MEERLOO, JOOST A. M. (b. 1903, The Hague, Netherlands). M.D., University of Leyden, Ph.D., University of Utrecht. Staff Psychiatrist and Private Practitioner of Psychotherapy and Psychoanalysis, 1928–34; served as psychiatric consultant to the Royal Court and to Dutch governmental agencies. In 1942 escaped to England and served as a Colonel, Chief of the Psychological Department of the Netherlands Army; 1944–46, High Commissioner for Welfare for the Netherlands Government and advisor to SHAEF and UNRRA. Distinguished Service Cross, 1943. Taught in several schools and conducted private practice of Psychotherapy and Psychoanalysis in New York. Honorary Member and Fellow of several professional societies and Associate Professor of Psychiatry, New York School of Psychiatry. Author of twenty books, among them *Total War and the Human Mind*; *Patterns of Panic*; *The Two Faces of Man*; *Two Studies on the Sense of Time and on Ambivalence*; *The Rape of the Mind*; *The Difficult Peace*; *The Dance*; *Suicide and Mass Suicide* and *Justice and Injustice*, and of three-hundred articles relating to the field of psychotherapy and psychoanalysis.

NAKAMURA, HAJIME (b. 1912, Matsue, Japan). D. Lit. Professor of Indian and Buddhist Philosophy, University of Tokyo, since 1943, and Dean of the Faculty of Letters, since 1964. Visiting Professor of Philosophy, Stanford University,

1951–52; Visiting Professor of Religion, University of Florida, 1961; Visiting Professor, 1959, 1964, and Senior Scholar, 1962 at University of Hawaii; Visiting Professor of World Religions, Harvard University, 1963–64; Delegate to the Conference on Freedom in Asia, Rangoon, 1955; State Guest by the Government of India, 1956 and 1960; Delegate to the Unesco-Pax-Ramana Conference, Manila, 1960; Senior Programme Member, The Third East-West Philosophers' Conference, 1959. Steering Committee Member, the Fourth East-West Philosophers' Conference, 1964; Director of the Japanese Association of Indian and Buddhist Studies. Honorary Fellow, Government Sanskrit College, Calcutta; Life Member of the Bhandarkar Oriental Institute, Poona. President of the Japan-India Society. Author of *Ways of Thinking of Eastern Peoples* (in English), (1960 and 1964); *Japan and Indian Asia* (in English) (1961); *History of Early Vendata Philosophy*, 4 vols. (in Japanese) (1951)f.; and other works. Associate Editor, *Philosophy East and West*; *Monumenta Nipponica*; *Philosophical Studies of Japan*; Corresponding Editor, *The Journal of the History of Ideas*.

NEEDHAM, JOSEPH (b. 1900). Sc.D., F.R.S. Cambridge University, specializing in biochemistry; in World War I was Surgeon Sub-Lieutenant R.N. Fellow of Gonville and Caius College, Cambridge, has held various University appointments in England and America; 1928–33, University Demonstrator in Biochemistry; 1933– , Sir William Dunn Reader in Biochemistry at Cambridge; 1929, Visiting Professor of Physiology at Stanford University; held visiting Lectureships at Yale, Cornell and Oberlin College. 1936, Oliver Sharpey Lecturer at the Royal College of Physicians, London; 1937, Herbert Spencer Lecturer at Oxford University, and lectured in the Universities of Warsaw, Lwow, Krakow and Wilno; 1939, Comte Memorial Lecturer in London; 1942–46, Head of the British Scientific Mission in China and Counsellor at the British Embassy, Chungking, directing the Sino-British Science Cooperation Office, the task of which was liaison between Chinese and Western scientists and engineers. At the invitation of the Soviet Government, took part in the commemoration of the 220th anniversary of the Moscow Academy of Sciences; Foreign Member of the Chinese National Academy (Academia Sinica), and the Societe Philomathique de Paris. Specialized in the borderline between biochemistry and embryology, but has also done much in the History and Philosophy of Science. Author of *Chemical Embryology* (3 vols., 1931), *Biochemistry and Morphogenesis* (1942), also a number of books of essays and addresses on general scientific and philosophical questions—*Man a Machine* (1927); *The Sceptical Biologist* (1929); *The Great Amphibium* (1932); *Time, The Refreshing River* (1940); *Order and Life* (1940); and *History is on our Side* (1945). Has edited a number of collaborative volumes, such as *Science, Religion and Reality* (1925); *Christianity and the Social Revolution* (1937); *Perspectives in Biochemistry* (1949). Now engaged on a comprehensive treatise on the history of science, scientific thought and technology in Chinese culture, *Science and Civilization in China* (1954–   10 vols. of which 5 have already appeared and one further in press); and byproduct monographs to the same, *The Development of Iron and Steel Technology in China* (1958), and *Heavenly Clockwork* (1960).

PIAGET, JEAN (b. 1896, Neuenburg, Switzerland). Since 1925 has held several Professorships in Psychology, Sociology and the Philosophy of Science at Geneva, Lausanne, the Sorbonne and other leading European Universities.

Was awarded honorary Doctorates by the Cambridge, Harvard, Chicago, McGill, the Sorbonne and other universities; participated in numerous conferences and held many positions of responsibility in international educational organizations. The author of about twenty books in the fields of child psychology and of about three hundred articles on zoology, child psychology, psychology of perception and related subjects.

RUSSELL, JOHN (b. 1906). Cambridge University, degree in Natural Science (part I) and Philosophy (part II) and subsequently obtained a Diploma in Agriculture and a Ph.D. for research on the colloidal properties of clay. Worked for two years in the Department of Soil Science at Oxford; entered the Society of Jesus in 1937. Since 1947, lectured at Heythrop Theological College on the Philosophy of Nature and on the History and Philosophy of Science. Author of *Science and Metaphysics* (1958).

SCHLEGEL, RICHARD (b. 1913). M.A. in philosophy, Ph.D. in physical chemistry. Professor of Physics, Michigan State University. Lecturer, Museum of Science and Industry, Chicago, 1937–40; Associate Physicist, University of Chicago; Instructor of Physics, Princeton University; Assistant Professor, Associate Professor, Professor since 1957 and Acting Head of Department of Physics, Michigan State University. Visiting Professor, University of California, Berkeley; Research Associate (on sabbatical leave from Michigan State University), Cavendish Laboratory, Cambridge University, 1954–55; Visiting Scholar, Department of History and Philosophy of Science, Cambridge University, 1961–62. Author of *Time and the Physical World* (1961).

STUTTERHEIM, CORNELIUS F. P. (b. 1903). Ph.D. Professor of Dutch Linguistics at the University of Leyden, Member of the Royal Dutch Academy of Sciences. Studies at the University of Amsterdam where he obtained his Ph.D. Degree, *cum laude*. Publications (in Dutch): *The Idea of 'Metaphor', a linguistical and philosophical inquiry*; *Introduction to the Philosophy of Language*; *Problems of the Science of Literature*; *Stylistics*; *Theory of Belles-Lettres*; *The Pianola, novel of a musician*, and over seventy essays concerning language, literature and philosophy.

WATANABE, SATOSI (b. 1910, Tokyo). B.S. and D.Sc. Tokyo. French Government Fellow, Institute Poincaré, Paris, France, D.Sc. University of Paris. Member of the Research Staff of the Institute für Theoretische Physik, Leipzig, 1937–39; Associate Professor at Tokyo University, 1942–45; Advisor to the Far East Command, Allied Powers, 1946–47; Professor and Chairman of Physics, St. Paul's University, 1949–50; Associate Professor at Wayne University, 1950–52; Professor at the U.S. Naval Postgraduate School, 1952–56; Senior Physicist at the IBM Research Center, Yorktown Heights, New York, 1956–64, now professor of applied science, Yale University. Adjunct Professor of Electrical Engineering, Columbia University; 1960–62 Fellow of American Physical Society elected a member of the International Academy of Philosophy of Science. Author of numerous articles in the *Physical Review, Reviews of Modern Physics, Zeitschrift für Physik, Revue de Métaphysique et de Morale* and elsewhere, on quantum field theory, philosophy of science, the problem of time and other subjects. Published ten books in Japanese and one book in French.

WHITROW, GERALD JAMES (b. 1912). M.A., Ph.D. Reader in Applied Mathematics in the University of London at the Imperial College of Science and Technology since 1951. Formerly Lecturer of Christ Church, Oxford. Member of Council of the Royal Astronomical Society; Membre Correspondant de l'Academie Internationale d'Histoire des Science; former Vice-Chairman of the British Society for the History of Science. Author of *The Structure of the Universe* (1949); *The Structure and Evolution of the Universe* (1959); *The Natural Philosophy of Time* (1961). Co-author of: *Atoms and the Universe* (1956); *Rival Theories of Cosmology* (1960). Former editor of *The Monthly Notices of the Royal Astronomical Society* and of *The Observatory Magazine*.

ZIMMERMAN, EDWARD JOHN (b. 1924). M.S., Ph.D. University of Illinois. Professor of Physics since 1960 and Chairman of the Department since 1962, University of Nebraska, Lincoln. Visiting Professor, University of Hamburg, 1957–58. Member A.P.S., AAPT and the British Society for the Philosophy of Science. Author of scientific articles on low energy nuclear and atomic physics in the *Physical Review* and on interpretations of physical theory in the *American Journal of Physics*.

GAONA, FRANCISCO (b. 1932). Ph.D. Assistant Professor of Spanish, Sonoma State College, California. Dissertation on *The Concept of Space in Cassirer's Philosophy of Symbolic Forms*. Author of *Alonso de Ercilla* for *Twayne's World Authors Series* (in preparation).

(Professor Gaona is the translator of Dr. Kümmel's essay from the German.)

MONTGOMERY, BETTY BROYLES, M.A. Education Specialist Ohio Department of Public Welfare, Division of Social Administration, Child Day Care Consultant: Consultation to teachers of very young children, 1954—present; directed five nursery programs in four different cities. M.A. in Child Care and Psychology, graduate study in psychology; work in learning in young children.

(Mrs. Montgomery is the translator of Professor Piaget's essay from the French.)

PARK, DAVID ALLEN (b. 1919). Ph.D. Research Associate in Physics, Radio Research Laboratories 1944–45; since 1950 Assistant Professor, Associate Professor and Professor of Physics, Williams College. Member, Institute for Advanced Studies, 1950–51; Fulbright Lecturer in the University of Ceylon, 1955–56; Department of Applied Mathematics and Theoretical Physics, University of Cambridge, 1962–63. Published articles and books on quantum mechanics, magnetic phenomena and the popularization of the results of contemporary research.

(Professor Park is the translator of Dr. Costa de Beauregard's essay from the French.)

690

RAWSKI, CONRAD H. (b. 1914). Ph.D. Professor of Library Science at Western Reserve University, Cleveland, Ohio. Studied at the University of Vienna and Austrian Institute for Historical Research, and Western Reserve University. Held a Faculty Fellowship of the Fund for the Advancement of Education, 1952–53; Visiting Fellow, Cornell University, Harvard University. Published a number of articles and papers on music and music theory in the Middle Ages. Edited Boismortier's *Flute Sonatas*, Op. 7, 2 vols., (1954–57). Among his translations are *Petrarch, Four Dialogues for Scholars*, (1964); *Knud Jeppesen, Italian Organ Music of the Early Sixteenth Century*; and *Notker's pertry* (forthcoming).

(Professor Rawski is the translator of Dr. Dürr's essay from the German.)

VERHOEFF, PIETER JOHANNES (b. 1930). Educated at Amsterdam University, he took a degree in English language and literature in 1957. In 1952 he received a Government grant for a year's study at Leeds University. From 1957–58 he worked in the American Institute of Amsterdam University, after which he was appointed lecturer in the newly-established English Department of the University of Utrecht, where he has since then taught phonetics, syntax, Old and Middle English. He has translated two volumes of Science Fiction stories besides various articles on linguistic subjects for the Dutch periodical *Lingua*. He is at present engaged on a Ph.D. dissertation, which will comprise an edition of two manuscripts of Robert Mannyng's *Chronicle*.

(Mr. Verhoeff is the translator of Professor C. F. P. Stutterheim's essay from the Dutch.)

Index

Index

Index

Index

Index